Keith Hohn (Ed.)

Feature Papers to Celebrate the Landmarks of Catalysts

MDPI

This book is a reprint of the Special Issue that appeared in the online, open access journal, *Catalysts* (ISSN 2073-4344) from 2014–2015 (available at: http://www.mdpi.com/journal/catalysts/special_issues/feature-paper-catalysts).

Guest Editor
Keith Hohn
Kansas State University
USA

Editorial Office
MDPI AG
Klybeckstrasse 64
Basel, Switzerland

Publisher
Shu-Kun Lin

Managing Editor
Zu Qiu

1. Edition 2016

MDPI • Basel • Beijing • Wuhan • Barcelona

ISBN 978-3-03842-220-4 (Hbk)
ISBN 978-3-03842-221-1 (PDF)

Table of Contents

List of Contributors

Cinthia Alegre Instituto de Carboquímica (CSIC), Miguel Luesma Castán 4, 50018 Zaragoza, Spain.

Hsunling Bai Institute of Environmental Engineering, National Chiao Tung University, Hsinchu 30010, Taiwan.

Miguel A. Bañares Catalytic Spectroscopy Laboratory, Instituto de Catalisis, ICP-CSIC, E-28049 Madrid, Spain.

Burcu Bayram William G. Lowrie Department of Chemical and Biomolecular Engineering, the Ohio State University, 151 W. Woodruff Avenue, Columbus, OH 43210, USA.

Pablo Beato Haldor Topsøe A/S, Nymøllevej 55, DK-2800 Kgs. Lyngby, Denmark.

Elisabeth Bordes-Richard CNRS UMR 8181, Unité de Catalyse et Chimie du Solide (UCCS), Université Lille 1 Sciences et Technologies, F-59655 Villeneuve d'Ascq Cedex, France; Ecole Nationale Supérieure de Chimie de Lille, ENSCL, F-59659 Villeneuve d'Ascq, France.

Maria Emma Borges Chemical Engineering Department, University of La Laguna, Avda. Astrofísico Fco. Sánchez s/n, La Laguna, Tenerife, Canary Island 38200, Spain.

Kyle M. Brunner Department of Chemical Engineering, Brigham Young University, Provo, UT, 84602, USA.

Sridhar Budhi Department of Chemistry and Geochemistry, Colorado School of Mines, Golden, CO 80401, USA.

Vanesa Calvino-Casilda Catalytic Spectroscopy Laboratory, Instituto de Catalisis, ICP-CSIC, E-28049 Madrid, Spain; Current Address: Departamento de Quimica Inorganica y Quimica Tecnica, UNED, 28040 Madrid, Spain.

Sebastiano Campisi Dipartimento di Chimica, Università degli Studi di Milano, via Golgi 19, I-20133 Milano, Italy.

Miguel Ángel Centeno Departamento de Química Inorgánica, Universidad de Sevilla e Instituto de Ciencias de Materiales de Sevilla Centro mixto US-CSIC, Avda. Américo Vespucio 49, 41092 Seville, Spain.

Hsiang-Yu Chan Department of Chemical Engineering, National Taiwan University, Taipei 10617, Taiwan.

Carine E. Chan-Thaw Dipartimento di Chimica, Università degli Studi di Milano, via Golgi 19, I-20133 Milano, Italy.

Jae-Soon Choi Oak Ridge National Laboratory, 1 Bethel Valley Road, Oak Ridge, TN 37831, USA.

M. Natália D. S. Cordeiro REQUIMTE, Departamento de Química e Bioquímica, Faculdade de Ciências, Universidade do Porto, P-4169-007 Porto, Portugal.

Mark Crocker University of Kentucky Center for Applied Energy Research, 2540 Research Park Drive, Lexington KY 40511, USA.

Bas de Bruin Van 't Hoff Institute for Molecular Sciences, University of Amsterdam, Science Park 904, 1098 XH, Amsterdam, The Netherlands.

Karla Herrera Delgado Institute for Chemical Technology and Polymer Chemistry, Karlsruhe Institute of Technology (KIT), 76128 Karlsruhe, Germany.

Olaf Deutschmann Institute for Catalysis Research and Technology; Institute for Chemical Technology and Polymer Chemistry, Karlsruhe Institute of Technology (KIT), 76128 Karlsruhe, Germany.

Vannia Cristina dos Santos Laboratory of Bioinorganic Chemistry and Catalysis, Federal University of Paraná, CP 19081, CEP 81531-990, Curitiba, Paraná, Brazil.

Franck Dumeignil CNRS UMR 8181, Unité de Catalyse et Chimie du Solide (UCCS), Université Lille 1 Sciences et Technologies, F-59655 Villeneuve d'Ascq Cedex, France; Institut Universitaire de France, IUF, Maison des Universités, 103 Bd St-Michel, F-75005 Paris, France.

Pedro Esparza Inorganic Chemistry Department, University of La Laguna, Avda. Astrofísico Fco. Sánchez s/n, La Laguna, Tenerife, Canary Island 38200, Spain.

José L. C. Fajín REQUIMTE, Departamento de Química e Bioquímica, Faculdade de Ciências, Universidade do Porto, P-4169-007 Porto, Portugal.

María Elena Gálvez Instituto de Carboquímica (CSIC), Miguel Luesma Castán 4, 50018 Zaragoza, Spain.

Dulce María García Chemical Engineering Department, University of La Laguna, Avda. Astrofísico Fco. Sánchez s/n, La Laguna, Tenerife, Canary Island 38200, Spain.

Preshit Gawade William G. Lowrie Department of Chemical and Biomolecular Engineering, the Ohio State University, 151 W. Woodruff Avenue, Columbus, OH 43210, USA.

Cao Huong Giang Vietnam Academy of Agricultural Science, Vinh Quynh, Thanh Tri, Ha Noi, Vietnam.

José R. B. Gomes CICECO, Departamento de Química, Universidade de Aveiro, 3810-193 Aveiro, Portugal.

Jan-Dierk Grunwaldt Institute of Catalysis Research and Technology (IKFT); Institute for Chemical Technology and Polymer Chemistry (ITCP), Karlsruhe Institute of Technology (KIT), Engesserstr. 20, 76131 Karlsruhe, Germany.

Yasemin Gumrukcu Van 't Hoff Institute for Molecular Sciences, University of Amsterdam, Science Park 904, 1098 XH, Amsterdam, The Netherlands.

Shaoli Guo Institute for the Study of Nanostructured Materials (ISMN)-CNR, Via Ugo La Malfa 153, 90146 Palermo, Italy; Department of Applied Physics, Northwestern Polytechnical University, Xi'an 710072, China.

Tahmina Haque Department of Chemistry, Tokyo Metropolitan University, 1-1 Minami Osawa, Hachioji, Tokyo 192-0397, Japan; Department of Chemistry, Jahangirnagar University, Savar, Dhaka 1342, Bangladesh.

William C. Hecker Department of Chemical Engineering, Brigham Young University, Provo, UT, 84602, USA.

Tania Hernández Chemical Engineering Department, University of La Laguna, Avda. Astrofísico Fco. Sánchez s/n, La Laguna, Tenerife, Canary Island 38200, Spain.

Keith L. Hohn Department of Chemical Engineering, Kansas State University, Manhattan, KS 66506, USA.

Martin Høj Department of Chemical & Biochemical Engineering, Technical University of Denmark (DTU), Søltofts Plads Building 229, DK-2800 Kgs. Lyngby, Denmark.

Baiyu Huang Department of Chemistry and Biochemistry, Brigham Young University, Provo, UT 84602, USA.

Svetlana Ivanova Departamento de Química Inorgánica, Universidad de Sevilla e Instituto de Ciencias de Materiales de Sevilla Centro mixto US-CSIC, Avda. Américo Vespucio 49, 41092 Seville, Spain.

Anker Degn Jensen Department of Chemical & Biochemical Engineering, Technical University of Denmark (DTU), Søltofts Plads Building 229, DK-2800 Kgs. Lyngby, Denmark.

Fangli Jing CNRS UMR 8181, Unité de Catalyse et Chimie du Solide (UCCS), Université Lille 1 Sciences et Technologies, F-59655 Villeneuve d'Ascq Cedex, France.

Benjamin Katryniok CNRS UMR 8181, Unité de Catalyse et Chimie du Solide (UCCS), Université Lille 1 Sciences et Technologies, F-59655 Villeneuve d'Ascq Cedex, France; Ecole Centrale de Lille, ECLille, F-59651 Villeneuve d'Ascq, France.

Wolfgang Kleist Institute of Catalysis Research and Technology (IKFT); Institute for Chemical Technology and Polymer Chemistry (ITCP), Karlsruhe Institute of Technology (KIT), Engesserstr. 20, 76131 Karlsruhe, Germany.

Ranjit T. Koodali Department of Chemistry, University of South Dakota, 414 E. Clark Street, Vermillion, SD 57069, USA.

Michael J. Lance Oak Ridge National Laboratory, 1 Bethel Valley Road, Oak Ridge, TN 37831, USA.

María Jesús Lázaro Instituto de Carboquímica (CSIC), Miguel Luesma Castán 4, 50018 Zaragoza, Spain.

Adam F. Lee European Bioenergy Research Institute, School of Engineering and Applied Sciences, Aston University, Birmingham B4 7ET, UK.

Angeliki A. Lemonidou Department of Chemical Engineering, Aristotle University of Thessaloniki, University Campus, GR-54124 Thessaloniki, Greece; Chemical Process and Energy Resources Institute (CERTH/CPERI), P.O. Box 60361 Thermi, Thessaloniki 57001, Greece.

Stephen E. Levine Mechanical and Aerospace Engineering, the Ohio State University, 201 W. 19th Street, N350 Scott Laboratory, Columbus, OH 43210, USA.

Samuel A. Lewis Sr. Oak Ridge National Laboratory, 1 Bethel Valley Road, Oak Ridge, TN 37831, USA.

Leonarda F. Liotta Institute for the Study of Nanostructured Materials (ISMN)-CNR, Via Ugo La Malfa 153, 90146 Palermo, Italy.

Lubow Maier Institute for Catalysis Research and Technology, Karlsruhe Institute of Technology (KIT), 76128 Karlsruhe, Germany.

Harry M. Meyer III Oak Ridge National Laboratory, 1 Bethel Valley Road, Oak Ridge, TN 37831, USA.

Rafael Moliner Instituto de Carboquímica (CSIC), Miguel Luesma Castán 4, 50018 Zaragoza, Spain.

Karren L. More Oak Ridge National Laboratory, 1 Bethel Valley Road, Oak Ridge, TN 37831, USA.

Tonya Morgan University of Kentucky Center for Applied Energy Research, 2540 Research Park Drive, Lexington, KY 40511, USA.

Van-Huy Nguyen Department of Chemical Engineering, National Taiwan University of Science and Technology, Taipei 106, Taiwan.

Kotohiro Nomura Department of Chemistry, Tokyo Metropolitan University, 1-1 Minami Osawa, Hachioji, Tokyo 192-0397, Japan; Advanced Catalytic Transformation for Carbon Utilization (ACT-C), Japan Science and Technology Agency (JST), Saitama 332-0012, Japan.

José Antonio Odriozola Departamento de Química Inorgánica, Universidad de Sevilla e Instituto de Ciencias de Materiales de Sevilla Centro mixto US-CSIC, Avda. Américo Vespucio 49, 41092 Seville, Spain.

Amin Osatiashtiani European Bioenergy Research Institute, School of Engineering and Applied Sciences, Aston University, Birmingham B4 7ET, UK.

Hayrani Oz Mechanical and Aerospace Engineering, the Ohio State University, 201 W. 19th Street, N350 Scott Laboratory, Columbus, OH 43210, USA.

Umit S. Ozkan William G. Lowrie Department of Chemical and Biomolecular Engineering, the Ohio State University, 151 W. Woodruff Avenue, Columbus, OH 43210, USA.

Robert Pace University of Kentucky Center for Applied Energy Research, 2540 Research Park Drive, Lexington, KY 40511, USA.

Laura Pastor-Pérez Laboratorio de Materiales Avanzados, Departamento de Química Inorgánica-Instituto Universitario de Materiales de Alicante, Universidad de Alicante, Apartado 99, E-03080 Alicante, Spain.

Sébastien Paul CNRS UMR 8181, Unité de Catalyse et Chimie du Solide (UCCS), Université Lille 1 Sciences et Technologies, F-59655 Villeneuve d'Ascq Cedex, France; Ecole Centrale de Lille, ECLille, F-59651 Villeneuve d'Ascq, France.

Maxime Perdu University of Kentucky Center for Applied Energy Research, 2540 Research Park Drive, Lexington, KY 40511, USA.

Laura Prati Dipartimento di Chimica, Università degli Studi di Milano, via Golgi 19, I-20133 Milano, Italy.

Fabrizio Puleo Institute for the Study of Nanostructured Materials (ISMN)-CNR, Via Ugo La Malfa 153, 90146 Palermo, Italy.

Joost N. H. Reek Van 't Hoff Institute for Molecular Sciences, University of Amsterdam, Science Park 904, 1098 XH, Amsterdam, The Netherlands.

Tomás Ramírez Reina Departamento de Química Inorgánica, Universidad de Sevilla e Instituto de Ciencias de Materiales de Sevilla Centro mixto US-CSIC, Avda. Américo Vespucio 49, 41092 Seville, Spain.

Juan Carlos Ruiz-Morales Inorganic Chemistry Department, University of La Laguna, Avda. Astrofísico Fco. Sánchez s/n, La Laguna, Tenerife, Canary Island 38200, Spain.

Eduardo Santillan-Jimenez University of Kentucky Center for Applied Energy Research, 2540 Research Park Drive, Lexington, KY 40511, USA.

Kirsten Schuh Institute for Chemical Technology and Polymer Chemistry (ITCP), Karlsruhe Institute of Technology (KIT), Engesserstr. 20, 76131 Karlsruhe, Germany.

Viviane Schwartz Oak Ridge National Laboratory, 1 Bethel Valley Road, Oak Ridge, TN 37831, USA.

Antonio Sepúlveda-Escribano Laboratorio de Materiales Avanzados, Departamento de Química Inorgánica-Instituto Universitario de Materiales de Alicante, Universidad de Alicante, Apartado 99, E-03080 Alicante, Spain.

Michael P. Snyder Mechanical and Aerospace Engineering, the Ohio State University, 201 W. 19th Street, N350 Scott Laboratory, Columbus, OH 43210, USA.

Hyuntae Sohn William G. Lowrie Department of Chemical and Biomolecular Engineering, the Ohio State University, 151 W. Woodruff Avenue, Columbus, OH 43210, USA.

Ilgaz I. Soykal William G. Lowrie Department of Chemical and Biomolecular Engineering, the Ohio State University, 151 W. Woodruff Avenue, Columbus, OH 43210, USA.

Henning Stotz Institute for Chemical Technology and Polymer Chemistry, Karlsruhe Institute of Technology (KIT), 76128 Karlsruhe, Germany.

Steffen Tischer Institute for Catalysis Research and Technology, Karlsruhe Institute of Technology (KIT), 76128 Karlsruhe, Germany.

Vanessa Trouillet Institute for Applied Materials (IAM) and Karlsruhe Nano Micro Facility (KNMF), Karlsruhe Institute of Technology (KIT), Hermann-von-Helmholtz-Platz 1, 76344 Eggenstein-Leopoldshafen, Germany.

Efterpi S. Vasiliadou Department of Chemical Engineering, Aristotle University of Thessaloniki, University Campus, GR-54124 Thessaloniki, Greece.

Alberto Villa Dipartimento di Chimica, Università degli Studi di Milano, via Golgi 19, I-20133 Milano, Italy.

Keith W. Waldron The Biorefinery Centre, Institute of Food Research, Norwich Research Park, Colney, Norwich NR4 7UA, UK.

Di Wang Institut für Nanotechnologie, Forschungszentrum Karlsruhe in der Helmholtz-Gemeinschaft Hermann-von-Helmholtz-Platz 1, 76344 Eggenstein-Leopoldshafen, Germany.

David R. Wilson The Biorefinery Centre, Institute of Food Research, Norwich Research Park, Colney, Norwich NR4 7UA, UK

Karen Wilson European Bioenergy Research Institute, School of Engineering and Applied Sciences, Aston University, Birmingham B4 7ET, UK.

Brian F. Woodfield Department of Chemistry and Biochemistry, Brigham Young University, Provo, UT, 84602, USA.

Chia-Ming Wu Department of Chemical Engineering, National Taiwan University, Taipei 10617, Taiwan.

Hongjing Wu Department of Applied Physics, Northwestern Polytechnical University, Xi'an 710072, China.

Jeffrey C.S. Wu Department of Chemistry, University of South Dakota, 414 E. Clark Street, Vermillion, SD 57069, USA.

Alexander Zellner Institute for Chemical Technology and Polymer Chemistry, Karlsruhe Institute of Technology (KIT), 76128 Karlsruhe, Germany.

Dan Zhao Department of Chemical Engineering, Ningbo University of Technology, Ningbo 315016, China.

About the Guest Editor

Keith L. Hohn obtained his PhD in Chemical Engineering at the University of Minnesota in 1995. He is currently the Honstead Professor of Chemical Engineering at Kansas State University. His research interests are in heterogeneous catalysis, with emphasis on energy applications. In particular, he has researched the conversion of hydrocarbons and alcohols to hydrogen and catalytic processes for converting biomass to fuels and chemicals. He is a member of the American Society of Engineering Educators, American Chemical Society (ACS), and American Institute of Chemical Engineers (AICHE). He has been active in ACS and AICHE as a session organizer and session chair at a variety of meetings.

Preface to "Feature Papers to Celebrate the Landmarks of *Catalysts*"

Keith L. Hohn

Reprinted from *Catalysts*. Cite as: Hohn, K.L. Feature Papers to Celebrate the Landmarks of *Catalysts*. *Catalysts* **2015**, 5, 2018–2023.

Catalysis is a critical scientific field that underpins much of the world's chemical industry. For example, it is often quoted that catalysis plays a role in 90% of all industrial chemical products. This importance has led to numerous academic journals and specialized conferences on the subject, as practitioners seek outlets to publish their cutting-edge research on catalysis.

Catalysts started in 2011 with the goal of providing an open-source outlet for outstanding research on catalysis. In its first few years, *Catalysts* has enlisted the help of an active editorial board and hard-working guest editors to establish itself as a high-quality forum for catalysis papers. A primary focus has been to get *Catalysts* included in the major citation indices in order to provide its authors with the visibility and impact they desire. Recently, that goal was achieved, as *Catalysts* was selected for inclusion in Scopus and Science Citation Index Expanded (SCIE). In addition, *Catalysts* received its first impact factor in 2015. These are major milestones for the journal, and are worthy of celebration.

This issue is being published in recognition of these achievements. It is an opportunity to celebrate *Catalysts*' beginnings, and a way to look forward to its future. Nineteen articles from pre-eminent scholars in catalysis have been compiled in this special issue, representing a wide range of topics. This diversity is representative of the field of catalysis in general, as it touches so many areas of scholarship. In this special issue, you can find articles on catalyst synthesis [1–4], modeling of reaction kinetics on catalysts [5], polymerization [6], photocatalysts [7,8], oxidation catalysis [9], biomass conversion [10–14], electrochemistry [15], fuel cell catalysis [16], computational catalysis [17,18], and hydrogen production [19].

Highlights include:

1. Budhi and coworkers describe the synthesis of a unique material, titanium dioxide nanoclusters dispersed on cubic MCM-48, via a room temperature synthesis method [1]. They report that their procedure leads to highly dispersed TiO_2 particles with Ti^{4+} ions mostly substituted in framework tetrahedral positions. The materials they synthesized were found to give 100% selectivity to cyclohexene oxide in cyclohexane oxidation at room temperature with tert-butylhydroperoxide as the oxidant.

2. Shuh and coworkers used a mild hydrothermal synthesis procedure to synthesize bismuth molybdate catalysts for the selective oxidation of propylene [2]. Notably, they found that the pH used during synthesis impacted the catalytic performance of the material. A pH of 9 led to a less active catalyst than pH values of 6 or 7. The catalyst performance of the bismuth molybdate catalysts was closely linked with the bismuth molybdate phases present and the catalyst surface area, variables that were impacted by the synthesis procedure.

3. Brunner and coworkers prepared iron Fischer-Tropsch catalysts by using solvent-deficient precipitation, focusing on the impacts of different catalyst preparation steps [3]. They found that the key steps in catalyst synthesis were whether the precursor catalyst was washed, how the promoted was added, and the drying condition. The most active Fischer-Tropsch catalysts were produced from an unwashed catalyst via a one-step method with the drying step at 100 °C.

4. Soykal and coworkers publish a unique study where they evaluate whether the microgravity environment on the International Space Station impacted the physical and chemical properties of cerium oxide synthesized by a controlled-precipitation method [4]. They found that in microgravity the solutions do not sediment and movement of the free particles is significantly easier. Microgravity changed the crystallization behavior where oriented aggregation of particles had a more pronounced effect. The samples prepared in space were generally larger with lower surface area and a broader pore size distribution when compared to the control samples synthesized on Earth. The samples prepared in space had lower surface area, lower pore volume, and a broader pore size distribution in the range from 30–600 Angstroms when compared to the control sample synthesized on Earth. To best of our knowledge this is the first ceria synthesis study conducted in space which demonstrates the effect of microgravity, as well as showing that ceria nanorods can be prepared at ambient temperature and pressure.

5. Delgado and coworkers report on their work to develop a microkinetic model to describe the catalytic conversion of methane on nickel catalysts in oxidative and reforming conditions [5]. A 52 elementary-step surface reaction mechanism is described that includes reactions of 14 surface and six gas-phase species. This model is evaluated by comparison to experimental data obtained over nickel catalysts for various reaction conditions, and is used to predict the performance of a fixed bed reactor.

6. Haque and Nomura review the recent work to develop an acyclic diene metathesis polymerization process that allows the synthesis of defect-free conjugated polymers with well-defined chain ends [6]. One of the most exciting implications of this work is that the polymer that results from this process

can further be functionalized to produce materials suitable for use as organic electronics. For instance, the authors demonstrate that the polymer can be treated by molybdenum-alkylidene complexes followed by addition of various aldehydes. Their results suggest that this polymerization method is attractive for synthesizing star conjugated polymers.

7. Chan and coworkers contribute to the state-of-the-art knowledge on photocatalytic processes by using NIR-Raman spectroscopy to monitor the photocatalytic epoxidation of cyclohexen over V-Ti/MCM-41 catalysts [7]. Using the *in situ* Raman spectroscopy to monitor characteristic bands, the authors were able to show, for the first time, that cyclohexene was directly photo-epoxidized to 1,2-epoxycyclohexane by t-BuOOH. They also found that both Ti and V were active in photo-epoxidation, and suggested that the two metals could work in concert to give enhanced reaction efficiency.

8. Borges and coworkers investigated photocatalysis for the removal of emerging contaminants from wastewater effluent [8]. The authors studied the reaction of paracetamol as a model contaminant model in both a stirred photoreactor and a packed bed photoreactor with a TiO_2 photocatalyst. The authors demonstrated high photocatalytic activities and suggested that a packed bed reactor system is an effective method for removing emerging pollutants.

9. Jing and coworkers describe how supported $(NH_4)_3HPMo_{11}VO_{40}$ catalysts change over time during the selective oxidation of isobutane by using a variety of characterization techniques [9]. Notably, they found that the thermal stability and reducibility of the Keggin units were enhanced by supporting 40% of the active phase on $Cs_3PMO_{12}O_{40}$. They described the molecular processes responsible for structural changes in these catalysts. For example, the authors describe the decomposition of ammonium cations that lead to formation of vacancies that can favor cationic exchanges between vanadium from the active phase and cesium from the support.

10. Santillan-Jimenez and coworkers investigated molybdenum carbide supported on different carbon supports (activated carbon, carbon nanofiber, and carbon nanotubes) for the hydrodeoxygenation of guaiacol [10]. The authors found that all three catalysts produced catechol and phenol as the main products in hydrodeoxygenation of guaiacol, suggesting that guaiacol was converted to phenol by sequential demethylation and hydrodeoxygenation. They showed that use of carbon nanofibers as the support was generally superior for selective production of phenol.

11. Choi and coworkers investigated molybdenum carbides for low-temperature hydroprocessing of acetic acid [11]. In particular, the authors investigated the structural changes in the molybdenum carbides that result from exposure to hot aqueous environments. They found that bulk Mo carbides are maintained during aging in hot liquid water, but that some of the carbidic Mo sites were

converted to oxide sites. Furthermore, they demonstrate that significant structural changes result from these reaction conditions, with changes in surface area and pore volume.

12. Guo and coworkers review the state-of-the-art in the use of B-site metal promoted $La_{1-x}Sr_xCo_{1-y}Fe_yO_{3-\sigma}$ perovskite oxides as cathodes for solid oxide fuel cells [16]. This review describes the impact of promoting perovskite oxides with Pd, Pt, Ag, Cu, Zn, and Ni, focusing on different synthetic methods and the effects of metal promotion on cathodic performance. A major contribution of the article is a discussion of the combined effects of the oxygen dissociation rate and the interfacial oxygen transfer rate between the metal and cathode phases.

13. Gurukcu, de Bruin, and Reek use computational methods to explore the mechanism of the allylicamination reaction between allyl alcohols and amines on a phosphoamidite palladium/1,3-diethylurea co-catalyst system [17]. This contribution uses DFT calculations to suggest that hydrogen bonding between the urea moiety and the hydroxyl group of the alcohol is key to facilitate the C-O oxidative addition step, which is the rate-limiting step in the reaction mechanism.

14. Pastor-Perez and coworkers explore the performance of a carbon-supported Ni-CeO$_2$ catalyst with a conventional Ni/CeO$_2$ bulk catalyst in the water–gas shift reaction [19]. They find that the carbon-supported catalyst is more active than the bulk catalyst due to its oxygen storage capacity. This study could lead to a new generation of Ni-ceria catalysts for hydrogen production via water–gas shift that have a reduced amount of ceria.

15. Alegre and coworkers contribute to the field of electrochemistry by studying the electro-oxidation of methanol on highly mesoporous carbon xerogel and Vulcan carbon black [15]. Their objective is to understand how the preparation procedure impacts catalytic performance. The authors show that the carbon xerogel gave higher catalytic activities towards CO and CH$_3$OH oxidation than the Vulcan support, mainly because of the higher mesoporosity of the carbon xerogel that allowed more facile diffusion of reactants within the catalyst.

16. Chan-Thaw and coworkers investigate AuPd nanoparticles for the selective oxidation of raw glycerol [12]. The impact of the support is investigated, and the authors demonstrate that activated carbon and nitrogen-functionalized carbon nanofibers gave the best catalytic results. The authors also note that glycerol obtained from the transesterification of rapeseed oil led to strong deactivation, in contrast with the results when pure glycerol was used.

17. Fajin and coworkers describe the use of the generalized-gradient approximation of the density functional theory to explore multicomponent catalysts for Fischer-Tropsch synthesis [18]. The authors describe how this

modeling technique allows them to separate the influence of different parameters in the global catalytic performance. A major conclusion of the article is that computational studies can now compete with modern experimental techniques because of the efficient computer codes that have been developed for use on powerful computers that allow realistic molecular systems to be simulated.

18. Vasiliadou and Lemonidou investigated the catalytic glycerol hydrodeoxygenation with ethanol as a hydrogen donor [13]. With this novel concept, the hydrogen necessary for hydrodeoxygenation is provided in the same reactor by ethanol reforming. The authors investigated Pt, Ni, and Cu catalysts, and found that Pt/Fe$_2$O$_3$-Al$_2$O$_3$ gave high selectivity to 1,2-propanediol. They also showed that milder reaction conditions should be applied to maximize 1,2-propanediol yield.

19. Giang and coworkers investigated the catalytic valorization of Vietnamese rice straw [14]. They propose that aqueous phase reforming of steam exploded rice straw hydrolysate and condensate could be used to produce platform chemicals. The authors show that tungstated zirconia catalysts could convert glucose to 5-hydroxymethylfurfural (HMF). They suggest that Lewis acid and/or base sites on the support catalyze glucose dehydration to fructose, which can then be dehydrated to HMF over Brönsted acidic tungsten oxide clusters.

I think practitioners in catalysis will find this Special Issue highly relevant and interesting. I want to thank all of the authors for their contributions, and the editorial staff at *Catalysts*, particularly Mary Fan, Senior Assistant Editor for their efforts. I hope you enjoy this Special Issue to commemorate the landmarks that *Catalysts* has achieved.

References

1. Budhi, S.; Wu, C.-M.; Zhao, D.; Koodali, R.T. Investigation of Room Temperature Synthesis of Titanium Dioxide Nanoclusters Dispersed on Cubic MCM-48 Mesoporous Materials. *Catalysts* **2015**, *5*, 1603–1621.

2. Schuh, K.; Kleist, W.; Høj, M.; Trouillet, V.; Beato, P.; Jensen, A.D.; Grunwaldt, J.-D. Bismuth Molybdate Catalysts Prepared by Mild Hydrothermal Synthesis: Influence of pH on the Selective Oxidation of Propylene. *Catalysts* **2015**, *5*, 1554–1573.

3. Brunner, K.M.; Huang, B.; Woodfield, B.F.; Hecker, W.C. Iron Fischer-Tropsch Catalysts Prepared by Solvent-Deficient Precipitation (SDP): Effects of Washing, Promoter Addition Step, and Drying Temperature. *Catalysts* **2015**, *5*, 1352–1374.

4. Soykal, I.I.; Sohn, H.; Bayram, B.; Gawade, P.; Snyder, M.P.; Levine, S.E.; Oz, H.; Ozkan, U.S. Effect of Microgravity on Synthesis of Nano Ceria. *Catalysts* **2015**, *5*, 1306–1320.

5. Delgado, K.H.; Maier, L.; Tischer, S.; Zellner, A.; Stotz, H.; Deutschmann, O. Surface Reaction Kinetics of Stean- and CO2-Reforming as Well as Oxidation of Methane over Nickel-Based Catalysts. *Catalysts* **2015**, *5*, 871–904.

6. Haque, T.; Nomura, K. Acyclic Diene Metathesis (ADMET) Polymerization for Precise Synthesis of Defect-Free Conjugated Polymers with Well-Defined Chain Ends. *Catalysts* **2015**, *5*, 500–517.

7. Chan, H.-Y.; Nguyen, V.-H.; Wu, J.C.S.; Calvino-Casilda, V.; Bañares, M.A.; Bai, H. Real-Time Raman Monitoring during Photocatalytic Epoxidation of Cyclohexene over V-Ti/MCM-41 Catalysts. *Catalysts* **2015**, *5*, 518–533.

8. Borges, M.E.; Garcia, D.M. Hernández, T.; Ruiz-Morales, J.C.; Esparza, P. Supported Photocatalyst for Removal of Emerging Contaminants from Wastewater in a Continuous Packed-Bed Photoreactor Configuration. *Catalysts* **2015**, *5*, 77–87.

9. Jing, F.; Katryniok, B.; Bordes-Richard, E.; Dumeignil, F.; Paul, S. Structural Evolution under Reaction Conditions of Supported $(NH_4)_3HPMo_{11}VO_{40}$ Catalysts for the Selective Oxidation of Isobutane. *Catalysts* **2015**, *5*, 460–477.

10. Santillan-Jimenez, E.; Perdu, M.; Pace, R.; Morgan, T.; Crocker, M. Activated Carbon, Carbon Nanofiber and Carbon Nanotube Supported Molybdenum Carbide Catalysts for the Hydrodeoxygenation of Guaiacol. *Catalysts* **2015**, *5*, 424–441.

11. Choi, J.-S.; Schwartz, V.; Santillan-Jimenez, E.; Crocker, M.; Lewis, S.A.; Lance, M.J.; Meyer, H.M.; More, K.L. Structural Evolution of Molybdenum Carbides in Hot Aqueous Environments and Impact on Low-Temperature Hydroprocessing of Acetic Acid. *Catalysts* **2015**, *5*, 406–423.

12. Chan-Thaw, C.E.; Campisi, S.; Wang, D.; Prati, L.; Villa, A. Selective Oxidation of Raw Glycerol Using Supported AuPd Nanoparticles. *Catalysts* **2015**, *5*, 131–144.

13. Vasiliadou, E.S.; Lemonidou, A.A. Catalytic Glycerol Hydrodeoxygenation under Inert Atmosphere: Ethanol as a Hydrogen Donor. *Catalysts* **2015**, *5*, 397–413.

14. Giang, C.H.; Osatiashtiani, A.; dos Santos, V.C.; Lee, A.L. Wilson, D.R.; Waldron, K.W.; Wilson, K. Valorisation of Vietnamese Rice Straw Waste: Catalytic Aqueous Phase Reforming of Hydrolysate from Steam Explosion to Platform Chemicals. *Catalysts* **2015**, *5*, 414–426.

15. Alegre, C.; Galvez, M.E.; Moliner, R.; Lazaro, M.J. Influence of the Synthesis Method for Pt Catalysts Supported on Highly Mesoporous Carbon Xerogel and Vulcan Carbon Black on the Electro-Oxidation of Methanol. *Catalysts* **2015**, *5*, 392–405.

16. Guo, S.; Wu, H.; Puleo, F.; Liotta, L. B-Site Metal (Pd, Pt, Ag, Cu, Zn, Ni) Promoted $La_{1-x}Sr_xCo_{1-y}Fe_yO_{3-\delta}$ Perovskite Oxides as Cathodes for IT-SOFCs. *Catalysts* **2015**, *5*, 366–391.

17. Gumrukcu, Y.; de Bruin, B.; Rook, J.N.H. A Mechanistic Study of Direct Activation of Allylic Alcohols in Palladium Catalyzed Amination Reactions. *Catalysts* **2015**, *5*, 349–365.

18. Fajin, J.L.C.; Natália, M.; Cordeiro, D.S.; Gomes, J.R.B. Fischer-Tropsch Synthesis on Multicomponent Catalysts: What Can We Learn from Computer Simulations? *Catalysts* **2015**, *5*, 3–17.

19. Pastor-Perez, L.; Reina, T.R.; Ivanova, S.; Centeno, M.A.; Odriozola, J.A.; Sepúlveda-Escribano, A. Ni-CeO$_2$/C Catalysts with Enhanced OSC for the WGS Reaction. *Catalysts* **2015**, *5*, 298–309.

Investigation of Room Temperature Synthesis of Titanium Dioxide Nanoclusters Dispersed on Cubic MCM-48 Mesoporous Materials

Sridhar Budhi, Chia-Ming Wu, Dan Zhao and Ranjit T. Koodali

Abstract: Titania containing cubic MCM-48 mesoporous materials were synthesized successfully at room temperature by a modified Stöber method. The integrity of the cubic mesoporous phase was retained even at relatively high loadings of titania. The TiO_2-MCM-48 materials were extensively characterized by a variety of physico-chemical techniques. The physico-chemical characterization indicate that Ti^{4+} ions can be substituted in framework tetrahedral positions. The relative amount of Ti^{4+} ions in tetrahedral position was dependent on the order of addition of the precursor. Even at relatively high loadings of titania, no distinct bulk phase of titania could be observed indicating that the titania nanoclusters are well dispersed on the high surface area mesoporous material and probably exist as amorphous nanoclusters. The TiO_2-MCM-48 materials were found to exhibit 100% selectivity in the cyclohexene oxidation at room temperature in the presence of *tert*-butylhydroperoxide (*t*-BHP) as the oxidant. The results suggest that room temperature synthesis is an attractive option for the preparation of TiO_2-MCM-48 materials with interesting catalytic properties.

Reprinted from *Catalysts*. Cite as: Budhi, S.; Wu, C.-M.; Zhao, D.; Koodali, R.T. Investigation of Room Temperature Synthesis of Titanium Dioxide Nanoclusters Dispersed on Cubic MCM-48 Mesoporous Materials. *Catalysts* **2015**, *5*, 1603–1621.

1. Introduction

The report of M41S series of periodic mesoporous materials in 1992 spurred excitement and brought a dramatic transformation in the field of porous materials [1,2]. The availability of these types of mesoporous materials, also helped push new frontiers in several interdisciplinary fields, notably catalysis [3], adsorption towards remediation of aqueous pollutants [4], and drug delivery [5]. The discovery has also led to the development of a series of periodic mesoporous materials from prominent groups worldwide, heralding a new era towards the development of porous materials with interesting pore geometries and topologies [6–10].

The M41S series of mesoporous materials comprises of three types, MCM-41, MCM-48, and MCM-50. Among these materials, the cubic MCM-48 material is an interesting material [11,12]. This is because MCM-48 consists of two continuous intersecting network of pores that leads to effective molecular trafficking of reactant(s) and product(s), thus minimizing clogging of pores and leading to enhanced catalytic reactivities in comparison to the uni-dimensional set of pores that occur in TiO_2-MCM-41 [13]. Although, MCM-48 is a favorable support material, literature reports regarding its use is lower in comparison to MCM-41. This is because the synthesis of the cubic MCM-48 phase is challenging and is formed only in a narrow range of conditions and is very sensitive to small deviations in the experimental conditions. In addition, the synthesis can be laborious and can take several days [14]. We had reported a facile method for the rapid and reproducible synthesis of MCM-48 [15]. The advantages of this method are that a readily available and common cationic surfactant, cetyltrimethylammonium bromide (CTAB) can be used as the surfactant, thus avoiding the need for the use of specialized gemini surfactants [16], the synthesis can be conducted at room temperature, and the cubic phase can be formed in thirty minutes.

Titanium supported mesoporous materials have attracted the attention of several researchers [17–19]. In particular, the oxidation of aromatics [20,21], oxidation of alkenes [22–28] and unsaturated alcohols [29], and oxidation of thioethers [30] have been successfully demonstrated using titania supported on mesoporous substrates such as MCM-41, SBA-15, MCM-48, etc. In contrast to the use of MCM-41 and SBA-15, studies involving MCM-48 as a support to disperse titania nanoclusters for oxidation reactions are relatively scarce.

Titania based MCM-48 materials have been utilized for selective oxidation of styrene [31], 2,6-di-tert-butylphenol [32,33], methyloleate [34], and cyclohexene [25,35–39]. TiO_2-MCM-48 prepared by a hydrothermal method was examined for the catalytic oxidation of cyclohexene. The results suggest that the corresponding alcohol, diol, ketone, and epoxide were formed with a selectivity of only 4.7% for cyclohexene oxide [35]. In another study, TiO_2-MCM-48 prepared by hydrothermal method and without any sodium ions exhibited higher activity (initial rate being four times higher) for the oxidation of cyclohexene in comparison to a material prepared using sodium ions [36]. However, the paper did not identify the products formed and the selectivity was not reported. TiO_2-MCM-48 was prepared by a hydrothermal method and the oxidation of cyclohexene was carried out under solvent free conditions in the presence of tert-butylhydroperoxide (t-BHP) as the oxidant. The results indicate that the predominant product was cyclohexene oxide [37]. In another study, TiO_2-MCM-48 was prepared by a hydrothermal method at 150 °C for 20 h and using a gemini surfactant. The oxidation of cyclohexene was chosen as a test reaction and the turnover frequency was found to be 5.1 (h^{-1}) [38]. The epoxidation of cyclohexene with aqueous hydrogen peroxide over mesostructured $Ti(Cp)_2Cl_2$-grafted TiO_2-MCM-48 was examined in another study [39]. It was observed that the TiO_2-MCM-48 catalyst exhibited higher activity in comparison to

2

TiO$_2$-MCM-41. Gemini surfactants were used for the preparation of high quality TiO$_2$-MCM-48 mesoporous materials by a hydrothermal method and TiO$_x$ layers were grafted onto the MCM-48 matrix. The turnover number was found to be ~27 after 2 h for the oxidation of cyclohexene [40].

The literature reports pertaining to TiO$_2$-MCM-48 indicate that synthesis methods involve long preparation times and all previous synthetic procedures were conducted under hydrothermal conditions at temperatures >110 °C. It is attractive to pursue synthetic methods for the preparation of TiO$_2$-MCM-48 mesoporous materials at room temperature. However, the synthesis of TiO$_2$-MCM-48 is quite challenging at room temperature. This is because the hydrolysis rates of the titania and silicon alkoxides are significantly different because of differences in the partial charge of the central metal atom, *i.e.*, Ti^{4+} and Si^{4+}. The partial charges of Ti^{4+} and Si^{4+} in titanium isopropoxide and tetraethyl orthosilicate are ~+0.61 and +0.32, respectively. This means that titanium isopropoxide undergo hydrolysis at almost twice the rate of tetraethyl orthosilicate. The differences in the rate of the hydrolysis lead to challenges in the reproducible preparation of TiO$_2$-MCM-48. Thus, the preparation of TiO$_2$-MCM-48 is extremely challenging in comparison to the siliceous form of MCM-48 (which in itself is extremely sensitive to the experimental parameters). In this work, we diligently examined the various factors affecting the synthesis, and optimized conditions for the reproducible synthesis of TiO$_2$-MCM-48. In a previous study, we reported the photocatalytic production of hydrogen using TiO$_2$-MCM-48 [41].

Hence, our present work is guided by the following factors: (i) use of a relatively facile method, *i.e.*, room temperature synthesis method for the preparation of TiO$_2$-MCM-48 catalysts thus minimizing synthesis procedures that are quite time-intensive; and (ii) lack of exploration of catalytic oxidation of organics using TiO$_2$-MCM-48 mesoporous catalysts prepared at room temperature. Thus, the present work is expected to provide guidance towards the reproducible room temperature method for the preparation of TiO$_2$-MCM-48 mesoporous materials with different titania loadings and their catalytic performance for oxidation of cyclohexene.

2. Results and Discussion

2.1. Physico-Chemical Characterization

As stated earlier, the reproducible and facile synthesis of titania based MCM-48 at room temperature is challenging and hence we examined four different methods for the preparation as discussed in the Experimental Section based on our previous experience [41]. The synthesis method is based on a modified Stöber synthesis developed by us for the preparation of MCM-48 [15]. Figure 1 shows the powder XRD data for the calcined TiO$_2$-MCM-48-200 materials prepared by the four different methods. The diffractograms show XRD patterns typical of the cubic phase with $Ia3d$ symmetry. The presence of a strong peak near $2\theta = 2.5°$ due to d_{211} diffraction planes and in particular the presence of a weak d_{220} reflection peak near $3.2°$ is indicative of the presence of the cubic phase. In addition, four additional reflections are seen in the 2θ range of $4°$ and $6°$, suggesting the high quality of the cubic phase. The results suggest that the rapid and facile synthesis method developed by us previous for the synthesis of MCM-48 [15] can be successfully extended to the preparation of TiO_2-MCM-48. One can also notice that there are some differences in the Full Width at Half Maxima (FWHM) in the four materials. The FWHM of the diffraction peak is a measure of the size distribution of the unit cell. The two materials, TiO_2-MCM-48-B-200 and TiO_2-MCM-48-D-200 show relatively smaller FWHM in comparison to the other two materials, TiO_2-MCM-48-A-200 and TiO_2-MCM-48-C-200. In addition to changes in the FWHM, one can also notice that the preparation methods A and B leads to a relative higher intensity of the d_{211} diffraction plane in comparison to methods, C and D. These differences can be attributed to variations in the order of addition of the precursors, which cause minor changes in the unit cell volume and in the FWHM (*i.e.*, size distribution of the unit cell). Figure 1B shows the long range XRD of all the TiO_2-MCM-48-200 materials prepared by the four methods. No peaks due to bulk titania are observed, indicating that the titania nanoclusters are well dispersed and perhaps amorphous and/or crystalline with small crystallite sizes (<3 nm), thus precluding their detection from XRD studies. A broad peak centered at 2θ near $25°$ can be observed in this material. This peak is assigned to the amorphous silica support.

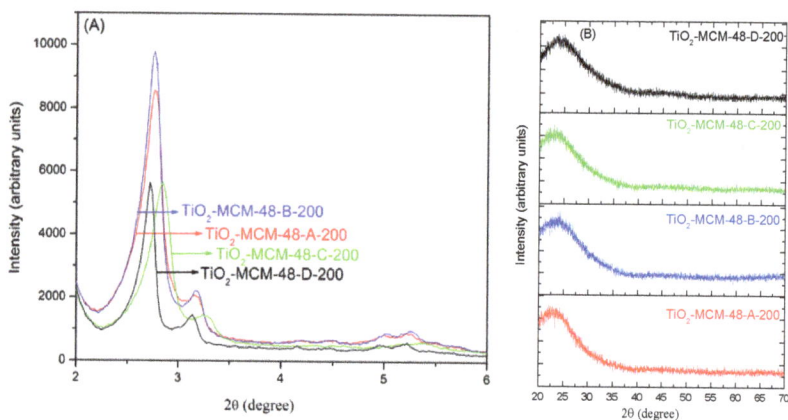

Figure 1. XRD patterns of TiO_2-MCM-48-200 materials (**A**) short and (**B**) long range.

In summary, the powder XRD studies indicate that high quality cubic phase can be formed at room temperature in as little as four hours. The absence of any peaks due to bulk titania indicate that the titania nanoclusters are amorphous and/or have small crystallite sizes (<3 nm).

We were also interested in preparing TiO$_2$-MCM-48 materials with different titania loadings in order to check if the room temperature synthesis method developed by us can be extended to high loadings of titania. For this purpose, method B (which exhibited the highest specific surface area) was adapted for the preparation of TiO$_2$-MCM-48 materials. The XRD data of TiO$_2$-MCM-48 materials with different loadings, ranging from Si/Ti = 200 to Si/Ti = 10 is shown in Figure 2. As shown in Figure 2, all materials form the cubic phase, evident from the presence of the small peak near $2\theta = 3.2°$, that is due to d_{220} diffraction planes. In addition, there are subtle differences in the FWHM and the relative intensity of the peak near $2\theta = 2.5°$, that is due to d_{211} diffraction planes. These differences can be attributed to the variation in the loading of titania that affects the volume and the size distribution of the unit cell. The inset in Figure 2 shows the long range XRD of TiO$_2$-MCM-48-B-10, i.e., the material with the highest titania loading. The results indicate that even at such high loadings, no peaks due to anatase phase of titania can be seen. This indicates that once again, that the titania nanoclusters are well dispersed on the MCM-48 support. Our results indicate that the intensity of the d_{211} diffraction plane does not monotonically decrease with titania loading suggesting that there are variations in the dispersion of titania in the MCM-48 matrix with increased loadings. The long range XRD of other materials are also similar and are not hence shown. In summary, the powder XRD results indicate that one can prepare TiO$_2$-MCM-48 mesoporous materials with relatively high loadings of titania (e.g., Si/Ti = 10) without destroying the periodicity of the cubic, MCM-48 mesophase even at room temperature. Also, the titania species are well dispersed and perhaps amorphous in nature. Raman studies of these materials also indicate the absence of peaks due to titania indicating the well dispersed and amorphous nature of the titania nanoclusters.

Nitrogen adsorption experiments were carried out for all the TiO$_2$-MCM-48 materials at 77 K. Figure 3 presents the nitrogen physisorption isotherms of all TiO$_2$-MCM-48-200 mesoporous materials. The isotherms belong to type IV class of porous materials. This shows the mesoporous nature of pores in these materials. At low relative pressure values ($P/P_0 < 0.2$), monolayer adsorption of nitrogen molecules takes place. At relative pressure (P/P_0) values in the range of 0.2 and 0.35, there is a relatively large inflection. This can be attributed to capillary condensation of nitrogen within the mesopores of the MCM-48 material and suggests the presence of highly ordered and periodic mesoporous nature in these materials. One can also notice that the position of the inflection in all the cubic phased mesoporous materials in this study is similar. This suggests that the pore sizes of these materials are nearly the same as indicated in Table 1. The specific surface area of the materials ranges from 1200 m^2/g to 1687 m^2/g whereas the pore diameter (estimated by using the

BJH equation to the desorption isotherm) is nearly 21 Å. The pore volume in this set of materials were found to be >0.7 cm^3/g.

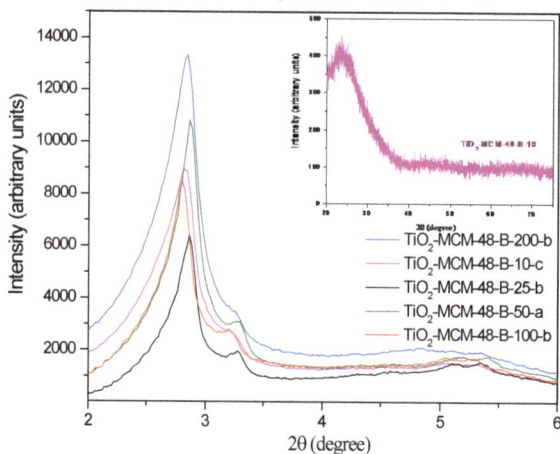

Figure 2. The short range powder XRD patterns of TiO$_2$-MCM-48 materials prepared by method B. The inset shows the long range XRD of TiO$_2$-MCM-48-B-10.

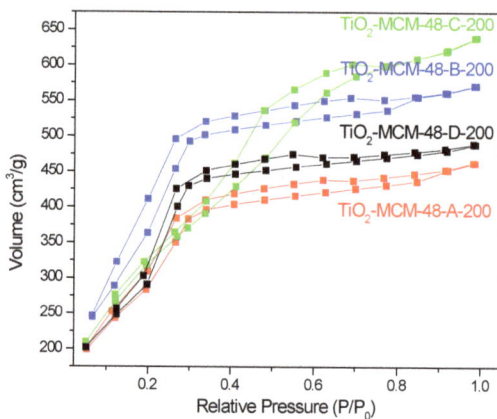

Figure 3. Nitrogen isotherms of TiO$_2$-MCM-48-200 materials.

The isotherms of TiO$_2$-MCM-48 mesoporous materials prepared with different titania loadings (Si/Ti = 100, 50, 25, and 10) were similar in nature to the materials with Si/Ti ratio of 200 and are hence not shown. The specific surface area of these materials are in general high, excepting for TiO$_2$-MCM-48-100, that reproducibly and consistently showed lower specific surface area (near 900 m^2/g) for reasons unknown at this moment and beyond the scope of this investigation. The pore volumes were in general high (>0.7 cm^3/g) excepting for TiO$_2$-MCM-48-100, whereas the pore sizes were fairly uniform and nearly 21 Å irrespective of the loading of titania. A summary of the textural properties is shown in Table 1.

The variations in the physico-chemical properties can be understood as follows. In method A, the titania precursor is added at the very end. In this method, hydrolysis and condensation of silica has already been initiated and a silica-surfactant mesophase has been pre-formed prior to the addition of the titania precursor. Since the composition of the synthesis mixture prior to addition of the titania precursor is an optimized and reported procedure by us previously [15], the addition of an ethanolic solution of the titania precursor at the very end does not significantly affect the quality of the cubic phase. Thus, as indicated previously, a relatively high intensity of the d_{211} plane can be observed for TiO$_2$-MCM-48-A-200 as shown in Figure 1. However, some of the titania nanoclusters can occlude the mesopores because they can precipitate (albeit slowly) under the experimental conditions employed. This results in a relatively lower specific surface area and pore volume. Method D is similar to method A, excepting that titanium alkoxide is used without any dilution. In such a situation, rapid hydrolysis of the titania precursor under the experimental conditions (of high pH) results in occlusion of pores by the titania nanoclusters. This lowers the phase contrast between the pores and the pore walls and hence the relative intensity of the d_{211} plane is lower as indicated in Figure 1. The pore volumes obtained in methods A and D seem similar with values of ~0.7 cm^3/g, while there seems to be differences in the specific surface areas, which are not fully clear at this moment. In method B, titanium alkoxide is added to the cationic surfactant first. This situation is akin to the Evaporation-Induced Self-Assembly (EISA) process. The interactions between the titania oligomers and the CTAB surfactant results in the formation of a hybrid titaniatropic phase of the type, Ti–OH$^+$ ... X$^-$ ... CTAB$^+$, where X$^-$ represents bromide ions in this work. The silica precursor (negatively charged since the solution pH is ~10.5) interacts with the already formed and well-organized titaniatropic mesophase through the cationic surfactant. Hence, in this method, a highly periodic mesostructure is formed as indicated by powder XRD studies and the specific surface areas are relatively large. In method C, the formation of the titaniatropic mesophase is presumed to be relatively slower in comparison to method B. This is because pre-hydrolysis of titanium alkoxides in polar solvents such as ethanol promotes solvation as explained by us previously [42]. It seems that the titaniatropic mesophase undergoes reorganization through evaporation of the solvent as explained by us in a recent review [43]. This results in a relatively lower quality cubic phase as noted by the fact that the intensity of the d_{211} plane is lower in the material prepared by method C in comparison to method B. In summary, the results suggest that pre-hydrolysis of titanium alkoxide in ethanol (methods B and D) is less preferable since they result in materials with relatively lower textural properties.

7

Table 1 lists the Si/Ti ratios in the synthesis gel and the actual Si/Ti ratios in the products determined by AAS analyses. The amounts of titania in materials prepared by methods A and C are higher in comparison to methods B and C, indicating that changes in the order of addition also cause variations in the actual amount of titania incorporated onto the cubic MCM-48 matrix. In summary, AAS results indicate that the Si/Ti ratios determined by AAS to be close to the theoretical values, indicating good incorporation of most of the titania precursor on the mesoporous silica support.

Table 1. Summary of physico-chemical properties of TiO_2-MCM-48 mesoporous materials.

Catalyst	Si/Ti ratio (Synthesis gel)	Si/Ti ratio (AAS)	Ratio of $Ti_{tet}^{4+}/Ti_{oct}^{4+}$	Specific Surface Area (m^2/g)	Pore Volume (cm^3/g)	Pore Diameter (Å)
TiO_2-MCM-48-A-200	200	189	1.61	1241	0.71	22
TiO_2-MCM-48-B-200	200	239	1.23	1687	0.98	21
TiO_2-MCM-48-C-200	200	177	1.51	1200	0.88	23
TiO_2-MCM-48-D-200	200	223	1.39	1436	0.75	21
TiO_2-MCM-48-B-100	100	87	1.60	898	0.53	20
TiO_2-MCM-48-B-50	50	47	1.23	1280	0.65	21
TiO_2-MCM-48-B-25	25	20	1.53	1563	0.82	21
TiO_2-MCM-48-B-10	10	13	0.91	1563	0.82	24

Transmission Electron Microscopic (TEM) studies were conducted for selected mesoporous materials and are shown in Figure 4.

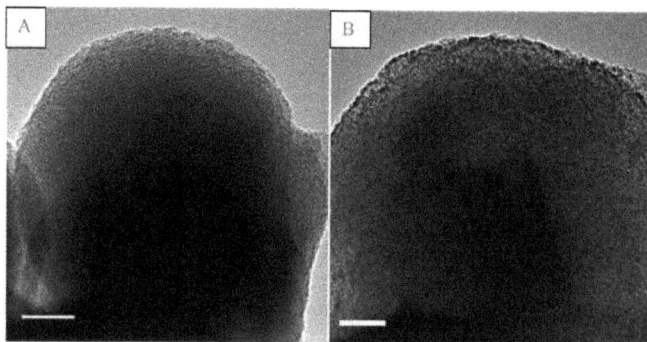

Figure 4. Representative TEM images of (**A**) TiO_2-MCM-48-B-200 (**left**) and (**B**) TiO_2-MCM-48-B-10 (**right**). The bar scale is 50 nm.

Figure 4 shows representative TEM images of TiO_2-MCM-48-B-200 and TiO_2-MCM-48-B-10 with Si/Ti ratios of 200 and 10. The materials exhibit long-range ordered cubic-typed pore structure of MCM-48. The incorporation of titania did not influence the morphology and the integrity of the pore structure even at high loadings, and is thus consistent with the results obtained from powder XRD studies.

A low magnification image of a representative cubic MCM-48 material is shown in Figure S1 indicating the nearly spherical shapes of the mesoporous materials. The bicontinuous and interpenetrating network of pores in MCM-48 poses unique and inherent challenges in imaging the cubic materials from our experiences. A relatively high magnification of a representative titania containing material is shown in Figure S2. The TEM images do not indicate the presence of any crystalline phase of titania in the TiO_2-MCM-48-B-10 material suggesting that the titania nanoclusters are perhaps amorphous. In addition, our TEM images do not indicate the presence of any large and aggregated titania particles in the TiO_2-MCM-48-200 set of materials. Our results are consistent with previous reports by Gies et al. [44]. In their work, despite relatively high loadings of titania, (13–16 wt. %), no large titania particles could be observed from their TEM studies. Our TEM data is consistent with UV-Vis Diffuse Reflectance Spectra (DRS) results that are discussed later. The absence of peaks near 370 nm due to bulk titania is a further confirmation of the absence of large titania crystals [44]. In summary, TEM studies indicate that the titania species are well dispersed on the cubic mesoporous support and probably are amorphous in nature.

Electron Spin Resonance (ESR) studies were conducted in order to better understand the geometry of the titania species. The TiO_2-MCM-48 mesoporous materials were illuminated in the presence of methanol as the sacrificial hole scavenger to suppress charge-carrier recombination. Figure 5 shows the ESR spectra of Ti-MCM-8 materials prepared by method B after UV irradiation at 77 K for 10 min. The ESR spectra were measured in the dark at 4.5 K. The samples were ESR silent prior to irradiation. A strong signal near $g = 2.004$ is observed in all spectra and is this is due to organic radicals (predominantly $•CH_2OH$ with a ratio of 1:2:1) formed by the reaction of the photogenerated holes in TiO_2-MCM-48 materials with methanol. The ESR spectra of TiO_2-MCM-48 mesoporous materials with Si/Ti ratio of 200 indicate the presence of Ti^{3+} species (Figure 5) with $g\perp = 1.951$ and $g\| = 1.910$. We have previously observed two sets of ESR signals (signal A and B) in TiO_2-MCM-48 materials prepared by first impregnating MCM-48 with titanium isopropoxide and then calcining at 550 °C [13]. Signal A with $g\perp = 1.952$ and $g\| = 1.902$ was attributed to titanium atoms in tetrahedral coordination in accordance with prior literature [45]. In addition, signal B, with $g\perp = 1.988$ and $g\| = 1.957$ was attributed to titanium atoms in distorted octahedral coordination. The presence of broad (albeit weak) signals in this study with g values similar to previous reports indicates the presence of titanium ions predominantly in tetrahedral coordination at low titania loadings (Si/Ti = 200) [46–48]. The presence of fairly broad and weak peaks precludes us from clearly observing the presence of titanium ions in octahedral coordination. Thus, we performed Diffuse Reflectance Spectroscopic (DRS) studies.

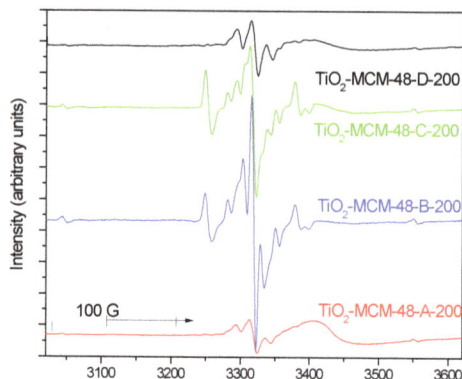

Figure 5.   ESR spectra of TiO_2-MCM-48-200 materials prepared by four different methods.

The UV-Vis DRS spectra of the TiO_2-MCM-48-200 materials are shown in Figure 6A–D. Peaks in the 210 to 230 nm range have been previously attributed to ligand-to-metal charge-transfer (LMCT) from O^{2-} to Ti^{4+} due to the presence of tetrahedrally coordinated Ti^{4+} ions [13,41], whereas the presence of octahedrally coordinated Ti^{4+} ions (or highly dispersed titania nanoparticles) can be discerned from the appearance of a peak in the 280 to 310 nm range. In order to identify the relative ratios of the Ti^{4+} ions in tetrahedral and octahedral coordination, the DRS bands were de-convoluted using Origin Pro 9.0 software (OriginLab Corporation, Northampton, MA, USA). The peak centers were fixed at 220 and 280 nm, respectively, for Ti^{4+} ions in tetrahedral and octahedral geometry. The relative ratios of Ti^{4+} ions in tetrahedral and octahedral coordination was then estimated by calculating the peak intensity from the baseline to the top of the peak center with width fixed at 220 ± 10 nm and 280 ± 10 nm in TiO_2-MCM-48-A-200, TiO_2-MCM-48-C-200, and TiO_2-MCM-48-D-200 as reported by us previously [49]. In TiO_2-MCM-48-B-200, the peak center with width was fixed at 220 ± 10 nm and 300 ± 10 nm. This suggests that slightly larger sized titania nanoparticles are formed in this material in comparison to the remaining three materials. Table 1 provides a summary of the ratio of $Ti_{tet}^{4+}/Ti_{oct}^{4+}$ for TiO_2-MCM-48-200 mesoporous materials. The ratio of $Ti_{tet}^{4+}/Ti_{oct}^{4+}$ in TiO_2-MCM-48-A-200, TiO_2-MCM-48-B-200, TiO_2-MCM-48-C-200, and TiO_2-MCM-48-D-200 was estimated to be 1.61, 1.23, 1.51, and 1.39, respectively, suggesting that the method of preparation modulates the amount of Ti^{4+} ions that can be incorporated into the framework tetrahedral positions in the cubic MCM-48 material.

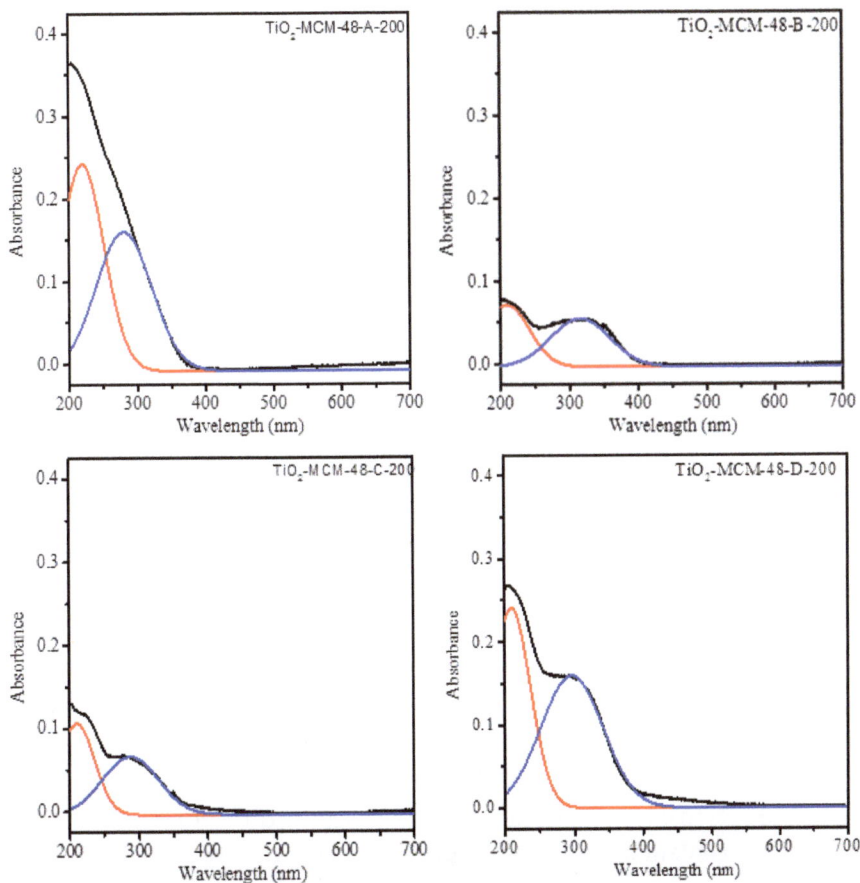

Figure 6. UV-Vis DRS of TiO_2-MCM-48-200 materials prepared by four different methods. The bold black line shows the original UV-Vis spectra whereas the red and blue bold lines indicate the de-convoluted spectra.

The UV-Vis DR spectra of TiO_2-MCM-48 with different loadings are shown in Figure 7 below. In general, one notices that there is a slight red shift in the onset of absorbance with increase in titania loading and also the obvious fact that there is also an increase in the absorbance values with increase in titania. De-convolution of the peaks indicate that the ratio of $Ti_{tet}^{4+}/Ti_{oct}^{4+}$ in TiO_2-MCM-48-10 is significantly lower in comparison to the rest of the four samples suggesting that formation of small sized titania clusters at such high loadings (*i.e.*, Si/Ti = 10). The titania clusters are fairly small in size and are probably amorphous too and are thus not seen in powder XRD studies.

11

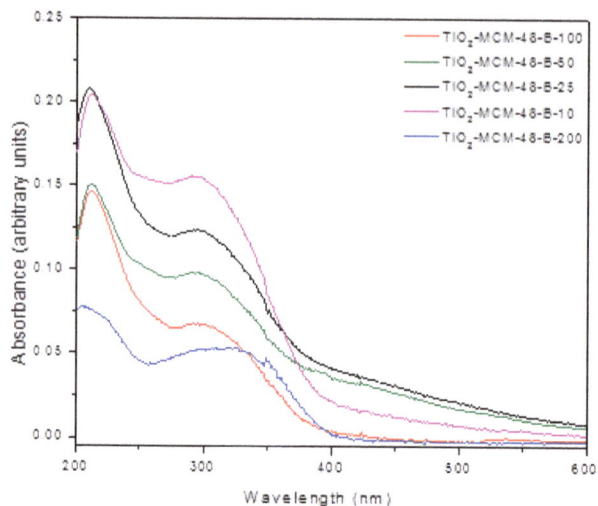

Figure 7. UV-Vis DR spectra of TiO_2-MCM-48-materials with different titania loadings.

2.2. Catalytic Studies

Oxidation reactions have been carried out using microporous TS-1 catalysts [22,46,47]. In contrast to crystalline TS-1 catalysts, the atomic ordering of the pore walls in mesoporous MCM-41 and MCM-48 type materials are amorphous. This often results in conversions and selectivities quite different in the mesoporous materials. TiO_2 containing mesoporous materials show catalytic activity for selective oxidation of organics using hydrogen peroxide or *tert*-butylhydroperoxide (*t*-BHP) under relatively mild conditions. In the present study, we examined the oxidation of cyclohexene using the TiO_2-MCM-48 materials prepared in this study. Preliminary catalytic experiments indicated that a molar ratio of cyclohexene: *t*-BHP = 1:1 was ideal. In addition, the optimal reaction temperature was found to be 25 ± 5 °C, and dichloroethane was the best solvent under the experimental conditions in this study. Control experiments with siliceous MCM-48 and a Degussa P25 indicate no conversion of cyclohexene even after 24 h of reaction indicating that well dispersed titania is essential for oxidation of cyclohexene.

Our initial catalytic experiments were carried out with the materials with low titania loadings since it has been reported that spatially isolated and tetrahedrally coordinated Ti^{4+} show relatively high activity [50]. We found that the catalytic activity for the oxidation of cyclohexene to be in the order, TiO_2-MCM-48-B-200 > TiO_2-MCM-48-A-200 > TiO_2-MCM-48-C-200 > TiO_2-MCM-48-D-200 as indicated in Figure 8.

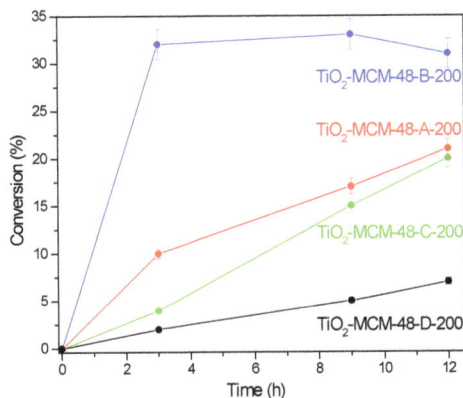

Figure 8. Conversion of cyclohexene over various TiO_2-MCM-48-200 catalysts.

Interestingly, all catalysts showed exclusive and 100% selectivity for the formation of cyclohexene oxide, contrary to a previous report in which other products have been identified [38]. The turnover number (moles of cyclohexene converted/mol Ti) after 12 h of reaction at room temperature was estimated to be 59, 32, 28, and 21 for TiO_2-MCM-48-B-200, TiO_2-MCM-48-A-200, TiO_2-MCM-48-C-200, and TiO_2-MCM-48-D-200 catalysts, respectively. The activity of TiO_2-MCM-48-B-200 (turnover number = 59 after 12 h) seems to be better than the activity of microporous TS-1 (turnover number = 47 after 24 h) previously reported under similar conditions [38], suggesting that the room temperature synthesis of TiO_2-MCM-48 materials is attractive for catalytic oxidation reactions. The ratio of $Ti_{tet}^{4+}/Ti_{oct}^{4+}$ was calculated to be 1.61, 1.23, 1.51, and 1.39 for TiO_2-MCM-48-B-200, TiO_2-MCM-48-A-200, TiO_2-MCM-48-C-200, and TiO_2-MCM-48-D-200, respectively, suggesting that the relative amount of Ti_{tet}^{4+} alone does not dictate the trends in catalytic activity. Indeed, Tuel and Hubert-Pfalzgraf have suggested that the coordination of the titania species may not be the only factor, and that other parameters such as dispersion and particle size of the titania nanoparticle may also play a vital role [51]. This is supported by the fact that Hutter *et al.* found that well dispersed titania nanodomains on TiO_2-SiO_2 are active for the epoxidation of cyclohexene [52]. Thus, our results are consistent with previous reports and that the presence of nanoclusters of titania do not necessarily decrease the catalytic performance (at least at low loadings of titania). Figure S3 shows a plot of specific surface area and ratio of $Ti_{tet}^{4+}/Ti_{oct}^{4+}$ as a function of turnover number. It seems that the high activity of TiO_2-MCM-48-B-200 is due to the relatively high specific surface area and pore volume of this material in comparison to the rest of the materials and hence the high dispersion of titania seems to be responsible for enhancing its activity. No strong correlation seems with the relative amount of Ti_{tet}^{4+} as stated previously.

13

The recycling studies with TiO_2-MCM-48-B-200 indicate a modest decrease in activity after two cycles. Powder XRD analysis of the spent catalyst indicates that the cubic phase is still preserved and the textural properties do not seem to be compromised. We also found that the turnover number of the catalyst decreased progressively with increase in the amount of TiO_2 loading. The turnover number was calculated to be 13, 3, 3, and 2.5 for TiO_2-MCM-48-B-100, TiO_2-MCM-48-B-50, TiO_2-MCM-48-B-25, and TiO_2-MCM-48-B-10, respectively, suggesting that the decrease in dispersion of titania was perhaps responsible for the decrease in catalytic activity. Interestingly, contrary to a previous report [53], we did not see any change in the selectivity towards the formation of cyclohexene oxide at high loadings of titania.

In summary, our catalytic results indicate that high selectivity and good activity can be obtained with titania nanoparticles dispersed on a cubic MCM-48 mesoporous support for the epoxidation of cyclohexene even at room temperature.

3. Experimental Section

3.1. Chemicals Used

The materials used for the synthesis of TiO_2-MCM-48 are ethanol (absolute 200 Proof, AAPER), tetraethyl orthosilicate ($Si(OEt)_4$, 98%, Alfa Aesar, Ward Hill, MA, USA), titanium(IV) isopropoxide (Ti-(^{i}OPr)$_4$, 98%, Alfa Aesar), cetyltrimethylammonium bromide (CTAB, 98%, Alfa Aaser), aq. Ammonia (Alfa Aesar). Deionized water was used throughout the studies. For the catalytic studies cyclohexene (99%, Acros, New Jersey, NJ, USA), dichloroethane (Acros), tert-butylhydroperoxide, and nonane (99%, Acros) as an internal standard were used.

3.2. Synthesis of TiO_2-MCM-48 Mesoporous Materials

A scheme for the four different methods for the preparation of TiO_2-MCM-48 is shown in Table 2 below.

Table 2. Preparation of TiO_2-MCM-48 by different methods.

Method A	Method B	Method C	Method D
CTAB (1.2 g)	CTAB (1.2 g)	CTAB (1.2 g)	CTAB (1.2 g)
H_2O (50 mL)	Ti-(^{i}OPr)$_4$	Ti-(^{i}OPr)$_4$ (Stock) *	H_2O (50 mL)
C_2H_5OH (25 mL)	H_2O (50 mL)	H_2O (50 mL)	C_2H_5OH (25 mL)
NH_3 (6 mL)	C_2H_5OH (25 mL)	C_2H_5OH (25 mL)	NH_3 (6 mL)
$Si(OEt)_4$ (1.8 mL)	NH_3 (6 mL)	NH_3 (6 mL)	$Si(OEt)_4$ (1.8 mL)
Ti-(^{i}OPr)$_4$ (Stock)*	$Si(OEt)_4$ (1.8 mL)	$Si(OEt)_4$ (1.8 mL)	Ti-(^{i}OPr)$_4$

* Preparation of stock solution: For Si/Ti = 200, 70 μL of Ti (^{i}OPr)$_4$ was dissolved in 10 mL of ethanol. This served as the stock solution. One milliliter of this stock solution was used for preparing TiO_2-MCM-48 with a Si/Ti ratio of 200.

In a 120 polypropylene beaker, 1.2 g (3.3 mmol) of CTAB were added to 50 mL of deionized water and CTAB was allowed to dissolve fully by stirring rapidly for a few min. Then, 25 mL of ethanol were added to the CTAB solution. To this mixture, 6 mL (0.09 mmol) of aqueous ammonia and the silica precursor, TEOS 1.8 mL were added. The titanium precursor (Ti-(^{i}OPr)$_4$ was dissolved in 10 mL of ethanol and served as the stock solution. In method A, 1 mL of the stock solution of Ti-(^{i}OPr)$_4$ was added at the very end as shown in Table 2. In method B, the required amount of Ti-(^{i}OPr)$_4$ solution (without dilution) was added to the CTAB solution as indicated in the Scheme below. In method C, the stock solution of Ti-(^{i}OPr)$_4$ was added after the addition of CTAB. In method D, Ti-(^{i}OPr)$_4$ (without dilution) was added to Si(OEt)$_4$ solution. After the addition of all the ingredients, the mixture was stirred for 4 h at 300 rpm at room temperature and the resulting TiO$_2$-MCM-48 was recovered by filtration, dried in an oven overnight, and then calcined at a heating rate of 3 °C/min at 550 °C for 6 h to remove the organic template. The mesoporous material, TiO$_2$-MCM-48 prepared with a Si/Ti ratio of 200 by method A is named as TiO$_2$-MCM-48-A-200. Other materials prepared in this work are named in a similar manner.

3.3. Characterization Techniques

Powder X-ray diffraction (XRD) measurements were done using a Rigaku Ultima IV X-ray diffractometer (Rigaku Corporation, Tokyo, Japan) with Cu Kα radiation (λ = 1.5408 Å) at room temperature. The diffractometer was typically operated at 1.76 kW (Voltage = 44 kV and Current = 40 mA) and scanned with a step size of 0.02°. The low angle regions were scanned from 2θ = 2°–6° with a step size of 0.02°. Wide angle XRD measurements were also made for selected mesoporous materials in the 2θ range of 20°–75°.

The nitrogen physisorption studies were done at 77 K using Nova 2200e analyzer (Quantachrome Instruments, Boynton Beach, FL, USA). The materials were dried in an oven at 373 K for overnight prior to the day of analysis. The materials were then degassed at 373 K extensively before the study. The Brunauer-Emmett-Teller (BET) equation was used to calculate the specific surface area from the adsorption data obtained at relative pressure P/P_0 values between 0.05–0.30. The total pore volume of TiO$_2$-MCM-48 mesoporous materials were calculated from the amount of nitrogen adsorbed at highest relative pressure ratio P/P_0 ~ 0.99. The pore size distribution was calculated by analyzing the N$_2$ isotherm using the Barrett-Joyner-Halenda (BJH) method and by applying the BJH equation to the desorption isotherm.

Transmission electron microscopic (TEM) images were acquired using a FEI Tecnai G^2 F30 instrument (FEI, Hillsboro, TX, USA) at an accelerating voltage of 200 kV. The materials for TEM studies were prepared by first sonicating ~2 mg of TiO$_2$-MCM-48 in 10 mL of ethanol for at least 30 min. One drop of this dispersion was carefully deposited on a carbon coated copper grid (200 mesh). The grid was allowed to dry at room temperature overnight before TEM analysis.

The Electron Spin Resonance (ESR) experiments were conducted in the X-band (continuous wave) using a Bruker Elexsys E580 spectrometer (Bruker, Billerica, MA, USA). This instrument was equipped with an Oxford CF935 helium flow cryostat and a ITC-5025 temperature controller (Oxford Instruments, Oxfordshire, UK). The mesoporous materials were dispersed in water-methanol mixture and were purged with argon gas to remove oxygen. The suspension was then illuminated at 77 K using a 300-W Xe UV lamp source (ILC) (Newport Corporation, Santa Clara, CA, USA). The EPR spectra were then recorded immediately at 5 K immediately after illumination. The g factors were calibrated by comparison to a coal standard, with $g = 2.00285 \pm 0.00005$.

The UV-Vis diffuse reflectance spectra were recorded using a Cary 100 Bio UV-Visible spectrophotometer (Varian Inc., Palo Alto, CA, USA). This instrument was equipped with a praying mantis diffuse reflection accessory (Harrick Scientific). A Si-MCM-48 material was used for baseline correction and the DR spectra were recorded at room temperature.

A Thermo Jarrell Ash Atomic Absorption (Thermo Jarrell Ash Corporation, Franklin, MA, USA) Spectrophotometer was used to determine the titanium content present in the calcined TiO_2-MCM-48 materials prepared by different methods and different Si/Ti ratios after dissolution of the materials in $HF:HNO_3$ solution. The silica was carefully filtered, and the supernatant solution was diluted to known volumes. The solutions were analyzed for Ti^{4+} ions.

The catalytic reactions were performed as described in the following manner. In a typical catalytic reaction, 19.3 mL of dichloroethane (DCE) was added to a single-neck round bottom flask. To this, *tert*-butylhydroperoxide (*t*-BHP, 0.32 mL, 2 mmol) in decane was added and reaction mixture was stirred for 15 min. Then, 0.06 g of the TiO_2-MCM-48 catalyst, nonane (0.2 mL), and cyclohexene (0.2 mL, 2 mmol) were added, respectively. The suspension was refluxed at room temperature ($25 \pm 5\,°C$) for various intervals of time. After the completion of the reaction, the catalyst was recovered, washed with acetone, and dried in an air oven at $80 \pm 10\,°C$. The supernatant solution (after separation of the catalyst) was injected into a GC-MS (Shimadzu QP 5000, Shimadzu Corporation, Kyoto, Japan). The GC was equipped with a silica column (J&W Scientific, Folsom, CA, USA, 122-5532, DB-5ms equivalent to a (5% phenyl)methyl polysiloxane, 30 m × 0.25 mm). The yield of the product (cyclohexene oxide) was determined using the internal standard method. After the catalytic reaction, the suspension was centrifuged and the catalyst was recovered. This was then washed with acetone and then subsequently dried in air, and tested for recycling experiments.

4. Conclusions

Room temperature synthesis of titania supported cubic MCM-48 mesoporous materials were successfully demonstrated. It was found that the order of addition of the titania precursor modulated the relative amount of Ti^{4+} ions in tetrahedral and octahedral coordination. The results indicate that the titania nanoclusters are well dispersed in the high surface area mesoporous matrix and no bulk phase of titania was detected even at relatively high titania loadings. The titania supported MCM-48 materials show excellent selectivity towards the formation of cyclohexene oxide from cyclohexene at room temperature using *tert*-butylhydroperoxide as the oxidant.

Acknowledgments: Ranjit T. Koodali thanks the South Dakota Center for Research and Development of Light-Activated Materials (CRDLM). We are thankful to Nada Dimitrijevic and Tijana Rajh, Argonne National Laboratory for help with ESR studies. Support from NSF-EPSCoR (EPS-0554609), NSF-CHE-0840507, and NSF-CHE-0722632 are gratefully acknowledged.

Author Contributions: All authors in this study were involved in the design of the experiments. Sridhar Budhi was primarily involved in the synthesis and catalytic studies; Dan Zhao and Chia-Ming Wu were involved in the characterization and analysis of data. All authors read and approved the final version of the manuscript.

Conflicts of Interest: The authors declare no conflict of interest.

References

1. Beck, J.S.; Vartuli, J.C.; Roth, W.J.; Leonowicz, M.E.; Kresge, C.T.; Schmitt, K.D.; Chu, C.T.W.; Olson, D.H.; Sheppard, E.W.; McCullen, S.B.; *et al.* A new family of mesoporous molecular-sieves prepared with liquid-crystal templates. *J. Am. Chem. Soc.* **1992**, *114*, 10834–10843.

2. Kresge, C.T.; Leonowicz, M.E.; Roth, W.J.; Vartuli, J.C.; Beck, J.S. Ordered mesoporous molecular-sieves synthesized by a liquid-crystal template mechanism. *Nature* **1992**, *359*, 710–712.

3. Schuth, F. Engineered porous catalytic materials. *Annu. Rev. Mater. Res.* **2005**, *35*, 209–238.

4. Nooney, R.I.; Kalyanaraman, M.; Kennedy, G.; Maginn, E.J. Heavy metal remediation using functionalized mesoporous silicas with controlled macrostructure. *Langmuir* **2001**, *17*, 528–533.

5. Vallet-Regi, M.; Balas, F.; Arcos, D. Mesoporous materials for drug delivery. *Angew. Chem. Int. Ed.* **2007**, *46*, 7548–7558.

6. Zhao, D.Y.; Feng, J.L.; Huo, Q.S.; Melosh, N.; Fredrickson, G.H.; Chmelka, B.F.; Stucky, G.D. Triblock copolymer syntheses of mesoporous silica with periodic 50 to 300 angstrom pores. *Science* **1998**, *279*, 548–552.

7. Huo, Q.S.; Margolese, D.I.; Ciesla, U.; Feng, P.Y.; Gier, T.E.; Sieger, P.; Leon, R.; Petroff, P.M.; Schuth, F.; Stucky, G.D. Generalized synthesis of periodic surfactant inorganic composite-materials. *Nature* **1994**, *368*, 317–321.

8. Tanev, P.T.; Pinnavaia, T.J. A neutral templating route to mesoporous molecular-sieves. *Science* **1995**, *267*, 865–867.

9. Joo, S.H.; Choi, S.J.; Oh, I.; Kwak, J.; Liu, Z.; Terasaki, O.; Ryoo, R. Ordered nanoporous arrays of carbon supporting high dispersions of platinum nanoparticles. *Nature* **2001**, *412*, 169–172.

10. Yanagisawa, T.; Shimizu, T.; Kuroda, K.; Kato, C. The preparation of alkyltrimethylammonium-kanemite complexes and their conversion to microporous materials. *Bull. Chem. Soc. Jpn.* **1990**, *63*, 988–992.

11. Vartuli, J.C.; Schmitt, K.D.; Kresge, C.T.; Roth, W.J.; Leonowicz, M.E.; McCullen, S.B.; Hellring, S.D.; Beck, J.S.; Schlenker, J.L.; Olson, D.H.; *et al.* Effect of surfactant silica molar ratios on the formation of mesoporous molecular-sieves: Inorganic mimicry of surfactant liquid-crystal phases and mechanistic implications. *Chem. Mater.* **1994**, *6*, 2317–2326.

12. Kresge, C.T.; Roth, W.J. The discovery of mesoporous molecular sieves from the twenty year perspective. *Chem. Soc. Rev.* **2013**, *42*, 3663–3670.

13. Peng, R.; Zhao, D.; Dimitrijevic, N.M.; Rajh, T.; Koodali, R.T. Room temperature synthesis of Ti-MCM-48 and Ti-MCM-41 mesoporous materials and their performance on photocatalytic splitting of water. *J. Phys. Chem. C* **2012**, *116*, 1605–1613.

14. Bronkema, J.L.; Bell, A.T. Mechanistic studies of methanol oxidation to formaldehyde on isolated vanadate sites supported on MCM-48. *J. Phys. Chem. C* **2007**, *111*, 420–430.

15. Boote, B.; Subramanian, H.; Ranjit, K.T. Rapid and facile synthesis of siliceous MCM-48 mesoporous materials. *Chem. Commun.* **2007**, 4543–4545.

16. Huo, Q.S.; Leon, R.; Petroff, P.M.; Stucky, G.D. Mesostructure design with gemini surfactants: Supercage formation in a 3-dimensional hexagonal array. *Science* **1995**, *268*, 1324–1327.

17. Tanev, P.T.; Chibwe, M.; Pinnavaia, T.J. Titanium-containing mesoporous molecular-sieves for catalytic-oxidation of aromatic-compounds. *Nature* **1994**, *368*, 321–323.

18. Wu, P.; Tatsumi, T.; Komatsu, T.; Yashima, T. Postsynthesis, characterization, and catalytic properties in alkene epoxidation of hydrothermally stable mesoporous Ti-SBA-15. *Chem. Mater.* **2002**, *14*, 1657–1664.

19. Anand, R.; Hamdy, M.S.; Gkourgkoulas, P.; Maschmeyer, T.; Jansen, J.C.; Hanefeld, U. Liquid phase oxidation of cyclohexane over transition metal incorporated amorphous 3D-mesoporous silicates M-TUD-1 (M = Ti, Fe, Co and Cr). *Catal. Today* **2006**, *117*, 279–283.

20. Eimer, G.A.; Casuscelli, S.G.; Ghione, G.E.; Crivello, M.E.; Herrero, E.R. Synthesis, characterization and selective oxidation properties of Ti-containing mesoporous catalysts. *Appl. Catal. A* **2006**, *298*, 232–242.

21. Vinu, A.; Srinivasu, P.; Miyahara, M.; Ariga, K. Preparation and catalytic performances of ultralarge-pore Ti-SBA-15 mesoporous molecular sieves with very high Ti content. *J. Phys. Chem. B* **2006**, *110*, 801–806.

22. Blasco, T.; Corma, A.; Navarro, M.T.; Pariente, J.P. Synthesis, characterization, and catalytic activity of Ti-MCM-41 structures. *J. Catal.* **1995**, *156*, 65–74.

23. Koyano, K.A.; Tatsumi, T. Synthesis of titanium-containing mesoporous molecular sieves with a cubic structure. *Chem. Commun.* **1996**, 145–146.

24. Chen, L.Y.; Chuah, G.K.; Jaenicke, S. Ti-containing mcm-41 catalysts for liquid phase oxidation of cyclohexene with aqueous H_2O_2 and tert-butyl hydroperoxide. *Catal. Lett.* **1998**, *50*, 107–114.

25. Pena, M.L.; Dellarocca, V.; Rey, F.; Corma, A.; Coluccia, S.; Marchese, L. Elucidating the local environment of Ti(iv) active sites in Ti-MCM-48: A comparison between silylated and calcined catalysts. *Microporous Mesoporous Mater.* **2001**, *44*, 345–356.

26. Ji, D.; Zhao, R.; Lv, G.M.; Qian, G.; Yan, L.; Suo, J.S. Direct synthesis, characterization and catalytic performance of novel Ti-SBA-1 cubic mesoporous molecular sieves. *Appl. Catal. A* **2005**, *281*, 39–45.

27. Berube, F.; Khadhraoui, A.; Janicke, M.T.; Kleitz, F.; Kaliaguine, S. Optimizing silica synthesis for the preparation of mesoporous Ti-SBA-15 epoxidation catalysts. *Ind. Eng. Chem. Res.* **2010**, *49*, 6977–6985.

28. Kumar, A.; Srinivas, D. Selective oxidation of cyclic olefins over framework Ti-substituted, three-dimensional, mesoporous Ti-SBA-12 and Ti-SBA-16 molecular sieves. *Catal. Today* **2012**, *198*, 59–68.

29. Bhaumik, A.; Tatsumi, T. Organically modified titanium-rich Ti-MCM-41, efficient catalysts for epoxidation reactions. *J. Catal.* **2000**, *189*, 31–39.

30. Kholdeeva, O.A.; Derevyankin, A.Y.; Shmakov, A.N.; Trukhan, N.N.; Paukshtis, E.A.; Tuel, A.; Romannikov, V.N. Alkene and thioether oxidations with H_2O_2 over Ti-containing mesoporous mesophase catalysts. *J. Mol. Catal. A* **2000**, *158*, 417–421.

31. Zhang, W.Z.; Pinnavaia, T.J. Transition metal substituted derivatives of cubic MCM-48 mesoporous molecular sieves. *Catal. Lett.* **1996**, *38*, 261–265.

32. Ahn, W.S.; Lee, D.H.; Kim, T.J.; Kim, J.H.; Seo, G.; Ryoo, R. Post-synthetic preparations of titanium-containing mesopore molecular sieves. *Appl. Catal. A* **1999**, *181*, 39–49.

33. Kang, K.K.; Ahn, W.S. Physiochemical properties of transition metal-grafted MCM-48 prepared using matallocene precursors. *J. Mol. Catal. A* **2000**, *159*, 403–410.

34. Guidotti, M.; Gavrilova, E.; Galarneau, A.; Coq, B.; Psaroa, R.; Ravasio, N. Epoxidation of methyl oleate with hydrogen peroxide. The use of Ti-containing silica solids as efficient heterogeneous catalysts. *Green Chem.* **2011**, *13*, 1806–1811.

35. Tatsumi, T.; Koyano, K.A.; Igarashi, N. Remarkable activity enhancement by trimethylsilylation in oxidation of alkenes and alkanes with H_2O_2 catalyzed by titanium-containing mesoporous molecular sieves. *Chem. Commun.* **1998**, 325–326.

36. Corma, A.; Kan, Q.B.; Rey, F. Synthesis of Si and Ti-Si-MCM-48 mesoporous materials with controlled pore sizes in the absence of polar organic additives and alkali metal ions. *Chem. Commun.* **1998**, 579–580.

37. Corma, A.; Serra, J.M.; Serna, P.; Valero, S.; Argente, E.; Botti, V. Optimisation of olefin epoxidation catalysts with the application of high-throughput and genetic algorithms assisted by artificial neural networks (softcomputing techniques). *J. Catal.* **2005**, *229*, 513–524.

19

38. Solberg, S.M.; Kumar, D.; Landry, C.C. Synthesis, structure, and reactivity of a new Ti-containing microporous/mesoporous material. *J. Phys. Chem. B* **2005**, *109*, 24331–24337.

39. Guidotti, M.; Pirovano, C.; Ravasio, N.; Lazaro, B.; Fraile, J.M.; Mayoral, J.A.; Coq, B.; Galarneau, A. The use of H_2O_2 over titanium-grafted mesoporous silica catalysts: A step further towards sustainable epoxidation. *Green Chem.* **2009**, *11*, 1421–1427.

40. Widenmeyer, M.; Grasser, S.; Kohler, K.; Anwander, R. TiO_x overlayers on MCM-48 silica by consecutive grafting. *Microporous Mesoporous Mater.* **2001**, *44*, 327–336.

41. Zhao, D.; Budhi, S.; Rodriguez, A.; Koodali, R.T. Rapid and facile synthesis of Ti-MCM-48 mesoporous material and the photocatalytic performance for hydrogen evolution. *Int. J. Hydrogen Energy* **2010**, *35*, 5276–5283.

42. Kibombo, H.S.; Zhao, D.; Gonshorowski, A.; Budhi, S.; Koppang, M.D.; Koodali, R.T. Cosolvent-induced gelation and the hydrothermal enhancement of the crystallinity of titania-silica mixed oxides for the photocatalytic remediation of organic pollutants. *J. Phys. Chem. C* **2011**, *115*, 6126–6135.

43. Mahoney, L.; Koodali, R.T. Versatility of evaporation-induced self-assembly (EISA) method for preparation of mesoporous TiO_2 for energy and environmental applications. *Materials* **2014**, *7*, 2697–2746.

44. Bandyopadhyay, M.; Birkner, A.; van den Berg, M.W.E.; Klementiev, K.V.; Schmidt, W.; Grunert, W.; Gies, H. Synthesis and characterization of mesoporous MCM-48 containing TiO_2 nanoparticles. *Chem. Mater.* **2005**, *17*, 3820–3829.

45. Gao, Y.L.; Konovalova, T.A.; Xu, T.; Kispert, L.A. Electron transfer of carotenoids imbedded in MCM-41 and Ti-MCM-41: EPR, ENDOR, and UV-Vis studies. *J. Phys. Chem. B* **2002**, *106*, 10808–10815.

46. Bal, R.; Chaudhari, K.; Srinivas, D.; Sivasanker, S.; Ratnasamy, P. Redox and catalytic chemistry of Ti in titanosilicate molecular sieves: An EPR investigation. *J. Mol. Catal. A* **2000**, *162*, 199–207.

47. Chaudhari, K.; Srinivas, D.; Ratnasamy, P. Reactive oxygen species in titanosilicates TS-1 and TiMCM-41: An *in situ* EPR spectroscopic study. *J. Catal.* **2001**, *203*, 25–32.

48. Chaudhari, K.; Bal, R.; Srinivas, D.; Chandwadkar, A.J.; Sivasanker, S. Redox behavior and selective oxidation properties of mesoporous titano- and zirconosilicate MCM-41 molecular sieves. *Microporous Mesoporous Mater.* **2001**, *50*, 209–218.

49. Wu, C.M.; Peng, R.; Dimitrijevic, N.M.; Rajh, T.; Koodali, R.T. Preparation of TiO_2-SiO_2 aperiodic mesoporous materials with controllable formation of tetrahedrally coordinated Ti^{4+} ions and their performance for photocatalytic hydrogen production. *Int. J. Hydrogen Energy* **2014**, *39*, 127–136.

50. Liu, Z.F.; Crumbaugh, G.M.; Davis, R.J. Effect of structure and composition on epoxidation of hexene catalyzed by microporous and mesoporous Ti-Si mixed oxides. *J. Catal.* **1996**, *159*, 83–89.

51. Tuel, A.; Hubert-Pfalzgraf, L.G. Nanometric monodispersed titanium oxide particles on mesoporous silica: Synthesis, characterization, and catalytic activity in oxidation reactions in the liquid phase. *J. Catal.* **2003**, *217*, 343–353.

52. Hutter, R.; Mallat, T.; Baiker, A. Titania-silica mixed oxides: II. Catalytic behavior in olefin epoxidation. *J. Catal.* **1995**, *153*, 177–189.

53. Dellarocca, V.; Marchese, L.; Pena, M.L.; Rey, F.; Corma, A.; Coluccia, S. Surface Properties of Mesoporous Ti-MCM-48 and Their Modifications Produced by Silylation. In Proceedings of Oxide-based Systems at the Crossroads of Chemistry—Second International Workshop, Como, Italy, 8–11 October 2000; Gamba, A., Colella, C., Coluccia, S., Eds.; Elsevier Science B.V.: Amsterdam, The Netherlands, 2001; pp. 209–220.

Bismuth Molybdate Catalysts Prepared by Mild Hydrothermal Synthesis: Influence of pH on the Selective Oxidation of Propylene

Kirsten Schuh, Wolfgang Kleist, Martin Høj, Vanessa Trouillet, Pablo Beato, Anker Degn Jensen and Jan-Dierk Grunwaldt

Abstract: A series of bismuth molybdate catalysts with relatively high surface area was prepared via mild hydrothermal synthesis. Variation of the pH value and Bi/Mo ratio during the synthesis allowed tuning of the crystalline Bi-Mo oxide phases, as determined by X-ray diffraction (XRD) and Raman spectroscopy. The pH value during synthesis had a strong influence on the catalytic performance. Synthesis using a Bi/Mo ratio of 1/1 at pH $\geq$ 6 resulted in γ-Bi_2MoO_6, which exhibited a better catalytic performance than phase mixtures obtained at lower pH values. However, a significantly lower catalytic activity was observed at pH = 9 due to the low specific surface area. γ-Bi_2MoO_6 synthesized with Bi/Mo = 1/1 at pH = 6 and 7 exhibited relatively high surface areas and the best catalytic performance. All samples prepared with Bi/Mo = 1/1, except samples synthesized at pH = 1 and 9, showed better catalytic performance than samples synthesized with Bi/Mo = 2/3 at pH = 4 and 9 and γ-Bi_2MoO_6 synthesized by co-precipitation at pH = 7. At temperatures above 440 °C, the catalytic activity of the hydrothermally synthesized bismuth molybdates started to decrease due to sintering and loss of surface area. These results support that a combination of the required bismuth molybdate phase and a high specific surface area is crucial for a good performance in the selective oxidation of propylene.

Reprinted from *Catalysts*. Cite as: Schuh, K.; Kleist, W.; Høj, M.; Trouillet, V.; Beato, P.; Jensen, A.D.; Grunwaldt, J.-D. Bismuth Molybdate Catalysts Prepared by Mild Hydrothermal Synthesis: Influence of pH on the Selective Oxidation of Propylene. *Catalysts* **2015**, *5*, 1554–1573.

1. Introduction

Since the development of bismuth molybdate catalysts for the oxidation and ammoxidation of propylene to acrolein or acrylonitrile by Sohio in 1959 [1,2], these mixed oxides have received strong attention and their catalytic properties have been studied in considerable detail [3–12]. Addition of further elements such as iron, cobalt or vanadium increased the acrolein and acrylonitrile yields and stabilized the catalyst during the reaction. Nevertheless, the surface layer of these multicomponent catalysts consists of mixed oxides based on Bi and Mo, and these two metals seem to form the key active sites [13]. The rate-determining step is the abstraction of hydrogen from propylene on bismuth or bismuth connected to molybdyl groups [14]. The addition of other transition metals and main group metals helps to increase the specific surface area of the catalyst [15], the extent of lattice oxygen participation [15,16] and the electronic conductivity [17]. α-$Bi_2Mo_3O_{12}$, β-$Bi_2Mo_2O_9$ and γ-Bi_2MoO_6 are the bismuth molybdate phases of significance for the selective oxidation of propylene to acrolein. Despite of intensive research there is still a debate in literature about the relative activity of these bismuth molybdate phases and the influence of the preparation route. Thus, the search for new synthesis approaches remains a key issue.

Carson *et al.* [18] stated that the catalytic activity for propylene oxidation decreases in the following order: $\alpha > \gamma > \beta$. On the other hand, γ-Bi_2MoO_6 was found to be the most active phase by Krenzke *et al.* [19] and Monnier *et al.* [20]. Brazdil *et al.* [21] even reported β-$Bi_2Mo_2O_9$ to be more active than the other two bismuth molybdate phases. We recently used flame spray pyrolysis for synthesis [22] and observed that the catalytic activity decreased in the following order: $\beta > \gamma > \alpha$. Furthermore, Snyder and Hill [23] claimed that the stability of the different bismuth molybdate phases, in particular β-$Bi_2Mo_2O_9$, depended on the temperature at which they were calcined.

The most common method to prepare bismuth molybdate catalysts is co-precipitation [9,10,18,19,24–26], but also solid-state routes [27], sol-gel synthesis [28], and spray drying of aqueous solutions [11,29] have been used. All of these methods require heating or calcination at temperatures >400 °C to yield crystalline materials, which may result in a decrease of the catalytic performance of the resulting phase due to bismuth enrichment of the catalyst surface [30]. Recently, γ-Bi_2MoO_6 with relatively high surface area (14 m^2/g) and good catalytic performance was reported using mesoporous carbon templates [31].

Alternatively, bismuth molybdates can be prepared by hydrothermal synthesis [32–38], which is a typical soft chemistry ("chimie douce") method and provides convenient access to advanced materials of high purity, controlled morphology, high crystallinity and good reproducibility [39]. α-$Bi_2Mo_3O_{12}$ (monoclinic, defective fluorite structure) and γ-Bi_2MoO_6 (orthorhombic, layered Aurivillius-type structure) have been successfully synthesized by a one-step synthesis under hydrothermal conditions applying different precursors and various synthesis conditions [32,33,36,40,41]. For the preparation of metastable, monoclinic β-$Bi_2Mo_2O_9$ the precursors were applied in the ratio Bi/Mo = 1/1 leading to a phase mixture which had to be further calcined at 560 °C to obtain crystalline β-bismuth molybdate [33,36]. The crystal growth and the formation of the various bismuth molybdate phases under hydrothermal conditions was studied *in situ* by time-resolved energy dispersive X-ray diffraction (EDXRD) and X-ray absorption spectroscopy (XAS) [33] and by combined EDXRD/XAS/Raman techniques [38]. Kongmark *et al.* [38] reported that $[MoO_4]^{2-}$ entities are required for the formation of γ-Bi_2MoO_6, demonstrating the importance of the pH value in the initial solution for the formation of the desired phase. Li *et al.* [36] successfully synthesized α- and γ-bismuth molybdate from bismuth nitrate and ammonium heptamolybdate with relatively high specific surface area (38–57 m^2/g) compared to other unsupported bismuth molybdate catalysts (1–4 m^2/g) [9,10,29,42] by variation of the pH between 1 and 9. Hydrothermally synthesized bismuth molybdates were mostly applied in photocatalysis under visible light irradiation [36,40,43], where, especially, γ-Bi_2MoO_6 is attractive due to its layered structure.

Although the controlled preparation of unsupported bismuth molybdate catalysts exposing a high surface area under mild conditions seems very attractive for the selective oxidation of olefins, hydrothermally prepared ones have hardly been applied in oxidation reactions. Hydrothermal synthesis as a preparation route for selective oxidation catalysts has been reported for example for iron molybdates [44], $La_{2-x}Sr_xCuO_4$ [45], wolframite type $MMoWO_x$ (with M = Ni and Co) [46] and more complex mixed oxides based on molybdenum and vanadium [47–49]. We have recently shown that hydrothermally synthesized bismuth molybdates were active catalysts for the oxidation of propylene to acrolein [50] and found that the use of nitric acid led to an improved catalytic performance. To further explore catalysts with higher surface area and altered structure, we have systematically studied the performance of hydrothermally synthesized bismuth molybdates. They were prepared under mild hydrothermal conditions by variation of the pH value, which should alter the structure of bismuth molybdates as reported by Li *et al.* [36] To understand the role of the composition and the surface area on the catalytic activity, the resulting product phases, the ratio of bismuth to molybdenum in the bulk and on the surface, as well as the specific surface area have been characterized and correlated to the performance in the selective oxidation of propylene to acrolein. This knowledge may in the future be applied to more complex systems, such as multicomponent molybdenum oxides.

2. Results and Discussion

2.1. Characterization of the Samples Synthesized with Bi/Mo = 1/1

The X-ray diffraction (XRD) patterns and Raman spectra of the samples synthesized with Bi/Mo = 1/1 revealed a strong influence of the pH value on the structure under hydrothermal conditions (Figure 1). At high pH values (pH $\geqslant$ 6) γ-Bi_2MoO_6 was formed, which is indicated by the reflections at $2\theta = 28.3°, 32.6°, 33.1°, 36.1°, 46.7°, 47.2°$ in the XRD patterns (Figure 1a; JCPDS card No. 77-1246). The corresponding Raman spectra showed bands at 848 cm^{-1} and 807 cm^{-1} as well as at 719 cm^{-1}, which were ascribed to γ-Bi$_2$MoO$_6$ [36,51]. Analysis of the bismuth and molybdenum concentration of the products by ICP-OES led to a Bi/Mo bulk ratio of 1.8–2.1 for pH = 6–9 (Table 1) suggesting that molybdenum partly remained in solution in the form of MoO$_4{}^{2-}$ [38,52], which was washed away during filtration, and, thus, did not precipitate at high pH values.

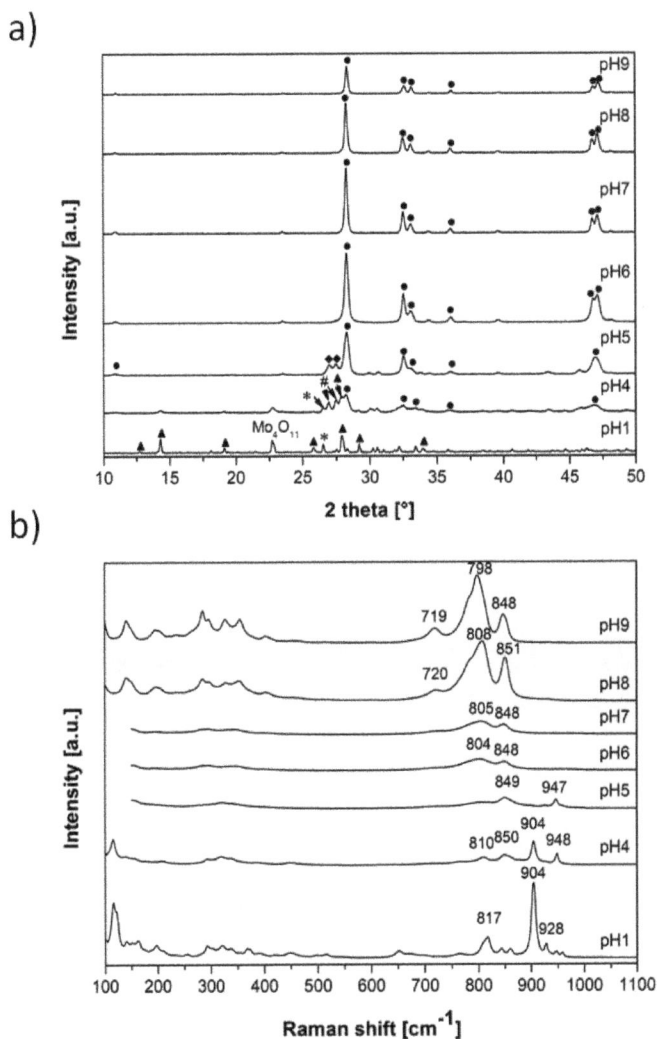

a)

b)

Figure 1. XRD patterns (**a**) and Raman spectra (**b**) of the samples with Bi/Mo = 1/1 at different pH values. The following symbols are used for the different phases: α-Bi$_2$Mo$_3$O$_{12}$ (▲), γ-Bi$_2$MoO$_6$ (●), MoO$_3$·2H$_2$O (◆), Bi$_2$O$_3$ (*) and Bi$_{26}$Mo$_{10}$O$_{69}$ (#).

Table 1. Characterization of the samples prepared by hydrothermal synthesis with different Bi/Mo ratios and at various pH values by X-ray diffraction measurements and Raman spectroscopy, as well as nitrogen physisorption measurements (BET), ICP-OES and XPS (main phases in bold letters).

Sample	Phases According to XRD	Phases According to Raman	Surface Area (BET) [m²/g]	Bi/Mo Ratio Bulk [a]	Bi/Mo Ratio Surface [b]
		Initial ratio Bi/Mo = 2/1			
Bi2Mo1_pH7	γ-Bi$_2$MoO$_6$	-	13	-	-
Bi2Mo1_pH8	γ-Bi$_2$MoO$_6$	-	7	-	-
		Initial ratio Bi/Mo = 1/1			
Bi1Mo1_pH1	**α-Bi$_2$Mo$_3$O$_{12}$,** Mo$_4$O$_{11}$, Bi$_2$O$_3$	α-Bi$_2$Mo$_3$O$_{12}$	3	1.0	-
Bi1Mo1_pH4	**γ-Bi$_2$MoO$_6$,** **α-Bi$_2$Mo$_3$O$_{12}$,** Bi$_2$O$_3$, Bi$_{26}$Mo$_{10}$O$_{69}$	α-Bi$_2$Mo$_3$O$_{12}$, γ-Bi$_2$MoO$_6$, β-Bi$_2$O$_3$	18	0.9	1.1
Bi1Mo1_pH5	**γ-Bi$_2$MoO$_6$,** MoO$_3$·2H$_2$O	γ-Bi$_2$MoO$_6$, Mo$_7$O$_{24}^{6-}$ or MoO$_3$(H$_2$O)$_2$	32	1.1	1.1
Bi1Mo1_pH6	γ-Bi$_2$MoO$_6$	γ-Bi$_2$MoO$_6$	26	1.8	1.7
Bi1Mo1_pH7	γ-Bi$_2$MoO$_6$	γ-Bi$_2$MoO$_6$	17	1.8	1.8
Bi1Mo1_pH8	γ-Bi$_2$MoO$_6$	γ-Bi$_2$MoO$_6$	10	1.9	2.5
Bi1Mo1_pH9	γ-Bi$_2$MoO$_6$	γ-Bi$_2$MoO$_6$	4	2.1	
		Initial ratio Bi/Mo = 2/3			
Bi2Mo3_pH1	**α-Bi$_2$Mo$_3$O$_{12}$,** H$_{0.68}$(NH$_4$)$_2$ Mo$_{14.16}$O$_{4.34}$·6.92H$_2$O	α-Bi$_2$Mo$_3$O$_{12}$	5	-	-
Bi2Mo3_pH4	**γ-Bi$_2$MoO$_6$,** MoO$_3$·2H$_2$O, Bi$_5$O$_7$NO$_3$	γ-Bi$_2$MoO$_6$, Mo$_7$O$_{24}^{6-}$ or MoO$_3$(H$_2$O)$_2$, NO$_3^-$	17	-	-
Bi2Mo3_pH9	γ-Bi$_2$MoO$_6$	γ-Bi$_2$MoO$_6$	5	-	-

[a] The calculated error accounts to around 10%. [b] Average of two XPS measurements at different spots.

In weakly acidic or basic aqueous solutions bismuth nitrate is "hydrolyzed" resulting in the formation of BiO$^+$ [53]. These bismuthyl subunits react with (MoO$_4$)$^{2-}$ and precipitate as γ-Bi$_2$MoO$_6$ [36,54]. X-ray absorption measurements at the Mo K-edge of the samples synthesized at pH = 6 and pH = 9 confirmed that the products contained only octahedral Mo (γ-Bi$_2$MoO$_6$; see Figure S1). In addition to γ-Bi$_2$MoO$_6$, a molybdenum oxide phase was formed at pH = 5 indicated by the reflections at 2θ = 26.9° and 27.5°, which could be assigned to MoO$_3$·2H$_2$O (JCPDS No. 39-363), and by the Raman band at 947 cm^{-1} assigned to the presence of Mo$_7$O$_{24}^{6-}$. At pH = 3–6 Mo$_7$O$_{24}^{6-}$ [38,55] and Mo$_2$O$_7^{2-}$ are the stable polymolybdate species in aqueous solution depending on the temperature [52], which agrees well with the Raman spectra of the solid phase. Bi1Mo1_pH5 exhibited a Bi/Mo bulk ratio of 1.1, which corresponds to the applied ratio and resulted in the largest surface area (32 m^2/g, Table 1) among the different bismuth molybdates. In contrast to Li *et al.* [36], γ-Bi$_2$MoO$_6$, and not α-Bi$_2$Mo$_3$O$_{12}$, was the most prominent phase as detected by X-ray diffraction (main reflection at 27.9°) or Raman spectroscopy (main band at 904 cm^{-1}) for the sample synthesized at pH = 5. The formation of the α-phase was observed at pH $\leqslant$ 4 and minor contributions from other phases were detected in addition to α- and γ-bismuth molybdate. Further decrease of the pH to 1 led to the disappearance of γ-Bi$_2$MoO$_6$ in the product and α-Bi$_2$Mo$_3$O$_{12}$ was the main phase observed by XRD (Figure 1a). In addition to the reflections at 2θ = 14.1°, 18.1°, 25.9°, 27.9° and 29.2°, which were assigned to α-Bi$_2$Mo$_3$O$_{12}$ (JCPDS No. 21-103), two reflections at 22.7° and 26.5° were observed. The reflection at 22.7° could be assigned to Mo$_4$O$_{11}$ (JCPDS No. 86-1269), but assignment of the reflection at 26.5° was not possible. The corresponding Raman spectra (Figure 1b) showed the characteristic bands of α-Bi$_2$Mo$_3$O$_{12}$ (928, 904, 861, 843 and 817 cm^{-1}). For all samples synthesized with a Bi/Mo ratio 1/1, β-Bi$_2$Mo$_2$O$_9$ was not detected, neither by X-ray diffraction (27.8°), nor by Raman spectroscopy (884 cm^{-1}), which agrees with literature [33,36] where calcination at 560 °C was required to form the β-phase. The samples synthesized at pH = 1–5 contained bismuth and molybdenum in the ratio 1/1 in the bulk (Table 1), corresponding to the applied ratio in the synthesis.

In addition to the characterization of the phase composition and the resulting Bi/Mo ratio in the bulk, the specific surface area and the Bi/Mo ratio on the surface of the products were determined (see Table 1). The XPS spectra showed one doublet for Bi 4f$_{7/2}$ at 159.4 eV and one for Mo 3d$_{5/2}$ at 232.7 eV (for representative spectra see Figure S2), which revealed an oxidation state of +3 for bismuth [56] and +6 for molybdenum [57]. The surface composition of the catalysts was calculated using the atom concentrations of both elements, which were determined from the peak area of the spectra. At pH = 4–7, the Bi/Mo ratio on the surface was not the same as the ratio in the bulk (Bi/Mo $\approx$ 1 for pH = 4 and 5, Bi/Mo $\approx$ 2 for pH = 6 and 7), whereas at pH = 8 a surface enrichment of bismuth (Bi/Mo = 2.5) was observed by XPS. High Bi/Mo surface ratios were also reported in the literature for γ-Bi$_2$MoO$_6$ prepared by different methods (spray drying: 2.4 [29], co-precipitation: 2.6 [10,58]), but the existence of surface Bi$_2$O$_3$ could neither be confirmed in literature nor in the present work. The sample synthesized at pH = 5 featured the highest specific surface area (32 m^2/g), whereas with increasing and decreasing pH value the surface area of the resulting product was reduced (Table 1) to 4 m^2/g and 3 m^2/g for pH = 9 and pH = 1, respectively. The product composition did not seem to be a determining factor for the specific surface area, as the two samples with the highest surface area showed different phases resulting in a different Bi/Mo ratio in the bulk and on the surface (Bi/Mo =1.1 for pH = 5 with 32 m^2/g and Bi/Mo =1.8 for pH = 6 with 26 m^2/g).

2.2. Variation of the Bi/Mo Ratio

To elucidate the influence of the applied Bi/Mo ratio on the phases, the amount of molybdenum was increased to Bi/Mo = 2/3. Despite of the higher initial Mo content, even at pH = 9 only γ-Bi₂MoO₆ with a high crystallinity was formed as evidenced by X-ray diffraction (Figure 2a) and Raman spectroscopy (Figure 2b). This indicated that also at low Bi/Mo ratio (*i.e.*, with an excess of molybdenum) bismuth-rich γ-Bi₂MoO₆ was formed at high pH values. The specific surface areas of the two samples synthesized at pH = 9 (Bi1Mo1_pH9 and Bi2Mo3_pH9) were similar ($4 \text{ m}^2/\text{g}$ and $5 \text{ m}^2/\text{g}$).

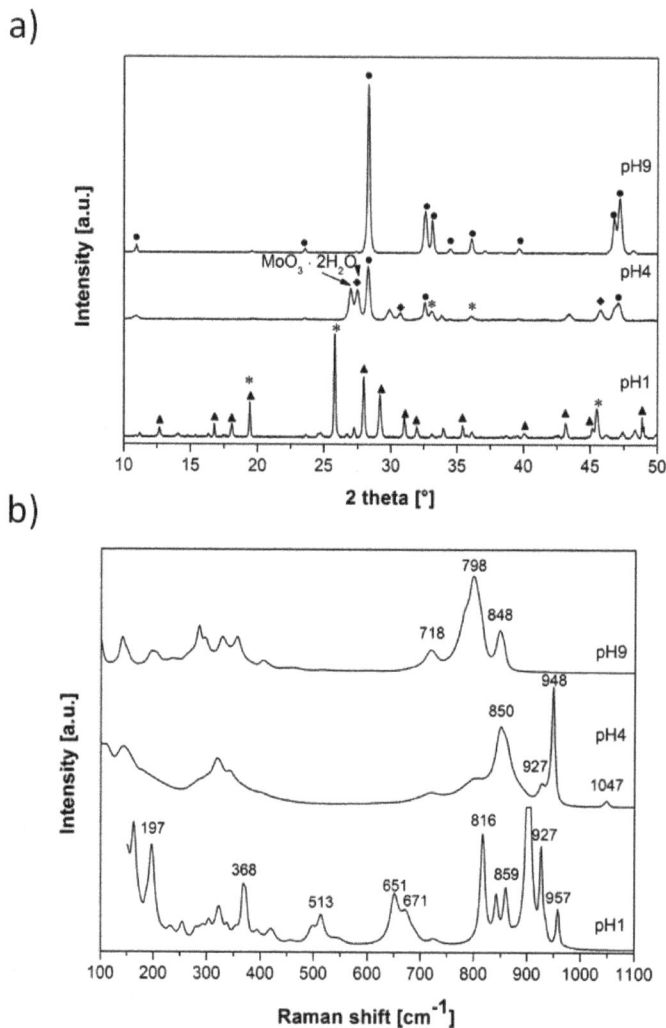

a)

b)

Figure 2. XRD patterns (**a**) and Raman spectra (**b**) of the samples with Bi/Mo = 2/3 at different pH values. The following symbols are used for the different phases: α-Bi₂Mo₃O₁₂ (▲), γ-Bi₂MoO₆ (●), H₀.₆₈(NH₄)₂Mo₁₄.₁₆O₄.₃₄·6.92H₂O (*) and Bi₅O₇NO₃ (◆).

Decreasing the pH to 4 led to the same phase composition as in sample Bi1Mo1_pH4 (γ-Bi$_2$MoO$_6$ and MoO$_3$·2H$_2$O), but additionally a nitrate-containing phase was found as indicated by the Raman band at 1047 cm^{-1} (Figure 2b). The reflections in the X-ray diffraction pattern (Figure 2a) at 27.6°, 30.7° and 45.7° could be assigned to Bi$_5$O$_7$NO$_3$ (JCPDS No. 51-525). Further decrease of the pH to 1 did not result in the formation of pure α-Bi$_2$Mo$_3$O$_{12}$, but an ammonium containing molybdenum oxide phase was additionally formed (see Figure 2 and Table 1).

The applied pH value and, thus, the resulting (poly)molybdate anions in the aqueous solution seem to have a stronger impact on the resulting phase than the applied Bi/Mo ratio. For all three samples synthesized with Bi/Mo = 2/3 the surface area determined by nitrogen physisorption was almost identical to the samples synthesized with Bi/Mo = 1/1 at the same pH value (Table 1).

Using an excess of bismuth and applying the initial ratio Bi/Mo = 2/1 at pH = 7 and pH = 8 led to the formation of highly crystalline γ-Bi$_2$MoO$_6$ (see XRD patterns in Figure S3). This is in agreement with Li et al. [36] and Zhang et al. [40], who also reported that hydrothermal synthesis with Bi/Mo = 2/1 always resulted in γ-Bi$_2$MoO$_6$ independent of the pH value (pH = 1–13).

2.3. Catalytic Performance in Propylene Oxidation to Acrolein

The samples synthesized with Bi/Mo = 1/1 were tested in propylene oxidation at 320 °C–520 °C to study the influence of the preparation parameters, i.e., the pH value and the corresponding structural properties on the catalytic performance of the hydrothermally synthesized bismuth molybdates. Figure 3 depicts the acrolein selectivity as a function of the propylene conversion at different flows (50, 80 and 120 NmL/min) leading to different catalyst contact times and, therefore, to a different propylene conversion and/or acrolein selectivity. At 320 °C (Figure 3a), the sample synthesized at pH = 6 showed the highest propylene conversion (19% and 31% propylene conversion, 74% and 73% acrolein selectivity for 120 NmL/min and 50 NmL/min, respectively), followed by the samples synthesized at pH = 7 ($X_{propylene}$ = 27% and $S_{acrolein}$ = 67% for 50 NmL/min) and pH = 8 ($X_{propylene}$ = 25% and $S_{acrolein}$ = 59% for 50 NmL/min).

Although the surface area of the sample synthesized at pH = 5 was higher, it converted less propylene ($X_{propylene}$ = 10%–21%). Both the samples prepared at pH = 5 and also at pH = 4, which performed even worse (8%–16% conversion of propylene), contained minor amounts of other phases additionally to γ-Bi$_2$MoO$_6$. This indicates that the samples which contained only γ-Bi$_2$MoO$_6$ were more active than the other samples. However, the sample synthesized at pH = 9, which also only contained crystalline γ-Bi$_2$MoO$_6$ but with lower specific surface area, reached propylene conversions of only 2%–4% at acrolein selectivities of 69%–49%. The sample synthesized at pH = 1, exhibiting a small surface area (3 m^2/g), was also nearly inactive, since the propylene conversion remained around 3% for 120–50 NmL/min total flow, while the acrolein selectivity strongly decreased from 80% to 38% with increasing flow.

The major by-products were CO_2 (selectivities around 10%–20%), acetaldehyde (7%–8%) and CO (1%–5%). Ethylene and hexadiene were only detected in very small amounts for the sample Bi1Mo1_pH9 at temperatures above 440 °C.

At 360 °C, the differences in activity between the various samples were less pronounced, but the samples synthesized at pH = 6 and pH = 7 still showed the best catalytic performance (propylene conversion of 40%–56% and 37%–54%, respectively) followed by the sample at pH = 8, which was slightly less selective (acrolein selectivity of 72%–80% compared to 78%–84%; see Figure 3b). All the samples featured significantly higher surface areas than in a previous study, where (i) the precursors were only dissolved in water; (ii) the precursor solutions were not prepared separately; and (iii) the pH was not adjusted by addition of acid or ammonia [50]. In line with previous studies, the use of nitric acid was beneficial and adjusting the pH by ammonia seemed to lead to a rather high surface area.

In Figure 4 the propylene conversion as well as the acrolein yield at 360 °C was correlated to the pH values during hydrothermal synthesis. Propylene conversion and acrolein yield showed volcano-like curves with respect to pH with a maximum at pH = 6. Interestingly, there was a steep increase in conversion from pH = 5 to pH = 6, although the specific surface area was higher at pH = 5. In addition, the samples at pH = 6 and pH = 7 featured a significantly higher Bi/Mo ratio of approximately 2 compared to the sample prepared at pH = 5, both in the bulk and on the surface (Table 1). Comparison of Table 1 and Figure 4 evidenced that a combination of the right bismuth molybdate phase (γ-Bi_2MoO_6) and a high surface area seemed to be important for high propylene conversion. It has been shown that oxygen diffusion through the lattice is most effective for γ-Bi_2MoO_6 compared to the other bismuth molybdate phases [59]. Thus, it may be rewarding in the future to also correlate the catalytic activity in relation to the nature of lattice oxygen.

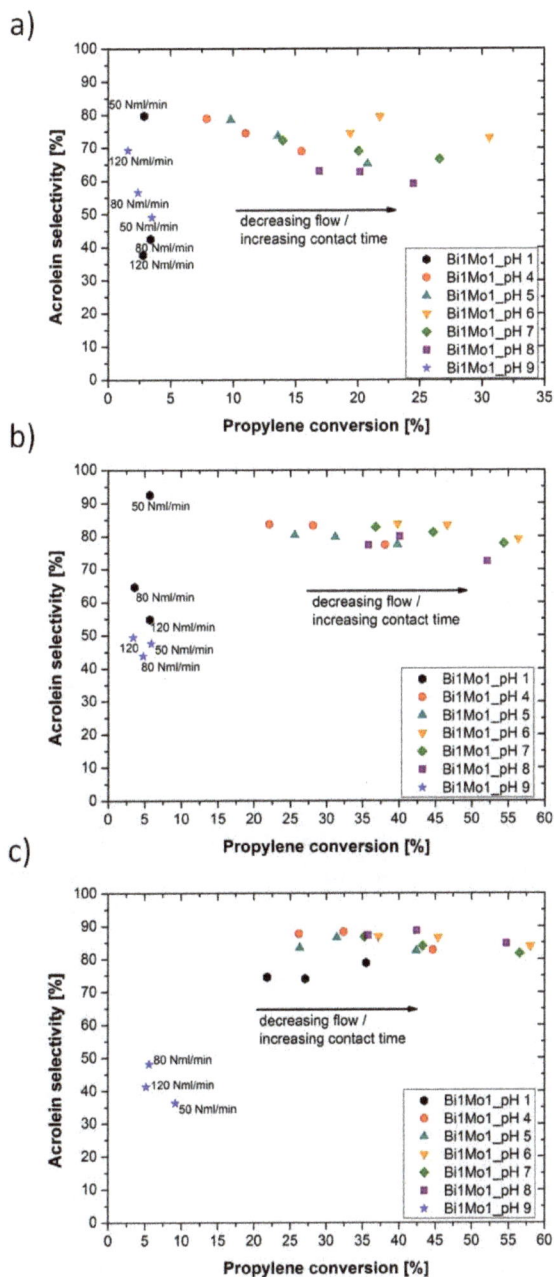

Figure 3. Catalytic performance of samples prepared with Bi/Mo = 1/1 at 320 °C (**a**); 360 °C (**b**) and 400 °C (**c**). The samples were dried in air at room temperature, crushed, sieved and pre-treated in the reactor in synthetic air at 300 °C. Reaction conditions: 500 mg catalyst; $C_3H_6/O_2/N_2 = 5/25/70$; 50, 80 and 120 NmL/min.

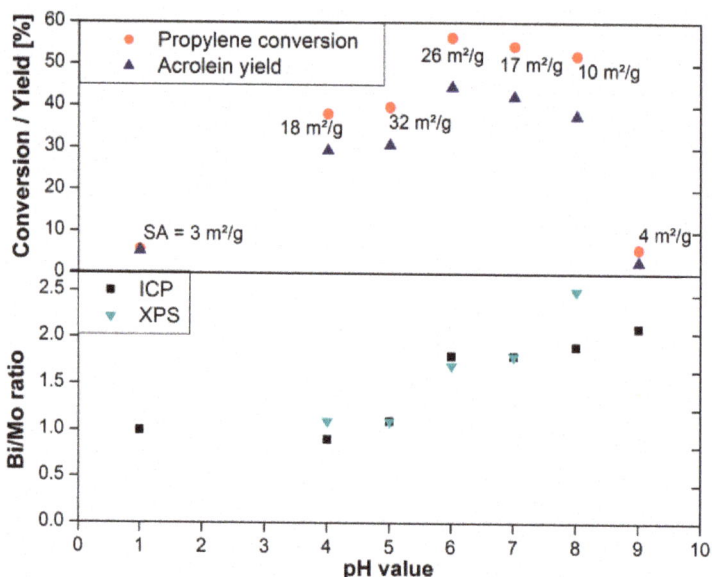

Figure 4. Correlation of the pH value in the initial solution, the Bi/Mo ratio determined in the bulk (ICP) and on the surface (XPS), and the catalytic performance at 360 °C using a flow of 50 NmL/min with a composition of $C_3H_6/O_2/N_2 =$ 5/25/70 and 500 mg of catalyst.

The sample synthesized at pH = 8 with a lower surface area (10 m^2/g) and bismuth excess on the surface led to slightly lower acrolein selectivity at 360 °C (Figure 3b). Propylene conversion of the bismuth molybdates prepared at pH = 1 and pH = 9 did not exceed 6% at this temperature. The catalyst synthesized at pH = 9 with Bi/Mo = 2/3 composed of γ-Bi$_2$MoO$_6$ gave similar activity and selectivity as Bi1Mo1_pH9 at 320 °C and 360 °C (see Figure 3; $X_{propylene} < 5\%$ and $X_{propylene} < 10\%$, respectively) in the oxidation of propylene to acrolein. Both samples synthesized at pH = 9 exhibited lower catalytic performance than the sample prepared by conventional co-precipitation CP_Bi2Mo1_450 consisting also of γ-Bi$_2$MoO$_6$ with a surface area of 6 m^2/g (this sample has been further characterized in Reference [50]). Bi2Mo3_pH4 showed a phase composition similar to Bi1Mo1_pH5 but with an additional nitrate phase and lower surface area (*cf.* Table 1) resulting in lower catalytic performance at 320 °C and 360 °C than the sample synthesized at pH = 4 with Bi/Mo = 1/1 (Figure 3). In general, the samples synthesized at pH = 4–8 with an initial Bi/Mo ratio of 1/1 yielded relatively high propylene conversion, along with a high selectivity for acrolein compared to the co-precipitated sample using Bi/Mo = 2/1 (compare Figure 3a,b with Figure 5).

a)

b)

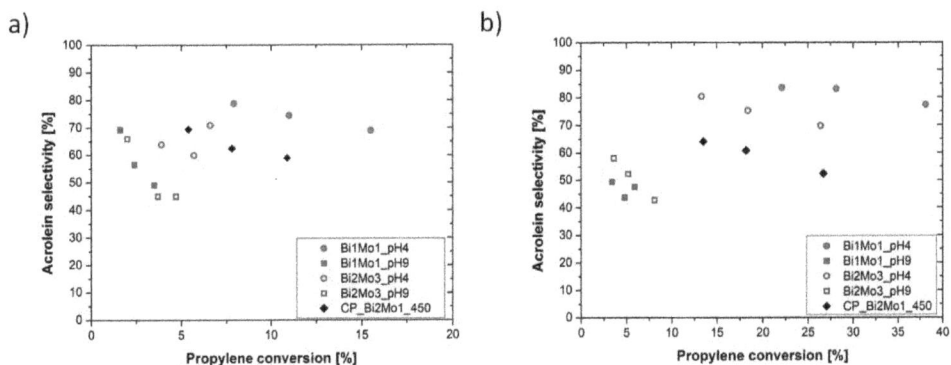

Figure 5. Comparison of the catalytic performance of the samples prepared at pH = 4 and pH = 9 with Bi/Mo = 1/1 and 2/3 and a reference sample synthesized by co-precipitation with Bi/Mo = 2/1 at 320 °C (**a**) and 360 °C (**b**). Reaction conditions: 500 mg catalyst; $C_3H_6/O_2/N_2$ = 5/25/70; 50, 80 and 120 NmL/min.

Bi1Mo1_pH9 showed low propylene conversion also at 400 °C (5%–9%) with acrolein selectivity below 50% and CO_2 selectivity of around 50% (Figure 3c), whereas the catalytic performance of the sample synthesized at pH = 1 increased strongly to propylene conversions of 22%–35% and acrolein selectivities of 74%–79%. For all the other samples the catalytic performance in propylene oxidation was only slightly increased when the temperature changed from 360 °C to 400 °C. The three samples synthesized at pH = 6–8 still exhibited the highest propylene conversion (35%–58%) of the tested samples with acrolein selectivities up to 92%. The sample synthesized at pH = 8, which was slightly less selective at 360 °C, gave the same acrolein selectivity as the samples synthesized at pH = 6–7 at 400 °C (Figure 3c). Further increase of the process temperatures to more than 400 °C did not result in an improvement of the catalytic activity. The corresponding full data set of propylene conversion and acrolein selectivities between 320 °C to 520 °C for the samples synthesized at pH = 5 (highest surface area) and pH = 6 (highest activity) as representative examples are given in the Supporting Information (Figure S4). An increase in temperature from 320 °C to 400 °C resulted in a better catalytic performance, whereas the activity at 440 °C was very similar to the one at 400 °C. At temperatures higher than 440 °C, the catalytic activity started to decrease and at 520 °C propylene conversion was similar to the values at 320 °C but with higher acrolein selectivities. The decrease in activity at temperatures higher than 440 °C was probably caused by a decrease in surface area for all samples. After application in propylene oxidation at 320 °C–520 °C the surface area of all used samples was ⩽1 m^2/g, whereas after 8 h of calcination at 360 °C the samples synthesized at pH = 5 and pH = 6 still exhibited surface areas of 15 m^2/g and 16 m^2/g, respectively. The decrease in surface area already started at 360 °C, but the decrease was more significant between 400 °C and 440 °C when catalyst deactivation started.

Figure 6 shows the X-ray diffraction patterns of the samples after their application in the selective oxidation of propylene at 320 °C–520 °C. The phase composition of the samples synthesized at pH = 6–9 (γ-Bi$_2$MoO$_6$) did not change during the catalytic activity tests, but the hydrothermally prepared materials synthesized at pH = 1–5 were all composed of a mixture of α-Bi$_2$Mo$_3$O$_{12}$ and γ-Bi$_2$MoO$_6$ after use. This indicates that the reduced surface area, and not the phase transformation, was the reason for the decreasing propylene conversion at temperatures above 440 °C. As already concluded earlier, a combination of the desired phase and a high surface area seemed to be crucial for propylene oxidation.

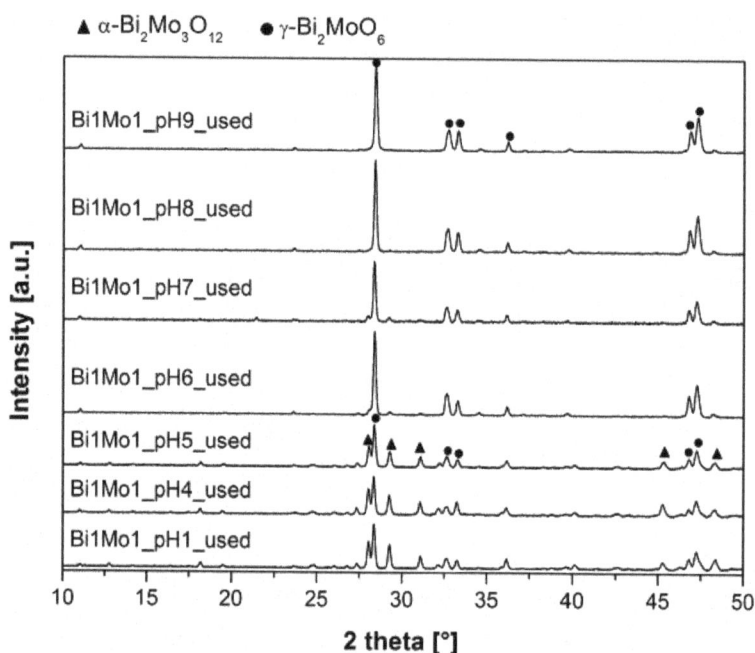

Figure 6. XRD patterns of the samples synthesized with Bi/Mo = 1/1 at different pH values after application in catalytic oxidation of propylene at temperatures up to 520 °C. The following symbols are used for marking the reflections of the different phases: α-Bi$_2$Mo$_3$O$_{12}$ (▲) and γ-Bi$_2$MoO$_6$ (●).

Jung et al. [60] tested the influence of the pH value during co-precipitation of γ-Bi$_2$MoO$_6$ (initial ratio Bi/Mo = 2/1) on the oxidative dehydrogenation of n-butylene and discovered that the sample synthesized at pH = 3 showed both the highest butylene conversion and 1,3-butadiene yield due to a high oxygen mobility of this sample. According to the group of Keulks [8,20], re-oxidation of the catalyst is the rate-determining step in propylene oxidation at temperatures below 400 °C, whereas abstraction of an α-hydrogen atom to form an allylic intermediate is the rate determining step at higher temperatures (>400 °C), in agreement with theory [14]. Recently, it was further reported that the reaction order in oxygen was zero at 340 °C and 400 °C for $Bi_2Mo_3O_{12}$ catalysts, *i.e.*, the reaction rate for acrolein formation, was independent of the partial pressure of oxygen, indicating that the transition temperature where re-oxidation of the catalyst becomes the rate determining step was lower than 340 °C [61,62]. The hydrothermally synthesized samples at pH = 4–8 exhibited high propylene conversion and also relatively high acrolein selectivities already at 360 °C. Therefore, oxygen mobility is not considered to be the decisive factor for the varying catalytic performance of the different samples depicted in Figure 3. The oxygen concentration used for the catalytic activity tests in this study was relatively high (C_3H_6/O_2 = 1/5) to guarantee a complete re-oxidation of the bismuth molybdate catalysts and to minimize the effect of decreasing oxygen partial pressure if the reaction order was not zero.

Aleshina *et al.* [42] prepared bismuth molybdates by co-precipitation with a Bi/Mo ratio of 2 at pH values between 0 and 7 and with surface areas of 1–3 m^2/g and tested them in propylene oxidation in the presence of steam. They detected Bi_2O_3 in the sample prepared at pH = 7 and claimed that with increasing pH value molybdenum dissolved and remained in solution leading to a bismuth rich product (mixture of Bi_2O_3 and γ-Bi_2MoO_6), which showed lower propylene conversion (73% compared to 86%) and very low acrolein selectivity compared to the sample containing only γ-Bi_2MoO_6. In contrast, Bi_2O_3 was not detected in the present work. The samples synthesized at pH = 8 and 9 contained mainly γ-Bi_2MoO_6 and also the formation of hexadiene during propylene conversion was hardly observed, although, in the literature, it was reported that hexadiene was formed in the presence of Bi_2O_3 [63,64]. Thus, the reason for the low activity of Bi1Mo1_pH9 and the lower selectivity of Bi1Mo1_pH8 may not be contamination of the product with bismuth oxide but, rather, the lower surface areas compared to the samples synthesized at pH = 6 and 7.

3. Experimental Section

3.1. Catalyst Preparation

The bismuth molybdate materials were synthesized by hydrothermal synthesis similar to the procedure reported by Li *et al.* [36]. All chemicals were analytical grade and used without further purification.

In a typical synthesis, 10 mmol $Bi(NO_3)_3 \cdot 5H_2O$ and stoichiometric amounts of $(NH_4)_6Mo_7O_{24} \cdot 4H_2O$ were dissolved in 20 mL 2.0 M nitric acid solution and 20 mL deionized water, respectively (for Bi/Mo = 1/1). The two solutions were mixed under vigorous stirring and the pH of the resulting mixture was adjusted to the desired value with an aqueous solution of 25 vol. % ammonia. For all samples addition of ammonia solution was necessary due to the low initial pH value. After stirring for 30 min the resulting solutions were heated to 180 °C in sealed 250 mL autoclaves (Berghof, Eningen, Germany) with Teflon inlays and kept in an oven for 24 h. Next, the solid product was separated by filtration at room temperature, and washed with 100 mL deionized water, 40 mL ethanol and finally 40 mL acetone. The resulting powder was dried at room temperature and ambient pressure. The samples are denoted as BixMoy_pH with Bi/Mo = x/y.

For comparison co-precipitation was used to synthesize a reference sample with Bi/Mo = 2/1 according to Carrazán *et al.* [65] using $(NH_4)_6Mo_7O_{24} \cdot 4H_2O$ dissolved in NH_4OH and $Bi(NO_3)_3 \cdot 5H_2O$ dissolved in HNO_3 at pH = 7. The resulting solid material was calcined at 450 °C to yield the γ-phase, and was referred to as CP_Bi2Mo1_450. This sample has been further studied and characterized in Reference [50].

3.2. Catalyst Characterization

X-ray diffraction (XRD) measurements were recorded using a Bruker D8 Advance powder diffractometer (Karlsruhe, Germany) in the range 2θ = 8°–80° (step size 0.016°) with Cu K$_\alpha$ radiation (Ni filter, 45 mA, 35 kV) on rotating sample holders. Raman spectra were recorded with a Horiba Jobin Yvon spectrometer (LabRam, Palaiseau, France) attached to an Olympus microscope (BX 40, Hamburg, Germany) using a 632.8 nm laser in the range 100–1100 cm^{-1} without pre-treatment of the sample on an object slide.

The specific surface area (SSA) was measured by nitrogen adsorption at its boiling point (Belsorp II mini, BEL Japan Inc., Osaka, Japan) using multipoint BET theory in the p/p_0 = 0.05–0.3 range.

Surface analysis by X-ray photoelectron spectroscopy (XPS) was performed with a K-Alpha spectrometer (ThermoFisher Scientific, Braunschweig, Germany) using a micro focused Al K$_\alpha$ X-ray source (400 μm spot size). Data acquisition and processing using Thermo Avantage software (Braunschweig, Germany) is described elsewhere [66]. Charge compensation during analysis was achieved using electrons of 8 eV energy and low energy argon ions to prevent any localized charging. Spectra were fitted with one or more Voigt profiles (binding energy uncertainty: ±0.2 eV) [12,67]. Scofield sensitivity factors were applied for quantification [68]. Charging was corrected by shifting to the binding energy of C 1s (C–C, C–H) at 285.0 eV. The energy scale was calibrated by means of the well-known photoelectron peaks of metallic Cu, Ag and Au.

The bulk composition of the catalysts was determined by optical emission spectrometry with inductively coupled plasma (ICP-OES, Agilent 720/725-ES,

Waldbronn, Germany). The plasma was created by a 40 MHz high-frequency generator and argon was applied as the plasma gas. For ICP-OES each sample was dissolved in 6 mL concentrated HNO_3, 2 mL concentrated HCl and 0.5 mL H_2O_2 in a microwave (at 600 W for 45 min).

X-ray absorption spectroscopy (XAS) was performed at beamline BM01B at the European Synchrotron Radiation Facility (ESRF, Grenoble, France). The samples were diluted with boron nitride and pressed to pellets for *ex situ* measurement in transmission mode at the Mo K edge (20.0 keV). XAS data were processed using the IFFEFIT software package (Chicago, IL, USA) [69].

3.3. Catalytic Tests

The catalytic performance was tested in a continuous flow fixed bed reactor, a U-shaped quartz reactor with 4 mm inner diameter. The catalyst powders were pressed into pellets, crushed and sieved to 150–300 μm particles, and 500 mg of sample were loaded in the reactor and stabilized with quartz wool. The quartz reactor was connected to a commercial test unit (ChimneyLab Europe, Hadsten, Denmark) with calibrated mass flow controllers (Brooks, Hatfield, PA, USA) and placed in an oven (Watlow, St. Louis, MO, USA) [70]. A thermocouple was placed inside the reactor just touching the catalyst bed to measure the reaction temperature and a pressure transducer placed upstream of the reactor measured the actual reaction pressure. The catalysts were pre-oxidized in dry air at 300 °C and activity tests were performed using a gas composition of $C_3H_6/O_2/N_2 = 5/25/70$ and flows of 50, 80, 120 NmL/min. The gas analysis was performed with a dual channel GC-MS (Thermo Fischer, Brauschweig, Germany) with a TCD detector for quantifying N_2, O_2, CO and CO_2 and a FID detector parallel to the MS for identification and quantification of saturated and unsaturated light hydrocarbons and oxygenated by-products. Before calculating the conversion of propylene and the selectivity for acrolein, the measured concentrations were corrected for expansion of the gas due to combustion using the nitrogen signal as internal standard.

4. Conclusions

In the present study different bismuth molybdate catalysts were synthesized using hydrothermal synthesis. The applied Bi/Mo ratio and the pH value had a strong influence on crystalline phases, specific surface area, bulk and surface composition and, thus, the catalytic performance. With increasing pH values from 1 to 9, the amount of more active γ-Bi_2MoO_6 compared to α-$Bi_2Mo_3O_{12}$ increased resulting in a higher catalytic activity.

The samples synthesized at pH = 4–7 exhibited relatively large surface areas (>20 m^2/g) compared to unsupported bismuth molybdates reported in literature. The higher surface area was beneficial for the catalytic activity in the selective oxidation of propylene under the applied conditions. The bismuth molybdates synthesized at pH = 6–7, which contained only γ-Bi$_2$MoO$_6$ and which exhibited high specific surface areas, yielded the highest propylene conversion at high acrolein selectivities. They were more active than samples with higher surface areas prepared at pH = 5, which also contained impurities of other phases. Very high (pH $\geqslant$ 9) or low (pH $\leqslant$ 1) pH values during synthesis led to catalysts with low catalytic activity and low specific surface area. This demonstrated that surface area is an important, but not the only, factor influencing activity, in agreement with literature.

Increasing the process temperature from 320 °C to 400 °C resulted in higher propylene conversion, while the catalysts deactivated at temperatures above 440 °C probably due to sintering and decreased specific surface area. Hence, especially at higher operating temperatures, stabilization of the surface area remains a key issue. Incorporation of further elements like Co, Fe or V, which are also included in industrial multicomponent bismuth molybdates, via hydrothermal synthesis could be beneficial for the stability of the prepared catalysts, as well as for the overall catalytic performance.

Acknowledgments: We thank the Danish Council for Strategic Research for financial support in the framework of the DSF proposal "Nanoparticle synthesis for catalysis" (grant No. 2106-08-0039). We gratefully acknowledge Hermann Köhler at the Institute of Catalysis Research and Technology (IKFT-KIT) for performing ICP-OES measurements and Angela Beilmann at Institute for Chemical Technology and Polymer Chemistry (ITCP-KIT) for the nitrogen physisorption measurements. The European Synchrotron Radiation Facility (ESRF) in Grenoble is acknowledged for providing XAS beamtime at the beamline BM01B. We thank Dmitry Doronkin and Hudson Carvalho (KIT) for the XAS measurements and Wouter van Beek (ESRF) for help and support during the beamtime.

Author Contributions: The experimental work and drafting of the manuscript was carried out by K.S., assisted by M.H. who performed catalytic measurements. V.T. and P.B. supported XPS and Raman experiments. W.K. participated in the interpretation of the scientific results and the preparation of the manuscript. J.-D.G. and A.D.J. supported the work and cooperation between DTU and KIT, supervised the experimental work, commented and approved the manuscript. The manuscript was written through comments and contributions of all authors. All authors have given approval to the final version of the manuscript.

Conflicts of Interest: The authors declare no conflict of interest.

References

1. Idol, J.D. Process for the Manufacture of Acrylonitrile. U.S. Patent 2,904,580, 15 September 1959.
2. Callahan, J.L.; Foreman, R.W.; Veatch, F. Attrition Resistant Oxidation Catalysts. U.S Patent 3,044,966, 17 July 1962.
3. Grasselli, R.K.; Burrington, J.D. Selective oxidation and ammoxidation of propylene by heterogeneous catalysis. *Adv. Catal.* **1981**, *30*, 133–163.
4. Snyder, T.P.; Hill, C.G. The mechanism of the partial oxidation of propylene over bismuth molybdate catalysts. *Catal. Rev. Sci. Eng.* **1989**, *31*, 43–95.

5. Hoefs, E.V.; Monnier, J.R.; Keulks, G.W. Investigation of the type of active oxygen for the oxidation of propylene over bismuth molybdate catalysts using Infrared and Raman spectroscopy. *J. Catal.* **1979**, *57*, 331–337.

6. Batist, P.A. Bismuth molybdates—Preparation and catalysis. *J. Chem. Technol. Biotechnol.* **1979**, *29*, 451–466.

7. German, K.; Grzybows, B.; Haber, J. Active centers for oxidation of propylene on Bi–Mo–O catalysts. *Acad. Pol. Sci. Chim.* **1973**, *21*, 319–325.

8. Krenzke, L.D.; Keulks, G.W. Catalytic oxidation of propylene 6. Mechanistic studies utilizing isotopic tracers. *J. Catal.* **1980**, *61*, 316–325.

9. Carson, D.; Coudurier, G.; Forissier, M.; Védrine, J.C.; Laarif, A.; Theobald, F. Synergy effects in the catalytic properties of bismuth molybdates. *J. Chem. Soc. Faraday Trans. 1* **1983**, *79*, 1921–1929.

10. Zhou, B.; Sun, P.; Sheng, S.; Guo, X. Cooperation between the α and γ phases of bismuth molybdate in the selective oxidation of propene. *J. Chem. Soc. Faraday Trans.* **1990**, *86*, 3145–3150.

11. Le, M.T.; van Well, W.J.M.; Stoltze, P.; van Driessche, I.; Hoste, S. Synergy effects between bismuth molybdate catalyst phases (Bi/Mo from 0.57 to 2) for the selective oxidation of propylene to arcrolein. *Appl. Catal. A* **2005**, *282*, 189–194.

12. Ayame, A.; Uchida, K.; Iwataya, M.; Miyamoto, M. X-ray photoelectron spectroscopic study on α- and γ-bismuth molybdate surfaces exposed to hydrogen, propene and oxygen. *Appl. Catal. A* **2002**, *227*, 7–17.

13. Hanna, T.A. The role of bismuth in the SOHIO process. *Coord. Chem. Rev.* **2004**, *248*, 429–440.

14. Getsoian, A.B.; Shapovalov, V.; Bell, A.T. DFT+U Investigation of Propene Oxidation over Bismuth Molybdate: Active Sites, Reaction Intermediates, and the Role of Bismuth. *J. Phys. Chem. C* **2013**, *117*, 7123–7137.

15. Morooka, Y.; Ueda, W. Multicomponent bismuth molybdate catalyst: A highly functionalized catalyst system for the selective oxidation of olefin. *Adv. Catal.* **1994**, *40*, 233–273.

16. Ueda, W.; Morooka, Y.; Ikawa, T.; Matsuura, I. Promotion effect of iron for the multicomponent bismuth molybdate catalysts as revealed by $^{18}O_2$ tracer. *Chem. Lett.* **1982**, 1365–1368.

17. Millet, J.M.M.; Ponceblanc, H.; Coudurier, G.; Herrmann, J.M.; Védrine, J.C. Study of multiphasic molybdate-based catalysts. 2. Synergy effect between bismuth molybdates and mixed iron cobalt molybdates in mild oxidation of propene. *J. Catal.* **1993**, *142*, 381–391.

18. Carson, D.; Forissièr, M.; Védrine, J.C. Kinetic study of the partial oxidation of propene and 2-methylpropene on different bismuth molybdate and on a bismuth iron molybdate phase. *J. Chem. Soc. Faraday Trans. 1* **1984**, *80*, 1017–1028.

19. Krenzke, L.D.; Keulks, G.W. The catalytic oxidation of propylene. 8. An investigation of kinetics over $Bi_2Mo_3O_{12}$, Bi_2MoO_6 and $Bi_3FeMo_2O_{12}$. *J. Catal.* **1980**, *64*, 295–302.

20. Monnier, J.R.; Keulks, G.W. The catalytic oxidation of propylene. 9. The kinetics and mechanism over β-$Bi_2Mo_2O_9$. *J. Catal.* **1981**, *68*, 51–66.

21. Brazdil, J.F.; Suresh, D.D.; Grasselli, R.K. Redox kinetics of bismuth molybdate ammoxidation catalysts. *J. Catal.* **1980**, *66*, 347–367.

22. Schuh, K.; Kleist, W.; Høj, M.; Trouillet, V.; Jensen, A.D.; Grunwaldt, J.-D. One-step synthesis of bismuth molybdate catalysts via flame spray pyrolysis for the selective oxidation of propylene to acrolein. *Chem. Commun.* **2014**, *50*, 15404–15406.

23. Snyder, T.P.; Hill, C.G. Stability of bismuth molybdate catalysts at elevated temperatures in air under reaction conditions. *J. Catal.* **1991**, *132*, 536–555.

24. Batist, P.A.; Bouwens, J.F.H.; Schuit, G.C.A. Bismuth molybdate catalysts—Preparation, characterization and activity of different compounds in Bi–Mo–O System. *J. Catal.* **1972**, *25*, 1–11.

25. Keulks, G.W.; Hall, J.L.; Daniel, C.; Suzuki, K. Catalytic oxidation of propylene. 4. Preparation and characterization of α-bismuth molybdate. *J. Catal.* **1974**, *34*, 79–97.

26. Soares, A.P.V.; Dimitrov, L.D.; de Oliveira, M.; Hilaire, L.; Portela, M.F.; Grasselli, R.K. Synergy effects between beta and gamma phases of bismuth molybdates in the selective catalytic oxidation of 1-butene. *Appl. Catal. A* **2003**, *253*, 191–200.

27. Rastogi, R.P.; Singh, A.K.; Shukla, C.S. Kinetics and mechanism of solid-state reaction between bismuth(III) oxide and molybdenum(VI) oxide. *J. Solid State Chem.* **1982**, *42*, 136–148.

28. Thang, L.M.; Bac, L.H.; van Driessche, I.; Hoste, S.; van Well, W.J.M. The synergy effect between gamma and beta phase of bismuth molybdate catalysts: Is there any relation between conductivity and catalytic activity? *Catal. Today* **2008**, *131*, 566–571.

29. Le, M.T.; van Craenenbroeck, J.; van Driessche, I.; Hoste, S. Bismuth molybdate catalysts synthesized using spray drying for the selective oxidation of propylene. *Appl. Catal. A* **2003**, *249*, 355–364.

30. Van Well, W.J.M.; Le, M.T.; Schiødt, N.C.; Hoste, S.; Stoltze, P. The influence of the calcination conditions on the catalytic activity of Bi_2MoO_6 in the selective oxidation of propylene to acrolein. *J. Mol. Catal. A* **2006**, *256*, 1–8.

31. Nell, A.; Getsoian, A.B.; Werner, S.; Kiwi-Minsker, L.; Bell, A.T. Preparation and Characterization of High-Surface-Area $Bi_{(1-x)/3}V_{1-x}Mo_xO_4$ Catalysts. *Langmuir* **2014**, *30*, 873–880.

32. Shi, Y.; Feng, S.; Cao, C. Hydrothermal synthesis and characterization of Bi_2MoO_6 and Bi_2WO_6. *Mater. Lett.* **2000**, *44*, 215–218.

33. Beale, A.M.; Sankar, G. *In situ* study of the formation of crystalline bismuth molybdate materials under hydrothermal conditions. *Chem. Mater.* **2003**, *15*, 146–153.

34. Yu, J.Q.; Kudo, A. Hydrothermal synthesis and photocatalytic property of 2-dimensional bismuth molybdate nanoplates. *Chem. Lett.* **2005**, *34*, 1528–1529.

35. Xie, H.; Shen, D.; Wang, X.; Shen, G. Microwave hydrothermal synthesis and visible-light photocatalytic activity of γ-Bi_2MoO_6 nanoplates. *Mater. Chem. Phys.* **2008**, *110*, 332–336.

36. Li, H.; Li, K.; Wang, H. Hydrothermal synthesis and photocatalytic properties of bismuth molybdate materials. *Mater. Chem. Phys.* **2009**, *116*, 134–142.

37. Gruar, R.; Tighe, C.J.; Reilly, L.M.; Sankar, G.; Darr, J.A. Tunable and rapid crystallisation of phase pure Bi_2MoO_6 (koechlinite) and $Bi_2Mo_3O_{12}$ via continuous hydrothermal synthesis. *Solid State Sci.* **2010**, *12*, 1683–1686.

38. Kongmark, C.; Coulter, R.; Cristol, S.; Rubbens, A.; Pirovano, C.; Loefberg, A.; Sankar, G.; van Beek, W.; Bordes-Richard, E.; Vannier, R.-N. A Comprehensive Scenario of the Crystal Growth of γ-Bi_2MoO_6 Catalyst during Hydrothermal Synthesis. *Cryst. Growth Des.* **2012**, *12*, 5994–6003.

39. Yoshimura, M.; Byrappa, K. Hydrothermal processing of materials: Past, present and future. *J. Mater. Sci.* **2008**, *43*, 2085–2103.

40. Zhang, L.; Xu, T.; Zhao, X.; Zhu, Y. Controllable synthesis of Bi_2MoO_6 and effect of morphology and variation in local structure on photocatalytic activities. *Appl. Catal. B* **2010**, *98*, 138–146.

41. Ren, J.; Wang, W.; Shang, M.; Sun, S.; Gao, E. Heterostructured Bismuth Molybdate Composite: Preparation and Improved Photocatalytic Activity under Visible-Light Irradiation. *ACS Appl. Mater. Interfaces* **2011**, *3*, 2529–2533.

42. Aleshina, G.I.; Joshi, C.; Tarasova, D.V.; Kustova, G.N.; Nikoro, T.A. Catalytic properties of Bi/Mo oxide catalysts prepared via precipitation. *React. Kinet. Catal. Lett.* **1984**, *26*, 203–208.

43. Guo, C.; Xu, J.; Wang, S.; Li, L.; Zhang, Y.; Li, X. Facile synthesis and photocatalytic application of hierarchical mesoporous Bi_2MoO_6 nanosheet-based microspheres. *Cryst. Eng. Commun.* **2012**, *14*, 3602–3608.

44. Beale, A.M.; Jacques, S.D.M.; Sacaliuc-Parvalescu, E.; O'Brien, M.G.; Barnes, P.; Weckhuysen, B.M. An iron molybdate catalyst for methanol to formaldehyde conversion prepared by a hydrothermal method and its characterization. *Appl. Catal. A* **2009**, *363*, 143–152.

45. Zhang, L.; Zhang, Y.; Dai, H.; Deng, J.; Wei, L.; He, H. Hydrothermal synthesis and catalytic performance of single-crystalline $La_{2-x}Sr_xCuO_4$ for methane oxidation. *Catal. Today* **2010**, *153*, 143–149.

46. Salamanca, M.; Licea, Y.E.; Echavarria, A.; Faro, A.C., Jr.; Palacio, L.A. Hydrothermal synthesis of new wolframite type trimetallic materials and their use in oxidative dehydrogenation of propane. *Phys. Chem. Chem. Phys.* **2009**, *11*, 9583–9591.

47. Ueda, W.; Oshihara, K. Selective oxidation of light alkanes over hydrothermally synthesized Mo–V–M–O (M = Al, Ga, Bi, Sb, and Te) oxide catalysts. *Appl. Catal. A* **2000**, *200*, 135–143.

48. Botella, P.; Garcia-Gonzalez, E.; Dejoz, A.; Nieto, J.M.L.; Vazquez, M.I.; Gonzalez-Calbet, J. Selective oxidative dehydrogenation of ethane on MoVTeNbO mixed metal oxide catalysts. *J. Catal.* **2004**, *225*, 428–438.

49. Sanfiz, A.C.; Hansen, T.W.; Girgsdies, F.; Timpe, O.; Rödel, E.; Ressler, T.; Trunschke, A.; Schlögl, R. Preparation of Phase-Pure M1 MoVTeNb Oxide Catalysts by Hydrothermal Synthesis—Influence of Reaction Parameters on Structure and Morphology. *Top. Catal.* **2008**, *50*, 19–32.

50. Schuh, K.; Kleist, W.; Høj, M.; Trouillet, V.; Beato, P.; Jensen, A.D.; Patzke, G.R.; Grunwaldt, J.-D. Selective oxidation of propylene to acrolein by hydrothermally synthesized bismuth molybdates. *Appl. Catal. A* **2014**, *482*, 145–156.

51. Hardcastle, F.D.; Wachs, I.E. Molecular structure of molybdenum oxide in bismuth molybdates by Raman spectroscopy. *J. Phys. Chem.* **1991**, *95*, 10763–10772.

52. Noack, J.; Rosowski, F.; Schlögl, R.; Trunschke, A. Speciation of Molybdates under Hydrothermal Conditions. *Z. Anorg. Allg. Chem.* **2014**, *640*, 2730–2736.

53. Briand, G.G. Bifunctional Ligands in Discerning and Developing the Fundamental and Medicinal Chemistry of Bismuth(III). Ph.D. Thesis, Dalhousie University, Halifax, NS, Canada, July 1999.

54. Trifiro, F.; Scarle, R.D.; Hoser, H. Relationships between structure and activity of mixed oxides as oxidation catalyst. 1. Preparation and solid state reactions of Bi-molybdates. *J. Catal.* **1972**, *25*, 12–24.

55. Dewangan, K.; Sinha, N.N.; Sharma, P.K.; Pandey, A.C.; Munichandraiah, N.; Gajbhiye, N.S. Synthesis and characterization of single-crystalline α-MoO$_3$ nanofibers for enhanced Li-ion intercalation applications. *Cryst. Eng. Commun.* **2011**, *13*, 927–933.

56. Chen, L.; Aarcon-Lado, E.; Hettick, M.; Sharp, I.D.; Lin, Y.; Javey, A.; Ager, J.W. Reactive Sputtering of Bismuth Vanadate Photoanodes for Solar Water Splitting. *J. Phys. Chem. C* **2013**, *117*, 21635–21642.

57. Choi, J.G.; Thompson, L.T. XPS study of as-prepared and reduced molybdenum oxides. *Appl. Surf. Sci.* **1996**, *93*, 143–149.

58. Herrmann, J.M.; el Jamal, M.; Forissier, M. Evidence by electrical conductivity for an excess of bismuth as Bi$^+$ interstitial at the surface of gamma-phase Bi$_2$MoO$_6$—Consequence for selectivity in propene catalytic oxidation. *React. Kinet. Catal. Lett.* **1988**, *37*, 255–260.

59. Ruckenstein, E.; Krishnan, R.; Rai, K.N. Oxygen depletion of oxide catalysts. *J. Catal.* **1976**, *45*, 270–273.

60. Jung, J.C.; Kini, H.; Choi, A.S.; Chung, Y.-M.; Kim, T.J.; Lee, S.J.; Oh, S.-H.; Song, I.K. Effect of pH in the preparation of γ-Bi$_2$MoO$_6$ for oxidative dehydrogenation of *n*-butene to 1,3-butadiene: Correlation between catalytic performance and oxygen mobility of γ-Bi$_2$MoO$_6$. *Catal. Commun.* **2007**, *8*, 625–628.

61. Zhai, Z.; Getsoian, A.B.; Bell, A.T. The kinetics of selective oxidation of propene on bismuth vanadium molybdenum oxide catalysts. *J. Catal.* **2013**, *308*, 25–36.

62. Zhai, Z.; Wang, X.; Licht, R.; Bell, A.T. Selective oxidation and oxidative dehydrogenation of hydrocarbons on bismuth vanadium molybdenum oxide. *J. Catal.* **2015**, *325*, 87–100.

63. Grzybowska, B.; Haber, J.; Janas, J. Interaction of allyl iodide with molybdate catalysts for selective oxidation of hydrocarbons. *J. Catal.* **1977**, *49*, 150–163.

64. Swift, H.E.; Bozik, J.E.; Ondrey, J.A. Dehydrodimerization of propylene using bismuth oxide as oxidant. *J. Catal.* **1971**, *21*, 212–224.

65. Carrazán, S.R.G.; Martin, C.; Rives, V.; Vidal, R. Selective oxidation of isobutene to methacrolein on multiphasic molybdate-based catalysts. *Appl. Catal. A* **1996**, *135*, 95–123.

66. Parry, K.L.; Shard, A.G.; Short, R.D.; White, R.G.; Whittle, J.D.; Wright, A. ARXPS characterisation of plasma polymerised surface chemical gradients. *Surf. Interface Anal.* **2006**, *38*, 1497–1504.

67. Grunwaldt, J.-D.; Wildberger, M.D.; Mallat, T.; Baiker, A. Unusual redox properties of bismuth in sol-gel Bi–Mo–Ti mixed oxides. *J. Catal.* **1998**, *177*, 53–59.

68. Scofield, J.H. Hartree-Slater subshell photoionization cross-sections at 1254 and 1487eV. *J. Electron Spectrosc. Relat. Phenom.* **1976**, *8*, 129–137.

69. Ravel, B.; Newville, M. ATHENA, ARTEMIS, HEPHAESTUS: Data analysis for X-ray absorption spectroscopy using IFEFFIT. *J. Synchrotron Radiat.* **2005**, *12*, 537–541.

70. Høj, M.; Jensen, A.D.; Grunwaldt, J.-D. Structure of alumina supported vanadia catalysts for oxidative dehydrogenation of propane prepared by flame spray pyrolysis. *Appl. Catal. A* **2013**, *451*, 207–215.

Iron Fischer-Tropsch Catalysts Prepared by Solvent-Deficient Precipitation (SDP): Effects of Washing, Promoter Addition Step, and Drying Temperature

Kyle M. Brunner, Baiyu Huang, Brian F. Woodfield and William C. Hecker

Abstract: A novel, solvent-deficient precipitation (SDP) method for catalyst preparation in general and for preparation of iron FT catalysts in particular is reported. Eight catalysts using a 2^3 factorial design of experiments to identify the key preparation variables were prepared. The catalysts were characterized by electron microprobe, N_2 adsorption, TEM, XRD, and ICP. Results show that the morphology of the catalysts, *i.e.*, surface area, pore volume, pore size distribution, crystallite sizes, and promoter distribution are significantly influenced by (1) whether or not the precursor catalyst is washed, (2) the promoter addition step, and (3) the drying condition (temperature). Consequently, the activity, selectivity, and stability of the catalysts determined from fixed-bed testing are also affected by these three variables. Unwashed catalysts prepared by a one-step method and dried at 100 °C produced the most active catalysts for FT synthesis. The catalysts of this study prepared by SDP compared favorably in activity, productivity, and stability with Fe FT catalysts reported in the literature. It is believed that this facile SDP approach has promise for development of future FT catalysts, and also offers a potential alternate route for the preparation of other catalysts for various other applications.

Reprinted from *Catalysts*. Cite as: Brunner, K.M.; Huang, B.; Woodfield, B.F.; Hecker, W.C. Iron Fischer-Tropsch Catalysts Prepared by Solvent-Deficient Precipitation (SDP): Effects of Washing, Promoter Addition Step, and Drying Temperature. *Catalysts* **2015**, *5*, 1352–1374.

1. Introduction

Demand for liquid fuel sources combined with political unrest in some of the world's regions most abundant in oil and natural gas, in addition to the recent natural gas boom from hydraulic fracturing have pushed global and domestic energy policies to focus on domestic production and sustainability. This drive for domestic supplies of fuel provides opportunities for industrial innovation in developing and improving alternative liquid fuel sources, including natural gas, biomass, and coal. The Fischer-Tropsch synthesis (FTS) is one commercially proven process for producing hydrocarbon products from carbon monoxide and hydrogen (syngas) and is a key step in gas to liquid (GTL), biomass to liquid (BTL), and coal to liquid (CTL) technologies [1].

Catalyst preparation is a complex process intended to produce desirable chemical, physical, and catalytic properties in the final catalyst through choice of materials (*i.e.*, metal, precursor, promoter, and support) and by manipulation of preparation variables and conditions (e.g., precipitation pH and temperature, washing, drying and calcination temperatures, and reduction environment and temperature) [2]. Promoters are added to catalysts to enhance their physical, chemical, and catalytic properties such as reduction temperature, rate of reaction, selectivity, or catalyst life. Supporting a catalyst on an oxide matrix (e.g., alumina, silica, ceria, or titania) requires extra preparation time and steps, but is desirable when the precursor is expensive (as is the case for cobalt or precious metal catalysts) or when structural enhancements (e.g., surface area, pore volume, or pore diameter) increase the dispersion, selectivity, or stability of the catalyst. Preparation variables largely determine the crystallite size, dispersion of the catalyst, and the pore and surface structure. After preparation, iron catalysts require pretreatment in order to create iron carbide, which is the active phase for FTS. Commercial iron FT catalysts at Sasol and at Synfuels China (currently the only commercial scale facilities using iron catalysts) are unsupported and precipitated catalysts promoted with copper or manganese (Synfuels China), potassium, and silica. This study focuses on catalysts promoted with copper, potassium and silica.

Good, extensive and detailed reviews of precipitated catalysts have been published [2,3]. A brief summary follows. Near the end of WWII and following, Ruhrchemie developed the first commercial iron catalyst which was used (and later modified) by Sasol [4]. Preparation involved pouring a near boiling solution of iron and copper nitrates (5 g Cu per 100 g Fe) into a hot solution of sodium carbonate while controlling pH around 6. After extensive washing in hot water to remove sodium, the precipitate was slurried and impregnated with potassium silicate (25 g SiO_2 per 100 g Fe) to improve thermal stability. Potassium was removed by adding nitric acid until the desired amount of potassium remained (5 g K per 100 g Fe). The effects of preparation method such as the precipitation order (acid into base or *vice versa*), final pH, temperature, viscosity of solvent at drying time, calcination temperature, and reduction temperature and environment (H_2 only or $H_2 + CO$) were known but not necessarily understood. Commercial catalysts used at Sasol and at Synfuels China are modifications of the Ruhrchemie catalyst. Although other methods and formulations have been developed, the fact that the Ruhrchemie preparation is still used in modified forms speaks to the robust nature and activity of these catalysts.

Improvements on the Ruhrchemie catalyst have been developed both by academic and industrial interests. Early work by Sasol showed: (1) lower required reduction temperature with increasing Cu content, (2) increased thermal stability and pore structure with increasing SiO_2 content, and (3) increased wax selectivity with increasing K content [4]. Later, Texas A&M studies spanning more than 10 years resulted in a continuous co-precipitation method to carefully control pH and temperature, resulting in highly active, selective, and stable catalysts for both FB and slurry reactor applications [5–8] which are among the most active iron catalysts reported in the literature to date [5–8]. Studies showed that K inhibits reducibility, increases WGS and iron FTS activity, and increases selectivity to olefins and heavy hydrocarbons. It was suggested that at higher K loadings, the apparent decrease in activity may be due to increased diffusion effects from pores filled with heavier molecular weight hydrocarbons. Increasing SiO_2 loading increased stability, but decreased reducibility, activity, and selectivity possibly due to diminishing effectiveness of Cu and K loadings as SiO_2 interacts with and neutralizes the effects of Cu and K. Cu increases reducibility and promotes WGS activity. Studies at UC Berkeley focused on promoter effects (K, Zn, Cu, and Ru) and on the effects of solvent viscosity on surface structure and promoter dispersion [9]. The goal was to increase the density of CO binding sites, thus bringing Fe catalysts to a level of activity comparable to that of Co catalysts. Promoters increased site density by encouraging smaller nucleation sites during reduction and reaction. Replacing water in precursor pores with low surface tension alcohols resulted in larger surface areas. Catalysts prepared by this method were of comparable activity to a low activity Co catalyst and had lower selectivities to CH_4. Work at the Chinese Academy of Science Institute of Coal Chemistry and by Synfuels China not only explored the effects of other metal promoters such as Mn but also produced highly active, stable and selective catalysts from sulfur-containing precursors which traditionally have been avoided due to the strong poisoning behavior of sulfur [10,11]. Recent studies showed that there could be an optimal range of K loading on Fe-based FTS catalysts, since higher K loading generally resulted in higher carburization rate, though too much K would lead to excessive carbon deposition [12,13].

Co-precipitation from aqueous solutions of precursors has been the principle preparation method for iron FT catalysts in the past; however, recent breakthroughs in solvent-deficient precipitation (SDP) of transition metal oxide nanoparticles by the BYU Chemistry Department and Cosmas Inc. provide exciting new possibilities for simpler, more uniform, and faster catalyst preparation methods [14–21]. Nanoparticle size distributions are tight and can be controlled over several orders of magnitude (1–10,000 nm) by simple factors such as adding water during the reaction, washing after reaction, and calcining at different temperatures. Successful synthesis of nanoparticle oxides has been achieved on over 20 different transition metals, rare earth metals, and groups I, II, and III of the periodic table including Ti [20], Al [16–18], Zr, Ni, Cu, Fe, Co, and La, all of which are common catalyst or support materials. To the best of our knowledge, the Cosmas SDP method had never been used for preparation of a catalyst prior to our current work. A report describing the preparation and performance of one Fe FT catalyst using the SDP method was published earlier this year [22]. It was also the subject of a dissertation [23].

In this paper, we report the preparation of eight iron FT catalysts and the use of a 2^3 factorial design of experiments to identify the key preparation variables. The catalysts are characterized by microprobe, N_2 adsorption, TEM, XRD, and ICP. Results show that the morphology of the catalysts, *i.e.*, surface area, pore volume, pore size distribution, crystallite sizes, and promoter distribution are significantly influenced by drying, washing, and the promoter addition step. Consequently, the activity, selectivity, and stability of the catalysts are also affected by these three variables. We also compare the catalysts in this study with other published, highly active iron FT catalysts. Results show that catalysts prepared via the SDP method have comparable or even better activity, productivity, and stability than traditional Fe FT catalysts. Thus, this facile SDP approach seems to have potential for the development of not only FT catalysts, but also for preparing heterogeneous catalysts for other applications as well.

2. Results and Discussion

2.1. Catalyst Characterizations

2.1.1. Catalyst Nomenclature Description

To investigate the effects of three key preparation variables—(1) timing of the addition of promoters potassium and silica, (2) presence or absence of a washing step for the precursor, and (3) drying temperature—two levels of each of these variables were chosen. Each level is represented by a single alphanumeric character, as shown in Scheme 1. The eight catalysts of the factorial design are designated by the three characters representing the variable levels. Levels for timing of promoter addition were "1 Step" (designated by a 1) in which potassium and silica promoters were added to the salts of iron and copper before precipitation, and the catalyst was created in a single step, and "2 Step" (designated by a 2) in which potassium and silica promoters were added in a separate step after precipitation and washing, if applicable. The levels for washing were unwashed (U) or washed (W) using 100 mL of deionized water five times immediately following precipitation. Levels for initial drying temperature were high (H) temperature (100 °C) and low (L) temperature (60 °C overnight, followed by 100 °C). The six alphanumeric characters for catalyst designation represent the levels of the promoter addition (1 or 2), washing (U or W), and drying steps (L or H). For example, 1UH designates a catalyst prepared in 1 Step (with potassium and silica added to salts before precipitation), not washed (unwashed) after precipitation, and dried at the high drying temperature condition (100 °C). 2WL denotes a catalyst prepared in two steps (with potassium and silica added after precipitation and washing), washed directly after precipitation, and dried first at the low drying temperature (60 °C) followed by drying at 100 °C.

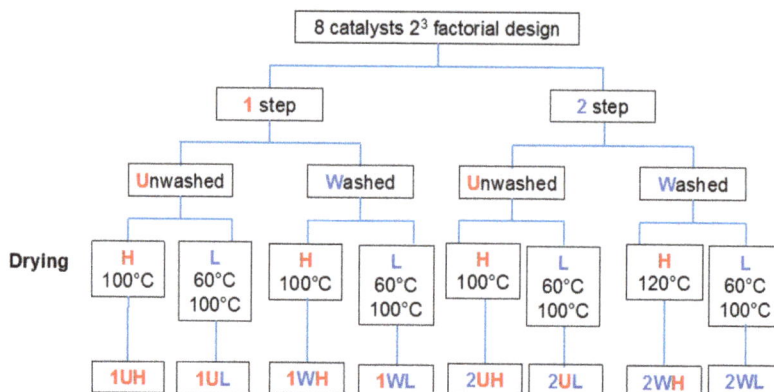

8 catalysts 2³ factorial design

1 step | 2 step

Unwashed | Washed | Unwashed | Washed

Drying
H 100°C | L 60°C 100°C | H 100°C | L 60°C 100°C | H 100°C | L 60°C 100°C | H 120°C | L 60°C 100°C

1UH | 1UL | 1WH | 1WL | 2UH | 2UL | 2WH | 2WL

Scheme 1. 2^3 Factorial design for eight catalysts of this study.

2.1.2. Catalyst Compositions and Metal Distributions

Catalyst elemental compositions determined by ICP analysis are shown in Table 1 and are in good agreement with nominal compositions, except for the 1 Step washed catalysts (1WH, 1WL). ICP analysis of Si content in all cases was unreliable due to the formation of volatile SiF_4 during digestion and therefore are not reported.

Table 1. Compositions and pore properties of Fe catalysts.

Catalyst	Nominal Composition (pbm) [a]	ICP composition [b] (pbm)	Fe Content [c] (Mass%)	Surface Area [d] (m²/g)	Pore Volume [e] (cm³/g)	Pore diameter (nm)	Pore Range [f] (nm)
1UH	100Fe/5.0Cu/4.0K/16.0SiO₂	100Fe/5.1Cu/3.6K	66.3	45.6	0.14	10.3	6.3–17.4
1UL	100Fe/5.0Cu/4.0K/16.0SiO₂	100Fe/4.8Cu/3.7K	72.1	51.7	0.13	8.1	4.2–15.5
1WH	100Fe/5.0Cu/4.0K/16.0SiO₂	100Fe/4.9Cu/0.3K	62.1	55.6	0.24	15.5	8.7–28.0
1WL	100Fe/5.0Cu/4.0K/16.0SiO₂	100Fe/5.1Cu/0.3K	70.2	37.2	0.24	34.2	16.1-70.7
2UH	100Fe/5.0Cu/4.0K/16.0SiO₂	100Fe/4.1Cu/3.6K	68.0	36.8	0.14	8.7	5.0–15.3
2UL	100Fe/5.0Cu/4.0K/16.0SiO₂	100Fe/5.5Cu/4.2K	62.8	54.7	0.15	13.1	4.0–43.0
2WH	100Fe/5.0Cu/4.0K/16.0SiO₂	100Fe/5.3Cu/5.2K	65.1	65.7	0.13	5.5	1.2–12.8
2WL	100Fe/5.0Cu/4.2K/16.0SiO₂	100Fe/5.1Cu/5.0K	72.8	23.6	0.20	46.1	16.6–127.8

[a] parts by mass equal to g per 100 g$_{Fe}$; [b] SiO₂ not quantified by ICP; [c] after reduction and passivation; [d] 95% confidence interval = $\pm$ 2.1 m²/g; [e] 95% confidence interval = $\pm$0.01 cm³/g; [f] $\pm$2 standard deviations of log normal pore diameter.

Figure 1 shows Fe, Cu, K, and SiO_2 distribution for 1UH, 2UH, and 2WH at the micron scale. Since Fe constitutes 60%–66% of these passivated catalysts, images of the other elements are compared to images of Fe to show contrasts in distributions. Figure 1a shows a microprobe image of a 50 μm agglomerate of 1UH. The distributions for Cu, K, and SiO_2 show uniform intensities and give the same details of cracks and boundaries as Fe. This indicates that the Cu, K, and Si distributions are uniform and that the promoters are in intimate association with Fe.

A similar 50 μm view of an agglomerate of 2UH (Figure 1b) shows a marked difference in elemental distributions of K and Si compared to 1UH. While Figure 1b shows uniform distributions of Fe and Cu, preferential distributions of K and Si along particle boundaries are observed. The center of the agglomerate shows almost no signal for K and Si, but since the signals lack sharp and well-defined edges, it appears that the promoters are distributed within outer structures and spaces and beginning to penetrate inward with the much smaller K^+ ions penetrating farther than the large clusters of SiO_2. This observation is supported by the fact that water was not added to 2UH during promoter addition as the precursor appeared moist enough to completely dissolve the $KHCO_3$ and SiO_2. The minimal amount of moisture distributed the promoters between particles, but the pores, already filled with a nearly saturated solution of NH_4NO_3 byproduct, presented some diffusional resistance for the promoters and prevented full penetration and uniform dispersion of the promoters.

A view of an agglomerate of 2WH at the same scale (Figure 1c) shows much better penetration of K and Si into the particles than Figure 1b. The K signal appears to reflect the structure shown by the Fe signal and shows no preferential accumulation at particle boundaries. The very center of the agglomerate shows some decrease, but not complete absence of signal. The Si signal clearly shows the structure and defects observed in the Fe distribution. The signal within the agglomerate is uniform despite the fact that the outside edges of the particle show well-defined edges and a stronger signal. Strong signal is also shown in spaces devoid of Fe signal between closely spaced particles. This is explained in part by the washing process and method of promoter addition. For 2WH, promoters were first dissolved in a minimal amount of water and then added to the precursor. Vacuum filtration following washing leaves the precursor relatively dry and the solution containing the promoters is readily absorbed into the pores, but there is still a tendency to accumulate on the outer edges of the particle, especially for SiO_2.

From this analysis, it is clear that the promoters in 1S preparations are more uniformly distributed than in 2S preparations. While the byproduct is present in both 1UH and 2UH, the uniform distribution of potassium and silica promoters in 1UH shows that the promoters are intimately mixed with the Fe and their distribution is unaffected by diffusion through concentrated solutions of the byproduct as opposed to 2UH in which the diffusion hinders full penetration of particles by the promoters. Promoter distribution in 2WH is uniform within particles (despite higher concentrations on the outside edges of particles) due to removing the byproduct (by washing) and much of the water (by vacuum) so that the solution of promoters is readily absorbed into pores, whereas in 2UH the promoters are more concentrated near particle boundaries.

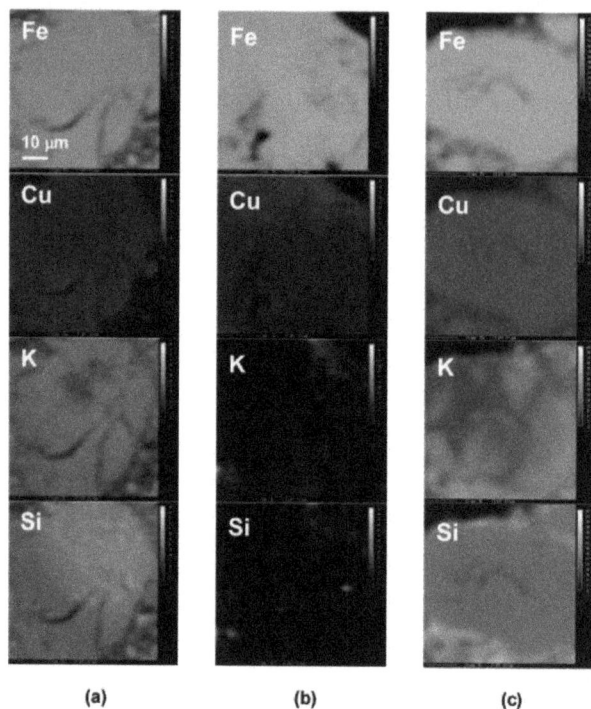

Figure 1. Electron microprobe images showing distributions of Fe, Cu, K, and Si of an agglomerate of (**a**) 1UH, (**b**) 2UH, and (**c**) 2WH.

2.1.3. Catalyst Pore Structure and Crystallite Size

BET data for reduced and passivated catalysts are summarized in Table 1. Surface areas for the four 1S catalysts are 37.2–55.6 m^2/g, whereas surface areas for the four 2S catalysts are 23.6–65.7 m^2/g. For all catalysts, surface area follows the same order of progression with WL < UH < UL < WH. One study on the effect of SiO_2 loading on precipitated iron catalyst surface area reported catalysts with no SiO_2 having reduced surface areas of 10 m^2/g or less (after reduction in H_2), whereas catalysts with 8 or 24 parts SiO_2 per 100 parts Fe have surface areas of 98 m^2/g and 150 m^2/g (after reduction in CO), respectively [24]. If reduction in CO rather than H_2 gives larger surface areas by a factor of 2–3 as it appears to do, then the surface areas of 1S and 2S catalysts are between reported surface areas for precipitated catalysts containing 8 and 24 parts SiO_2, as expected. Notable exceptions are 2UH and 2WL, which have surface areas less than 40 m^2/g. These low surface area values support the previous supposition that the distribution of SiO_2 in these catalysts is not uniform.

Pore volumes of all reduced catalysts are 0.13–0.24 cm^3/g with washed catalysts giving larger pore volumes (0.13–0.24 cm^3/g compared with 0.13–0.15 cm^3/g) except 2WH (0.13 cm^3/g), which is a washed catalyst dried at 120 °C rather than 100 °C. The smaller pore volumes of the unwashed catalysts again suggest a different precursor structure than is found in the structure of the washed catalysts.

Pore diameters for the catalysts follow log-normal distributions with average pore diameters between 5.5 and 46.1 nm (see Table 1 and Figure 2). Figure 2a shows pore size distributions (PSD) for 1S catalysts. The pore size distributions for 1UH and 1UL are very similar at 4–200 nm, whereas the distributions for 1WH and 1WL are broader at 4–600 nm. Average diameters for 1UH and 1UL are 10.3 nm and 8.1 nm, respectively, whereas 1WH is 50–100% larger at 15.5 nm, and 1WL is 220–320% larger at 34.2 nm. Thus, for the 1S preparation, unwashed catalysts have smaller pores and narrower distributions of pores.

Further study of PSDs indicates that drying temperature has a combined effect with the presence of water such that lower drying temperatures correspond to larger pores (and larger particles). BET measurements on 1S catalysts were taken after calcination and again after reduction. 1S PSDs after calcination are nearly identical and show only pores smaller than 9 nm with averages between 2 nm and 4 nm; however, the PSDs after reduction are very different, as described above and shown in Figure 2a. Since all of the 1S catalyst pore structures are nearly identical after calcination, differences after reduction would not be expected. These differences after reduction could be due to the extent that the structures after drying lend themselves to particle growth (sintering) and to reduction, *i.e.* degree of crystallinity and contact between particles. It has been reported that the degree of crystallinity of ferrihydrite (precursor) increases as a function of temperature and time in the presence of water [25]. One theory of agglomeration states that linkages between molecules of water on the surfaces of smaller particles cause the particles to agglomerate. When the water is driven off, the ordered linkages lend themselves to phase transition from ferrihydrite to Fe_2O_3 as larger agglomerates [26]. Thus, pore size (which is related to particle size) increases with time and temperature in the presence of water. The lower drying temperature for 1WL may have driven water off slowly enough and provided enough time to increase the agglomeration sufficiently and create larger linked agglomerates which resulted in the largest pore size distribution of the 1S catalysts. The higher drying temperature of 1WH afforded the creation of some linkages (more so than in 1UH or 1UL, which had no water added), but drove the water off too quickly to create the extensive linking in the larger agglomerates of 1WL. Thus, the 1WH PSD is larger than 1UH and 1UL, but slightly smaller than 1WL.

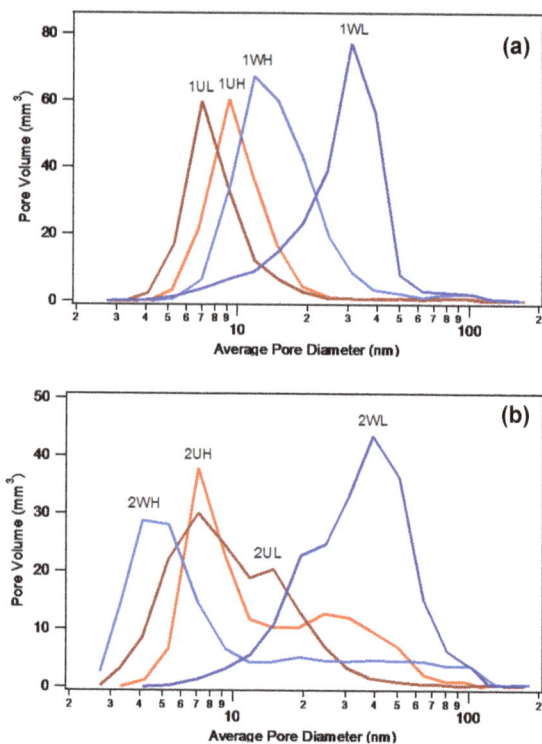

Figure 2. Pore size distributions of (a) 1S and (b) 2S catalysts after reduction and passivation.

Since the PSDs of 1UH and 1UL are very similar, it follows that drying temperature alone has very little effect on PSD. The reason may be that in these unwashed precursors, the formation of inter-particle linkages is inhibited. As shown in Equations 1 and 2 the products of reaction provide 10 molecules of H_2O for every atom of Fe in the preparation. This is more than adequate for hydration and coordination of the precursor, yet the PSDs of 1UH and 1UL are almost identical. Even though there is plenty of water present, the water is nearly saturated with byproduct NH_4NO_3. To the original agglomeration theory related above (see Figure 3), this work adds that the byproduct NH_4NO_3 may interfere with the weak water linkages between particles, making it difficult for larger agglomerates to form. Once dry, the particles are no longer free to form linkages and agglomerate, resulting in more narrow PSDs and smaller average diameters.

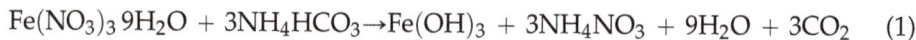

$$Fe(NO_3)_3\, 9H_2O + 3NH_4HCO_3 \rightarrow Fe(OH)_3 + 3NH_4NO_3 + 9H_2O + 3CO_2 \quad (1)$$

$$Fe(OH)_3 \rightarrow FeOOH + H_2O \quad (2)$$

NH$_4^+$ and NO$_3^-$
interference

2 nm
ferrihydrite

2 nm
ferrihydrite

2 nm
ferrihydrite

particle
growth

2 nm
ferrihydrite

2 nm
ferrihydrite:
91 units of
Fe$_5$O$_{12}$H$_9$

dehydration

dehydroxylation   surface
water linkage

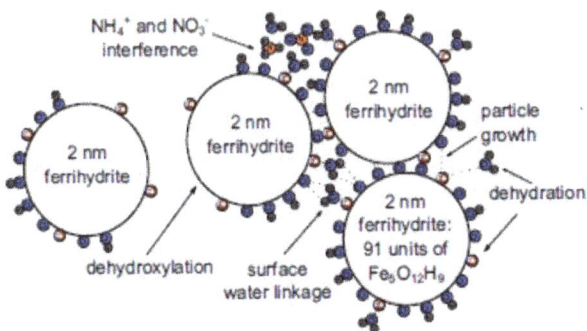

Figure 3. Illustration of agglomerate formation from ferrihydrite precursor particles. Large circles represent cross-sections of 2 nm spheres of ferrihydrite, whereas smaller circles represent 1.5 Å atomic cross-sections. Adapted from Ref. [24].

PSDs for the four 2S catalysts are shown in Figure 2b. The PSD for 2UH shows a bimodal distribution with the first average pore size at 8.7 nm and a +/−2σ range from 5.0 nm to 15.3 nm representing only 60.8% of the total pore volume as given in Table 1. The second average pore size at 34.6 nm has a +/−2σ range from 12.5 nm to 99.1 nm and accounts for 38.2% of the pore volume. The PSD for 2UL is fit as a single peak, even though it has a slight shoulder representing a second peak, giving a volume weighted average diameter of 13.1 nm and a range of 4.0–43.0 nm. The average pore diameter of 2WL is 46.1 nm and the range is 16.6–127.8 nm. It is interesting to note that PSDs for these three 2S catalysts cover much broader ranges of diameters than their 1S counter parts. 2WH is the exception and is discussed below. Like 1WL for 1S catalysts, 2WL has the largest average pore diameter and PSD of 2S catalysts. In contrast to the 1U catalysts, the PSDs for 2UH and 2UL are quite broad. The primary PSD for 2UH is in the expected narrow range and of smaller average diameter for an unwashed catalyst as explained above, but the secondary PSD is much larger. A comparison of PSDs of 2S catalysts after calcination with PSD after reduction indicates that this secondary range may be the result of adding precursors in a second step. The PSDs after calcination for all 2S catalysts have bimodal distributions with 50–75% of the pore volume in pores less than 9 nm. The balance of the pore volume for all of these catalysts spans the pore sizes described above. This shows that while the primary PSDs increased with reduction, the secondary PSDs for these catalysts did not. This behavior is expected if the secondary PSDs are filled with and supported by SiO$_2$ which was added in the second preparation step. For 2UL, to which more than an adequate amount of water was added during promoter addition, the extra water may have diluted the byproduct enough so that it did not interfere with the formation of linkages between particles leading to a larger PSD than 1UL and also to a larger PSD than the primary PSD of 2UH.

In contrast to the other three 2S catalysts, 2WH has smaller average pore diameters and narrower PSDs than all other catalysts. With other catalysts dried at the lower temperature, linkages formed between particles produced larger pore sizes and broad PSDs after reduction, but for catalysts dried at 120 °C, any linkages that may have formed appear to have been destroyed. This suggests that at 120 °C an energy barrier is surpassed which does more than simply dehydrate inter-particle linkages. One DSC/TGA study on ferrihydrite shows an endothermic transformation around 105–125 °C and also suggests that the surface of the ferrihydrite particles are covered in hydroxide groups [27]. Others show that drying at 130 °C results in partially dehydroxylating the ferrihydrite [28]. These two observations may be related, meaning that the endothermic transformation may actually be a removal of the surface hydroxides. A particle undergoing partial dehydroxylation would lose its surface hydroxides at temperatures lower than its interior hydroxides, leaving an oxide surface with a protected oxyhydroxide core. Surface hydroxides are the likely structures to form linkages with surface water between particles. As described above, slowly dehydrating the linked structures leads to larger agglomerates that grow into single particles of Fe_2O_3. At 120 °C, the water linkages may dehydrate at the same time that the surface hydroxides decompose which would destroy any linkages between particles and result in smaller pore diameters and very narrow PSDs.

Dispersion and average crystallite diameters of iron particles for all catalysts are listed in Table 2. Diameters estimated from chemisorption follow the trends for pore diameter and PSD and particle size is attributed to the agglomeration theory discussed above. Estimated average crystallite diameters for 1UH, 1UL, and 1WH agree amazingly well with their average pore diameters (within ±0.8 nm), while 1WL is within 5 nm (14%) of its average pore diameter (see Figure 4). Estimated diameters for 2UH, 2UL, and 2WL are roughly half of their pore diameters whereas the crystallite diameter for 2WH is within 1 nm (20%) of its pore diameter. Diameters from XRD data for Fe crystals in the passivated catalysts are larger than the diameters calculated from H_2 chemisorption. Diameters for Fe_3O_4 crystals are also larger than chemisorption diameters, except for 1WH, 1WL, 2UL, and 2WL.

Table 2. H_2 and CO uptakes, extents of reduction, dispersions, and crystallite diameters of catalysts after re-reduction in H_2 at 300 °C for 6 h.

Catalyst	Uptake [a]					Crystallite diameter		
	H_2 [b] μmol/g	CO [c] μmol/g	CO/H [d] %	EOR %	Disp nm	H_2 [a] nm	Fe [e] nm	Fe_3O_4 [e,f] nm
1UH	159	583	1.85	21.5	12.5	9.8	19.7	20.5
1UL	146	893	3.05	15.5	14.8	8.3	19.8	22.6
1WH	203	746	1.85	46.1	8.0	15.3	26.3	15.2
1WL	131	675	2.60	50.6	4.2	29.5	36.4	21.2
2UH [g]	158	145	0.46	9.3	31.1	3.9	-	24.5
2UL	118	264	1.10	12.1	17.4	7.1	17.8	6.6
2WH	164	242	0.75	14.8	19.0	6.5	18.3	16.4
2WL	61	166	1.35	14.2	6.7	18.5	35.0	16.2

[a] after re-reduction for 6 h at 300 °C; [b] 95% confidence interval $=\pm25\%$; [c] 95% confidence interval $=\pm 23\%$; [d] assume H concentration is double of H_2 concentration; [e] from XRD peak broadening; [f] Fe_3O_4 indistinguishable from FeO; [g] 2UH showed 26.8 nm Fe_2O_3, but no Fe.

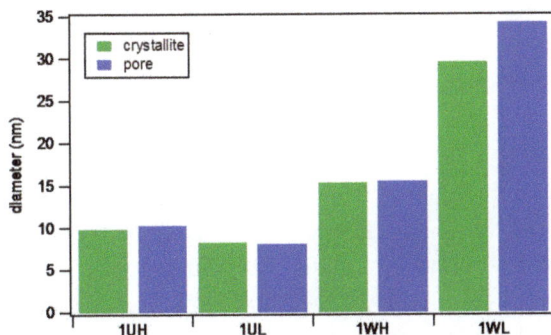

Figure 4. Comparison of average pore diameters from BET with average Fe crystallite diameters calculated from H_2 uptake for 1S catalysts.

TEM images of catalysts confirm the presence of crystallites with particle diameters equal to those estimated from chemisorption and XRD analysis. Figure 5 shows TEM images of 300 nm agglomerates of 1UH and 1WH, respectively. Figure 5a shows fine grains (5–10 nm) clumped together in 20–30 nm agglomerates, some of which have begun to merge into continuous particles, but which retain textures of the fine grains. This confirms the chemisorption crystallite diameter estimate of 9.8 nm for 1UH and also the 20 nm diameter XRD estimates of Fe and Fe_3O_4 crystals. The TEM image of 1WH (Figure 5b) shows much coarser single grains 10–30 nm in size. In contrast with 1UH, these larger particles appear to be continuous without smaller constituent pieces or textures. This also confirms the chemisorption diameter estimate of 15.3 nm as well as the XRD estimates of Fe (26.3 nm) and Fe_3O_4 (14.1 nm) crystallite diameters.

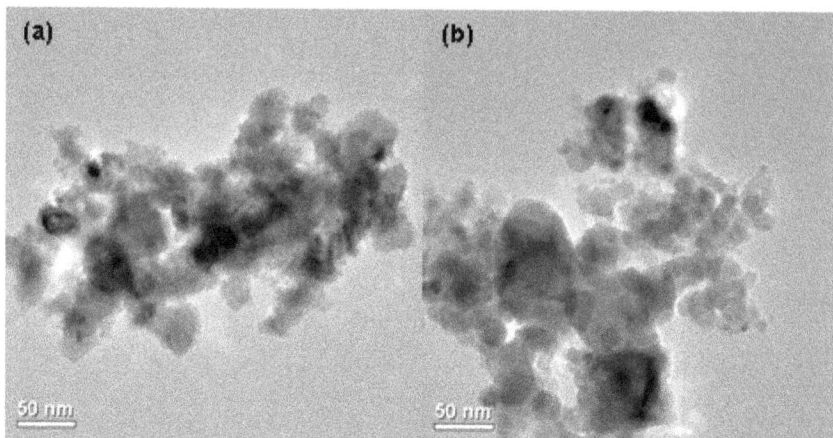

Figure 5. TEM micrograph of (**a**) 1UH and (**b**) 1WH catalysts.

2.2. Catalytic Performance

2.2.1. Catalyst Activity

Figure 6 shows experimental rate constant (k) values as functions of temperature for all the catalysts of this study. The solid lines were obtained using the Arrhenius equation and "best-fit" A and E_A values obtained from the k data for each catalyst. The most active catalyst is 2UH, followed by 1WL, 1UL, 1UH, and 2UL at 250 °C. Excluding 1WL and 1WH which have low K loadings, all 1U and 2U catalysts are more active than W catalysts at 250 °C. The activity for these catalysts may be explained by the iron carbide phase after reaction. As shown in Figure 7, the tallest and broadest peak for all catalysts (except 1WH) corresponds to Fe_2C, which may be a candidate for the active phase of the catalyst. 1UH, 1WL, 2UH, and 2UL (which are the most active catalysts) show the least signal (or none at all) for Fe_3O_4 and Fe_5C_2. 1UL, 1WH, 2WH and 2WL have strong Fe_5C_2 peaks, whereas 1UL and 1WH also show peaks for Fe_3O_4. As stated previously, these results do not prove which phase is active (or which is most active for FTS); however, it is interesting that the most active catalysts show the weakest signal for Fe_3O_4 and Fe_5C_2 and the strongest signal for Fe_2C.

Figure 8 illustrates the effects of the factorial variables (U/W and H/L) on rate of reaction at 250 °C. The 1W catalysts were excluded to avoid confusion over compounding effects from the lower potassium loadings of these two catalysts. The figure shows higher rates for unwashed catalysts than for washed catalysts. For example, the 2U catalysts show an average 60% increase (14.0 mmol/g/h) in rate compared to 2W catalysts. In addition, for each pair of 1U, 2U, and 2W catalysts, H catalysts show an average 19% increase (5.2 mmol/g/h for 2S catalysts).

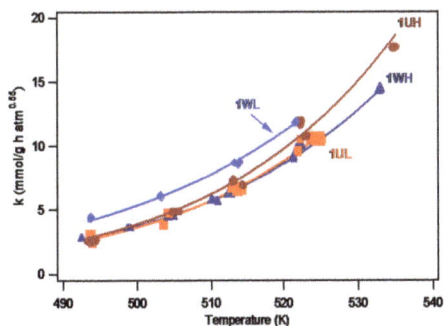

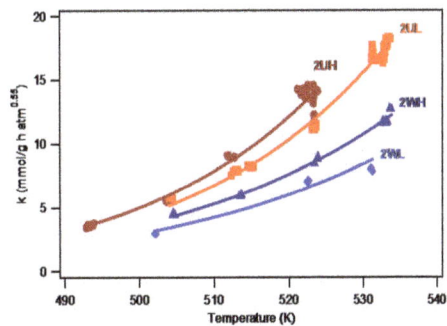

(a)   (b)

Figure 6. Rate constant as a function of temperature for (**a**) 1S and (**b**) 2S catalyst kinetic data sets.

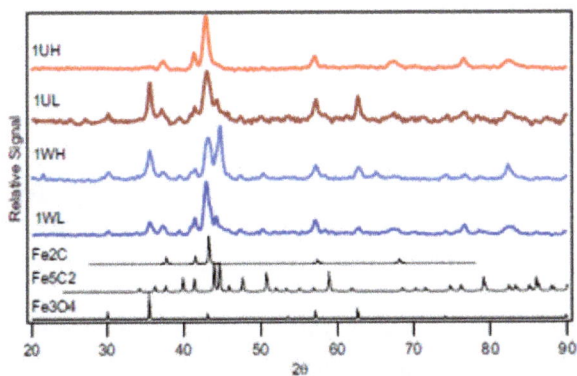

(a)

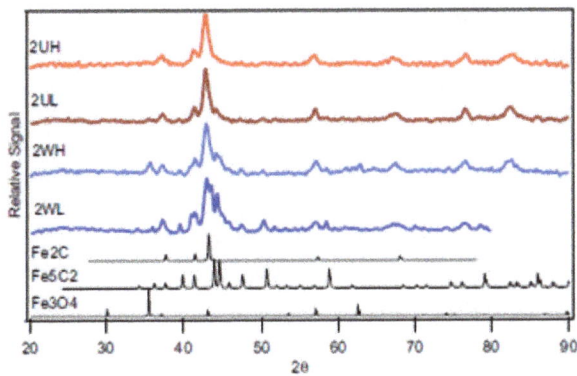

(b)

Figure 7. XRD spectra of (**a**) 1S catalysts and (**b**) 2S catalysts after FTS reaction.

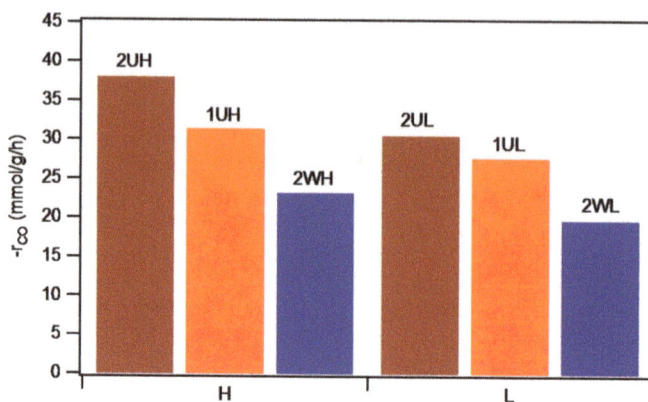

Figure 8. Effects of washing and drying on rate of reaction for 1S and 2S catalysts at 250 °C. 1WH and 1WL are not shown to avoid confusion over effects of low potassium loading.

These activity results indicate that the washing step effects a fundamental change in the nature of the active sites. Site activity is related to binding energy. Sites with low binding energies, like planar sites, can adsorb and desorb reactants quickly and may not allow for sufficient time and contact for the reaction to proceed efficiently. Sites with very high binding energies, like corner sites or defect sites, may bind reactants too strongly and prevent desorption of products. Sites with intermediate binding energies, like edge sites found at the boundaries between crystallites, may provide the majority of the turnover and activity. As discussed in the PSD section, washing removes NH_4NO_3 from between particles and allows them to form larger, more ordered agglomerates as seen in the TEM images of 1WH (Figure 5b). The image of 1WH shows larger, smoother particles than does the image of 1UH (Figure 5a). The surface reorganizations resulting from washing and subsequent grain growth during heat treatments appear to eliminate boundaries between crystallites, resulting in fewer edges and corners therefore eliminating active binding sites and resulting in lower catalyst activities.

Figure 9 shows rate of reaction at 250 °C *versus* crystallite size and pore diameter for 1U and 2S catalysts. 1W catalyst with low K loading are excluded from these charts. It is clear that rate of reaction increases with decreasing particle and pore diameters. U and WH catalysts have particle and pore diameters mainly between 10 and 20 nm which may be ideal for FTS, whereas WL catalysts show larger diameters and lower rates.

2.2.2. Selectivity

For H_2-deficient feed stocks like coal or biomass, the water gas shift (WGS) activity of Fe FT catalysts is important. Table 3 reports the mole percent of CO converted to CO_2 as an indicator of the WGS activity of each catalyst in this study at several temperatures. WGS selectivity increases with increasing temperature for all catalysts in this study. This suggests that the activation energy for the WGS reaction is larger than for the FT reactions.

Table 3. Selectivity of catalysts at different temperatures.

Catalyst	CO to CH$_4$ (mole%) [a,b]			CO to CO$_2$ (mole%) [c]		
	230 °C	240 °C	250 °C	230 °C	240 °C	250 °C
1UH	4.7	5.3	6.3	41.7	43.1	45.2
1UL	4.0	4.6	5.0	38.2	43.3	45.5
1WH	14.8	15.5	15.1	10.7	14.4	18.9
1WL	10.8	12.0	12.5	9.8	14.2	18.0
2UH	6.9	7.2	8.4	30.5	31.2	32.6
2UL	3.6	4.9	6.0	23.2	29.5	34.0
2WH	3.1	3.8	4.9	24.0	28.7	33.9
2WL	5.1	5.5	7.2	31.2	32.5	37.1

[a] CO_2-free basis; [b] 95% confidence interval less than ±0.7%; [c] 95% confidence interval less than ±1.5%.

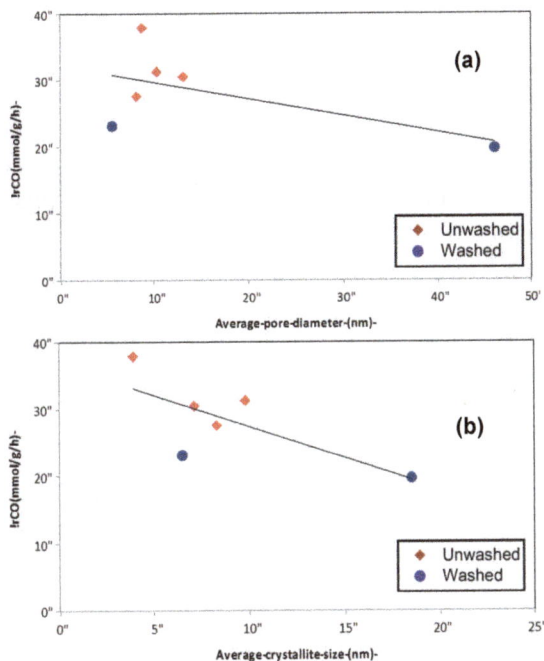

Figure 9. Rate of reaction at 250 °C *versus* (**a**) average crystallite diameter after reduction and (**b**) average pore diameter after passivation.

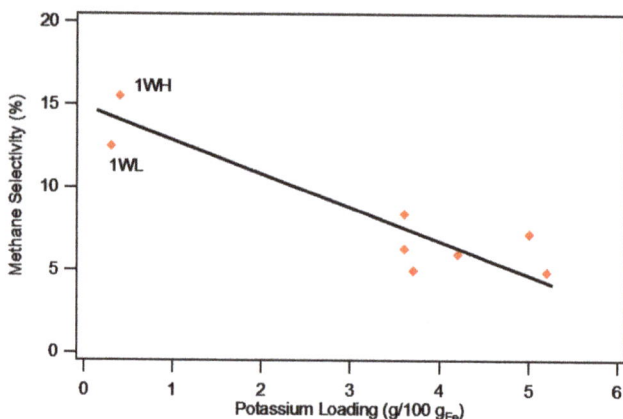

Figure 10. CH$_4$ selectivity *versus* potassium loading for SDP catalysts.

CH$_4$ selectivity increases with increasing temperature also, as demonstrated in Table 3. Selectivities for the eight 1S and 2S catalysts appear to increase at about the same rate with temperature suggesting that K loading has little or no effect on the activation energy of the methane reaction. As shown in Figure 10, 1W catalysts (0.3 pbm K) have selectivities 50–200% higher than catalysts with nominal potassium loading (3.6–5.2 pbm K). Decreasing CH$_4$ selectivity with increasing K loading is well documented in the literature [4,24,29].

The 1U catalysts exhibit the highest WGS activity with 38.2–45.5% selectivity to CO$_2$. 1W shows the lowest WGS activity of the eight catalysts with 9.8–18.9% selectivity. 2S catalysts fall between these levels. The differences in selectivities may be due to the content and dispersion of K in the catalysts. The higher WGS activities of 1U catalysts may be due to more even distribution and greater dispersion of K, whereas the low WGS activities of the 1W catalysts are probably due to almost complete loss of K during washing (K loadings < 0.4 pbm). The intermediate WGS activities of 2S catalysts reflect K distributions less uniform than 1S catalysts from the microprobe results.

2.2.3. Stability

In addition to activity and selectivity, catalyst stability is a key metric for discerning between catalysts. Both catalyst activity and selectivity can change with time. Figure 11 shows the reaction rate constant for 1UH and 2UH as a function of time-on-stream. As shown in Figure 11a, data of 1UH span more than 500 hours and show little or no downward trend with time, thereby suggesting that this catalyst has excellent activity stability. The variation (scatter) in apparent rate constant measurements may be due in part to inaccurate orders of reaction used to calculate k. In contrast, 2UH shows an apparent constant rate of decrease

60

in activity (deactivation). For illustration purposes, a linear slope and intercept are regressed to the data recorded at 250 °C, excluding the temperature and partial pressure experiments, and are represented by the dark line in Figure 11b. The cause of the deactivation is probably carbon deposition on the catalyst surface where the concentration of K appears to be higher. As shown in microprobe images for promoter distribution (Figure 1), 1S catalysts show more uniform distribution than 2S catalysts. Therefore, in 2S catalysts, the carbon is more likely to deposit on the grain boundaries, where K concentration is higher, and induce catalyst deactivation.

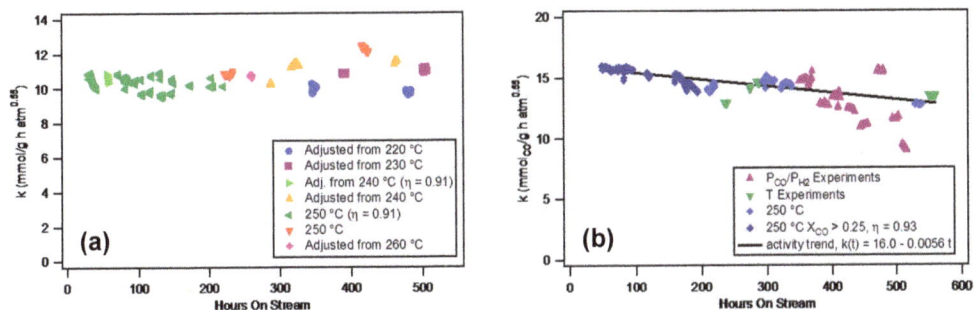

Figure 11. Experimental rate constant values at 250 °C as a function of time-on-stream for **(a)** 1UH and **(b)** 2UH catalysts. Legends represent data converted from different reaction temperatures.

2.2.4. Comparison to Published Catalysts

The catalysts from this study compare well with data on published catalysts in the literature. The activity, productivity, selectivity of 1UH, 2UH, and published, highly active iron catalysts [8] are listed in Table 4. The activity and productivity for 1UH and 2UH are comparable with TAMU catalysts and higher than Mobil catalyst. For the selectivity, although 1UH and 2UH show higher CH_4 selectivity, their C_{2+} selectivity are comparable or even higher than TAMU and Mobil catalysts. While acknowledging that the TAMU catalyst was tested in a slurry phase CSTR, and the Mobil catalyst was tested in a slurry bubble reactor, nevertheless using 1UH and 2UH as proxy, catalysts made by the solvent-deficient precipitation method show great potential for having activity, selectivity, and productivity comparable to those of some of the most active and selective catalysts in the literature. It is important to note that the two published catalysts were subjected to an optimized pretreatment process that significantly enhanced the catalyst activity and productivity. The BYU catalyst pretreatment has not been optimized yet and improvements in catalyst preparation techniques and pretreatment are expected to improve catalytic properties significantly.

Table 4. Comparison of 1UH and 2UH with published, highly active iron catalysts [a].

Catalyst	Activity (mmol/g h MPa)	Productivity (g_{HC}/g_{Fe} h)	Selectivity			
			H₂/CO	CH₄	C₂₊	CO₂
1UH	125	0.55	1.0	0.040	0.49	0.47
2UH	134	0.71	1.0	0.063	0.61	0.34
TAMU [b]	102-180	0.5-0.8	0.67	0.014	0.50	0.48
Mobil [b]	102	0.39	0.73	0.012	0.49	0.50

[a] Data from ref. [7]; [b] TAMU catalyst operated at 260 °C and 22 atm, Mobil catalyst operated at 257 °C and 15 atm activity in mmol of CO per g Fe per MPa H_2.

3. Experimental Section

3.1. Catalyst Preparation

As illustrated in Scheme 1 and described above, the effects of three key preparation variables—(1) timing of the addition of promoters potassium and silica, (2) presence or absence of a washing step for the precursor, and (3) drying temperature—were investigated by preparing eight catalysts in a 2^3 factorial design. Levels for timing of promoter addition were "1 Step" in which potassium and silica promoters were added to the salts of iron and copper before precipitation,, and "2 Step" in which potassium and silica promoters were added in a separate step after precipitation and washing, if applicable. The levels for washing were unwashed (U) or washed (W) using 100 mL of deionized water five times immediately following precipitation. Levels for initial drying temperature were high (H) temperature (100 °C) and low (L) temperature (60 °C overnight, followed by 100 °C).

The eight catalysts were prepared with target compositions designated as 100 Fe/5 Cu/4 K/16 SiO₂ which indicates relative mass values of each component. In the 1 Step preparation, the silica promoter was added to the iron and copper salts (acid) and the potassium promoter was added to ammonium bicarbonate (base) before combining the two mixtures and stirring vigorously. In a typical preparation, 4.822 g fumed SiO₂ (Cab-O-Sil) were added to 217.005 g Fe(NO₃)₃·9H₂O and 5.485g Cu(NO₃)₂·2.5H₂O and mixed well with a pestle in a large glass bowl. In a separate container, 3.071 g KHCO₃ were mixed with 128.989 g NH₄HCO₃ followed by adding the bicarbonate mixture to the metal salt mixture. The combined mixture of powders was vigorously mixed with the pestle, leading to waters of hydration being released. Mixing continued for approximately 20 min until precipitation of Fe and Cu hydroxides, referred to as the precursor, was complete as indicated by cessation of CO₂ release [15]. Then, the precursor was split into portions which were treated according to their code, respectively, and pelletized, crushed and sieved to −30/+60 mesh. The resulting four catalysts are collectively identified as 1S catalysts.

For the "2 Step" preparations, the Fe and Cu metal salts were precipitated according to the SDP method. The K and Si promoters and enough water to make a thick clay were added to the damp precipitate before drying. For 2UH and 2UL, the promoters were added immediately after complete precipitation and before drying. For 2WH and 2WL, the promoters were added immediately after washing. 2WH was dried at 120 °C instead of 100 °C, but all other precursors were dried at 100 °C for 16–48 hours or at 60 °C for 16–48 hours followed by 100 °C as designated. The dried precursors were pelletized, crushed and sieved to −30/+60 mesh. These catalysts are collectively identified as 2S catalysts.

3.2. Characterization Techniques

3.2.1. N_2 Adsorption

Pore properties were measured by nitrogen adsorption using a Micromeritics TriStar 3000 BET analyzer. Sample sizes were typically 0.3–0.5 g and degassed at 200 °C overnight prior to measurements. Surface areas (SA) were calculated using the BET model using P/P_0 from 0.05 to 0.2 [30]. Pore volume (Vpore) was measured at a single point at P/P_0 of 0.995. Average pore diameter and PSD were calculated using a split pore geometry (SPG) model [31].

3.2.2. Thermal Gravimetric Analysis (TGA)

Temperature programmed oxidation (TPO), temperature programmed reduction (TPR), isothermal oxygen titration and CO chemisorption were performed on 10–40 mg samples in a Mettler Toledo TGA/DSC 1 equipped with an automated GC 200 gas controller. Gas flow rates of H_2, CO, and O_2 (or air depending on the source) were set by rotameters, but gas switching during all experiments was controlled by the GC 200 controller and the TGA software.

TPO experiments were used to design temperature programs for bulk calcination. The rate of mass loss during a constant temperature ramp of 3 °C /min from ambient temperature to 700 °C in 100 mL/min of 70–80% air/He was analyzed to determine appropriate temperature ramps and soaks for controlling byproduct decomposition at low rates.

TPR experiments were used to determine temperature programs for bulk reduction. Again, the rate of mass loss during a straight temperature ramp of 3 °C/min from ambient temperature to 700 °C in 100 mL/min 10% H_2/He was analyzed to determine appropriate temperature ramps and soaks for reduction to proceed at an acceptable rate without producing high partial pressures of H_2O.

Isothermal oxygen titration experiments were used to determine the extent of reduction (EOR) to Fe metal following reduction. EOR was calculated from O_2 uptake during oxidation at 400 °C after the re-reduction of the previously passivated catalyst for 6 hours at 300 °C in 10% H_2/He.

Gravimetric carbon monoxide adsorption was used as a relative measurement of chemisorption site density. CO uptake was measured at 25 °C in 10% CO/He following re-reduction of passivated catalysts at 300 °C for 6 h in 10% H_2/He and a one-hour purge in 100% He at 290 °C.

3.2.3. Hydrogen Chemisorption

Dispersion and crystallite diameters of reduced catalysts were calculated from hydrogen chemisorption uptake measurements. Hydrogen was chemisorbed on reduced catalysts in a flowthrough adsorption system using procedures and equipment developed in the BYU catalysis lab [32]. Passivated catalysts were re-reduced at 300 °C in 100% H_2 for 6 hours followed by purging in Ar at 280 °C to remove residual H_2 from the reduction. Hydrogen was adsorbed at 100 °C before purging at dry ice/acetone temperatures (77 K) to remove physisorbed molecules. Finally, hydrogen was desorbed during a temperature ramp up to 600 °C. Dispersion (*Disp*) and average crystallite diameter (d_c) estimates were calculated from the hydrogen uptake and extent of reduction data using Equations 3 and 4, respectively [2].

$$Disp = 1.12 \times \frac{Hydrogen\ Uptake}{EOR \times WeightLoading} \tag{3}$$

$$d_c = \frac{123}{Disp} \tag{4}$$

Crystallite diameter (d_c) is in nm, H_2 uptake is in μmol/g, EOR is the fraction of Fe in the metallic state, and weight loading is the mass percent of Fe in the catalyst.

3.2.4. X-ray Diffraction (XRD)

Crystalline phases of catalysts after drying, calcining, reducing, and carbiding (a result of FB FTS testing) were identified using X-ray diffraction (XRD) patterns in order to understand phase changes at each of these steps and to estimate crystallite diameters. XRD patterns were collected using a PANalytical X'Pert Pro diffractometer with a Cu source and a Ge monochromator tuned to the Cu-Kα1 wavelength (λ = 1.540598 Å). Samples were scanned from 20 to 90° 2θ using a step size of 0.016° at scan rates between 100 and 400 s/step. Diffraction patterns were compared to standard patterns in the International Center for Diffraction Data (ICDD) database. Average crystallite size (d_c) was calculated by Scherrer's formula.

3.2.5. Electron Microscope Analysis

To determine the uniformity of promoter distributions at the micron level, macro elemental distributions of Fe, Cu, K, and Si in 50–400 μm catalyst agglomerates were imaged using a Cameca SX50 electron microprobe at 15 kV and 20–30 nA.

Confirmation of the presence of crystallite diameters estimated from other techniques was attempted with TEM imaging. Crystallites and agglomerates were imaged on a FEI TF30 TEM operating at 300 keV or on a FEI TF20 Ultra-twin TEM/STEM operating at 200 keV.

3.2.6. Inductively Coupled Plasma (ICP) Analysis

To confirm the elemental content of prepared catalysts, digested catalysts and catalyst washes were analyzed in a Perkin Elmer Optima 2000 DV ICP analyzer. Catalyst samples (20–45 mg) were digested in hydrofluoric acid, dried, and then dissolved in 10 mL 3% nitric acid. 0.5–1 mL of digested sample was diluted with 20–40 mL of 3% nitric acid, giving final analyte concentrations of about 100 mg_{cat}/L. Samples of catalyst washes (3–5 mL) were diluted with 15 mL of 3% nitric acid before analysis. Analyte wavelengths were 238.2 nm (Fe), 327.4 nm (Cu), and 766.5 nm (K).

3.3. Activity Tests

Fischer-Tropsch Synthesis (FTS) was conducted in a fixed-bed reactor (stainless steel, 3/8 inch OD) described previously [22,23]. Each sample to be tested (0.25 g, 250–590 μm) was diluted with quartz sand (−50/+70 mesh) to improve temperature stability in the catalytic zone. Before FTS, the sample was reduced *in situ* at 280–320 °C in 10% H_2/He for 10 h followed by 100% H_2 for 6 h. Then, the catalyst was cooled to 180 °C and the system was pressurized to 2.1 MPa in flowing syngas (H_2:CO = 1). The catalyst was then heated (still in syngas) and activated at 280 °C and a CO conversion level of ~60% for 48–90 h. These activation conditions were chosen based on studies by Bukur *et al.* that showed that Fe FTS catalysts activated under syngas had lower deactivation rates compared to those activated under CO and had lower methane selectivity compared to those activated under H_2 [33,34]. After activation, the reactor conditions were changed to those to be tested which included temperatures from 220 °C to 260 °C and H_2:CO ratios from 0.66 to 1.0. In addition to CO and H_2, the feed gas contained 20–40% helium used as a diluent to keep the total pressure the same for all runs. The exit gas and liquid effluent passed through a hot trap (110 °C) and a cold trap (0 °C) to collect heavy hydrocarbons and liquid products. The effluent gaseous product was analyzed using an HP 6890 gas chromatograph equipped with a thermal conductivity detector and 60/80 carboxene-1000 column. CO and H_2 conversions and product selectivities were determined with the aid of an Ar tracer.

The rate of reaction (CO depletion) was determined by operating at low CO conversions ($X_{CO} \leqslant 0.25$), assuming differential reactor conditions, and thus using the resulting reactor performance equation:

$$-r_{co} = \frac{F^0_{co} X_{co}}{W_{cat}} \tag{5}$$

where W_{cat} is the mass of catalyst and $F_{CO}{}^0$ is the inlet molar CO flow rate. Experimental rate constant (k) values were determined for every data point from the rate data using Equation 6 by assuming the P_{CO} and P_{H2} dependencies proposed by Eliason [35]. Values of P_{H2} and P_{CO} are simple averages of the inlet and outlet partial pressures of each gas.

$$k = \frac{-r_{co}}{P_{CO}^{-0.05} P_{H_2}^{0.6}} \tag{6}$$

4. Conclusions

In summary, we have successfully prepared active and stable iron FT catalysts via a facile, solvent-deficient precipitation method. Key preparation variables, i.e., drying, washing, and promoter addition step, were identified by a 2^3 factorial experimental design. Results in this study show that (1) unwashed catalysts are statistically more active (60% on average) than washed catalysts of the same potassium loading; (2) catalysts dried initially at higher temperature (100 °C) are more active (~20%) than catalysts dried initially at lower temperature (60 °C); (3) 1S (one-step preparation) catalysts appear to have superior properties to 2S catalysts under reaction conditions when considering activity, selectivity, and stability together; (4) catalysts with average particle diameters and average pore sizes both between 5 and 15 nm are up to 300% more active than catalysts with larger average particle (17–23 nm) and pore diameters (45–50 nm). In addition, 1UH and 2UH in this study show comparable or even better activity, selectivity, and stability than those of published, highly active catalysts, though consideration should be given to the different reactor geometries used in the various catalyst tests. As a simple and time-effective method, solvent-deficient precipitation not only contributes to the development of iron FT catalysts, but also offers promising potential for preparation of other catalysts used in different applications.

Acknowledgments: We thank Jeff Farrer and BYU microscopy lab for their assistance with TEM imaging. The financial support for this work was provided by members of the Brigham Young University Fischer-Tropsch Consortium and the University of Wyoming Clean Coal Technologies program.

Author Contributions: Kyle M. Brunner prepared the catalysts, performed the experiments and wrote the original manuscript. Baiyu Huang helped prepare the final manuscript. William

C. Hecker supported the catalyst preparation, the experiments and revised the final version of paper. Brian F. Woodfield was instrumental in the development of the SDP method.

Conflicts of Interest: The authors declare no conflict of interest.

References

1. Schulz, H. *Fischer-Tropsch Synthesis in a Modern Perspective*; Wiley-VCH Verlag GmbH & Co. KGaA: WeinHeim, Germany, 2012; pp. 301–324.
2. Bartholomew, C.H.; Farrauto, R.J. *Fundamentals of Industrial Catalytic Processes*; Wiley: Hoboken, U.S.A., 2011.
3. Steynberg, A.P. Introduction to Fischer-Tropsch technology. *Fischer-Tropsch Technol.* **2004**, *152*, 1–63.
4. Dry, M.E. Chapter 4; *The Fischer-Tropsch Synthesis*; Springer: Berlin, Germany, 1981; pp. 159–255.
5. Bukur, D.B.; Lang, X.; Mukesh, D.; Zimmerman, W.H.; Rosynek, M.P.; Li, C.P. Binder Support Effects on the Activity and Selectivity of Iron Catalysts in the Fischer-Tropsch Synthesis. *Ind. Eng. Chem. Res.* **1990**, *29*, 1588–1599.
6. Bukur, D.B.; Koranne, M.; Lang, X.S.; Roa, K.R.P.M.; Huffman, G.P. Pretreatment Effect Studies with a Precipitated Iron Fischer-Tropsch Catalyst. *Appl. Catal. A* **1995**, *126*, 85–113.
7. Bukur, D.B.; Lang, X.S. A precipitated iron Fischer-Tropsch catalyst for synthesis gas conversion to liquid fuels. *Natural Gas Convers. V* **1998**, *119*, 113–118.
8. Bukur, D.B.; Lang, X.S. Highly active and stable iron Fischer-Tropsch catalyst for synthesis gas conversion to liquid fuels. *Ind. Eng. Chem. Res.* **1999**, *38*, 3270–3275.
9. Li, S.Z.; Krishnamoorthy, S.; Li, A.W.; Meitzner, G.D.; Iglesia, E. Promoted iron-based catalysts for the Fischer-Tropsch synthesis: Design, synthesis, site densities, and catalytic properties. *J. Catal.* **2002**, *206*, 202–217.
10. Wu, B.; Bai, L.; Xiang, H.; Li, Y.-W.; Zhang, Z.; Zhong, B. An active iron catalyst containing sulfur for Fischer–Tropsch synthesis. *Fuel* **2004**, *83*, 205–212.
11. Yang, Y.; Xiang, H.W.; Zhang, R.L.; Zhong, B.; Li, Y.W. A highly active and stable Fe-Mn catalyst for slurry Fischer-Tropsch synthesis. *Catal. Today* **2005**, *106*, 170–175.
12. Ma, W.; Jacobs, G.; Graham, U.M.; Davis, B.H. Fischer-Tropsch Synthesis: Effect of K Loading on the Water-Gas Shift Reaction and Liquid Hydrocarbon Formation Rate over Precipitated Iron Catalysts. *Top. Catal.* **2014**, *57*, 561–571.
13. Park, J.-Y.; Lee, Y.-J.; Jun, K.-W.; Bae, J.W.; Viswanadham, N.; Kim, Y.H. Direct conversion of synthesis gas to light olefins using dual bed reactor. *J. Ind. Eng. Chem.* **2009**, *15*, 847–853.
14. Liu, S.F.; Liu, Q.Y.; Boerio-Goates, J.; Woodfield, B.F. Preparation of a wide array of ultra-high purity metals, metal oxides, and mixed metal oxides with uniform particle sizes from 1 nm to bulk. *J. Adv. Mater.* **2007**, *39*, 18–23.
15. Bartholomew, C.H.; Woodfield, B.F.; Huang, B.; Olsen, R.E.; Astle, L. Method for making highly porous, stable metal oxide with a controlled pore structure. WO2011119638A2, March 2011.

16. Huang, B.; Bartholomew, C.H.; Smith, S.J.; Woodfield, B.F. Facile solvent-deficient synthesis of mesoporous γ-alumina with controlled pore structures. *Microporous Mesoporous Mater.* **2013**, *165*, 70–78.

17. Huang, B.; Bartholomew, C.H.; Woodfield, B.F. Facile structure-controlled synthesis of mesoporous γ-alumina: Effects of alcohols in precursor formation and calcination. *Microporous Mesoporous Mater.* **2013**, *177*, 37–46.

18. Huang, B.; Bartholomew, C.H.; Woodfield, B.F. Facile synthesis of mesoporous γ-alumina with tunable pore size: The effects of water to aluminum molar ratio in hydrolysis of aluminum alkoxides. *Microporous Mesoporous Mater.* **2014**, *183*, 37–47.

19. Khosravi Mardkhe, M.; Woodfield, B.; Bartholomew, C. Facile one-pot synthesis of a thermally stable silica-doped alumina having a large pore volume and large bimodal pores. In Proceedings of 247th ACS National Meeting, Dallas, TX, USA, 16–20 March 2014.

20. Olsen, R.E.; Bartholomew, C.H.; Huang, B.Y.; Simmons, C.; Woodfield, B.F. Synthesis and characterization of pure and stabilized mesoporous anatase titanias. *Microporous Mesoporous Mater* **2014**, *184*, 7–14.

21. Smith, S.J.; Page, K.; Kim, H.; Campbell, B.J.; Boerio-Goates, J.; Woodfield, B.F. Novel synthesis and structural analysis of ferrihydrite. *Inorg. Chem.* **2012**, *51*, 6421–6424.

22. Brunner, K.M.; Harper, G.E.; Keyvanloo, K.; Woodfield, B.F.; Bartholomew, C.H.; Hecker, W.C. Preparation of Unsupported Iron Fischer-Tropsch Catalyst by Simple, Novel, Solvent Deficient Precipitation (SDP) Method. *Energy Fuels* **2015**, *29*, 1972–1977.

23. Brunner, K.M. Novel Iron Catalyst and Fixed-Bed Reactor Model for the Fischer-Tropsch Synthesis. Ph.D. Thesis, Brigham Young University, August 2012.

24. Bukur, D.B.; Mukesh, D.; Patel, S.A.; Rosynek, M.P.; Zimmerman, W.H. Development and Process Evaluation of Improved Fischer-Tropsch Slurry Catalysts. Available online: http://www.fischer-tropsch.org/DOE/DOE_reports/80011/80011-1/80011-1_toc.htm (accessed on 23 July 2015).

25. Ristić, M.; De Grave, E.; Musić, S.; Popović, S.; Orehovec, Z. Transformation of low crystalline ferrihydrite to α-Fe₂O₃ in the solid state. *J. Mol. Struct.* **2007**, *834–836*, 454–460.

26. Feng, Z.; Zhao, J.; Huggins, F.E.; Huffman, G.P. Agglomeration and Phase-Transition of a Nanophase Iron-Oxide Catalyst. *J. Catal.* **1993**, *143*, 510–519.

27. Eggleton, R.A.; Fitzpatrick, R.W. New Data and a Revised Structural Model for Ferrihydrite. *Clays Clay Miner.* **1988**, *36*, 111–124.

28. Zhao, J.M.; Huggins, F.E.; Feng, Z.; Huffman, G.P. Ferrihydrite - Surface-Structure and Its Effects on Phase-Transformation. *Clays Clay Miner.* **1994**, *42*, 737–746.

29. Arakawa, H.; Bell, A.T. Effects of Potassium Promotion on the Activity and Selectivity of Iron Fischer-Tropsch Catalysts. *Ind. Eng. Chem. Process Des. Dev.* **1983**, *22*, 97–103.

30. Gregg, S.J.; Sing, K.S.W. *Adsorption, Surface Area, and Porosity*; Academic Press: San Diego, USA, 1991.

31. Huang, B.; Bartholomew, C.H.; Woodfield, B.F. Improved calculations of pore size distribution for relatively large, irregular slit-shaped mesopore structure. *Microporous Mesoporous Mater.* **2014**, *184*, 112–121.

32. Jones, R.D.; Bartholomew, C.H. Improved Flow Technique for Measurement of Hydrogen Chemisorption on Metal-Catalysts. *Appl. Catal.* **1988**, *39*, 77–88.

33. Bukur, D.B.; Nowicki, Z.; Manne, R.K.; Lang, X. Activation studies with a precipitated iron catalysts for Fischer-Tropsch synthesis. II. Reaction studies. *J. Catal.* **1995**, *155*, 366–375.

34. Bukur, D.B.; Okabe, K.; Rosynek, M.P.; Li, C.; Wang, D.; Rao, K.R.P.M.; Huffman, G.P. Activation studies with a precipitated iron catalyst for Fischer-Tropsch synthesis. I. Characterization studies. *J. Catal.* **1995**, *155*, 353–365.

35. Eliason, S.A.; Bartholomew, C.H. Kinetics and Deactivation of Unpromoted and Potassium-Promoted Iron Fischer-Tropsch Catalysts at Low Pressure (1 Atm). *Stud. Surf. Sci. Catal.* **1991**, *68*, 211–218.

Effect of Microgravity on Synthesis of Nano Ceria

Ilgaz I. Soykal, Hyuntae Sohn, Burcu Bayram, Preshit Gawade,
Michael P. Snyder, Stephen E. Levine, Hayrani Oz and Umit S. Ozkan

Abstract: Cerium oxide (CeO_2) was prepared using a controlled-precipitation method under microgravity at the International Space Station (ISS). For comparison, ceria was also synthesized under normal-gravity conditions (referred as control). The Brunauer-Emmett-Teller (BET) surface area, pore volume and pore size analysis results indicated that the ceria particles grown in space had lower surface area and pore volume compared to the control samples. Furthermore, the space samples had a broader pore size distribution ranging from 30–600 Å, whereas the control samples consisted of pore sizes from 30–50 Å range. Structural information of the ceria particles were obtained using TEM and XRD. Based on the TEM images, it was confirmed that the space samples were predominantly nano-rods, on the other hand, only nano-polyhedra particles were seen in the control ceria samples. The average particle size was larger for ceria samples synthesized in space. XRD results showed higher crystallinity as well as larger mean crystal size for the space samples. The effect of sodium hydroxide concentration on synthesis of ceria was also examined using 1 M and 3 M solutions. It was found that the control samples, prepared in 1 M and 3 M sodium hydroxide solutions, did not show a significant difference between the two. However, when the ceria samples were prepared in a more basic medium (3 M) under microgravity, a decrease in the particle size of the nano-rods and appearances of nano-polyhedra and spheres were observed.

Reprinted from *Catalysts*. Cite as: Soykal, I.I.; Sohn, H.; Bayram, B.; Gawade, P.; Snyder, M.P.; Levine, S.E.; Oz, H.; Ozkan, U.S. Effect of Microgravity on Synthesis of Nano Ceria. *Catalysts* **2015**, *5*, 1306–1320.

1. Introduction

Cerium oxide (CeO_2) is a widely used catalyst support due to its high oxygen storage capacity and high oxygen ion conductivity. The high oxygen mobility of ceria leads to a change in its oxidation state between Ce^{3+} and Ce^{4+}, which facilitates its use as an oxygen buffer for catalytic redox reactions [1,2]. In addition, ceria has been shown to have catalytic activity itself for various reactions, such as oxidation and reforming [3,4]. Moreover, the high thermal stability of ceria is attractive for reactions which are conducted at high temperatures [2]. Therefore, ceria has been extensively utilized in three-way catalysts [5–7] and fuel cells [8,9], as well as investigated for its

70

activity in catalytic reactions such as CO oxidation [10,11], steam reforming [12–17], and water gas shift (WGS) [18,19].

A variety of methods to synthesize ceria have been reported until now, including precipitation [20], spray pyrolysis [21,22], thermal hydrolysis [23], micro-emulsion [24], combustion [25], sonochemical [26], hydrothermal [27], solvothermal [28,29], and sol-gel [30] techniques. It is a well-established fact that changing the synthesis parameters, such as temperature, pressure, pH, and crystallization time, produces different particle sizes and morphologies, which, in turn, alter the physical and chemical properties of ceria, as well as the catalytic performance of the ceria-supported catalysts. For example, an increase in oxygen vacancies with decreased particle size of ceria has been reported in many studies [31–33]. Deshpande *et al.* [32] obtained 17% to 44% increase in Ce^{3+} concentration as ceria particles size was reduced from 30 nm to 3 nm, which was prepared by micro-emulsion. This was attributed to the generation of oxygen vacancies of ceria [32]. Flytzani-Stephanopoulos and colleagues synthesized nano-polyhedra, rods and cubic ceria particles using hydrolysis method and examined its catalytic activity for various reactions [19,34–37]. They demonstrated that different shapes of ceria nanoparticles exposing different crystal planes on the catalyst surface significantly influenced the Au dispersion on the ceria support.

Although there are many studies about the ceria synthesis process, very little is understood regarding the effect of microgravity on the characteristics of ceria. It is expected that gravitational forces would affect the crystallization process and lead to different chemical properties of a solid especially when it is prepared in a heterogeneous solution. This can be attributed to the gravity-driven disturbance in the solution (convection), sedimentation of the solid phase, and the frequency of collisions between particles during the reaction. In the literature, several microgravity studies have been conducted on synthesis of zeolites [38–41]. Coker *et al.* [38] obtained larger average crystal dimensions, less surface area of the Zeolite Socony Mobil (ZSM)-5 under microgravity conditions. The formation of larger crystals of ZSM-5 zeolite was explained by lack of convection and sedimentation of zeolite particles during crystallization. Sano and coworkers [40] prepared ZSM-5 zeolite in space and observed significantly better uniformity of shape and size. Tsuchida *et al.* calculated the formation rates for colloidal crystallization of silica spheres. It was shown that rate constants were significantly smaller in microgravity [42]. Smith *et al.* reported that formation of silica crystals to be slower under microgravity environment due to diffusion-limited crystallization where mass transfer dominates the aggregation of particles rather than reaction kinetics [43]. The effect of microgravity on synthesis of microporous tin(IV) sulfides was examined by Ahari and coworkers [44]. Their results indicated a higher degree of crystallinity, as well as smoother crystal surface of the sample. Additionally, the Langmuir isotherm showed 60% larger pore volume [44]. Similarly, a significant reduction of number of pores were reported from aluminum-copper alloy by Fujii *et al.* [45].

Previously, we have reported on synthesis of ceria with different morphologies (polyhedra, rods, cubes) and different particle sizes (3 nm–25 nm). These materials were examined both as supports, as well as catalysts, in ethanol steam reforming to understand the effect of particle size and morphology [3,46]. In this article, we report the effect of microgravity on the synthesis of ceria.

Ceria samples were synthesized using cerium nitrate and sodium hydroxide solutions. Solutions with different concentrations were taken separately to the International Space Station and mixed under microgravity. The samples were launched on 22 January 2011, and the experiment was performed on 21 February 2011. After staying at microgravity for 23 days, the samples were returned to Earth and delivered to our laboratories. Similar synthesis experiments were also performed on Earth as control experiments.

Brunauer-Emmett-Teller (BET) surface area, pore volume and pore size distribution were measured using nitrogen physisorption technique for both sets of samples. Transmission electron microscopy (TEM) images were taken to obtain particle size and shape. X-ray Diffraction (XRD) was also used to determine crystallite size and exposed surface planes.

2. Results and Discussion

2.1. BET Surface Area, Pore Size and Pore Volume

BET surface analysis was conducted on the ceria samples to determine the effect of microgravity on their surface characteristics. Table 1 compares the surface areas and pore volumes of the various samples tested. The control samples exhibited higher surface area and higher pore volume than the space samples. Regarding the effect of pH on the pore volume and surface area, higher surface areas and pore volumes were observed with increase in basicity of the medium for space samples whereas the opposite trend was exhibited by the control samples.

Table 1. Surface area and pore volume of CeO_2 sample measured using N_2 physisorption.

Samples	BET Surface Area (m^2/g)	Pore Volume (cm^3/g)
Control 1 M	138	0.3
Control 3 M	118	0.23
Space 1 M	47	0.18
Space 3 M	69	0.19

There was a considerable difference between control and space ceria in terms of the pore size distribution, as observed in Figure 1. The control ceria had a narrower pore size distribution with most of the pores being the 50–100 Å range and no pores larger than ~200 Å. On the other hand, for the space samples, a wide range of pore diameters ranging from 30–600 Å were observed. The lower pore volume and surface area can be due to larger particles size of the space samples compared to control samples. This was further confirmed in the TEM section. With respect to the effect of NaOH concentration, broader pore size distribution was obtained in case of 1 M control ceria than 3 M. The highest pore volume was obtained at a pore diameter of 83 Å and 62 Å for 1 M and 3 M, respectively. For space-grown ceria, there were no significant changes in the pore size distribution.

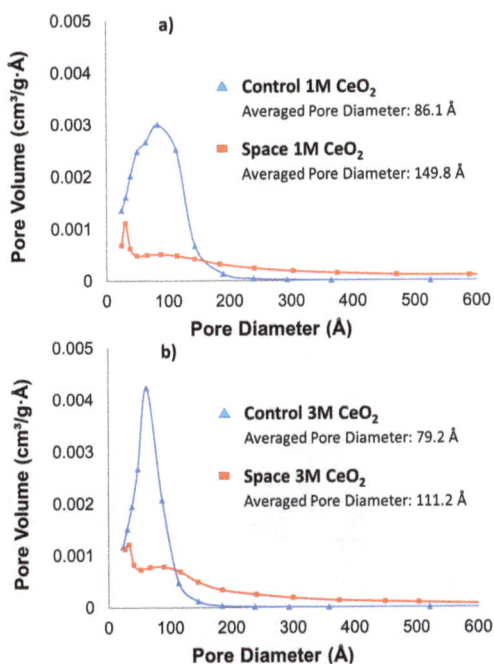

Figure 1. Barrett-Joyner-Halenda (BJH) Pore size distribution of CeO_2 sample measured using N_2 physisorptionprepared (**a**) with 1M NaOH (**b**) with 3M NaOH.

2.2. Transmission Electron Microscopy (TEM)

TEM images were obtained to confirm the ceria morphology and particle size (Figures 2 and 3). The particle size measurement was done by random sampling using multiple micrographs. Additionally, different batches of space and control samples are used for analysis.

In Figure 2a–c, TEM images of control ceria prepared with 1 M NaOH solution are shown. These samples consisted of polyhedron-shaped particles with an average particle size of 3.8 nm, fairly uniform particle size and shape throughout the sample.

Similarly, control samples prepared at more basic conditions (Figure 2d–f) contained smaller nanopolyhedra with an average particle size of 2.9 nm.

TEM images of the space samples are presented in Figure 3. The samples had mostly nano-rods, with 2–13 nm in width and 10–50 nm in length. Samples prepared with 1 M-solution were seen to give larger and more uniform particles on the average. The 3 M-space samples had some nano-spheres as well as rods. The particle size distributions for all four ceria samples are shown in Figures 4 and 5.

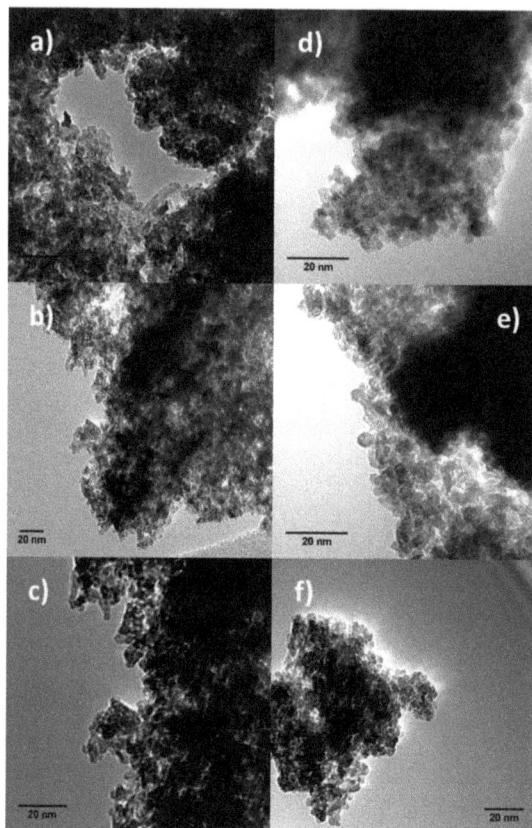

Figure 2. TEM image of control CeO$_2$ sample: 1 M (**a–c**) and 3 M (**d–f**).

These results are significant in showing that ceria nanorods can be synthesized at room temperature and atmospheric pressure. To the best of our knowledge, there have not been any publications for the synthesis of nanorods under such conditions. In most cases, studies have concluded that ceria nanorods are formed at much higher temperatures and pressures [27,46–49]. In order to understand the formation of ceria nanorods under microgravitiy condition, two crystal growth

74

processes, Oswald ripening and oriented attachment, can be considered. These crystal growth mechanisms have been utilized to demonstrate the increase in particle size and change in morphology of the ceria particles [50–52]. The Ostwald ripening process follows the dissolution and reprecipitation of solution ions (Ce^{4+}, OH^-) from the ceria nuclei minimizing the surface energy state [53]. On the other hand, the oriented attachment mechanism predominantly depends on the collision between nuclei during reaction. This process allows ceria nano-crystal to grow into a certain direction [53]. Under microgravity condition, the lack of sedimentation of ceria particles in the solution may increase the residence time for nucleation of ceria crystals [40]. Additionally, the diffusion and/or natural convection of ceria particles can be somewhat enhanced. These factors can significantly affect the crystal growth through oriented attachment, which results in more frequent collision between nuclei, thus forming larger particles and nanorods. There have been several articles showing evidence that oriented attachment crystal growth is the main route for the formation of nanorods of ceria [51], as well as various metal oxides [54].

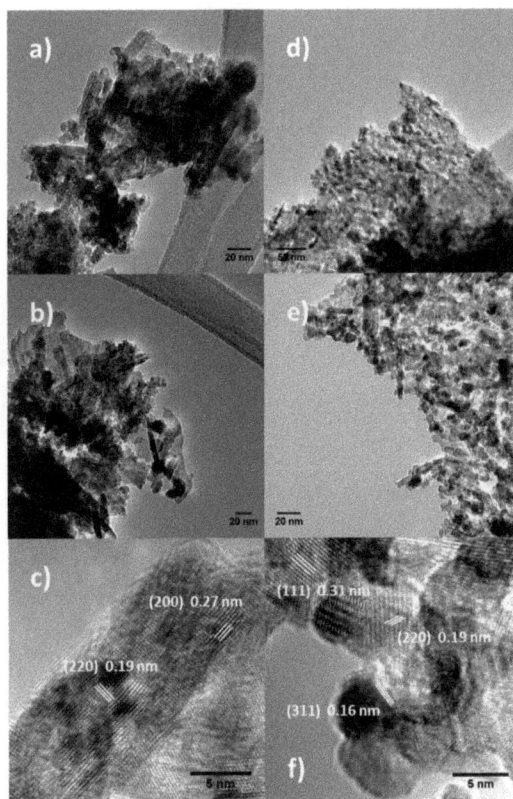

Figure 3. TEM image of Space CeO_2 sample: 1 M (**a–c**) and 3 M (**d–f**).

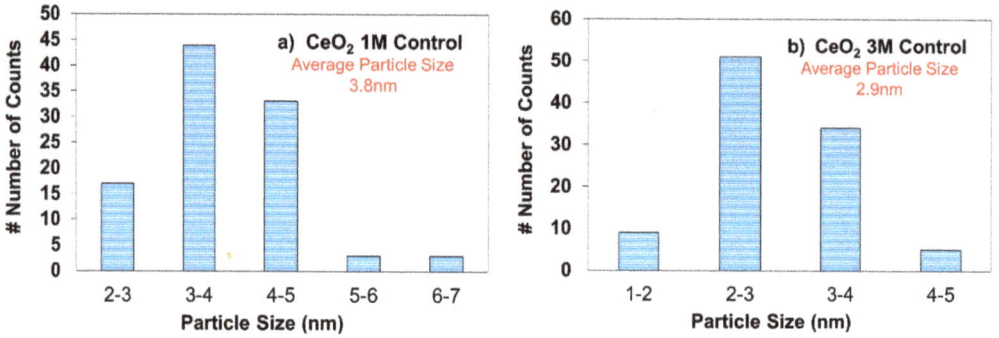

Figure 4. Particle size distribution histogram of Control CeO$_2$ sample: (**a**) 1 M; (**b**) 3 M.

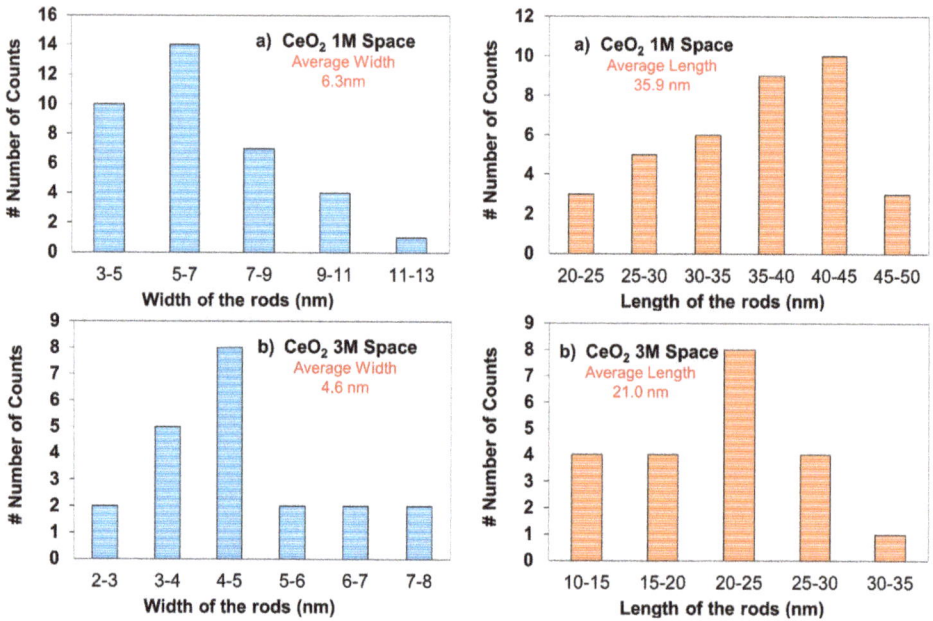

Figure 5. Width and length distribution histogram of Space CeO$_2$ sample: (**a**) 1 M; (**b**) 3 M.

In the literature, numerous studies on understanding the effect of pH on ceria particles size and shape have been conducted, however, depending on the synthesis method and choice of ceria precursor, different results have been reported. Tok *et al.* [50] observed similar behavior where smaller grains were seen in a more basic environment. This was because the Ce(OH)$_4$ precipitate is basic; thus, the solubility of Ce(OH)$_4$ in a more basic medium limits further agglomeration to form larger particles [50]. However, an opposite trend was reported by Yang and coworkers [51], where changing the concentration of NaOH from 1 M to 5 M, increased the particle size of the ceria cubes from 20 nm to 25–40 nm. Mai *et al.* [52] prepared ceria using cerium nitrate precursor at different NaOH concentrations of 1, 3, 6 and 9 M, at a fixed reaction temperature of 100 °C and reaction time of 24 h. It was observed that polyhedra particles were formed using 1 M NaOH, but as the solution became more basic, only rods were obtained.

In this study, the effect of pH on the ceria morphology and particle size was more apparent in the case of the space samples. It was found that the length of the nanorods were longer for the 1 M-space samples whereas smaller dimensions were observed for the 3 M-space samples. Moreover, the rods were prominent the former, while the number of rods were smaller in the 3 M-space samples, with nano-polyhedra and spheres also being observed. It should also be noted that the control ceria samples were under reaction condition of 25 °C and 1 atm. For example, Wu *et al.* [53] prepared ceria powders in acidic, basic and neutral reaction mediums indicating that the effect of pH on particles size was insignificant at room temperature. The similarities between the particles size and shape in the 1 M and 3 M-control samples are in good agreement with their results. Therefore, it can be concluded that under microgravity, the pH affects the ceria morphology and particle size, whereas the effect is less prominent under normal gravity at room temperature.

Lastly, identification of the diffracted planes were conducted. D-spacing was measured using Image J software. Figure 5c shows a ceria rod including (220) and (200) planes with 0.19 and 0.27 nm d-spacing [34,46,52]. Figure 5f indicates ceria nanopolyhedra and spheres with (311) and (220) planes [54].

2.3. X-ray Diffraction (XRD) Analysis

The XRD patterns of the ceria samples are shown in Figure 6. All patterns exhibited five distinct peaks which ensured that pure ceria was formed. Each peak was assigned to a particular crystal plane. The peak at a 2θ value of 28.5°, which showed the highest intensity, was identified as (111) plane of the cubic cerianite phase (ICDD 81-792). Others at 2θ values of 33°, 47.5°, 56°, and 69.4° were identified as (200), (220), (311), and (400) planes of ceria. For comparison, sharper diffraction lines were obtained for space samples, which suggest higher crystallinity of the structure.

The full width at half maximum (FWHM) values are shown in Figure 7. These values were used to calculate the mean crystal size which is listed in Table 2. It appears that the (111) peak is broader for control samples, resulting smaller crystal size than those that were prepared in space. Generally, the crystal size estimated from XRD is in good agreement with the particle size observed in TEM. In the control samples, however, an increase of crystal size from 31.6 to 39.6 Å was obtained with an increasing NaOH concentration, whereas almost no change in terms of particle size was concluded from TEM results. In contrast, the effect of pH on the crystal size was clear for space samples, where a significant decrease of crystal size from 67.9 to 53.5 Å was seen, which follows the same trend as that of the TEM results.

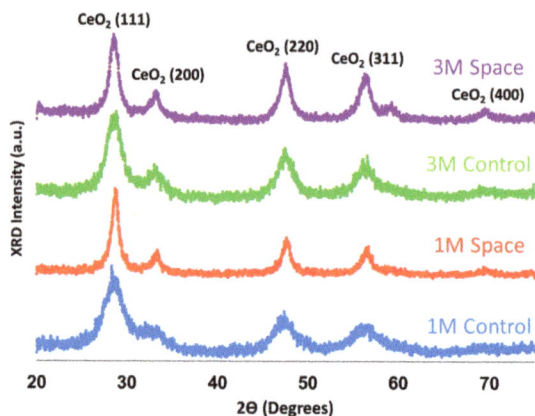

Figure 6. XRD pattern of the CeO_2 sample.

Table 2. Mean crystal sizes of CeO_2 sample.

	Mean Crystal Size (nm)
1 M Control	3.2
3 M Control	4.0
1 M Space	6.8
3 M Space	5.3

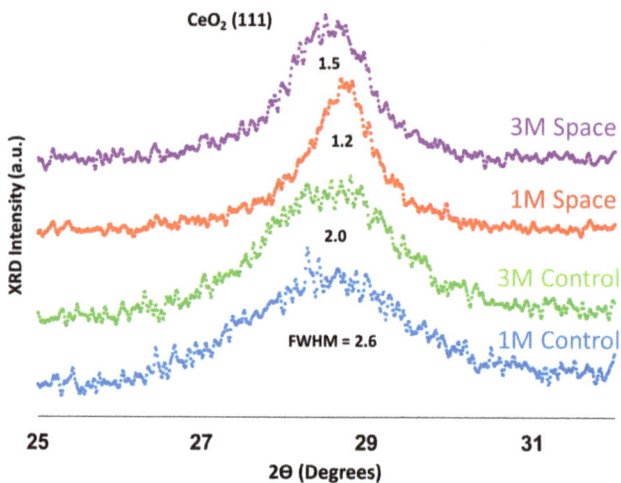

Figure 7. Comparison of FWHM of (111) plane for CeO_2 sample.

3. Experimental Methods

3.1. Catalyst Preparation

A CubeLab container (Nanoracks, Houston, TX, USA), shown in Figure 8, with the dimensions of 10 cm × 10 cm × 20 cm was utilized to protect the experimental containers during take-off and landing on route to and back from the International Space Station (ISS). Therefore, the catalyst preparation apparatus had to fit into a single CubeLab container.

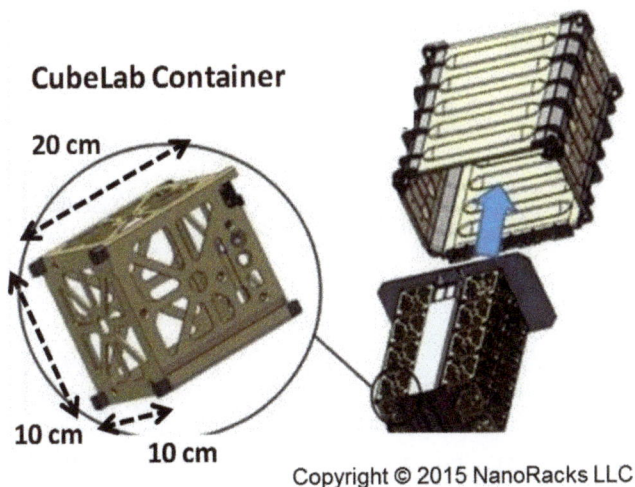

Copyright © 2015 NanoRacks LLC

Figure 8. Schematic and dimensions of CubeLab and NanoRack Modules.

Ceria nanoparticles were prepared by precipitation method, where 2.1 g $Ce(NO_3)_3 \cdot 6H_2O$ (Sigma-Aldrich, 99.999%, St. Louis, MO, USA) was dissolved in 13.5 mL water. In separate containers, caustic solutions with 66.5 mL of 1 M and 3 M NaOH were prepared. The solutions were loaded into concentric cylindrical rods. The innermost tube contained the cerium nitrate solution. Surrounding the inner glass tube was the sodium hydroxide solution. The astronauts were instructed to bend the assembly and thereby break the innermost glass tubing, allowing the solutions to mix. The assembly was then shaken to allow for a more thorough mixing. It should be noted that the liquid stays spread out through the length of the tube in microgravity, thereby showing different precipitation behavior. The concentric tubes and the protective cover layers around the catalyst preparation assembly are shown in Figure 9.

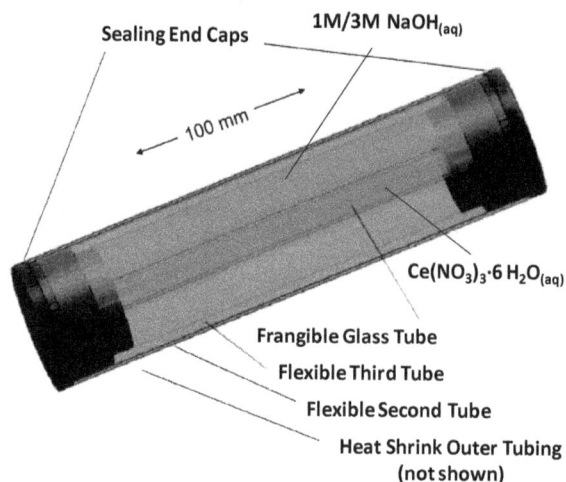

Figure 9. Schematic for the catalyst preparation assembly.

Twenty-four rods containing different molarities of NaOH solutions were loaded into the Cubelab. Another batch of 24 rods were prepared and kept in our laboratories as control. The CubeLab sample holder was then tested for vibration, vacuum, and off-gassing [55], verification as per the NASA space-flight qualification requirements and flight safety regulations, and was sent to the International Space Station (ISS). The samples were not exposed to air during this process. It should be noted that ISS was kept at 25 °C and 1 atm [55]. On approximately the same day, the control samples were treated under similar reaction conditions and the solutions were mixed in our laboratories. No further treatment was done until the samples prepared at ISS were delivered to our laboratories.

Following the delivery of the ISS samples, the samples were tested for ruptures. No serious containment breaches were found. Following the inspection, all samples including the control were filtered and washed with deionized and distilled water until a pH of 7 is reached. The samples were then dried in an oven at 90 °C overnight and were ready for characterization. Throughout the text, the ceria samples prepared in normal-gravity and microgravity are referred to as control and space samples, respectively.

3.2. Surface Analysis

BET surface area and pore volume measurements were carried out on a Micromeritics ASAP 2020 (Norcross, GA, USA) accelerated surface area and porosimetry instrument. The samples were degassed at 130 °C for 12 h prior to analysis under vacuum ($<2\ \mu m \cdot Hg$). Adsorption/desorption isotherms were collected at liquid nitrogen temperature where the desorption branch of the isotherm was used to calculate Barrett-Joyner-Halenda (BJH) pore size distributions.

3.3. Transmission Electron Microscopy (TEM)

A Phillips Tecnai F20 TEM (Hillsboro, OR, USA) equipped with a Field Emission Gun (FEG) and operated at 200 kV was used for imaging the particles. All exposures were collected in the bright field. The samples were suspended in ethanol and were sonicated for ~10 min to improve dispersion. Following sonication, the resulting mixture was immediately deposited on a 200-mesh copper grid coated with lacey carbon in order to prevent agglomeration of the particles. Image J open-source processing and analysis software was used to estimate the particle sizes.

3.4. X-ray Diffraction (XRD)

XRD powder patterns were collected using a Rigaku diffractometer (Tokyo, Japan) with Cu Kα radiation with λ = 1.5418 Å operated at 40 kV and 25 mA. The patterns were collected in the 2θ range of $20°$–$75°$. International Center for Diffraction Data (ICDD) database was used for identification of the crystalline phases. The crystal size of the samples were calculated by Scherrer equation $L = K\lambda/\beta\cos\theta$ where K is the Scherrer constant, λ is the wavelength of the X-ray, β is the full width half maximum (FWHM), and θ is the diffraction angle.

4. Conclusions

The effect of microgravity on ceria synthesis was investigated. Ceria was synthesized by mixing cerium nitrate and sodium hydroxide solutions at room temperature, under both microgravity (space) and normal-gravity (control) for comparison. The ceria samples that were prepared in space showed lower surface area and pore volume compared to the Earth samples. BJH pore size analysis indicated that the space ceria samples consisted of larger pores with a broader pore size distribution. The morphology of the ceria particles were observed from TEM images. It was concluded that the space samples were mostly nano-rods, whereas control samples contained nano-polyhedra. The average particle size observed was larger for space ceria. From XRD results, space ceria samples were more crystalline than the control samples. The mean crystal size obtained using Scherer equation was larger in case of space samples than control. The difference in particle size and shape between space and control ceria is thought to be due to improved oriented attachment crystal growth process where more collision of ceria particle occurs. This can be attributed to lack of sedimentation of the ceria particles under microgravity condition. Lastly, the effect of pH was examined by changing the concentration of sodium hydroxide solution (1 M and 3 M). The ceria samples synthesized under normal-gravity at room temperature showed minimal differences between 1 M and 3

81

M. However, a significant decrease in particle size was observed for space samples which were prepared in a more basic medium, with a number of nano-polyhedra and spheres also being present.

Acknowledgments: We gratefully acknowledge the funding from the US Department of Energy for the Grant DE-FG36-05GO15033.

Author Contributions: Ilgaz I. Soykal worked primarily on characterization of the samples, data analysis and co-wrote the manuscript. Hyuntae Sohn worked on sample characterization and co-wrote the manuscript. Burcu Bayram helped design the experiments. She was the liaison between the Department of Chemical and Biomolecular Engineering and Mechanical and Aerospace Engineering. Preshit Gawade worked on sample preparation and synthesis. Michael P. Snyder, Stephen E. Levine and Hayrani Oz coordinated the CubeLab experimental rack and organized the portion of the experimentation that took place in the International Space Station. Umit S. Ozkan is the Project Leader leading the project. She conceived and designed the experiments, oversaw characterization and co-wrote the manuscript.

Conflicts of Interest: The authors declare no conflict of interest.

References

1. Sun, C.; Li, H.; Chen, L. Nanostructured ceria-based materials: Synthesis, properties, and applications. *Energy Environ. Sci.* **2012**, *5*, 8475–8505.
2. Trovarelli, A. Catalytic Properties of Ceria and CeO_2-Containing Materials. *Catal. Rev. Sci. Eng.* **1996**, *38*, 439–520.
3. Soykal, I.I.; Sohn, H.; Ozkan, U.S. Effect of Support Particle Size in Steam Reforming of Ethanol over Co/CeO_2 Catalysts. *ACS Catal.* **2012**, *2*, 2335–2348.
4. Soykal, I.I.; Sohn, H.; Singh, D.; Miller, J.T.; Ozkan, U.S. Reduction Characteristics of Ceria under Ethanol Steam Reforming Conditions: Effect of the Particle Size. *ACS Catal.* **2014**, *4*, 585–592.
5. Kaspar, J.; Fornasiero, P.; Graziani, M. Use of CeO_2-based oxides in the three-way catalysis. *Catal. Today* **1999**, *50*, 285–298.
6. Diwell, A.; Rajaram, R.; Shaw, H.; Truex, T. The Role of Ceria in Three-Way Catalysts. *Stud. Surf. Sci. Catal.* **1991**, *71*, 139–152.
7. Whittington, B.; Jiang, C.; Trimm, D. Vehicle exhaust catalysis: I. The relative importance of catalytic oxidation, steam reforming and water-gas shift reactions. *Catal. Today* **1995**, *26*, 41–45.
8. Murray, E.P.; Tsai, T.; Barnett, S. A direct-methane fuel cell with a ceria-based anode. *Nature* **1999**, *400*, 649–651.
9. Steele, B.C.; Heinzel, A. Materials for fuel-cell technologies. *Nature* **2001**, *414*, 345–352.
10. Carrettin, S.; Concepcion, P.; Corma, A.; Nieto, J.M.L.; Puntes, V.F. Nanocrystalline CeO_2 Increases the Activity of Au for CO Oxidation by Two Orders of Magnitude. *Angew. Chem.* **2004**, *43*, 2538–2540.
11. Diagne, C.; Idriss, H.; Pearson, K.; Gómez-García, M.A.; Kiennemann, A.R. Efficient hydrogen production by ethanol reforming over Rh catalysts. Effect of addition of Zr on CeO_2 for the oxidation of CO to CO_2. *C.R. Chim.* **2004**, *7*, 617–622.

12. Biswas, P.; Kunzru, D. Steam reforming of ethanol on Ni–CeO$_2$–ZrO$_2$ catalysts: Effect of doping with copper, cobalt and calcium. *Catal. Lett.* **2007**, *118*, 36–49.

13. Cai, W.; Wang, F.; van Veen, A.C.; Provendier, H.; Mirodatos, C.; Shen, W. Autothermal reforming of ethanol for hydrogen production over an Rh/CeO$_2$ catalyst. *Catal. Today* **2008**, *138*, 152–156.

14. Cai, W.; Zhang, B.; Li, Y.; Xu, Y.; Shen, W. Hydrogen production by oxidative steam reforming of ethanol over an Ir/CeO$_2$ catalyst. *Catal. Commun.* **2007**, *8*, 1588–1594.

15. Frusteri, F.; Freni, S.; Chiodo, V.; Donato, S.; Bonura, G.; Cavallaro, S. Steam and auto-thermal reforming of bio-ethanol over MgO and CeO$_2$/Ni supported catalysts. *Int. J. Hydrogen Energy* **2006**, *31*, 2193–2199.

16. Laosiripojana, N.; Assabumrungrat, S. Catalytic steam reforming of ethanol over high surface area CeO$_2$: The role of CeO$_2$ as an internal pre-reforming catalyst. *Appl. Catal.* **2006**, *66*, 29–39.

17. Song, H.; Ozkan, U.S. Changing the Oxygen Mobility in Co/Ceria Catalysts by Ca Incorporation: Implications for Ethanol Steam Reforming. *J. Phys. Chem. A* **2010**, *114*, 3796–3801.

18. Jacobs, G.; Chenu, E.; Patterson, P.M.; Williams, L.; Sparks, D.; Thomas, G.; Davis, B.H. Water-gas shift: Comparative screening of metal promoters for metal/ceria systems and role of the metal. *Appl. Catal. A* **2004**, *258*, 203–214.

19. Yi, N.; Si, R.; Saltsburg, H.; Flytzani-Stephanopoulos, M. Active gold species on cerium oxide nanoshapes for methanol steam reforming and the water gas shift reactions. *Energy Environ. Sci.* **2010**, *3*, 831–837.

20. Anupriya, K.; Vivek, E.; Subramanian, B. Facile synthesis of ceria nanoparticles by precipitation route for UV blockers. *J. Alloys Compd.* **2014**, *590*, 406–410.

21. Shih, S.J.; Tzeng, W.L.; Kuo, W.L. Fabrication of ceria particles using glycine nitrate spray pyrolysis. *Surf. Coat. Technol.* **2014**, *259*, 302–309.

22. Christensen, J.; Deiana, D.; Grunwaldt, J.D.; Jensen, A. Ceria Prepared by Flame Spray Pyrolysis as an Efficient Catalyst for Oxidation of Diesel Soot. *Catal. Lett.* **2014**, *144*, 1661–1666.

23. Hirano, M.; Inagaki, M. Preparation of monodispersed cerium(IV) oxide particles by thermalhydrolysis: Influence of the presence of urea and Gd doping on their morphology and growth. *J. Mater. Chem.* **2000**, *10*, 473–477.

24. Bumajdad, A.; Zaki, M.I.; Eastoe, J.; Pasupulety, L. Microemulsion-Based Synthesis of CeO$_2$ Powders with High Surface Area and High-Temperature Stabilities. *Langmuir* **2004**, *20*, 11223–11233.

25. Charojrochkul, S.; Nualpaeng, W.; Laosiripojana, N.; Assabumrungrat, S. *Materials Challenges and Testing for Supply of Energy and Resources*; Springer: Berlin, Germany, 2012; pp. 189–202.

26. Derakhshandeh, P.G.; Soleimannejad, J.; Janczak, J. Sonochemical synthesis of a new nano-sized cerium(III) coordination polymer and its conversion to nanoceria. *Ultrason. Sonochem.* **2015**, *26*, 273–280.

27. Cui, M.Y.; He, J.X.; Lu, N.P.; Zheng, Y.Y.; Dong, W.J.; Tang, W.H.; Chen, B.Y.; Li, C.R. Morphology and size control of cerium carbonate hydroxide and ceria micro/nanostructures by hydrothermal technology. *Mater. Chem. Phys.* **2010**, *121*, 314–319.

28. Zawadzki, M. Preparation and characterization of ceria nanoparticles by microwave-assisted solvothermal process. *J. Alloys Compd.* **2008**, *454*, 347–351.

29. Zhang, D.; Niu, F.; Li, H.; Shi, L.; Fang, J. Uniform ceria nanospheres: Solvothermal synthesis, formation mechanism, size-control and catalytic activity. *Powder Technol.* **2011**, *207*, 35–41.

30. Abbasi, Z.; Haghighi, M.; Fatehifar, E.; Rahemi, N. Comparative synthesis and physicochemical characterization of CeO_2 nanopowder via redox reaction, precipitation and sol-gel methods used for total oxidation of toluene. *Asia Pac. J. Chem. Eng.* **2012**, *7*, 868–876.

31. Zhou, X.D.; Huebner, W. Size-induced lattice relaxation in CeO_2 nanoparticles. *Appl. Phys. Lett.* **2001**, *79*, 3512–3514.

32. Deshpande, S.; Patil, S.; Kuchibhatla, S.V.N.T.; Seal, S. Size dependency variation in lattice parameter and valency states in nanocrystalline cerium oxide. *Appl. Phys. Lett.* **2005**, *87*, 133113.

33. Zhang, F.; Chan, S.W.; Spanier, J.E.; Apak, E.; Jin, Q.; Robinson, R.D.; Herman, I.P. Cerium oxide nanoparticles: Size-selective formation and structure analysis. *Appl. Phys. Lett.* **2002**, *80*, 127–129.

34. Si, R.; Flytzani-Stephanopoulos, M. Shape and Crystal-Plane Effects of Nanoscale Ceria on the Activity of Au–CeO_2 Catalysts for the Water-Gas Shift Reaction. *Angew. Chem. Int. Ed.* **2008**, *47*, 2884–2887.

35. Yi, N.; Si, R.; Saltsburg, H.; Flytzani-Stephanopoulos, M. Steam reforming of methanol over ceria and gold-ceria nanoshapes. *Appl. Catal. B* **2010**, *95*, 87–92.

36. Lee, Y.; He, G.; Akey, A.J.; Si, R.; Flytzani-Stephanopoulos, M.; Herman, I.P. Raman Analysis of Mode Softening in Nanoparticle $CeO_{2-\delta}$ and Au–$CeO_{2-\delta}$ during CO Oxidation. *J. Am. Chem. Soc.* **2011**, *133*, 12952–12955.

37. Boucher, M.B.; Goergen, S.; Yi, N.; Flytzani-Stephanopoulos, M. "Shape effects" in metal oxide supported nanoscale gold catalysts. *Phys. Chem. Chem. Phys.* **2011**, *13*, 2517–2527.

38. Coker, E.N.; Jansen, J.C.; DiRenzo, F.; Fajula, F.; Martens, J.A.; Jacobs, P.A.; Microporous, A.S., Jr. Zeolite ZSM-5 synthesized in space: Catalysts with reduced external surface activity. *Microporous Mesoporous Mater.* **2001**, *46*, 223–236.

39. Coker, E.N.; Jansen, J.C.; Martens, J.A.; Jacobs, P.A.; DiRenzo, F.; Fajula, F.; Microporou, A.S., Jr. The synthesis of zeolites under micro-gravity conditions: A review. *Microporous Mesoporous Mater.* **1998**, *23*, 119–136.

40. Sano, T.; Mizukami, F.; Kawamura, M.; Takaya, H.; Mouri, T.; Inaoka, W.; Toida, Y.; Watanabe, M.; Toyoda, K. Crystallization of ZSM-5 zeolite under microgravity. *Zeolites* **1992**, *12*, 801–805.

41. Warzywoda, J.; Baç, N.; Rossetti, G.A., Jr.; van der Puil, N.; Jansen, J.C.; Bekkum, H.V.; Sacco, A., Jr. Synthesis of high-silica ZSM-5 in microgravity. *Microporous Mesoporous Mater.* **2000**, *38*, 423–432.

42. Tsuchida, A.; Taguchi, K.; Takyo, E.; Yoshimi, H.; Kiriyama, S.; Okubo, T.; Ishikawa, M. Microgravity experiments on colloidal crystallization of silica spheres in the presence of sodium chloride. *Colloid Polym. Sci.* **2000**, *278*, 872–877.

43. Smith, D.D.; Sibille, L.; Cronise, R.J.; Hunt, A.J.; Oldenburg, S.J.; Wolfe, D.; Halas, N.J. Effect of Microgravity on the Growth of Silica Nanostructures. *Langmuir* **2000**, *16*, 10055–10060.

44. Ahari, H.; Bedard, R.L.; Bowes, C.L.; Coombs, N.; Dag, O.; Jiang, T.; Ozin, G.A.; Petrov, S.; Sokolov, I.; Verma, A.; *et al.* Effect of microgravity on the crystallization of a self-assembling layered material. *Nature* **1997**, *388*, 857–860.

45. Fujii, H.; Nogi, K. Formation and disappearance of pores in aluminum alloy molten pool under microgravity. *Sci. Technol. Adv. Mater.* **2004**, *5*, 219–223.

46. Soykal, I.I.; Bayram, B.; Sohn, H.; Gawade, P.; Miller, J.T.; Ozkan, U.S. Ethanol steam reforming over Co/CeO_2 catalysts: Investigation of the effect of ceria morphology. *Appl. Catal. A* **2012**, *449*, 47–58.

47. Ho, C.M.; Yu, J.C.; Kwong, T.; Mak, A.C.; Lai, S.Y. Morphology-Controllable Synthesis of Mesoporous CeO_2 Nano- and Microstructures. *Chem. Mater.* **2005**, *17*, 4514–4522.

48. Huang, P.X.; Wu, F.; Zhu, B.L.; Gao, X.P.; Zhu, H.Y.; Yan, T.Y.; Huang, W.P.; Wu, S.H.; Song, D.Y. CeO_2 Nanorods and Gold Nanocrystals Supported on CeO_2 Nanorods as Catalyst. *J. Phys. Chem. B* **2005**, *109*, 19169–19174.

49. Liu, L.; Yao, Z.; Deng, Y.; Gao, F.; Liu, B.; Dong, L. Morphology and Crystal-Plane Effects of Nanoscale Ceria on the Activity of CuO/CeO_2 for NO Reduction by CO. *ChemCatChem* **2011**, *3*, 978–989.

50. Tok, A.I.Y.; Boey, F.Y.C.; Dong, Z.; Sun, X.L. Hydrothermal synthesis of CeO_2 nano-particles. *J. Mater. Process. Technol.* **2007**, *190*, 217–222.

51. Yang, Z.; Zhou, K.; Liu, X.; Tian, Q.; Lu, D.; Yang, S. Single-crystalline ceria nanocubes: Size-controlled synthesis, characterization and redox property. *Nanotechnology* **2007**, *18*, 185606.

52. Mai, H.X.; Sun, L.D.; Zhang, Y.W.; Si, R.; Feng, W.; Zhang, H.P.; Liu, H.C.; Yan, C.H. Shape-Selective Synthesis and Oxygen Storage Behavior of Ceria Nanopolyhedra, Nanorods, and Nanocubes. *J. Phys. Chem. B* **2005**, *109*, 24380–24385.

53. Wu, N.C.; Shi, E.W.; Zheng, Y.Q.; Li, W.J. Effect of pH of Medium on Hydrothermal Synthesis of Nanocrystalline Cerium(IV) Oxide Powders. *J. Am. Ceram. Soc.* **2002**, *85*, 2462–2468.

54. Li, J.G.; Ikegami, T.; Wang, Y.; Mori, T. Reactive Ceria Nanopowders via Carbonate Precipitation. *J. Am. Ceram. Soc.* **2002**, *85*, 2376–2378.

55. Levine, S.; Snyder, M. Microgravity Cube Lab Experiment Design: Setting New Precedents for Micro-Gravity Testing. In Proceedings of the AIAA SPACE 2012 Conference & Exposition, Pasadena, CA, USA, 11–13 September 2012.

Surface Reaction Kinetics of Steam- and CO₂-Reforming as Well as Oxidation of Methane over Nickel-Based Catalysts

Karla Herrera Delgado, Lubow Maier, Steffen Tischer, Alexander Zellner, Henning Stotz and Olaf Deutschmann

Abstract: An experimental and kinetic modeling study on the Ni-catalyzed conversion of methane under oxidative and reforming conditions is presented. The numerical model is based on a surface reaction mechanism consisting of 52 elementary-step like reactions with 14 surface and six gas-phase species. Reactions for the conversion of methane with oxygen, steam, and CO_2 as well as methanation, water-gas shift reaction and carbon formation via Boudouard reaction are included. The mechanism is implemented in a one-dimensional flow field description of a fixed bed reactor. The model is evaluated by comparison of numerical simulations with data derived from isothermal experiments in a flow reactor over a powdered nickel-based catalyst using varying inlet gas compositions and operating temperatures. Furthermore, the influence of hydrogen and water as co-feed on methane dry reforming with CO_2 is also investigated.

Reprinted from *Catalysts*. Cite as: Delgado, K.H.; Maier, L.; Tischer, S.; Zellner, A.; Stotz, H.; Deutschmann, O. Surface Reaction Kinetics of Steam- and CO₂-Reforming as Well as Oxidation of Methane over Nickel-Based Catalysts. *Catalysts* **2015**, 5, 871–904.

1. Introduction

In recent years, much attention has been paid towards the reforming of light hydrocarbons to produce synthesis gas (H_2/CO), which is an important intermediate in the chemical industry for manufacturing valuable basic chemicals and synthetic fuels, via methanol synthesis, oxo synthesis, and Fischer-Tropsch synthesis [1–6]. Hydrogen separated from the synthesis gas is largely used in the manufacturing of ammonia, a variety of petroleum hydrogenation processes, and for power generation [7–9]. Manufacturing syngas constitutes a significant portion of the investments in large-scale gas conversion plants based on natural gas [4].

Processes such as steam reforming (SR), partial oxidation (POX), autothermal reforming (ATR) and dry reforming (DR) are the most common catalytic technologies for converting natural gas to synthesis gas in various compositions [2].

Since 1930, the most important industrial method to produce syngas has been the steam reforming of methane Equation (1) [10,11]. Conventional steam reformers deliver relatively high concentrations of hydrogen at high fuel conversion [2,4,11,12]. The reaction is highly endothermic, and requires a large efficient external energy supply; also the efficiency of the process is severely affected by the catalyst deactivation due to carbon formation [4].

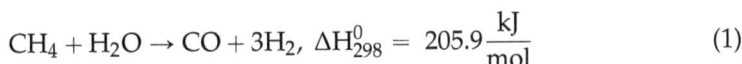

$$CH_4 + H_2O \rightarrow CO + 3H_2, \ \Delta H^0_{298} = 205.9 \frac{kJ}{mol} \qquad (1)$$

Due to the increasing environmental concerns about global warming and oil depletion, methane reforming with CO_2 Equation (2) has gained considerable attention in the field of catalysis, because it offers the opportunity to convert greenhouse gases (CH_4 and CO_2) as carbon-containing materials synthesis gas. The reforming of methane with CO_2 has been proposed for the production of H_2/CO with lower ratios, which is more suited for stream processes such as oxo synthesis of the aldehydes, syntheses of methanol, and acetic acid [13]. However, one of the main challenges in dry reforming of methane, especially at industrial conditions, is the formation of carbon, which causes catalyst deactivation [2,14].

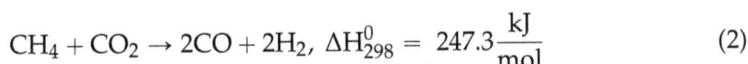

$$CH_4 + CO_2 \rightarrow 2CO + 2H_2, \ \Delta H^0_{298} = 247.3 \frac{kJ}{mol} \qquad (2)$$

The catalytic partial oxidation Equation (3) of methane over nickel-based catalysts has been deeply studied as a promising alternative to the endothermic reforming processes [15–20], because no additional steam or heat are required. However, the process is complicated; different pre-treatment conditions affect the state of the catalyst surface and may change the reaction mechanism. Consequently, many studies have been carried out in order to elucidate the kinetics behind this reaction [15–21].

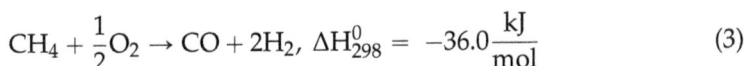

$$CH_4 + \frac{1}{2}O_2 \rightarrow CO + 2H_2, \ \Delta H^0_{298} = -36.0 \frac{kJ}{mol} \qquad (3)$$

Similarly to the reforming processes, the catalytic oxidative conversion of methane at elevated temperatures and pressures suffers from coke formation as well. Coke deposition on catalysts and reactor pipe walls are serious problems in many industrial reactors that involve methane as fuel; in some cases it leads to the blocking of reactor tubes as well as the physical disintegration of the catalyst structure [22–27].

Noble metals have been found to be less prone to coke formation under oxidation and reforming conditions [28]. However, their high prices make them economically unsustainable. Ni-based catalysts are preferred in industrial applications due to fast turnover rates, good availability and low costs, but the use is limited by their higher tendency towards coke formation [29–31].

In order to optimize both the processes for the catalytic oxidation and reforming of methane, it is necessary to achieve a better understandings of the elementary steps involved in the reaction mechanism at a the molecular level and, along with the deactivation kinetics of coke formation. Here, micro-kinetic modeling is the approach of choice to cover different scales and varying conditions [32,33]. The kinetic model then needs to be is coupled with mass and heat transport models in order to numerically simulate the behavior of reforming reactors.

Reforming and oxidation of methane have been studied using several techniques. Different reaction mechanisms and corresponding kinetic models have been proposed. However, despite all reported experimental and theoretical studies, the detailed path for the conversion of methane to syngas and carbon remains a controversial issue [3]. In a pioneering work, Xu and Froment [11] proposed a reaction mechanism for the steam reforming of methane accompanied by water-gas shift reactions on a $Ni/MgAl_2O_4$ catalyst. Bradford and Vannice [13] studied the mechanism and kinetics of dry reforming over Ni catalysts with different supports. The authors proposed a model for CH_4-CO_2 kinetics based on CH_4 activation to form CH_x and CH_xO. Aparicio [10] proposed an overall model that described CH_4-H_2O kinetics over Ni/MgO-$MgAl_2O_4$ catalysts. The rate constants of surface elementary reactions were extracted from transient isotopic experimental data by fitting the measured response curves to micro-kinetic models [10]. Chen $et\ al.$ [31,32] modified Aparicio's micro-kinetic model for methane reforming with CO_2 and deactivation by carbon formation. Furthermore, they extended a hierarchical multiscale approach developed by Mhadeshwar and Vlachos [34] on Rh to a multiscale model of an industrial reformer for steam reforming on supported nickel catalysts. Wei and Iglesia [35] proposed a common sequence of elementary steps for CH_4 decomposition and water-gas shift reactions on Ni/MgO catalysts. Isotopic studies and reaction rate measurements showed a mechanistic equivalence among all CH_4 reactions [35]. Blaylock $et\ al.$ [36,37] developed a micro-kinetic model for methane steam reforming using thermodynamic data from plane wave density functional theory (DFT) over nickel crystals. The model developed in this study predicts overall SR rates that are approximately 3 orders of magnitude slower than experimentally measured reforming rates using commercial supported catalysts. However, the comparison of calculated parameters with single-crystal Ni(111) high-resolution electron energy loss spectroscopy (HREELS) data shows the accuracy of the computational methods employed here at least for the case of methane decomposition. Wang $et\ al.$ [38] investigated the reaction pathways of methane reforming with CO_2 on Ni (111) by using DFT calculations. The authors developed a surface reaction mechanism on the basis of computed energy barriers. There, the CH_4 dissociative adsorption is the rate determining step, and adsorbed CHO species are considered as key intermediates on the surface. The reaction paths for partial oxidation of methane and its kinetics over platinum and rhodium have been widely studied by several groups [34,39–43]. A review on catalytic partial oxidation of methane to synthesis gas with emphasis on reaction mechanisms over transition metal catalysts was published by Enger and co-workers [44]. De Groote and Froment [45] proposed a one-dimensional adiabatic model for partial oxidation of methane to syngas in a fixed bed reactor on nickel catalysts. The authors considered total and partial oxidation of methane, steam reforming, and water-gas shift.

Despite all the kinetic studies performed over nickel-based catalysts, the development of a detailed mechanism for simultaneous modeling of partial oxidation, steam and dry reforming of methane over nickel-based catalysts, as well as the sub systems behind these reactions (e.g., H_2 and CO oxidation, water-gas shift, and its reverse reaction) have not been described yet.

In this paper, we present an experimental and modeling study for catalytic conversion of methane under oxidative and reforming conditions over nickel-based catalysts. Therefore, in the present work a previously developed kinetic model for methane steam reforming over Ni/alumina catalysts [46] has been modified and extended. The newly developed thermodynamically consistent reaction mechanism includes carboxyl species (COOH) as an intermediate on the surface and new reaction paths for carbon formation. The mechanism is tested by the comparison of the simulation results with experimentally obtained data for partial oxidation, steam and dry reforming of methane over a powdered nickel-based catalyst in a temperature range of 373–1123 K.

2. Experimental Setup

Methane partial oxidation and reforming are studied in a fixed bed reactor as depicted in Figure 1. Details of the experimental setup can be found elsewhere [47]. The reactor consists of a quartz tube with an inner diameter of 10 mm filled with 20 mg of a nickel-based catalyst with a reaction zone of 27 mm length surrounded by a quartz frit and glass wool. The nickel-based catalyst was synthesized by BASF as part of the BMWI "DRYREF project" (reference FKZ0327856A) and has a particle size between 500 and 1000 μm.

The experiments were carried out at 4 slpm (standard liters per minute, $T = 298.15$ K and $p = 1.01325$ bar) and 1 bar total pressure. The reaction temperatures were increased from 373 K to 1173 K at a rate of 15 K/min. Two separate thermocouples were used to measure the temperature of the gas-phase during the reaction: type K in front of the catalyst, and type N behind the catalyst, respectively. The entire reactor was surrounded by a furnace for heating and thermal insulation. Table 1 shows the different inlet compositions investigated in this study. The dosage of the gases (H_2, CO, O_2, CH_4, CO_2, and N_2) was controlled by mass flow controllers produced by Bronkhorst Hi-Tec. Water was provided by a liquid flow controller from a water reservoir. After evaporation, the water steam was mixed directly into the reactant gas stream.

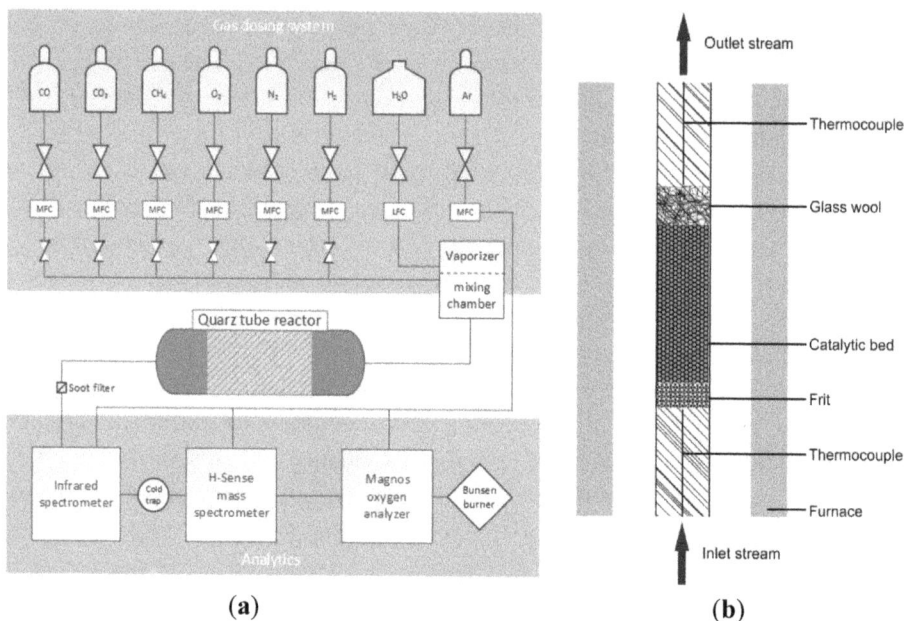

(a)   (b)

Figure 1. Schematic diagram of the experimental setup used for catalytic oxidation and reforming experiments over a nickel bed catalyst (**a**) and schematic drawing of the packed bed reactor (**b**).

Table 1. Investigated inlet gas mixtures on which the studies have been performed.

Fuel composition	CH_4 (vol.%)	O_2 (vol.%)	CO_2 (vol.%)	H_2 (vol.%)	H_2O (vol.%)	N_2 (vol.%)
CH_4/O_2	1.33	0.81	-	-	-	97.86
CH_4/H_2O	1.60	-	-	-	2.00	96.40
CH_4/CO_2	2.00	-	2.00	-	-	96.00
$CH_4/CO_2/H_2$	1.62	-	2.08	1.80	-	94.50
$CH_4/CO_2/H_2O$	1.67	-	2.13	-	2.13	94.07

The product composition was measured by means of an online Fourier Transform Infrared Spectrometer (FT-IR), a Mass spectrometer (H_2-MS) and paramagnetic oxygen detection (O_2-analyser, ABB, Zurich, Switzerland). Remaining water in the product stream was removed by a cold trap after passing the FT-IR to protect the downstream analytics. Argon was exclusively used as purifier gas for the analytics. Before all experiments, the powdered nickel-based catalyst was conditioned with 20 vol.% O_2 diluted in nitrogen at 673 K for 30 min and then reduced with 10 vol.% H_2 diluted in nitrogen with a total flow of 4 slpm at 873 K, over 60 min; then the reactor was cooled down to 373 K.

3. Mathematical Model and Numerical Simulation

The numerical simulation was performed using the software package DETCHEM™ [48], a program specifically designed for numerical simulation of the flow field coupled with detailed gas- phase and surface kinetics in chemical reactors at laboratory and technical scale.

All experiments used in the model development were carried out in a packed bed reactor at isothermal conditions.

3.1. Modeling the Flow Field in the Packed Bed Reactor

For simulations of the packed bed reactor the computer code DETCHEM^{PACKEDBED} was used. The code uses an one-dimensional heterogeneous model that assumes no radial variations in the flow properties and axial diffusion of any quantity is negligible relative to the corresponding convective term [48].

The one-dimensional isothermal fixed-bed reactor model is based on the following set of equations:

Continuity equation

$$\frac{d\,(\rho u)}{dz} = a_v \sum_{i=1}^{N_g} \dot{s}_i M_i \tag{4}$$

Species conservation

$$\rho u \frac{d\,(Y_i)}{dz} + Y_i a_v \sum_{i=1}^{N_g} \dot{s}_i M_i = M_i \left(a_v \dot{s}_i + \dot{w}_i \varepsilon \right) \tag{5}$$

and equation of state

$$pM = \rho RT \tag{6}$$

Here, ρ = density, u = velocity, a_v = catalytic area to volume ratio, ε = porosity, A_c = area of cross section of the channel, N_g = number of gas-phase species, N_s = number of surface species, $\dot{s}_i$ = molar rate of production of species i by surface reaction, $\dot{w}_i$ = molar rate of production of species i by the gas-phase reaction, M_i = molecular mass of species i, Y_i = mass fraction of species , p = pressure, and M = average molecular weight. The porosity of the catalytic bed was calculated to be $\varepsilon = 0.42$, using a statistical method proposed by Pushnov [49]. The active catalytic area to volume ratio (a_v) was calculated by

$$a_v = D_{Ni} \cdot \frac{m_{Ni}}{M_{Ni}} \cdot \frac{1}{\Gamma} \cdot \frac{1}{V_{bed}} \tag{7}$$

The catalyst dispersion D_{Ni} was experimentally determined by chemisorption measurements. The surface-site density was chosen to be $\Gamma = 2.66 \times 10^{-5}\text{mol m}^{-2}$ [12,46]. In Equation (7), M_{Ni} represents the molar mass of nickel (58.7 g/mol), V_{bed} the total volume of the catalytic bed (m³), and m_{Ni} the amount of catalyst loading (g). A value of $(a_v) = 9.85 \times 10^6 \text{ m}^{-1}$ was derived for the nickel-based catalyst used in this work.

3.2. Modeling the Surface Reaction Kinetics

The mean-field approximation is used to model the surface reaction kinetics. The approximation is related to the sizes of the computational cells in the flow field simulation, assuming that the local state of the active surface can be represented by means of values for this cell such as coverages (Θ_i) and temperature. [50,51]. A surface reaction is expressed as:

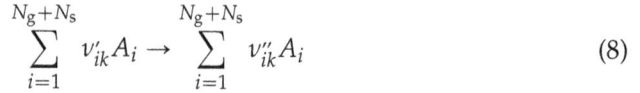

$$\sum_{i=1}^{N_g+N_s} v'_{ik} A_i \rightarrow \sum_{i=1}^{N_g+N_s} v''_{ik} A_i \tag{8}$$

where N_g is the number of gas-phase species, N_s is the number of surface species, with $v_{ik} = v''_{ik} - v'_{ik}$ are the stoichiometric coefficients and A_i denotes the species i. The concentration of the absorbed species can be expressed in terms of a surface coverage (Θ_i), according to the relation

$$\Theta_i = \frac{c_i \sigma_i}{\Gamma} \tag{9}$$

where c_i is the concentration of adsorbed species, which are given, e.g., in mol/m², σ_i represents the number of surface sites that are occupied by species i.

The total molar production rate $\dot{s}_i$ of surface species i on the catalyst was calculated in analogy to gas-phase reactions as a product of rate coefficients and concentrations determined by

$$\dot{s}_i = \sum_{k=1}^{K_s} v_{ik} k_{fk} \prod_{j=1}^{N_g+N_s} c_j^{v'_{jk}} \tag{10}$$

Thus, locally resolved reaction rates depend on the local gas-phase concentrations, surface coverage, and temperature.

The temperature dependence of the rate coefficients k_{fk} is described by a modified Arrhenius expression where additional coverage dependencies of the activation energy ε_{ik} are taken into account:

$$k_{fk} = A_k T^{\beta_k} \exp\left[\frac{-E_{ak}}{RT}\right] \prod_{i=1}^{N_s} \exp\left[\frac{\varepsilon_{ik} \Theta_i}{RT}\right] \tag{11}$$

Here, A_k is the pre-exponential factor, β_k the temperature exponent and E_{a_k} the activation energy. The rate for adsorption reactions were calculated using sticking coefficients $S_{0,i}$.

$$\dot{s}_i^{ads} = S_{0,i}\sqrt{\frac{RT}{2\pi M_i}}c_i \cdot \prod_{j=1}^{N_s} \Theta_j^{v'_j} \qquad (12)$$

3.3. Sensitivity Analysis

Sensitivity analysis is used to identify the rate determining steps of the mechanism and their key parameters. In its simplest form we consider a perfectly mixed reactor at a constant temperature with only surface reactions. Then the change of amount n_i of species i is given by

$$\frac{dn_i}{dt} = A_{cat}\dot{s}_i \qquad (13)$$

where A_{cat} is the catalytic surface area. We define a time dependent sensitivity coefficient $E_{i,k}(t)$ as the change of amount of species i with respect to a relative change of rate coefficient k_{fk}, i.e.,

$$E_{i,k}(t) = \frac{\partial n_i(t)}{\partial \ln k_{fk}} . \qquad (14)$$

Inserting equations Equation (10) and Equation (13), we can solve for the time development of the sensitivity coefficient

$$\frac{dE_{i,k}(t)}{dt} = A_{cat}v_{ik}k_{fk}\prod_{j=1}^{N_g+N_s}c_j^{v'_{jk}} + A_{cat}\sum_{l=1}^{K_s}v_{il}k_{fl}\prod_{j=1}^{N_g+N_s}c_j^{v'_{jl}}\left(\sum_{j=1}^{N_g+N_s}v'_{jl}\frac{E_{j,l}}{n_j}\right) \qquad (15)$$

Thus, the sensitivity coefficient describes the contribution of the kth reaction on the production of species i. Equation (15) can be integrated in time along with the solution of the conservation equations of each species. Since we are only interested in the relative contributions of all reactions on the products, we finally rescale the sensitivity coefficients for a given i such that the largest absolute value $|E_{i,k}|$ becomes unity.

3.4. Reaction Flow Analysis

Reaction flow analysis was used to identify the main pathway for the conversion of the reactants and the formation of the products. Here we look at individual contributions of a reaction k of the mechanism to the production or depletion of a chemical species i. The time-dependent production rate of species i by reaction k is

$$\frac{d\tilde{E}_{i,k}(t)}{dt} = \nu_{ik} k_{fk} \prod_{j=1}^{N_g+N_s} c_j^{\nu'_{jk}} \tag{16}$$

Only those $\tilde{E}_{i,k}$ for which species i is an immediate product of reaction k are of interest. Since all these $\tilde{E}_{i,k}$ are non-negative, they can be seen as weights in a directed graph that connects reactants and products along edges of elementary-step reactions. The $\tilde{E}_{i,k}$ are scaled such that the sum of the weights originating in a common root node becomes unity. Usually, only the most significant reactions are shown in the figures.

3.5. Thermodynamic Consistency

One of the requirements of a micro-kinetic model is that in the limit of infinite time the thermodynamic equilibrium is properly predicted. As a consequence, we need to guarantee that each elementary step is microkinetically reversible. Consider a pair of reversible reactions

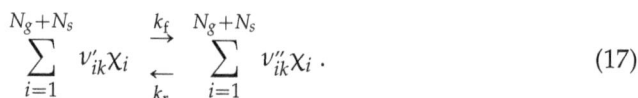

$$\sum_{i=1}^{N_g+N_s} \nu'_{ik} X_i \underset{k_r}{\overset{k_f}{\rightleftharpoons}} \sum_{i=1}^{N_g+N_s} \nu''_{ik} X_i \, . \tag{17}$$

For given temperature dependent rate coefficients $k_f(T)$ and $k_r(T)$, the equilibrium condition yields

$$\frac{k_f(T)}{k_r(T)} = K_c(T) \tag{18}$$

where $K_c(T)$ is the equilibrium constant with respect to concentrations. On the other hand, thermodynamic equilibrium is expressed by

$$K_p(T) = \exp\left(-\frac{\Delta_R G(T)}{RT}\right) \tag{19}$$

Here, $K_p(T)$ is the equilibrium constant with respect to pressures (or more precise: activities), $\Delta_R G(T)$ the change of Gibbs energy of the reaction, and R the universal gas constant. The two equilibrium constants can be converted by the factor

$$F_{c/p} = \frac{K_c(T)}{K_p(T)} = \prod_{i=1}^{N_g+N_s} (c_i^{\ominus})^{\nu_i} \tag{20}$$

where $c_i^{\ominus}$ denotes the reference concentration of species i at standard pressure, *i.e.*, $c_i^{\ominus} = \frac{p^{\ominus}}{RT}$ for ideal gas species and $c_i^{\ominus} = \frac{\Gamma}{\sigma_i}$ for surface species. Thus, we can link the reaction rates to the thermodynamic properties by

$$\frac{k_f(T)}{k_r(T)} = F_{c/p} \cdot \exp\left(-\frac{\Delta_R G(T)}{RT}\right) \tag{21}$$

The change of Gibbs energy of a reaction can be written in terms of the Gibbs energies of each species.

$$\Delta_R G(T) = \sum_{i=1}^{N_g+N_s} \nu_i G_i(T) = \sum_{i=1}^{N_g+N_s} \nu_i (H_i(T) - T \cdot S_i(T)) \tag{22}$$

A reaction mechanism is considered thermodynamically consistent if Equation (21) is fulfilled for all reactions. The straight forward conclusion would be to use this equation to calculate the rate of the reverse reactions $k_r(T)$ for every forward rate $k_f(T)$. However, in the development of surface reaction mechanisms, the thermodynamic properties of surface species are often unknown. Therefore the rate coefficients for forward and reverse reactions need to be given explicitly. Nevertheless, thermodynamic consistency is satisfied for a reaction mechanism if a set of $G_i(T)$ for all surface species can be found such that Equation (21) is simultaneously satisfied for all pairs of reversible reactions.

A detailed surface reaction mechanism usually contains more pairs of reversible reactions than unknown thermodynamic functions $G_i(T)$. Thus, there are more equations of type Equation (21) than there are degrees of freedom to satisfy these equations (the latter is in most cases the number of surface species minus one). So, $k_f(T)$ and $k_r(T)$ cannot be chosen independently for all reactions.

In a previous publication [46] we described a method to ensure thermodynamic consistency in such a case. There, no assumptions about the functions $G_i(T)$ were made, just their existence was required. However, it turned out that by doing so, the adjusted mechanisms may have required some species with negative heat capacity. Obviously such a mechanism cannot be claimed to be realistic in detail. Therefore, the algorithm has been improved by considering physically meaningful temperature dependencies of the functions $G_i(T)$. We will only limit to the case of constant heat capacities c_{pi} for surface species. Then the enthalpy must be a linear function of temperature

$$H_i(T) = H_{0i} + c_{pi} \cdot (T - T_0) \tag{23}$$

and the entropy contains a logarithmic dependency

$$S_i(T) = S_{0i} + c_{pi} \cdot \ln\left(\frac{T}{T_0}\right) \tag{24}$$

Thus, the Gibbs energy becomes

$$G_i(T) = H_{0i} + c_{pi} \cdot (T - T_0) - TS_{0i} + c_{pi}T \cdot \ln\left(\frac{T}{T_0}\right) \tag{25}$$

Taking the logarithm of Equation (21) and separating the known (*i.e.*, for gas-phase species) and the unknown (*i.e.*, for surface species) thermodynamic variables, we get

$$\begin{aligned}\ln k_f - \ln k_r &= \ln F_{c/p} - \sum_{i=1}^{N_g} v_i \frac{G_i(T)}{RT} \\ &\quad - \sum_{i=N_g+1}^{N_g+N_s} \frac{v_i}{R} \left[\frac{H_{0i} - c_{pi}T_0}{T} + c_{pi}(1 - \ln T_0) - S_{0i} + c_{pi}\ln T \right]\end{aligned} \tag{26}$$

The temperature dependent term of the rate coefficients is given according to Equation (11) by

$$k_{f,r}(T) = A_{f,r} \cdot T^{\beta_{f,r}} \cdot \exp\left(-\frac{E_{af,r}}{RT}\right) \tag{27}$$

Adsorption reactions can be treated in the same way by converting expressions based on sticking coefficients into the same functional form.

Obviously the terms $\ln k_f$, $\ln k_r$ and $\frac{G_i(T)}{RT}$ for the surface species with unknown thermodynamic properties belong to a set of functions

$$F = \left\{ y(T) = a + b\ln T + \frac{c}{T} \mid a, b, c \in \mathbb{R} \right\} \tag{28}$$

with constant coefficients a, b and c. The objective to make a reaction mechanism thermodynamically consistent is now to find functions $x_k(T) \in F$ and $y_i(T) \in F$ such that

$$(\ln k_f + x_f(T)) - (\ln k_r - x_r(T)) = \ln F_{c/p} - \sum_{i=1}^{N_g} v_i \frac{G_i(T)}{RT} - \sum_{i=N_g+1}^{N_g+N_s} v_i \cdot y_i(T) \tag{29}$$

is fulfilled for all reactions. Here $x_k(T)$ shall denote the necessary changes in the rate coefficients of a proposed reaction mechanism. With good agreement over a wide temperature range, we can approximate the given terms by a function $z(T) \in F$

$$z(T) = \ln F_{c/p} - \sum_{i=1}^{N_g} v_i \frac{G_i(T)}{RT} - \ln k_f + \ln k_r \tag{30}$$

96

Therefore, the problem is transformed to solve a system of equations for pairs of reversible reactions k:

$$x_{fk}(T) - x_{rk}(T) = z_k(T) - \sum_{i=N_g+1}^{N_g+N_s} v_{ik} \cdot y_i(T) \qquad (31)$$

It is not necessary to correct all reactions with terms $x_k(T)$. Some of these terms may be omitted, which means that the rate coefficients shall not be changed in the adjustment. The method proposed by Mhadeshvar and Vlachos [52], for instance, keeps a maximum number of reactions unchanged and adjusts exactly one reaction for each so-called thermodynamic cycle. In our iterative reaction mechanism development process, we fix the parameters of reactions that were identified to be most sensitive, leaving an underdetermined system of linear Equation (31).

For this underdetermined system of equations we are interested in the solution that minimizes the necessary changes to the effective rate constants in terms of an objective function

$$\Phi = \sum_k w_k \cdot ||x_k||_2 \rightarrow \min \qquad (32)$$

where w_k are selectable non-negative weights and

$$||x_k||_2 = \int_{T_1}^{T_2} (x_k(T))^2 \, dT \qquad (33)$$

We can use the method of Lagrange multipliers to minimize the objective function Φ. As result we get correction terms $x_k(T)$ and thermodynamic functions $y_i(T)$.

This algorithm can also easily be extended to coverage dependent rate expressions as in Equation (11) by formulation of the minimization problem for each species i with non-zero coverage dependent activation energies ε_{ik} by assuming coverage-dependent Gibbs Free Energies

$$G_i(T) = H_{0i} + c_{pi} \cdot (T - T_0) - T S_{0i} + c_{pi} T \cdot \ln\left(\frac{T}{T_0}\right) + \sum_j \varepsilon_{ij} \Theta_j \qquad (34)$$

Finally, the adjustment algorithm yields a surface reaction mechanism with minimum changes in the Arrhenius rate constants and coverage dependent activation energies. The resulting mechanism is thermodynamically consistent in the sense that thermodynamic functions with correct temperature dependency exist for all species. This procedure needs to be applied during mechanism development every time a rate coefficient has been changed manually.

97

4. Surface Reaction Mechanism

The conversion of methane to syngas includes different molecular paths considered as a combination of the following overall reactions (Table 2).

Table 2. Overall reactions (OR_x) in the methane reforming and oxidation system.

OR_x	Reaction	Reaction Enthalpy
	Methane Steam Reforming	
OR_1	$CH_4 + H_2O \leftrightarrow CO + 3H_2$	$\Delta H°_{298} = 205.9 \text{ kJ/mol}$
OR_2	$CH_4 + 2H_2O \leftrightarrow CO_2 + 4H_2$	$\Delta H°_{298} = 164.7 \text{ kJ/mol}$
	Methane dry reforming	
OR_3	$CH_4 + CO_2 \leftrightarrow 2CO + 2H_2$	$\Delta H°_{298} = 247.3 \text{ kJ/mol}$
	Methane partial oxidation	
OR_4	$CH_4 + 1/2O_2 \leftrightarrow CO + 2H_2$	$\Delta H°_{298} = -35.6 \text{ kJ/mol}$
	Methane total oxidation	
OR_5	$CH_4 + 2O_2 \leftrightarrow CO_2 + 2H_2O$	$\Delta H°_{298} = -880 \text{ kJ/mol}$
	Water-gas shift	
OR_6	$CO + H_2O \leftrightarrow CO_2 + H_2$	$\Delta H°_{298} = -41.2 \text{ kJ/mol}$
	Methanation	
OR_7	$CO + 3H_2 \leftrightarrow CH_4 + H_2O$	$\Delta H°_{298} = -206 \text{ kJ/mol}$
OR_8	$2CO + 2H_2 \leftrightarrow CH_4 + CO_2$	$\Delta H°_{298} = -247 \text{ kJ/mol}$
	Boudouard reaction	
OR_9	$2CO \leftrightarrow {}^*C + CO_2$	$\Delta H°_{298} = -172.4 \text{ kJ/mol}$
	Methane cracking	
OR_{10}	$CH_4 \leftrightarrow {}^*C + 2H_2$	$\Delta H°_{298} = 74.9 \text{ kJ/mol}$
	Gasification of carbon	
OR_{11}	${}^*C + H_2O \leftrightarrow CO + H_2$	$\Delta H°_{298} = 131.3 \text{ kJ/mol}$
OR_{12}	${}^*C + O_2 \leftrightarrow CO_2$	$\Delta H°_{298} = -393.5 \text{ kJ/mol}$

* C as graphite.

A previously established model for steam reforming on nickel [46] served as basis for the development of a kinetic scheme for the extended region of operating conditions covering all the ways from total oxidation to pyrolysis. The unity bond index-quadratic exponential potential (UBI-QEP) approach [53–55] was applied to determine heats of adsorption of adsorbed species, reaction enthalpy changes, and the original activation barriers for all relevant steps of the mechanism. Here, the reaction scheme is extended by adding new reaction paths involving carboxyl species as intermediate together with carbon formation paths.

The thermodynamically consistent surface reaction mechanism presented in Table 3 was developed using several sets of experiments for oxidation and reforming of methane. The predictive behavior of the overall reactor model was assessed by adaption of the kinetics through iterative comparisons of numerically predicted and experimentally determined species concentrations. The new model consists of 52 reactions with 6 gas-phase species and 14 surface species.

The kinetic model presented in Table 3 involves adsorption and desorption steps of all reactants and products, as well as surface reaction steps. Figure 2 describes the main pathways for methane reforming and oxidation processes on the nickel catalyst for syngas production. The availability of adsorbed atomic oxygen O(s), produced via dissociative adsorption of O_2, H_2O or CO_2, plays an important role to determine the reaction rate of these reactions. The importance of this common reaction intermediate is supported by TPR experimental results obtained by Qin et al. [56], in situ isotope transient experiments performed by Aparicio [10], and DFT studies carried out by Zu et al. [14]. Sticking coefficients were used to model adsorption reactions (H_2, CH_4, CO, CO_2, O_2, and H_2O). The initial values of the sticking coefficients were taken from the previous kinetic model [46].

The initial kinetic data are based on the surface science literature, which will be summarized below for individual steps.

4.1. H_2 on Ni Surface

Hydrogen desorption from nickel was studied by many authors, the activation energy was found to be between 90 and 97 kJ/mol for a single crystal surface. Bartholomew [57] and Weatherbee [58] reported heats of adsorption in a range of approximately 82–89 kJ/mol. Katzer et al. [59] calculate a H_2 desorption energy of 96 kJ/mol. Chen et al. [32] estimated the activation energy for H_2 desorption to be 97 kJ/mol applying the UBI-QEP method; this value is close to the parameter obtained by Zhu and White [60], using static secondary ion mass spectroscopy (SSIMS) on Ni(100) of 95 kJ/mol and the activation energy of 96 kJ/mol reported by Aparicio [10]. Bengaard et al. [61] determined the activation energies of desorption of H_2 from Ni(111), Ni(100), Ni(110) and Ni(445) to be 96, 90, 89, and 87 kJ/mol, respectively.

The activation energy for hydrogen desorption of 81.21 kJ/mol used in our previous model was estimated by the UBI-QEP method. However, this value was too low in comparison with the results reported by the previous mentioned references using different theoretical and experimental methods. In the new model, the activation energy for hydrogen desorption was modified from 81.21 kJ/mol to 95 kJ/mol, nevertheless the model predicts former and the currently presented experimental data. This modification shows an improvement in the prediction for hydrogen production via water-gas shift reaction (WGS), especially at high temperatures.

Table 3. Surface reaction mechanism for methane oxidation and reforming over Ni-based catalyst.

Rx	Reaction	$A/(\text{cm}^2/\text{mol}\cdot\text{s})/S_0$	β	$E_a/(\text{kJ/mol})$	ε_1 (kJ/mol)
R1	$H_2 + Ni(s) + Ni(s) \rightarrow H(s) + H(s)$	3.00×10^{-2}	0.000	5.0	
R2	$H(s) + H(s) \rightarrow Ni(s) + Ni(s) + H_2$	$2.54 \times 10^{+20}$	0.000	95.2	
R3	$O_2 + Ni(s) + Ni(s) \rightarrow O(s) + O(s)$	4.36×10^{-2}	-0.206	1.5	
R4	$O(s) + O(s) \rightarrow Ni(s) + Ni(s) + O_2$	$1.18 \times 10^{+21}$	0.823	468.9	
R5	$H_2O + Ni(s) \rightarrow H_2O(s)$	1.00×10^{-1}	0.000	0.0	
R6	$H_2O(s) \rightarrow H_2O + Ni(s)$	$3.73 \times 10^{+12}$	0.000	60.7	
R7	$CO_2 + Ni(s) \rightarrow CO_2(s)$	7.00×10^{-6}	0.000	0.0	
R8	$CO_2(s) \rightarrow CO_2 + Ni(s)$	$6.44 \times 10^{+7}$	0.000	25.9	
R9	$CO + Ni(s) \rightarrow CO(s)$	5.00×10^{-1}	0.000	0.0	
R10	$CO(s) \rightarrow CO + Ni(s)$	$3.56 \times 10^{+11}$	0.000	111.2	$-50.0\ \theta_{CO(s)}$
R11	$CH_4 + Ni(s) \rightarrow CH_4(s)$	8.00×10^{-3}	0.000	0.0	
R12	$CH_4(s) \rightarrow CH_4 + Ni(s)$	$8.70 \times 10^{+15}$	0.000	37.5	
R13	$CH_4(s) + Ni(s) \rightarrow CH_3(s) + H(s)$	$1.54 \times 10^{+21}$	0.087	55.8	
R14	$CH_3(s) + H(s) \rightarrow CH_4(s) + Ni(s)$	$1.44 \times 10^{+22}$	-0.087	63.4	
R15	$CH_3(s) + Ni(s) \rightarrow CH_2(s) + H(s)$	$1.54 \times 10^{+24}$	0.087	98.1	
R16	$CH_2(s) + H(s) \rightarrow CH_3(s) + Ni(s)$	$3.09 \times 10^{+23}$	-0.087	57.2	
R17	$CH_2(s) + Ni(s) \rightarrow CH(s) + H(s)$	$3.70 \times 10^{+24}$	0.087	95.2	
R18	$CH(s) + H(s) \rightarrow CH_2(s) + Ni(s)$	$9.77 \times 10^{+24}$	-0.087	81.0	
R19	$CH(s) + Ni(s) \rightarrow C(s) + H(s)$	$9.88 \times 10^{+20}$	0.500	21.9	
R20	$C(s) + H(s) \rightarrow CH(s) + Ni(s)$	$1.70 \times 10^{+24}$	-0.500	157.9	
R21	$CH_4(s) + O(s) \rightarrow CH_3(s) + OH(s)$	$5.62 \times 10^{+24}$	-0.101	92.7	
R22	$CH_3(s) + OH(s) \rightarrow CH_4(s) + O(s)$	$2.98 \times 10^{+22}$	0.101	25.8	
R23	$CH_3(s) + O(s) \rightarrow CH_2(s) + OH(s)$	$1.22 \times 10^{+25}$	-0.101	134.6	
R24	$CH_2(s) + OH(s) \rightarrow CH_3(s) + O(s)$	$1.39 \times 10^{+21}$	0.101	19.0	
R25	$CH_2(s) + O(s) \rightarrow CH(s) + OH(s)$	$1.22 \times 10^{+25}$	-0.101	131.3	
R26	$CH(s) + OH(s) \rightarrow CH_2(s) + O(s)$	$4.40 \times 10^{+22}$	0.101	42.4	
R27	$CH(s) + O(s) \rightarrow C(s) + OH(s)$	$2.47 \times 10^{+21}$	0.312	57.7	
R28	$C(s) + OH(s) \rightarrow CH(s) + O(s)$	$2.43 \times 10^{+21}$	-0.312	118.9	
R29	$H_2O(s) + Ni(s) \rightarrow H(s) + OH(s)$	$3.67 \times 10^{+21}$	-0.086	92.9	
R30	$H(s) + OH(s) \rightarrow H_2O(s) + Ni(s)$	$1.85 \times 10^{+20}$	0.086	41.5	
R31	$H(s) + O(s) \rightarrow OH(s) + Ni(s)$	$3.95 \times 10^{+23}$	-0.188	104.3	
R32	$OH(s) + Ni(s) \rightarrow H(s) + O(s)$	$2.25 \times 10^{+20}$	0.188	29.6	
R33	$OH(s) + OH(s) \rightarrow H_2O(s) + O(s)$	$2.34 \times 10^{+20}$	0.274	92.3	
R34	$H_2O(s) + O(s) \rightarrow OH(s) + OH(s)$	$8.14 \times 10^{+24}$	-0.274	218.4	
R35	$C(s) + O(s) \rightarrow CO(s) + Ni(s)$	$3.40 \times 10^{+23}$	0.000	148.1	
R36	$CO(s) + Ni(s) \rightarrow C(s) + O(s)$	$1.75 \times 10^{+13}$	0.000	116.2	
R37	$CO(s) + H(s) \rightarrow C(s) + OH(s)$	$3.52 \times 10^{+18}$	-0.188	105.4	
R38	$C(s) + OH(s) \rightarrow H(s) + CO(s)$	$3.88 \times 10^{+25}$	0.188	62.5	$-50.0\ \theta_{CO(s)}$
R39	$CO(s) + CO(s) \rightarrow C(s) + CO_2(s)$	$1.62 \times 10^{+14}$	0.500	241.7	$-50.0\ \theta_{CO(s)}$
R40	$CO_2(s) + C(s) \rightarrow CO(s) + CO(s)$	$7.29 \times 10^{+28}$	-0.500	239.2	$-100.0\ \theta_{CO(s)}$
R41	$CO(s) + O(s) \rightarrow CO_2(s) + Ni(s)$	$2.00 \times 10^{+19}$	0.000	123.6	
R42	$CO_2(s) + Ni(s) \rightarrow CO(s) + O(s)$	$4.64 \times 10^{+23}$	-1.000	89.3	$-50.0\ \theta_{CO(s)}$
R43	$CO(s) + OH(s) \rightarrow COOH(s) + Ni(s)$	$6.00 \times 10^{+21}$	0.213	97.6	
R44	$COOH(s) + Ni(s) \rightarrow CO(s) + OH(s)$	$1.46 \times 10^{+24}$	-0.213	54.3	
R45	$CO_2(s) + H(s) \rightarrow COOH(s) + Ni(s)$	$6.25 \times 10^{+24}$	-0.475	117.2	
R46	$COOH(s) + Ni(s) \rightarrow CO_2(s) + H(s)$	$3.73 \times 10^{+20}$	0.475	33.6	
R47	$CO(s) + H(s) \rightarrow HCO(s) + Ni(s)$	$4.00 \times 10^{+20}$	-1.000	132.2	$-50.0\ \theta_{CO(s)}$
R48	$HCO(s) + Ni(s) \rightarrow CO(s) + H(s)$	$3.71 \times 10^{+21}$	0.000	0.0	
R49	$HCO(s) + Ni(s) \rightarrow CH(s) + O(s)$	$3.79 \times 10^{+14}$	0.000	81.9	$+50.0\theta_{CO(s)}$
R50	$CH(s) + O(s) \rightarrow HCO(s) + Ni(s)$	$4.59 \times 10^{+20}$	0.000	109.9	
R51	$H(s) + COOH(s) \rightarrow HCO(s) + OH(s)$	$6.00 \times 10^{+22}$	-1.163	104.8	
R52	$HCO(s) + OH(s) \rightarrow COOH(s) + H(s)$	$2.28 \times 10^{+20}$	0.263	15.9	

The rate coefficients are given in the form of $k = AT^\beta \exp(-E_a/RT)$; adsorption kinetics is given in form of sticking coefficients; the surface site density is $\Gamma = 2.66 \times 10^{-9}\ \text{mol}\cdot\text{cm}^{-2}$ [12,46]. Rx represents the reaction number.

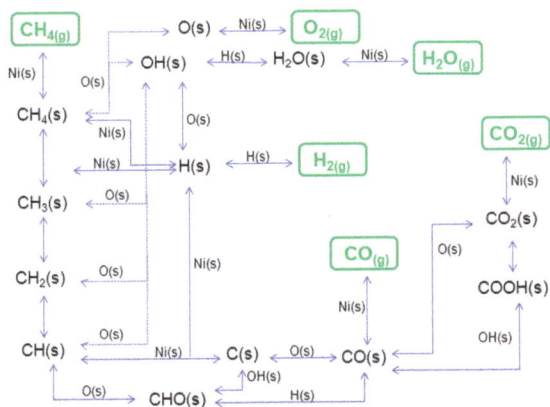

Figure 2. Reaction pathways for methane oxidation and reforming.

4.2. O_2 on Ni Surface

The interaction of oxygen with nickel was subject of a large number of surface science studies using a wide range of techniques. Stuckless *et al.* [62] listed the initial heats of adsorption of oxygen for the three low index crystal planes of Ni. The activation energies for desorption are 520 kJ/mol, 470 kJ/mol, and 485 kJ/mol for the Ni(100), Ni(111) and Ni(110) planes, respectively. These values are similar to the *ab-initio* theoretical results obtained by Siegbahn and Wahlgren [63]; in their work 540 kJ/mol was estimated for Ni(100) and 480 kJ/mol for Ni(111). In our model, the activation energy for oxygen desorption R4 of 468.9 kJ/mol was estimated by UBI-QEP method; being close to the activation energy for Ni(111) proposed by Stuckless *et al.* [62].

4.3. H_2O on Ni Surface

The desorption of H_2O has been studied by Stulen *et al.* [64] at a low temperature on clean Ni(111) using thermal desorption spectrometry (TDS) and electron-simulated desorption (ESD). They reported an activation energy for H_2O desorption of 41 kJ/mol. Pache *et al.* [65] reported an activation energy for H_2O desorption of 57 ± 5 kJ/mol measured for a clean Ni(111) surface. Chen *et al.* [31] considered desorption energies for H_2O of 64.4 kJ/mol in their macro-kinetic model. Aparicio [10] performed detailed micro-kinetic studies of the hydrogen exchange reaction of H_2O with D_2 on Ni/MgAl$_2$O$_3$ and reports an activation energy of 64.4 kJ/mol for H_2O desorption. Zakharov *et al.* [66] performed theoretical cluster calculations on Ni(111); leading to activation energies between 51.4 kJ/mol and 67.1 kJ/mol. The value of 60.8 kJ/mol used in the present work for H_2O desorption is comparable with the experimental and theoretical values reported in the literature.

4.4. CO on Ni Surface

Estimated values for CO desorption found in theoretical studies using DFT calculations are in the range of 120 ± 10 kJ/mol [32,36,67,68]. Using temperature-programmed desorption (TPD) Bjørgum *et al.* [69] derived a CO desorption energy of 119 kJ/mol for low CO surface coverages. Al-Sarraf *et al.* [70] estimated an initial heat of CO adsorption of 122 ± 4 kJ/mol for Ni(100) using a single crystal micro-calorimeter. In our work, the heat of CO adsorption of 111.27 kJ/mol was estimated from the UBI-QEP method and is close to the value of 115 kJ/mol as reported by Aparicio [10]. Coverage dependency for CO is included into the kinetic model to describe the lateral interaction of adsorbed species. The values were estimated based on the comparison of model predictions and experimental results.

4.5. CH_4 on Ni Surface

The reference values for the activation energies of methane dehydrogenation (R11–R19), oxygen assisted methane dehydrogenation (R21–R27), water activation (R29–R34), and CO dissociation (R35–R36) are taken from our former kinetic study for steam reforming on nickel [46]; such values were slightly modified due to the computational algorithm that is applied to guarantee thermodynamic consistency [71].

4.6. C on the Ni Surface

The surface reaction mechanism includes reaction paths to describe the formation of up to one monolyer of carbon from adsorbed CO species. The first path is through CO(s) dissociation (R35–R36); the reference values for activation barriers of this reaction are taken from our former mechanism [46]. In the current mechanism, we introduce new reaction paths for carbon formation (R37–R40). The reaction steps R37–R38 were estimated by the UBI-QEP method. The Boudouard reaction (R39–R40) is also included in the model; the activation energies of these reactions are based on the kinetic data reported by Blaylock *et al.* [37] from DFT calculations. However, such data were modified within the enthalphic consistency.

4.7. COOH on Ni Surface

Two reaction paths are proposed for CO oxidation. The first path is the direct oxidation of CO(s) with O(s) (R41–R42). The second one takes place through the surface intermediate carboxyl (COOH) formed from CO(s) and OH(s). It is assumed that COOH(s) binds to the surface by its carbon atom. Formate (HCOO) as additional reactive surface intermediate has been discussed in the literature [72–74]; it is bonded to the surface through its oxygen atom. Carboxyl and formate intermediates are isomers and both have been detected experimentally over different transition metals [73,75–79]. Tibiletti *et al.* [78] identified formate, carbonate and carboxyl species at the surface of a Pt/CeO_2 catalyst during the forward water-gas shift (WGS) and the reverse reaction (RWGS). Shido and Iwasawa [79] also studied the formation of formate species on Rh/CeO_2 during WGS. Despite all studies performed, there is still no overall agreement on the nature of the rate-determining intermediate, as to whether it is a carboxyl or a formate species. Theoretical calculations favor the formation of carboxyl species, while the surface formate is considered as a spectator species [80,81]. Lin *et al.* [82] performed mechanistic studies using density functional theory (DFT) to clarify the mechanism of WGS reaction on transition metals. The authors proposed that the WGS reaction involves three mechanisms: (i) redox; (ii) carboxyl; and (iii) formate. Lin *et al.* [82] showed that the formate path is energetically unfavorable in comparison to the redox and carboxyl mechanisms. The carboxyl mechanism is supported by Boisen *et al.* [83], who assumed that the carboxyl species plays an important role especially on supports containing CeO_2, regardless of the metal type, that the extraction of the first hydrogen from water is a slow step and that the subsequent reaction of CO and OH results in a carboxyl (COOH) intermediate which decomposes into CO_2 and H_2. Grabow *et al.* [84] presented a micro-kinetic model as well as experimental data for the low-temperature water-gas shift (WGS) reaction catalyzed by Pt at temperatures ranging from 523 K to 573 K for various gas compositions at a pressure of 1 atm. The authors concluded that the most significant reaction channel proceeds via the carboxyl (COOH) intermediate. Gokhale *et al.* [80] used self-consistent density functional theory (DFT-GGA) calculations to investigate the WGS mechanism on Cu(111) and identified carboxyl, as central intermediate. In a recent study, Karakaya *et al.* [85] developed a kinetic model for the WGS reaction over Rh. In this model, the main path for CO_2 formation is concluded to be the direct oxidation of CO with O species at high temperatures, whereas the formation of the carboxyl group is significant at low temperatures.

During the development of our kinetic model, sensitivity and reaction flow analysis were performed to determine the significance of both reactions paths. The analysis showed that the carboxyl path was more sensitive especially for conditions were water was added as co-feed. However, the results did not show a significant participation of the formate intermediate in any of the cases studied. Therefore, we do not further consider the formate path in the reaction mechanism. The DFT calculation data presented by Blaylock *et al.*[36,37] for steam reforming on nickel are used as a reference for the reaction paths and enthalpy values in (R43–R46) (Table 3).

4.8. HCO on Ni surface

It has been proposed by several authors that formyl species (HCO) are coordinated through carbon and acts as intermediate during the reforming and oxidation of methane over transition metals [31,32,37,38,86,87].

Pistonesi *et al.* [88] have studied HCO(s) on Ni(111) and proposed based on DFT calculations that during methane steam reforming on nickel, dissociation of HCO(s) to CO(s) and H(s) is favored. Zhou *et al.* [14] concluded from their DFT studies that HCO is a key intermediate during dry reforming of methane to CO and H_2. Blaylock *et al.* [36] studied the kinetics of steam reforming over Ni(111). Based on DFT calculations they obtain an activation energy of 150 kJ/mol for the $CH(s) + O(s) \rightarrow HCO(s)$ step. The authors also conclude that the formation of HCO(s) is an important step during steam reforming of methane. Chen *et al.* [32] also includes the HCO(s) surface species as intermediate in their micro-kinetic model for methane reforming.

The implementation of the HCO(s) in the kinetic model presented in this work is supported experimentally by TPRS and TR-FTIR experiments [89] and theoretically by the UBI-QEP method [54,90] as well as DFT studies [36–38]. Our previous mechanism [46] is used as a reference for the kinetic data for the reaction steps R47 and R50, where HCO(s) is produced from CO(s) and from CH(s). In addition to these reactions, the formation of HCO(s) through carboxyl intermediate (R51–R52) is also included. The kinetic data presented by Chen *et al.* [32] are used as a reference for R51–R52 in our model.

5. Results and Discussion

5.1. Methane Catalytic Partial Oxidation (CPOX)

Catalytic partial oxidation of methane has been extensively studied [5,15–20,39–44,91–94]. Nevertheless, the mechanism of syngas production is still controversial. Two main paths have been suggested for partial oxidation of methane. On the one hand, the direct oxidation mechanisms where H_2 is directly originated from methane decomposition. Further interaction of adsorbed hydrocarbon species CH_x (x = 0, 1, 2, 3) with adsorbed atomic oxygen produces carbon monoxide [5,42,91,94,95]. On the other hand, the indirect route where methane is totally oxidized to CO_2 and H_2O, as long as oxygen is present close to the catalyst surface, and then the remaining CH_4 is reformed with steam or CO_2 to H_2 and CO [15,39,41,43,96–98].

Figure 3 shows the experimentally measured species concentrations in comparison with the numerical predictions as a function of the temperature in our fixed bed reactor. Methane conversion starts at 723 K leading to CO_2 and H_2O formation only. No significant amounts of H_2 or CO were detected up to a temperature of 880 K. At temperatures above 880 K, H_2 and CO formation increases while, CO_2 and H_2O concentrations decrease leading almost to the equilibrium composition at given operating conditions (Figure 3b).

CO_2 and H_2O concentrations are above the ones obtained at equilibrium in the medium temperature range. At equilibrium more H_2 and CO is predicted. This behavior is an indication for the indirect way to H_2 and CO via formation of H_2O and CO_2.

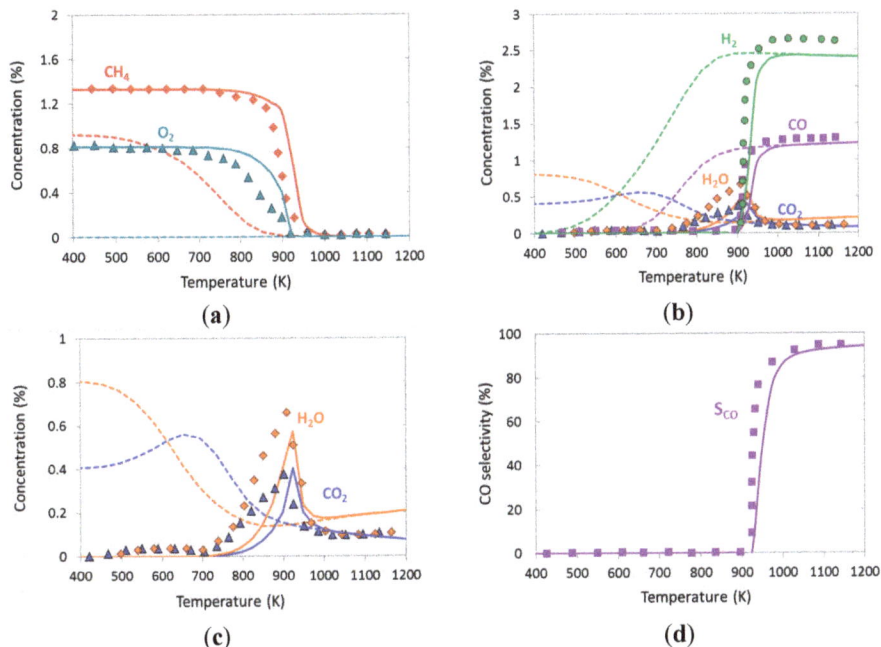

Figure 3. Comparison of experimentally determined (symbols) and numerically predicted (lines) concentrations as a function of temperature for catalytic partial oxidation of methane in a fixed bed reactor: (**a**) reactants; (**b**) products; (**c**) zoom-in of CO_2 and H_2O formation; (**d**) selectivity of CO; inlet gas composition of CH_4/O_2 = 1.6 in N_2; 1 bar; T_{inlet} = 373 K; total flow rate of 4 slpm; dashed lines = equilibrium composition at given temperature.

Numerically predicted concentrations of O_2 and CH_4 (Figure 3a) as well as H_2 and CO (Figure 3b) concentrations agree well with the experimentally derived data in the whole temperature range under investigation. The selectivity of CO is presented in Figure 3d.

The computed concentrations of the gas-phase species along the catalytic bed at 973 K after ignition (Figure 4) also support the indirect route of H_2 and CO formation. At the beginning of the catalytic bed CO_2 and H_2O are first formed by total oxidation of methane, whereas H_2 and CO are produced through steam and some dry reforming of methane further downstream when oxygen is completely consumed. The very distinct transition for total oxidation and reforming at an axial position of 14 mm is also caused by the fact that a one-dimensional packed-bed model is used, *i.e.*, mass transfer limitations are neglected.

There is a competitive adsorption between CH_4 and O_2 species on metallic nickel sites during partial oxidation of methane. However, O_2 adsorption is stronger than that of CH_4 [21]. Before ignition, the surface is mainly covered by oxygen; the system is controlled by surface reaction kinetics. Figure 5 shows the computed surface coverage at 973 K after ignition. Oxygen coverage is dominating the entrance region of the catalytic bed where oxygen is available in the gas flow. At an axial position of 14 mm, the oxygen coverage decreases fast, producing free nickel sites together with carbon monoxide and adsorbed hydrogen. The species OH(s) and $H_2O(s)$ have a maximum concentration on the surface at the transition point (Figure 5b). Further downstream some carbon C(s) increases on the surface.

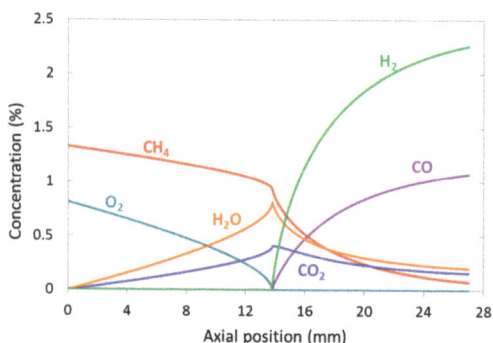

Figure 4. Computed concentration of the gas-phase species: O_2, CH_4, H_2, H_2O, CO and CO_2 along the catalytic bed after the ignition for partial oxidation of methane over nickel at 973 K, $CH_4/O_2 = 1.6$ and 1 bar.

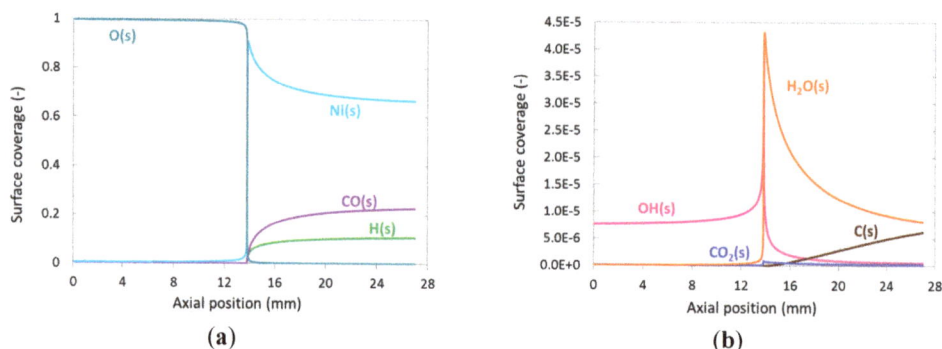

Figure 5. Computed surface coverage of adsorbed species: (a) O(s), CO(s), H(s), Ni; (b) OH(s), $H_2O(s)$, $CO_2(s)$, and C(s) along the catalytic bed at 973 K, 4 slpm $CH_4/O_2 = 1.6$ and 1 bar.

106

Sensitivity analysis of the reaction mechanism was carried out at two different temperatures: 723 K (before ignition) and 973 K (after ignition) with $CH_4/O_2 = 1.6$ in nitrogen dilution. The sensitivities of the gas-phase concentrations of CH_4, CO_2, H_2O, H_2 and CO at the outlet of the bed were analyzed by perturbing the pre-exponentials of each reaction. Results for the CO mole fraction, presented in Figure 6, show that for both temperatures the system is highly sensitive to methane adsorption/desorption (R11–R12) along with $CH_4(s)$ dissociation with oxygen assistance (R21) and CO_2 desorption (R8). At 723K, where the total oxidation takes place, the reactions to produce OH(s) and $H_2O(s)$ species (R30–R32) observed to become sensitive due to the availability surface oxygen compared to high temperature region, where oxygen is consumed. The reactions of CO_2 dissociation (R42) and CO desorption (R10) are the most sensitive and are rate determining steps for the production of CO at the high temperature.

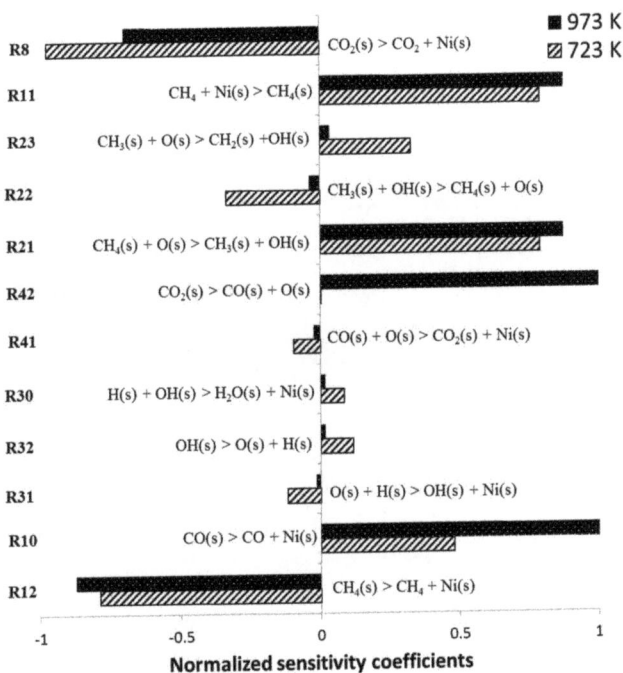

Figure 6. Sensitivity analysis of CO gas-phase concentration for Catalytic Partial Oxidation (CPOX) reaction at 723 K and 948 K.

5.2. Methane Steam Reforming (SR)

A kinetic study of steam reforming of methane over Ni/Al_2O_3 coated monoliths was presented in our previous work including the establishment of a multi-step reaction mechanism [46]. This SR mechanism was successfully applied in studies of steam reforming of methane over, Ni/YSZ anodes of solid oxide fuels cells [99–104], and experimental results from literature [105]. Despite the introduction of new reaction paths, together with adjustments of the kinetic parameters for partial oxidation and dry reforming of methane, the current mechanism extension is still able to predict the experimental results for methane steam reforming. For the evaluation of the newly developed kinetic model, additional experiments were conducted in a fixed bed reactor.

Experimentally determined and computed species concentrations are shown as function of the temperature in a fixed bed reactor in Figure 7. A small amount of CO_2 as a product was observed at temperatures between 623 and 973 K due to the availability of OH(s) species on the surface, originating from the dissociation of water (Figure 7c).

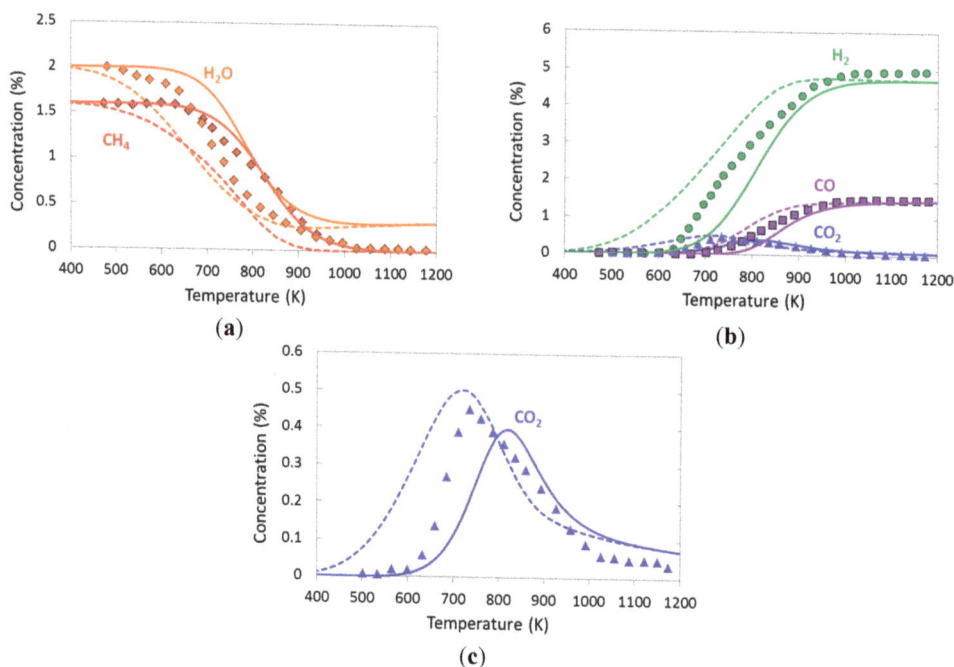

Figure 7. Comparison of experimentally determined (symbols) and numerically predicted (lines) concentrations as a function of temperature for catalytic steam reforming of methane: (**a**) reactants; (**b**) products; (**c**) zoom-in of CO_2 formation; C/S = 0.8 in N_2, 1 bar; T_{inlet} = 373 K; total flow rate of 4 slpm, dashed lines = equilibrium composition at given temperature.

For the sake of clarity we investigate on the validity of the chosen one-dimensional reactor model and the computed reactor outlet concentrations, by roughly estimating the influence of the radial temperature gradient based on the

Mears criterion [106] for radial isothermicity. To estimate this effect it is necessary to calculate the radial heat-transfer over the bed cross section. We apply this criterion for the steam reforming inlet gas mixture, which imposes the most critical condition of all experiments in this study, due to the high endothermic value of the heat of reaction amongst the processes considered herein. In reality these harsh conditions are diminished by the coupled water-gas shift reaction which is exothermic in nature.

For a reactor feed temperature of 800 K we estimate a radial temperature difference of about 28.5 K. It can be concluded that the simplification of the used one-dimensional packed-bed model assuming isothermal conditions over the bed cross section introduces an error in the simulated species concentrations. This error gives an explanation for the deviation of simulated and measured concentrations as shown in Figure 7. However, since the kinetics of the SR mechanism has been tested separately against experimental data from a monolithic reactor configuration under steam reforming conditions [46], it remains reasonably and simulated concentrations are expected to match experimentally measured ones closer, if instead of a one-dimensional model a two-dimensional packed-bed model is used.

In the present study, reaction flow analysis is performed at 823 K were the maximum CO_2 formation is observed. It can be seen that CO_2 production is preferred through the direct oxidation of CO(s) with O(s) in comparison with the carboxyl path (Figure 8).

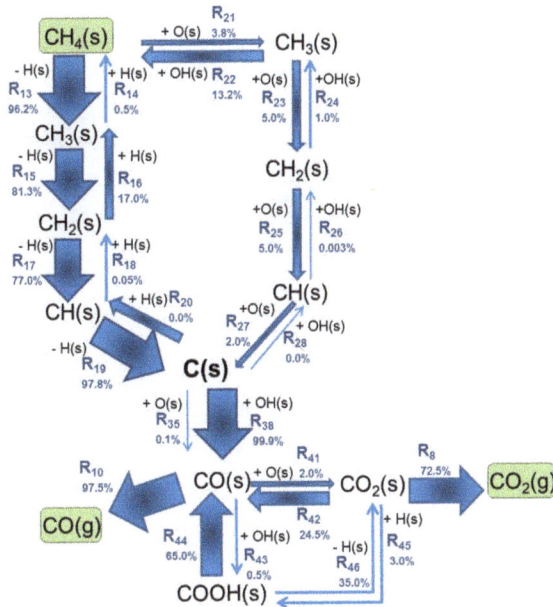

Figure 8. Reaction flow analysis for steam reforming of methane on nickel at 823 K, 1 bar, and a ratio of $CH_4/H_2O = 0.8$.

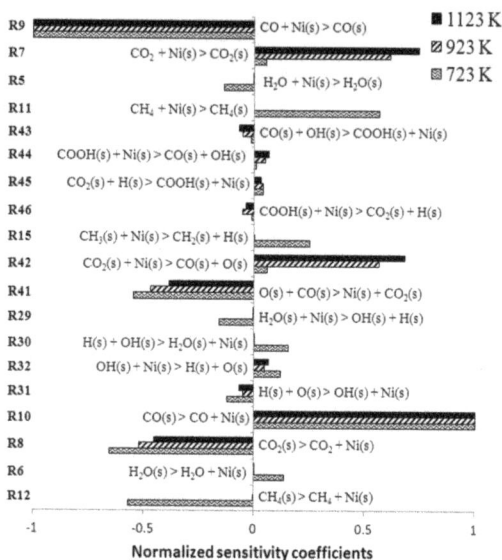

Figure 9. Sensitivity analysis of CO gas-phase concentration for steam reforming methane at different temperatures for $CH_4/H_2O = 0.8$ and 1 bar.

Figure 9 shows the normalized sensitivity coefficients on the CO yield at 723 K, 923 K, and 1123 K for an inlet methane to steam feed ratio (CH_4/H_2O) of 0.8 and at 1 bar. It can be observed that the CO adsorption (R9) and desorption (R10) are highly sensitive as expected for all the reported temperatures. Additionally, at low temperatures (723 K), the CO yield is sensitive to both H_2O adsorption (R5) and desorption (R6) as well as elementary reaction steps of the WGS reaction (R29) and (R30). Besides, adsorption (R11), desorption (R12) and dehydrogenation of methane (R15) play a sensitive role in CO formation. CO_2 dissociation (R42) becomes a sensitive step at high temperatures (823 and 923 K).

5.3. Methane Dry Reforming (DR)

Methane reforming with CO_2 was also studied in the fixed bed reactor. The experimentally measured and computed concentrations of the gas-phase species (CH_4, CO_2, CO, H_2, and H_2O) at the reactor outlet are close to equilibrium for the entire temperature range (Figure 10).

Figure 11 shows the most sensitive reactions for methane conversion by dry reforming, steam reforming and partial oxidation with respect to the rate constant at 1073 K. It can be seen that methane adsorption (R11) and desorption (R12) reactions are the most sensitive reactions for DR and CPOX, while oxygen-assisted dehydrogenation of CH_4 (R21) becomes the most sensitive step for steam reforming. The results indicate that pyrolytic methane dehydrogenation (R13) is sensitive only for dry reforming. However, oxygen-assisted dehydrogenation of methane (R21) is a sensitive step for all processes at 1073 K, in particular for partial oxidation.

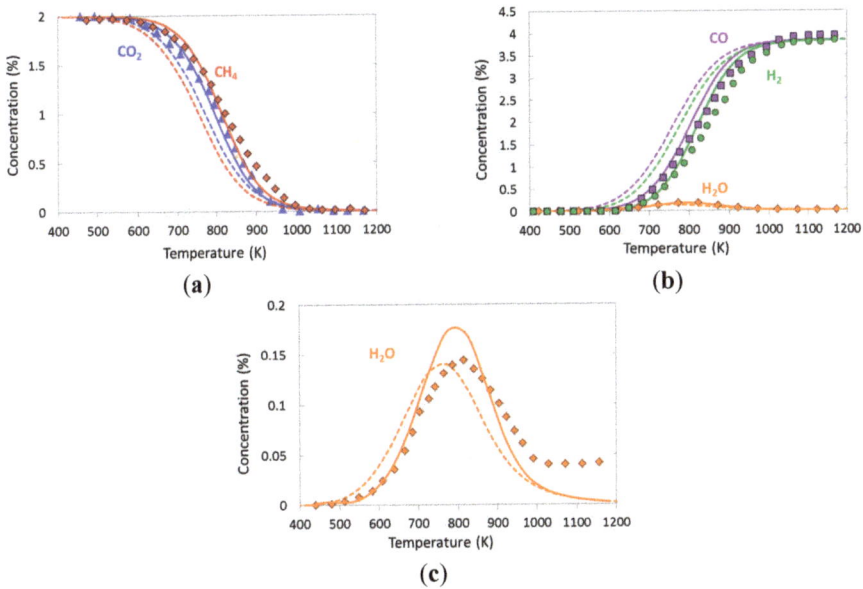

(a)

(b)

(c)

Figure 10. Comparison of experimentally determined (symbols) and numerically predicted (lines) concentrations as a function of temperature for catalytic reforming of methane with CO_2: (**a**) reactants; (**b**) products; (**c**) zoom-in H_2O formation; for CH_4/CO_2 = 1 in N_2; 1 bar; T_{inlet} = 373 K; total flow rate of 4 slpm, dashed lines = equilibrium composition at given temperature.

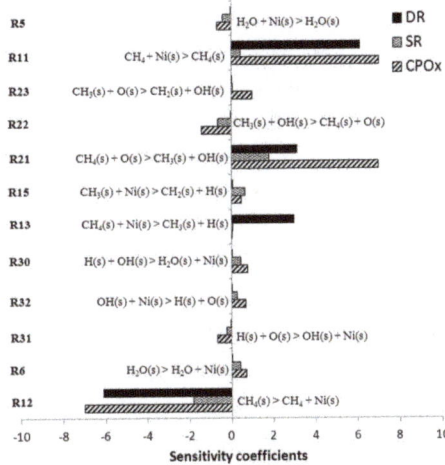

Figure 11. Sensitivity coefficients for CH_4 at 1073 K at DR, SR and CPOX conditions.

5.4. Influence of H_2 and H_2O on Methane Reforming with CO_2

As mentioned above, one of the main problems in many industrial reactors that involve methane as fuel is coke deposition on catalysts and the reactor pipe walls. The formation of coke can lead to lower catalytic activity and even complete catalyst deactivation, depending on the amount of solid carbon deposited on the surface [22–26]. Hydrogen and water have recently been studied as inhibitors of coke formation in dry reforming of methane at higher pressure and temperatures [26]. The influence of hydrogen and water in DR of methane was studied in the fixed bed reactor at atmospheric pressure, 4 slpm flow rate, residence time of 0.013 s, temperatures between 373 and 1173 K and inlet mixture of 1.6% CH_4, 2.1% CO_2 and 1.8% H_2 in N_2 dilution. Hydrogen addition led to an increase of water at a lower temperature compared to dry reforming (Figure 12). This water was produced through the RWGS reaching a maximum at 673 K (Figure 12b). As the temperature increases, the water was consumed together with unconverted methane due to the steam reforming reaction.

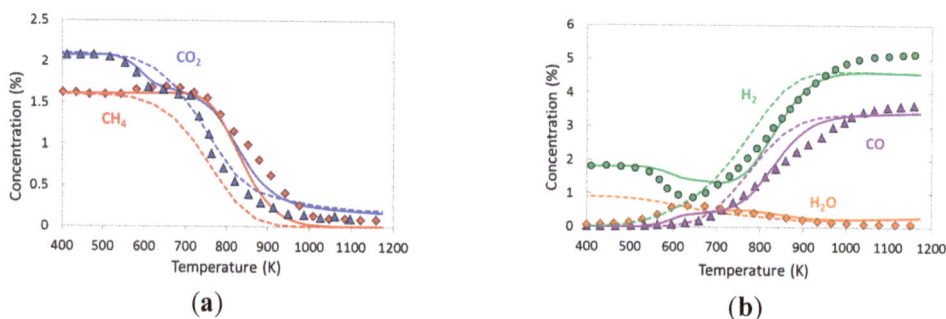

(a)   (b)

Figure 12. Comparison of experimentally determined (symbols) and numerically predicted (lines) concentrations as a function of temperature for catalytic dry reforming of methane with co-feed H_2: (a) CH_4 and CO_2; (b) H_2O, CO, H_2, inlet gas composition of 1.6 vol.% CH_4, 2.1 vol.% CO_2, 1.8 vol.% H_2 in N_2; 1 bar; T_{inlet} = 373 K; total flow rate of 4 slpm; dashed lines = equilibrium composition at given temperature.

The computed surface coverage reveals a high coverage with hydrogen and CO at low and medium temperatures respectively (Figure 13). The maximum formation of carbon C(s) is a temperatures between 373 and 573 K, mainly produced by the reaction between CO(s) and H(s) (R37) at low temperature. At higher temperature the total coverage with adsorbed species is rather low.

In Figure 14, the experimental and numerical results are given for dry reforming with steam as co-feed. At temperatures between 723 and 823 K, some additional CO_2 was produced, by the WGS reaction.

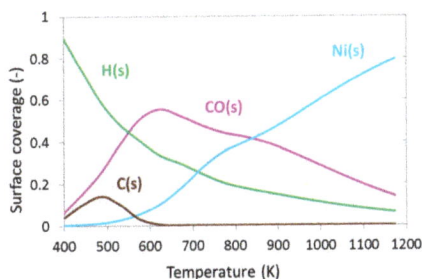

Figure 13. Computed surface coverage of adsorbed species as function of the temperature for methane dry reforming with H_2 co-feed: inlet gas composition of 1.6 vol.% CH_4, 2.1 vol.% CO_2, 1.8 vol.% H_2 in N_2; 1 bar; total flow rate 4 slpm.

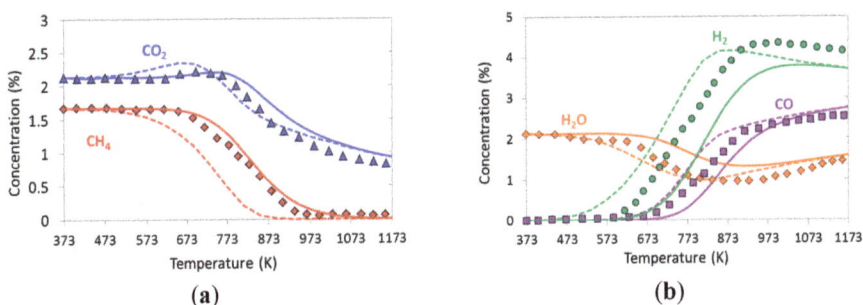

(a) (b)

Figure 14. Comparison of experimentally determined (symbols) and numerically predicted (lines) concentrations as a function of temperature for catalytic dry reforming of methane with co-feed H_2O: (a) CH_4, and CO_2; (b) H_2O, CO, H_2, inlet gas composition of 1.7 vol.% CH_4, 2.1 vol.% CO_2, 2.1 vol.% H_2O in N_2; 1 bar; $T_{inlet} = 373$ K, total flow rate of 4 slpm, dashed lines = equilibrium composition at given temperature.

In the intermediate temperature range, the outlet stream has not reached the equilibrium composition yet. At temperatures below 523 K, the surface was mainly covered by oxygen coming from $H_2O(s)$ dissociation in the simulation (Figure 15). As the temperature increases the oxygen coverage decreases rapidly leading to vacant nickel sites and H(s) and CO(s) become the most abundant surface intermediates. No significant concentration of surface carbon was observed at low temperatures in comparison with the previous results using H_2 as co-feed (Figure 15).

During the experiments performed in this study, no formation of visible carbon deposits on the surface was observed at all operations conditions. Nevertheless, a comparison of the computed carbon coverage at the three different conditions studied for DR reveals the inhibition of C(s) formation by H_2 and H_2O (Figure 16).

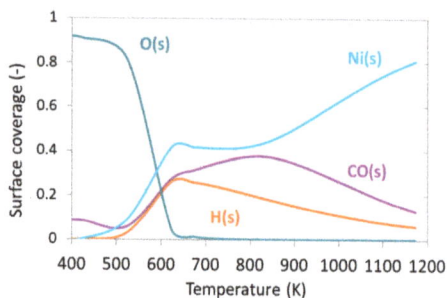

Figure 15. Computed surface coverage of adsorbed species as function of the temperature for methane dry reforming with H_2O co-feed: 1.7 vol.% CH_4, 2.1 vol.% CO_2, 2.1 vol.% H_2O in N_2; 1 bar; T_{inlet} = 373 K; total flow rate of 4 slpm.

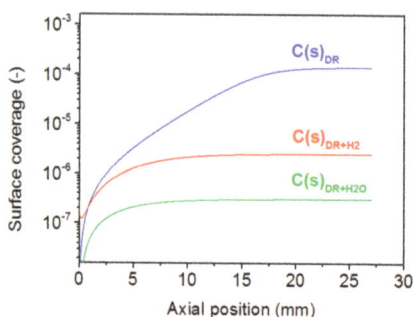

Figure 16. Computed surface coverage for carbon along the catalytic bed for methane reforming with CO_2 at 1123 K: (**a**) comparison of methane dry reforming without and with H_2, and H_2O co-feed respectively.

Figure 17 describes the most sensitive reactions for carbon formation during dry reforming in presence of H_2 and H_2O at 1123 K. Addition of both H_2O and H_2 make the reactions (R37) and (R38) highly sensitive. It can be seen that adsorption/desorption of CO_2 (R7, R8) are very sensitive reactions in the presence of H_2O as well as CO_2 dissociation (R42) and CO_2 formation thought the direct oxidation of CO(s) with O(s) (R41). Besides, direct oxidation of CO(s) (R41), CO(s) is also oxidized with OH(s) (R43) to produce CO_2(s) involving the COOH(s) path (R46). The results also indicate that in the presence of H_2 as co-feed for the DR, water adsorption (R5) and desorption (R6) become sensitive reactions as well as H_2O(s) formation (R30) and dissociation (R29). In both cases, either by the addition of hydrogen or water, the formation (R31) and dissociation of OH(s) (R32) becomes rate determining steps. This is the case, because OH(s) is the main oxidant for C(s) in the system.

114

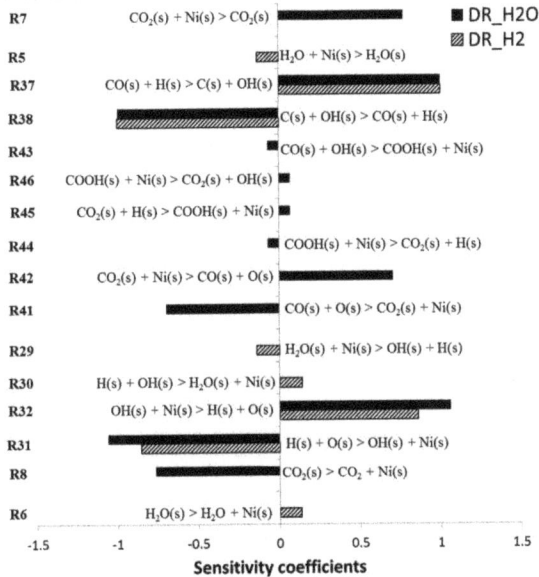

Figure 17. Sensitivity analysis of C(s) formation for dry reforming in the presence of H_2 and H_2O at 1123 K.

6. Summary

The kinetics of oxidation and reforming of methane over nickel were studied experimentally and numerically. Experimental investigations for partial oxidation dry and steam reforming of methane were performed in a flow reactor over a powdered Ni-based catalyst. A detailed reaction mechanism for the catalytic conversion of methane under oxidative and reforming conditions was developed and evaluated by comparison of experimentally derived and numerically predicted conversion and selectivity. The mechanism was implemented into a one-dimensional flow field description of a fixed bed reactor. A new numerical adjustment procedure was applied through the development process to ensure the overall thermodynamic consistency of the mechanism.

The newly developed kinetic model makes it possible to predict the product distribution for partial oxidation and reforming of methane as well as the impact of the co-feed of products such as hydrogen and water on the dry reforming process. Furthermore, the simulation tools developed allow the numerical simulation of chemical species profiles and surface coverage within the catalytic bed.

Experimental and numerical results for synthesis gas production via partial oxidation of methane are consistent with the indirect path where the total oxidation of methane takes place, first producing CO_2 and H_2O, which react with the remaining methane through reforming reactions to produce H_2 and CO.

115

Experimental results where H_2O and H_2 are co-fed together with methane and CO_2 show that both H_2O and H_2 act as inhibitors for coke deposition. The numerical simulations reveal a significant decrease in surface carbon concentration during reaction conditions where H_2 and H_2O were added as co-reactants. However, H_2O provides a better inhibition effect than H_2 on the catalytic surface.

The model developed can be extended for industrial applications, quantitatively predicting the effect of inlet compositions, operating conditions and undesirable transient modifications of the active catalytic phase, e.g., by deactivation and coking, which are the main challenge in industrial catalytic reformers. Future work will focus on the implementation of a coking model in the surface reaction scheme to describe transient carbon deposition on the surface.

Acknowledgments: The authors are grateful for the financial support of the German Federal Ministry of Economics and Technology (No. FKZ 0327856A) and our project partners BASF, Linde AG, hte AG, Technische Universität München, and Universität Leipzig for fruitful discussions on the dry reforming of methane. Personal funding (K.H.D.) and a cost-free academic license of DETCHEM™ by the Steinbeis GmbH & Co. KG für Technologietransfer (STZ 240 Reaktive Strömung) are gratefully acknowledged.

Author Contributions: K. Herrera Delgado did the simulation and experimental work, did the development of new extended version of the mechanism, wrote the paper, answered the questions of the reviewers, made editing and improvement corrections of manuscript; L. Maier proposed the idea, developed the first version of the mechanism and supervised their extension and revision, proposed the experiments for mechanism evaluation, answered the questions of the reviewers, made editing and improvement corrections of manuscript; S. Tischer developed the procedure and computer code for thermodynamic consistency, wrote appropriate part of the manuscript; A. Zellner did technical support on experimental layout, performed some experiments; H. Stotz made the estimation of the isothermcity of the reactor, wrote appropriate statement to the model description, helped in addressing the reviewer's questions, made editing and corrections of the manuscript; O. Deutschmann is a supervisor, worked on all topics listed above.

Conflicts of Interest: The authors declare no conflict of interest.

References

1. Rostrup-Nielsen, J.; Dybkjaer, I.; Aasberg-Petersen, K. Synthesis gas for large scale fischer-tropsch synthesis. *Am. Chem. Soc. Div. Pet. Chem. Prepr.* **2000**, *45*, 186–189.
2. Rostrup-Nielsen, J.R.; Sehested, J.; Nørskov, J.K. Hydrogen and synthesis gas by steam- and CO_2 reforming. In *Advances in Catalysis*; Elsevier: Amherst, MA, USA, 2002; Volume 47, pp. 65–139.
3. Iglesia, E. Design, synthesis, and use of cobalt-based fischer-tropsch synthesis catalysts. *Appl. Catal. A* **1997**, *161*, 59–78.
4. Rostrup-Nielsen, J.R. New aspects of syngas production and use. *Catal. Today* **2000**, *63*, 159–164.
5. Hickman, D.A.; Schmidt, L.D. Production of syngas by direct catalytic oxidation of methane. *Science* **1993**, *259*, 343–346.

6. Peña, M.A.; Gómez, J.P.; Fierro, J.L.G. New catalytic routes for syngas and hydrogen production. *Appl. Catal. A* **1996**, *144*, 7–57.

7. Shao, Z.; Haile, S.M. A high-performance cathode for the next generation of solid-oxide fuel cells. *Nature* **2004**, *431*, 170–173.

8. Park, S.; Vohs, J.M.; Gorte, R.J. Direct oxidation of hydrocarbons in a solid-oxide fuel cell. *Nature* **2000**, *404*, 265–267.

9. Singhal, S.C. Advances in solid oxide fuel cell technology. *Solid State Ionics* **2000**, *135*, 305–313.

10. Aparicio, L.M. Transient isotopic studies and microkinetic modeling of methane reforming over nickel catalysts. *J. Catal.* **1997**, *165*, 262–274.

11. Xu, J.; Froment, G.F. Methane steam reforming, methanation and water-gas shift: I. Intrinsic kinetics. *AIChE J.* **1989**, *35*, 88–96.

12. Rostrup-Nielsen, J. Catalytic Steam Reforming. In *Catalysis*; Anderson, J., Boudart, M., Eds.; Springer: Berlin/Heidelberg, Germany, 1984; Volume 5, pp. 1–117.

13. Bradford, M.C.J.; Vannice, M.A. CO_2 reforming of ch_4. *Catal. Rev.* **1999**, *41*, 1–42.

14. Zhu, Y.-A.; Chen, D.; Zhou, X.-G.; Yuan, W.-K. DFT studies of dry reforming of methane on Ni catalyst. *Catal. Today* **2009**, *148*, 260–267.

15. Dissanayake, D.; Rosynek, M.P.; Kharas, K.C.C.; Lunsford, J.H. Partial oxidation of methane to carbon monoxide and hydrogen over a Ni/Al_2O_3 catalyst. *J. Catal.* **1991**, *132*, 117–127.

16. Zhu, T.; Flytzani-Stephanopoulos, M. Catalytic partial oxidation of methane to synthesis gas over $Ni–CeO_2$. *Appl. Catal. A* **2001**, *208*, 403–417.

17. Tang, S.; Lin, J.; Tan, K.L. Partial oxidation of methane to syngas over Ni/MgO, Ni/CaO and Ni/CeO_2. *Catal. Lett.* **1998**, *51*, 169–175.

18. Goula, M.A.; Lemonidou, A.A.; Grünert, W.; Baerns, M. Methane partial oxidation to synthesis gas using nickel on calcium aluminate catalysts. *Catal. Today* **1996**, *32*, 149–156.

19. Liu, Z.-W.; Jun, K.-W.; Roh, H.-S.; Park, S.-E.; Oh, Y.-S. Partial oxidation of methane over nickel catalysts supported on various aluminas. *Korean J. Chem. Eng.* **2002**, *19*, 735–741.

20. Vernon, P.F.; Green, M.H.; Cheetham, A.; Ashcroft, A. Partial oxidation of methane to synthesis gas. *Catal. Lett.* **1990**, *6*, 181–186.

21. Li, C.; Yu, C.; Shen, S. Role of the surface state of Ni/Al_2O_3 in partial oxidation of CH_4. *Catal. Lett.* **2000**, *67*, 139–145.

22. Rostrup-Nielsen, J.R. Sulfur-passivated nickel-catalysts for carbon-free steam reforming of methane. *J. Catal.* **1984**, *85*, 31–43.

23. Trimm, D.L. Formation and removal of coke from nickel-catalyst. *Catal. Rev. Sci. Eng.* **1977**, *16*, 155–189.

24. Bartholomew, C.H. Carbon deposition in steam reforming and methanation. *Catal. Rev.* **1982**, *24*, 67–112.

25. Blekkan, E.A.; Myrstad, R.; Olsvik, O.; Rokstad, O.A. Characterization of tars and coke formed during the pyrolysis of methane in a tubular reactor. *Carbon* **1992**, *30*, 665–673.

26. Kahle, L.C.S.; Roussière, T.; Maier, L.; Herrera Delgado, K.; Wasserschaff, G.; Schunk, S.A.; Deutschmann, O. Methane dry reforming at high temperature and elevated pressure: Impact of gas-phase reactions. *Ind. Eng. Chem. Res.* **2013**, *52*, 11920–11930.

27. Wang, S.; Lu, G.Q.M. CO_2 reforming of methane on Ni catalysts: Effects of the support phase and preparation technique. *Appl. Catal. B* **1998**, *16*, 269–277.

28. Ross, J.R.H.; vanKeulen, A.N.J.; Hegarty, M.E.S.; Seshan, K. The catalytic conversion of natural gas to useful products. *Catal. Today* **1996**, *30*, 193–199.

29. Ginsburg, J.M.; Pina, J.; El Solh, T.; de Lasa, H.I. Coke formation over a nickel catalyst under methane dry reforming conditions: Thermodynamic and kinetic models. *Ind. Eng. Chem. Res.* **2005**, *44*, 4846–4854.

30. Guo, J.; Lou, H.; Zheng, X.M. The deposition of coke from methane on a $Ni/MgAl_2O_4$ catalyst. *Carbon* **2007**, *45*, 1314–1321.

31. Chen, D.; Lodeng, R.; Anundskas, A.; Olsvik, O.; Holmen, A. Deactivation during carbon dioxide reforming of methane over ni catalyst: Microkinetic analysis. *Chem. Eng. Sci.* **2001**, *56*, 1371–1379.

32. Chen, D.; Lødeng, R.; Svendsen, H.; Holmen, A. Hierarchical multiscale modeling of methane steam reforming reactions. *Ind. Eng. Chem. Res.* **2010**, *50*, 2600–2612.

33. Kunz, L.; Maier, L.; Tischer, S.; Deutschmann, O. Modeling the rate of heterogeneous reactions. In *Modeling and Simulation of Heterogeneous Catalytic Reactions*; Wiley-VCH Verlag GmbH & Co. KGaA: Weinheim, Germany, 2011; pp. 113–148.

34. Mhadeshwar, A.B.; Vlachos, D.G. Hierarchical multiscale mechanism development for methane partial oxidation and reforming and for thermal decomposition of oxygenates on Rh. *J. Phys. Chem. B* **2005**, *109*, 16819–16835.

35. Wei, J.; Iglesia, E. Isotopic and kinetic assessment of the mechanism of reactions of CH_4 with CO_2 or H_2O to form synthesis gas and carbon on nickel catalysts. *J. Catal.* **2004**, *224*, 370–383.

36. Blaylock, D.W.; Zhu, Y.-A.; Green, W. Computational investigation of the thermochemistry and kinetics of steam methane reforming over a multi-faceted nickel catalyst. *Top. Catal.* **2011**, *54*, 828–844.

37. Blaylock, D.W.; Ogura, T.; Green, W.H.; Beran, G.J.O. Computational investigation of thermochemistry and kinetics of steam methane reforming on Ni(111) under realistic conditions. *J. Phys. Chem. C* **2009**, *113*, 4898–4908.

38. Wang, S.-G.; Liao, X.-Y.; Hu, J.; Cao, D.-B.; Li, Y.-W.; Wang, J.; Jiao, H. Kinetic aspect of CO_2 reforming of CH_4 on Ni(111): A density functional theory calculation. *Surf. Sci.* **2007**, *601*, 1271–1284.

39. Schwiedernoch, R.; Tischer, S.; Correa, C.; Deutschmann, O. Experimental and numerical study on the transient behavior of partial oxidation of methane in a catalytic, monolith. *Chem. Eng. Sci.* **2003**, *58*, 633–642.

40. Nogare, D.D.; Degenstein, N.J.; Horn, R.; Canu, P.; Schmidt, L.D. Modeling spatially resolved data of methane catalytic partial oxidation on Rh foam catalyst at different inlet compositions and flowrates. *J. Catal.* **2011**, *277*, 134–148.

41. Diehm, C.; Deutschmann, O. Hydrogen production by catalytic partial oxidation of methane over staged Pd/Rh coated monoliths: Spatially resolved concentration and temperature profiles. *Int. J. Hydrog. Energy* **2014**, *39*, 17998–18004.

42. Horn, R.; Williams, K.A.; Degenstein, N.J.; Bitsch-Larsen, A.; Dalle Nogare, D.; Tupy, S.A.; Schmidt, L.D. Methane catalytic partial oxidation on autothermal Rh and Pt foam catalysts: Oxidation and reforming zones, transport effects, and approach to thermodynamic equilibrium. *J. Catal.* **2007**, *249*, 380–393.

43. Donazzi, A.; Beretta, A.; Groppi, G.; Forzatti, P. Catalytic partial oxidation of methane over a 4% Rh/α-Al$_2$O$_3$ catalyst: Part i: Kinetic study in annular reactor. *J. Catal.* **2008**, *255*, 241–258.

44. Christian Enger, B.; Lødeng, R.; Holmen, A. A review of catalytic partial oxidation of methane to synthesis gas with emphasis on reaction mechanisms over transition metal catalysts. *Appl. Catal. A* **2008**, *346*, 1–27.

45. De Groote, A.M.; Froment, G.F.; Kobylinski, T. Synthesis gas production from natural gas in a fixed bed reactor with reversed flow. *Can. J. Chem. Eng.* **1996**, *74*, 735–742.

46. Maier, L.; Schädel, B.; Herrera Delgado, K.; Tischer, S.; Deutschmann, O. Steam reforming of methane over nickel: Development of a multi-step surface reaction mechanism. *Top. Catal.* **2011**, *54*, 845–858.

47. Hartmann, M.; Maier, L.; Minh, H.D.; Deutschmann, O. Catalytic partial oxidation of iso-octane over rhodium catalysts: An experimental, modeling, and simulation study. *Combust. Flame* **2010**, *157*, 1771–1782, ™.

48. Deutschmann, O.; Tischer, S.; Kleditzsch, S.; Janardhanan, V.M.; Correa, C.; Chatterjee, D.; Mladenov, N.; Minh, H.D. DETCHEM™ software package. Available online: www.detchem.com (accessed on 26 May 2015).

49. Pushnov, A.S. Calculation of average bed porosity. *Chem. Petrol. Eng.* **2006**, *42*, 14–17.

50. Deutschmann, O. Computational Fluid Dynamics Simulation of Catalytic Reactors. In *Handbook of Heterogeneous Catalysis*; Erlt, H.K.G., Schüth, F., Weitkamp, J., Eds.; Wiley-VCH Verlag GmbH & Co. KGaA: Weinheim, Germany, 2008; pp. 1811–1828.

51. Kee, R.J.; Coltrin, M.E.; Glarborg, P. Heterogeneous chemistry. In *Chemically Reacting Flow*; John Wiley & Sons, Inc.: Hoboken, NJ, USA, 2005; pp. 445–486.

52. Mhadeshwar, A.B.; Vlachos, D.G. A thermodynamically consistent surface reaction mechanism for co oxidation on pt. *Combust. Flame* **2005**, *142*, 289–298.

53. Lee, M.B.; Yang, Q.Y.; Tang, S.L.; Ceyer, S.T. Activated dissociative chemisorption of CH$_4$ on Ni(111): Observation of a methyl radical and implication for the pressure gap in catalysis. *J. Chem. Phys.* **1986**, *85*, 1693.

54. Shustorovich, E. *The Bond-Order Conservation Approach to Chemisorption and Heterogeneous Catalysis: Applications and Implications*; WILEY-VCH Verlag GmbH & Co. KgaA: Weinheim, Germany, 1991; Volume 22, pp. 1522–2667.

55. Sellers, H. The generalized ubi-qep method for modeling the energetics of reactions on transition metal surfaces. *Surf. Sci.* **2003**, *524*, 29–39.

56. Qin, D.; Lapszewicz, J.; Jiang, X. Comparison of partial oxidation and steam-CO$_2$ mixed reformingof CH$_4$ to syngas on MgO-supported metals. *J. Catal.* **1996**, *159*, 140–149.

57. Bartholomew, C. Hydrogen adsorption on supported cobalt, iron, and nickel. *Catal. Lett.* **1990**, *7*, 27–51.

58. Weatherbee, G.D.; Bartholomew, C.H. Effects of support on hydrogen adsorption/desorption kinetics of nickel. *J. Catal.* **1984**, *87*, 55–65.

59. Kratzer, P.; Hammer, B.; Norskov, J.K. A theoretical study of CH_4 dissociation on pure and gold-alloyed Ni(111) surfaces. *J. Chem. Phys.* **1996**, *105*, 5595–5604.

60. Zhu, X.Y.; White, J.M. Hydrogen interaction with nickel (100): A static secondary ion mass spectroscopy study. *J. Phys. Chem.* **1988**, *92*, 3970–3974.

61. Bengaard, H.S.; Nørskov, J.K.; Sehested, J.; Clausen, B.S.; Nielsen, L.P.; Molenbroek, A.M.; Rostrup-Nielsen, J.R. Steam reforming and graphite formation on Ni catalysts. *J. Catal.* **2002**, *209*, 365–384.

62. Stuckless, J.T.; Wartnaby, C.E.; Al-Sarraf, N.; Dixon-Warren, S.J.B.; Kovar, M.; King, D.A. Oxygen chemisorption and oxide film growth on Ni{100}, {110}, and {111}: Sticking probabilities and microcalorimetric adsorption heats. *J. Chem. Phys.* **1997**, *106*, 2012–2030.

63. Siegbahn, P.E.M.; Wahlgren, U. A theoretical study of atomic oxygen chemisorption on the Ni(100) and Ni(111) surfaces. *Int. J. Quantum Chem.* **1992**, *42*, 1149–1169.

64. Stulen, R.H.; Thiel, P.A. Electron-stimulated desorption and thermal desorption spectrometry of H_2O on nickel (111). *Surf. Sci* **1985**, *157*, 99–118.

65. Pache, T.; Steinrück, H.P.; Huber, W.; Menzel, D. The adsorption of H_2O on clean and oxygen precovered Ni(111) studied by arups and tpd. *Surf. Sci.* **1989**, *224*, 195–214.

66. Zakharov, I.I.; Avdeev, V.I.; Zhidomirov, G.M. Non-empirical cluster model calculations of the adsorption of H_2O on Ni(111). *Surf. Sci.* **1992**, *277*, 407–413.

67. Shustorovich, E.; Sellers, H. The UBI-QEP method: A practical theoretical approach to understanding chemistry on transition metal surfaces. *Surf. Sci. Rep.* **1998**, *31*, 1–119.

68. Shustorovich, E. Reaction Energetics on Transition Metal Surfaces: A Bond-Order Conservation Approach. In *Quantum Chemistry Approaches to Chemisorption and Heterogeneous Catalysis*; Ruette, F., Ed.; Springer: Dordrecht, The Netherlands, 1992; Volume 6, pp. 231–252.

69. Bjorgum, E.; Chen, D.; Bakken, M.G.; Christensen, K.O.; Holmen, A.; Lytken, O.; Chorkendorff, I. Energetic mapping of ni catalysts by detailed kinetic modeling. *J. Phys. Chem. B* **2005**, *109*, 2360–2370.

70. Al-Sarraf, N.; Stuckless, J.T.; Wartnaby, C.E.; King, D.A. Adsorption microcalorimetry and sticking probabilities on metal single crystal surfaces. *Surf. Sci.* **1993**, *283*, 427–437.

71. Chan, D.; Tischer, S.; Heck, J.; Diehm, C.; Deutschmann, O. Correlation between catalytic activity and catalytic surface area of a Pt/Al_2O_3 doc: An experimental and microkinetic modeling study. *Appl. Catal. B* **2014**, *156–157*, 153–165.

72. Jacobs, G.; Graham, U.M.; Chenu, E.; Patterson, P.M.; Dozier, A.; Davis, B.H. Low-temperature water-gas shift: Impact of Pt promoter loading on the partial reduction of ceria and consequences for catalyst design. *J. Catal.* **2005**, *229*, 499–512.

73. Jacobs, G.; Patterson, P.M.; Williams, L.; Chenu, E.; Sparks, D.; Thomas, G.; Davis, B.H. Water-gas shift: *In situ* spectroscopic studies of noble metal promoted ceria catalysts for co removal in fuel cell reformers and mechanistic implications. *Appl. Catal. A* **2004**, *262*, 177–187.

74. Jacobs, G.; Patterson, P.M.; Graham, U.M.; Sparks, D.E.; Davis, B.H. Low temperature water-gas shift: Kinetic isotope effect observed for decomposition of surface formates for Pt/ceria catalysts. *Appl. Catal. A* **2004**, *269*, 63–73.

75. Jacobs, G.; Patterson, P.M.; Graham, U.M.; Crawford, A.C.; Davis, B.H. Low temperature water gas shift: The link between the catalysis of WGS and formic acid decomposition over Pt/ceria. *Int. J. Hydrog. Energy* **2005**, *30*, 1265–1276.

76. Jacobs, G.; Williams, L.; Graham, U.; Sparks, D.; Davis, B.H. Low-temperature water-gas shift: *In-situ* drifts-reaction study of a Pt/CeO$_2$ catalyst for fuel cell reformer applications. *J. Phys. Chem. B* **2003**, *107*, 10398–10404.

77. Kalamaras, C.M.; Olympiou, G.G.; Efstathiou, A.M. The water-gas shift reaction on Pt/γ-Al$_2$O$_3$ catalyst: Operando ssitka-drifts-mass spectroscopy studies. *Catal. Today* **2008**, *138*, 228–234.

78. Tibiletti, D.; Goguet, A.; Reid, D.; Meunier, F.C.; Burch, R. On the need to use steady-state or operando techniques to investigate reaction mechanisms: An *in situ* drifts and ssitka-based study example. *Catal. Today* **2006**, *113*, 94–101.

79. Shido, T.; Iwasawa, Y. Reactant-promoted reaction mechanism for water-gas shift reaction on Rh-doped ceo2. *J. Catal.* **1993**, *141*, 71–81.

80. Gokhale, A.A.; Dumesic, J.A.; Mavrikakis, M. On the mechanism of low-temperature water gas shift reaction on copper. *J. Am. Chem. Soc.* **2008**, *130*, 1402–1414.

81. Chen, Y.; Wang, H.; Burch, R.; Hardacre, C.; Hu, P. New insight into mechanisms in water-gas-shift reaction on Au/CeO$_2$(111): A density functional theory and kinetic study. *Faraday Discuss.* **2011**, *152*, 121–133.

82. Lin, C.-H.; Chen, C.-L.; Wang, J.-H. Mechanistic studies of water–gas-shift reaction on transition metals. *J. Phys. Chem. C* **2011**, *115*, 18582–18588.

83. Boisen, A.; Janssens, T.V.W.; Schumacher, N.; Chorkendorff, I.; Dahl, S. Support effects and catalytic trends for water gas shift activity of transition metals. *J. Mol. Catal. A* **2010**, *315*, 163–170.

84. Grabow, L.C.; Gokhale, A.A.; Evans, S.T.; Dumesic, J.A.; Mavrikakis, M. Mechanism of the water gas shift reaction on Pt: First principles, experiments, and microkinetic modeling. *J. Phys. Chem. C* **2008**, *112*, 4608–4617.

85. Karakaya, C.; Otterstätter, R.; Maier, L.; Deutschmann, O. Kinetics of the water-gas shift reaction over rh/Al$_2$O$_3$ catalysts. *Appl. Catal. A* **2014**, *470*, 31–44.

86. Inderwildi, O.R.; Jenkins, S.J.; King, D.A. An unexpected pathway for the catalytic oxidation of methylidyne on Rh(111) as a route to syngas. *J. Am. Chem. Soc.* **2007**, *129*, 1751–1759.

87. Inderwildi, O.R.; Jenkins, S.J.; King, D.A. Mechanistic studies of hydrocarbon combustion and synthesis on noble metals. *Angew. Chem. Int. Ed.* **2008**, *47*, 5253–5255.

88. Pistonesi, C.; Juan, A.; Irigoyen, B.; Amadeo, N. Theoretical and experimental study of methane steam reforming reactions over nickel catalyst. *Appl. Surf. Sci.* **2007**, *253*, 4427–4437.

89. Weng, W.Z.; Chen, M.S.; Yan, Q.G.; Wu, T.H.; Chao, Z.S.; Liao, Y.Y.; Wan, H.L. Mechanistic study of partial oxidation of methane to synthesis gas over supported rhodium and ruthenium catalysts using *in situ* time-resolved FTIR spectroscopy. *Catal. Today* **2000**, *63*, 317–326.

90. Hei, M.J.; Chen, H.B.; Yi, J.; Lin, Y.J.; Lin, Y.Z.; Wei, G.; Liao, D.W. CO_2-reforming of methane on transition metal surfaces. *Surf. Sci.* **1998**, *417*, 82–96.

91. Horn, R.; Williams, K.A.; Degenstein, N.J.; Schmidt, L.D. Syngas by catalytic partial oxidation of methane on rhodium: Mechanistic conclusions from spatially resolved measurements and numerical simulations. *J. Catal.* **2006**, *242*, 92–102.

92. Deutschmann, O.; Schmidt, L.D. Two-dimensional modeling of partial oxidation of methane on rhodium in a short contact time reactor. *Symp. (Int.) Combust.* **1998**, *27*, 2283–2291.

93. Deutschmann, O.; Schmidt, L.D. Modeling the partial oxidation of methane in a short-contact-time reactor. *AIChE J.* **1998**, *44*, 2465–2477.

94. Li, C.; Yu, C.; Shen, S. Isotopic studies on the mechanism of partial oxidation of CH_4 to syngas over a Ni/Al_2O_3 catalyst. *Catal. Lett.* **2001**, *75*, 183–189.

95. Tang, S.; Lin, J.; Tan, K.L. Pulse-ms studies on CH_4/Cd_4 isotope effect in the partial oxidation of methane to syngas over Pt/α-Al_2O_3. *Catal. Lett.* **1998**, *55*, 83–86.

96. Wang, H.Y.; Ruckenstein, E. Catalytic partial oxidation of methane to synthesis gas over γ-Al_2O_2-supported Rhodium catalysts. *Catal. Lett.* **1999**, *59*, 121–127.

97. Kim, S.-B.; Kim, Y.-K.; Lim, Y.-S.; Kim, M.-S.; Hahm, H.-S. Reaction mechanism of partial oxidation of methane to synthesis gas over supported Ni catalysts. *Korean J. Chem. Eng.* **2003**, *20*, 1023–1025.

98. Choudhary, V.R.; Rajput, A.M.; Prabhakar, B. Nonequilibrium oxidative conversion of methane to CO and H_2 with high selectivity and productivity over Ni/Al_2O_3 at low temperatures. *J. Catal.* **1993**, *139*, 326–328.

99. Menon, V.; Janardhanan, V.M.; Tischer, S.; Deutschmann, O. A novel approach to model the transient behavior of solid-oxide fuel cell stacks. *J. Power Sources* **2012**, *214*, 227–238.

100. Menon, V.; Fu, Q.; Janardhanan, V.M.; Deutschmann, O. A model-based understanding of solid-oxide electrolysis cells (soecs) for syngas production by H_2O/CO_2 co-electrolysis. *J. Power Sources* **2015**, *274*, 768–781.

101. Janardhanan, V.M.; Deutschmann, O. CFD analysis of a solid oxide fuel cell with internal reforming: Coupled interactions of transport, heterogeneous catalysis and electrochemical processes. *J. Power Sources* **2006**, *162*, 1192–1202.

102. Bessler, W.G.; Gewies, S.; Vogler, M. A new framework for physically based modeling of solid oxide fuel cells. *Electrochim. Acta* **2007**, *53*, 1782–1800.

103. Hecht, E.S.; Gupta, G.K.; Zhu, H.; Dean, A.M.; Kee, R.J.; Maier, L.; Deutschmann, O. Methane reforming kinetics within a ni–ysz sofc anode support. *Appl. Catal. A* **2005**, *295*, 40–51.

104. Zhu, H.; Kee, R.J.; Janardhanan, V.M.; Deutschmann, O.; Goodwin, D.G. Modeling elementary heterogeneous chemistry and electrochemistry in solid-oxide fuel cells. *J. Electrochem. Soc.* **2005**, *152*, A2427–A2440.

105. Ryu, J.-H.; Lee, K.-Y.; La, H.; Kim, H.-J.; Yang, J.-I.; Jung, H. Ni catalyst wash-coated on metal monolith with enhanced heat-transfer capability for steam reforming. *J. Power Sources* **2007**, *171*, 499–505.

106. Mears, D.E. Diagnostic criteria for heat transport limitations in fixed bed reactors. *J. Catal.* **1971**, *20*, 127–131.

Acyclic Diene Metathesis (ADMET) Polymerization for Precise Synthesis of Defect-Free Conjugated Polymers with Well-Defined Chain Ends

Tahmina Haque and Kotohiro Nomura

Abstract: This accounts introduces unique characteristics by adopting the acyclic diene metathesis (ADMET) polymerization for synthesis of conjugated polymers, poly(arylene vinylene)s, known as promising molecular electronics. The method is more suitable than the other methods in terms of atom efficiency affording defect-free, stereo-regular (exclusive *trans*) polymers with well-defined chain ends; the resultant polymers possess better property than those prepared by the conventional methods. The chain ends (vinyl group) in the resultant polymer prepared by ruthenium-carbene catalyst(s) can be modified by treating with molybdenum-alkylidene complex (olefin metathesis) followed by addition of various aldehyde (Wittig type cleavage), affording the end-functionalized polymers exclusively. An introduction of initiating fragment, the other conjugated segment, and one-pot synthesis of end-functionalized block copolymers, star shape polymers can be achieved by adopting this methodology.

Reprinted from *Catalysts*. Cite as: Haque, T.; Nomura, K. Acyclic Diene Metathesis (ADMET) Polymerization for Precise Synthesis of Defect-Free Conjugated Polymers with Well-Defined Chain Ends. *Catalysts* **2015**, *5*, 500–517.

1. Introduction

Organic electronics are important emerging technologies, and conjugated polymers like poly(*p*-arylene vinylene)s are promising materials as novel class of organic semiconductors [1–4]. It has been known that the properties for their device efficiency are generally influenced by their structural regularity, chemical purity, and supramolecular order [1–4]. Both high temperatures (180–300 °C) and vacuum conditions are employed to convert nonconjugated precursor polymers into poly(*p*-phenylene vinylene)s (PPVs, Scheme 1(1)) [1,5,6]; the synthetic procedure is very sensitive to trace amounts of oxygen during the conversion step that generates oxidation products [5,6]. These structural defects reduce the luminescence quantum efficiency of the final PPV films. There are still several concerns, such as impurities (halogen, sulfur, *etc.*) and structural (stereo-, regio-) irregularity in other conventional methods (shown in Scheme 1 such as (3) Horner-Wittig-Emmons (HWE) reactions or (4) Heck coupling, *etc.*) [1–6]. Therefore, considerable attention has thus been

124

paid to a study for synthesis of structurally regular, chemically pure polymers by development of new synthetic methods.

Conventional Synthetic Route

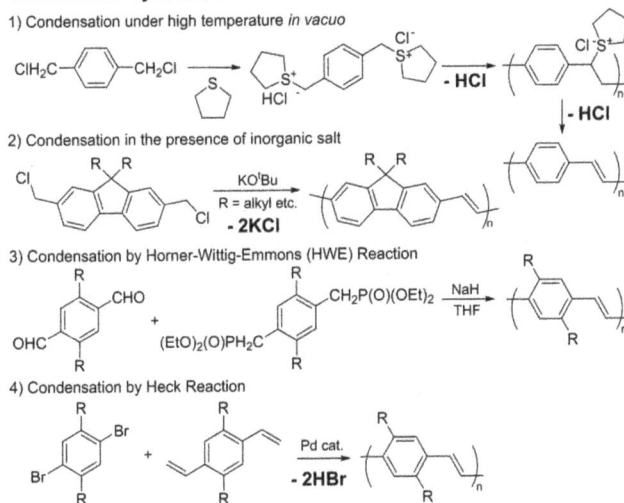

1) Condensation under high temperature *in vacuo*

2) Condensation in the presence of inorganic salt

3) Condensation by Horner-Wittig-Emmons (HWE) Reaction

4) Condensation by Heck Reaction

Scheme 1. Conventional routes for synthesis of poly(arylene vinylene)s [1–6].

Recently, it has been demonstrated that acyclic diene metathesis (ADMET) polymerization [7–11] offers a new possibility as the efficient route of π-conjugated materials [12–24]. We first demonstrated a synthesis of high molecular weight, all *trans* poly(9,9-di-*n*-octyl-fluorene-2,7-vinylene) (PFV) by ADMET polymerization of 2,7-divinyl-9,9-di-*n*-octyl-fluorene using Schrock type molybdenum-alkylidene [21]. Ru(CHPh)(Cl)$_2$(IMesH$_2$)(PCy$_3$) (Cy = cyclohexyl, IMesH$_2$ = 1,3-bis(2,4,6-trimethylphenyl)-2-imidazolidinylidene), was also effective for syntheses of high molecular weight poly(2,5-dialkyl-phenylene-1,4-vinylene)s (PPVs) [22], PFVs and poly(N-alkylcarbazole-2,7-vinylene)s (PCVs) [23] (Scheme 2). The fact introduced an interesting contrast, because the initial attempts by this approach afforded oligomer mixtures [12–20]. It turned out that an optimization of the reaction conditions (catalyst, monomer/catalyst molar ratios, initial monomer concentration, *etc.*), especially removal of ethylene by-produced from the reaction medium should be prerequisite for obtainment of the high molecular weight polymers in this condensation polymerization [21–23].

The promising characteristics by adopting this approach can be summarized as follows: (1) the resultant polymers are *defect-free* (without termination of the conjugated units, without containing any negative impurities such as halogen, sulfur), (2) the resultant polymers (oligomers) possessed highly *trans* olefinic double bonds (because, as initially proposed by Thorn-Csányi *et al.* [7,8,11–14], the

125

reaction proceeds via metallacycle intermediate, Scheme 3). Moreover, the resultant polymers prepared by Ru catalyst possessed well-defined polymer chain ends (as vinyl group) [22,25–30]; therefore, modification of the conjugated materials can be demonstrated by utilization of the chain ends [25–30].

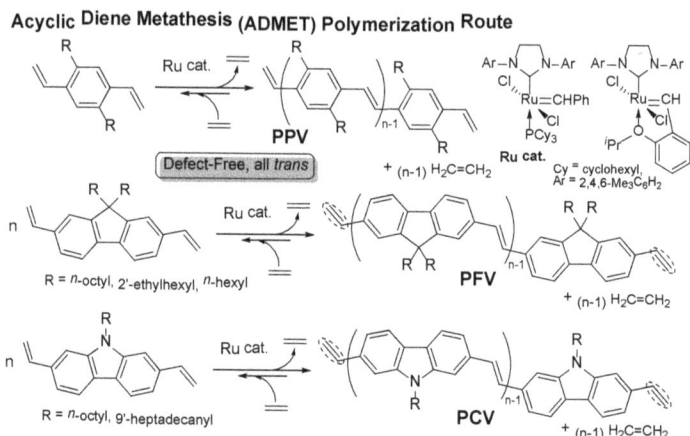

Scheme 2. Acyclic diene metathesis (ADMET) (right) [21–23] routes for synthesis of poly(arylene vinylene)s.

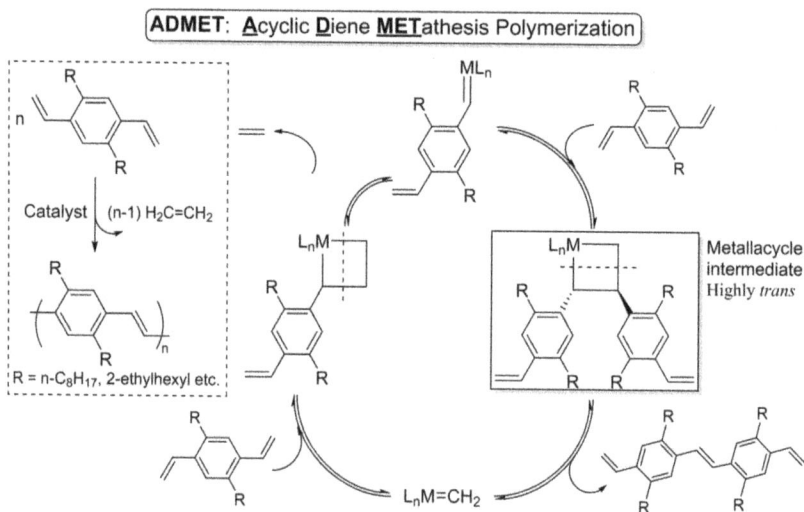

Scheme 3. Proposed mechanism for ADMET polymerization of 2,5-dialkyl-1,4-divinylbenzene [7,8,11–14,21–23].

In this accounts, unique characteristics by adopting the ADMET polymerization methods in terms of properties in the resultant polymers as well as synthesis of various advanced materials especially by the end modifications, including the points that have to be taken into consideration for the efficient synthesis, have been introduced.

2. Synthesis of Conjugated Polymers by ADMET Polymerization

Table 1 summarizes the results for ADMET polymerization of 2,7-divinyl-9,9-dialkylfluorenes by ruthenium-carbene catalysts (Scheme 4). Synthesis of high molecular weight PFVs with uniform molecular weight distributions could be achieved by adopting this method under optimization of the reaction conditions (catalyst, monomer/catalyst molar ratios, initial monomer concentration, *etc.*), although the perfect control of the repeating units cannot be obtained in this condensation (step growth) polymerization (M_w/M_n = *ca.* 2). Continuous removal of ethylene by-product from the reaction medium should be prerequisite for obtainment of the high molecular weight polymers in the present polymerization due to an equilibrium shown in Scheme 3 [21–23]. The resultant polymers possess all-*trans* internal olefinic double bonds with *defect-free* nature (without termination of conjugation).

Table 1. Acyclic diene metathesis (ADMET) polymerization of 2,7-divinyl-9,9-dialkylfluorene using various Ru-carbene complex catalysts [23][a].

Alkyl in C9 R	conc.[b]	Ru cat. (equiv)[c]	solvent (mL)	Temp/°C	Time/h	$M_n{}^d \times 10^{-4}$	$M_w/M_n{}^d$	Yield[e]/%
n-C8H17 (1)	90	A (20)	toluene (1.0)	50	8	-	-	-[f]
n-C8H17 (1)	90	B (40)	toluene (1.0)	50	7.5	1.84	1.8	75
n-C8H17 (1)	180	B (40)	toluene (1.0)	50	8	2.75	2.0	90
n-C8H17 (1)	150	B (40)	CH2Cl2 (1.2)	40	5	2.58	2.2	89
n-C8H17 (1)	180	D (80)	CH2Cl2 (1.0)	40	5	0.45	1.6	>99
n-C8H17 (1)	180	D (40)	C6H5Br (1.0)	90	3.5	0.33	1.8	75
2'-ethylhexyl (2)	180	B (40)	toluene (1.0)	50	3	2.10	2.1	75
2'-ethylhexyl (2)	270	B (40)	toluene (1.0)	50	3	3.30	2.2	82
2'-ethylhexyl (2)	180	B (40)	toluene (1.0)	50	8	3.00	2.6	79
n-C6H13 (3)	100	B (40)	toluene (1.0)	50	3	2.30	2.5	93
n-C6H13 (3)	200	C (40)	toluene (1.0)	50	5	2.10	2.1	82
n-C6H13 (3)	103	B (35)	CH2Cl2 (2.0)	40	8	3.20	2.0	>99

[a] Conditions: solvent 1.0–3.0 mL, RuCl2(CHPh)(PCy3)2 (Ru(A)), RuCl2(CHPh)(IMesH2)(PCy3) (Ru(B), IMesH2 = 1,3-bis(2,4,6-trimethylphenyl)-2-imidazolidinylidene), RuCl2(CH-2-O^iPr-C6H4)(IMesH2) (Ru(C)), RuCl2(CHPh)(IMesH2)(3-BrC5H4N)2 (Ru(D)); [b] Initial monomer concentration in μmol/mL; [c] Initial molar ratio based on monomer/Ru; [d] GPC data in THF vs polystyrene standards; [e] Isolated yields; [f] No polymers (oligomers) were formed.

The polymerizations of 2,7-divinyl-9,9-di-*n*-octyl-fluorene using Ru(CHPh)(Cl)2(IMesH2)(PCy3) (Ru(B)) in toluene proceeded under the optimized conditions, affording rather high molecular weight polymers with unimodal molecular weight distributions. Dichloromethane could also be used as the solvent, especially in terms of improvement of the solubility in the resultant PFV. The similar results were observed in the ADMET polymerization of the 2'-ethylhexyl analogue. The polymerization did not complete if the *n*-hexyl analogue was used as the monomer under the similar conditions, probably due to precipitation of the resultant polymer after a certain period; the polymerization under low initial monomer concentration conditions or in CH_2Cl_2 was effective.

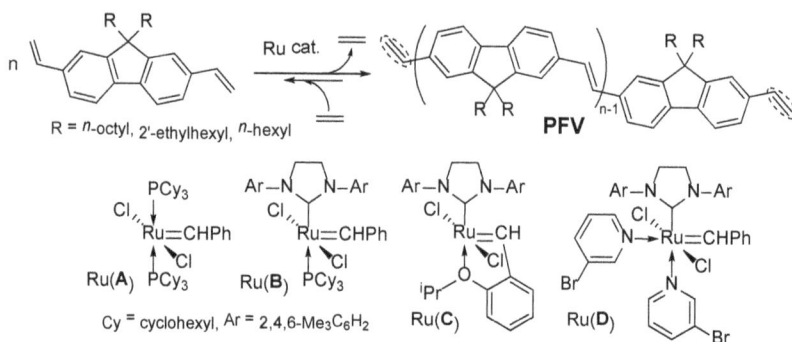

Scheme 4. Acyclic diene metathesis (ADMET) polymerization of 2,7-divinyl-9,9-dialkylfluorenes by ruthenium-carbene catalysts [23].

The ADMET polymerization of the *n*-octyl analogue using $RuCl_2(CHPh)(PCy_3)_2$ [Ru(**A**)] did not proceed, whereas the polymerization by Ru(**B**) afforded high molecular weight polymers under the same conditions. High molecular weight polymers could also be obtained by $RuCl_2(CH-2-O^iPr-C_6H_4)(IMesH_2)$ (Ru(**C**)). However, the polymerization using $RuCl_2(CHPh)(IMesH_2)(3-BrC_5H_4N)_2$ (Ru(**D**)) afforded low molecular weight oligomers even if the reaction was conducted at high temperature (90 °C). This would be explained as due to a formation of the pyridine-coordinated dormant species (probably after formation of the methylidene species, Ru=CH$_2$). Taking into account of these results, it is clear that Ru(**B**) and Ru(**C**) are suited to this polymerization [23].

Figure 1 shows typical ^{1}H NMR spectrum in the resultant polymer prepared by the ADMET polymerization using ruthenium-carbene catalyst [25]. Olefinic double bonds in the resultant PFVs possessed exclusive *trans* regularity confirmed by ^{1}H NMR spectra [21,23], because the reaction proceeds via metallacyclobutane intermediate (Scheme 3). Moreover, the resultant PFVs prepared by Ru catalysts possessed vinyl groups at the both polymer chain ends [22,23,25]. These are remarkable contrasts with that prepared by the previous (precursor) method, especially two broad resonances ascribed to protons at 4.5–4.8 ppm in addition to rather broad resonances ascribed to aromatic protons in the polymer prepared by the precursor method [23].

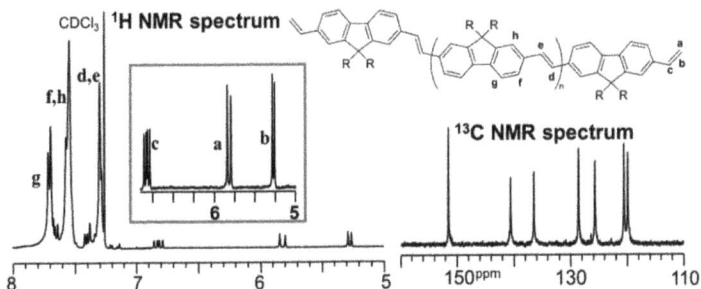

Figure 1. ^{1}H and ^{13}C NMR spectrum (in CDCl$_3$ at 25 °C) of poly(9,9-di-*n*-octyl-fluorene vinylene) (PFV) prepared by ADMET polymerization using ruthenium catalyst [25].

Figure 2a shows UV-vis spectra (in THF, 1.0×10^{-5} M at 25 °C) for PFVs (*n*-hexyl analogue) prepared by the ordinary method [available from Aldrich, M_n = 24,100, M_w/M_n = 1.70 and M_n = 1720, M_w/M_n = 1.11, prepared from the sulfonium salt precursor monomer under high-temperature (180–300 °C) and vacuum conditions (10^{-6} mbar)], and by the ADMET polymerization using Ru(**B**) (M_n = 32,000, M_w/M_n = 2.0). The spectrum prepared by the ADMET polymerization displays two absorption bands at 427 and 455 nm, which can also be attributed to π-π* transitions of the conjugated backbone [31]. Three absorption peaks at 455, 427, and 400 nm are attributed to 0-0, 0-1, and 0-2 transitions [32], respectively, with corresponding emission peaks at 465, 496, and 530 nm (described below, Figure 2b). The sample prepared by the ADMET polymerization showed a rather sharp contrast to that prepared by the Hormer-Emmons (HWE) route [reaction of 2,7-bis(methylenediethyl phosphate)-9,9-di-*n*-octylfluoene with 9,9-di-*n*-octylfluorene-2,7-dicarbaldehyde in the presence of KO^{t}Bu] [31], probably because that the resultant polymer possessed higher stereo-regularity (all *trans* in the internal olefinic double bond) than that prepared by the HWE route (mixture of *cis*/*trans*). Note that the spectrum prepared by the ordinary method (the precursor route) shows a broad absorption band with λ_{max} = 414 nm; this band can be attributed to π-π* transitions of the conjugated backbones with a shoulder at *ca.* 380 nm; the bands for the samples prepared by the ADMET polymerization are intensified and red-shifted when compared to that for the sample prepared by the precursor method. The observed difference is due to presence of defect, leading to an enhanced conjugation length [31].

Figure 2b shows the fluorescence spectra (in THF, 1.0×10^{-6} M at 25 °C, excitation wavelength at 390 nm or 426 nm). The spectrum of PFV prepared by the ADMET method showed a strong emission band at 465 nm with a shoulder at 496 nm along with a slight shoulder at 530 nm. It is explained that the shoulder arises from coupling between the fluorene and vinylene units to form a new electronic state with a lower energy. No significant differences were observed in the spectra for the

129

other alkyl analogues [23]. The spectrum of PFV prepared by the precursor route also showed a rather strong emission band at 464 nm with a shoulder at 494 nm, but remarkable differences in their intensities were seen. Moreover, a shoulder observed at 450 nm in the sample prepared by the HWE route was not observed in the samples prepared by the ADMET route and the emission band at 465 nm possessed narrow half-width. The observed difference would be probably due to that the resultant polymer possessed high stereo-regularity (all *trans* olefinic double bond) without any defects. Therefore, it is clear that PFVs prepared by the ADMET polymerization are defect-free polymers with high stereo-regularity as well as with better optical property compared with samples prepared by the other methods.

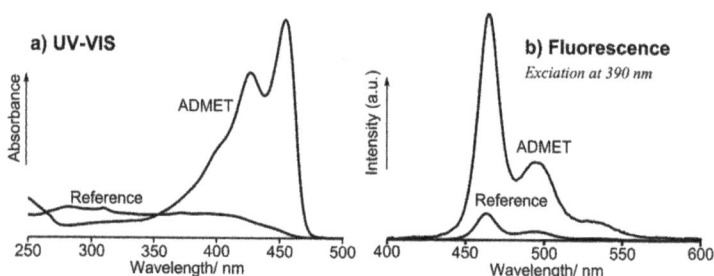

Figure 2. (a) UV-vis (conc. 1.0×10^{-5} M) and (b) fluorescent (conc. 1.0×10^{-6} M) spectra for PFVs in THF at 25 °C ($R = n\text{-}C_6H_{13}$; ADMET: $M_n = 32,000$, $M_w/M_n = 2.0$; Reference (Aldrich, St. Louis, MO, USA): $M_n = 24,100$, $M_w/M_n = 1.70$ and $M_n = 1720$, $M_w/M_n = 1.11$) [23]. The M_n and the M_w/M_n values were measured by GPC *vs.* polystyrene standards.

3. Modification of Conjugated Polymers by Exclusive End-Functionalization

Since the resultant polymers prepared by the ADMET polymerization using Ru catalyst possess well-defined chain ends as vinyl group [22,23], a facile, exclusive end-functionalization can be achieved by treating the vinyl groups with molybdenum-alkylidene, $Mo(CHCMe_2Ph)(N\text{-}2,6\text{-}Me_2C_6H_3)[OCMe(CF_3)_2]_2$ (**Mo** cat.) followed by Wittig-type cleavage with aldehyde [25–30,33–35].

A facile, exclusive end-functionalization of PFV prepared by ADMET polymerization using Ru(**B**) has been achieved by treating the vinyl groups in PFV chain ends with **Mo** cat. followed by Wittig-type cleavage with $4\text{-}Me_3SiOC_6H_4CHO$; precise synthesis of ABA type amphiphilic triblock copolymers could be accomplished by grafting PEG into both the PFV chain ends (Scheme 5) [25].

The $SiMe_3$ group in the resultant polymers (**PFV-OTMS**) was easily cleaved by treating with HCl aq. to afford **PFV-OH** in high yields. No significant changes in the M_n values were seen before/after the procedure. The OH groups in the PFV chain ends were then treated with KH in THF, and the subsequent reaction with mesylated poly(ethylene glycol) (**PEGMs2**) gave ABA-type amphiphilic triblock copolymers in rather high yields (69%–90%). The resultant copolymers possessed uniform molecular weight distributions; the copolymers were identified by ^{1}H (Figure 3) and ^{13}C NMR spectra and confirmed that no residual PEG was seen in GPC traces for the isolated polymer(s). The M_n values in the resultant triblock copolymers estimated based on the integration ratios with methylene protons of the PEG segment (Figure 3b) were very close to those estimated by both GPC [$M_{n(calc.)} = M_{n(GPC)}/1.6$] and the value in the starting PEG [25].

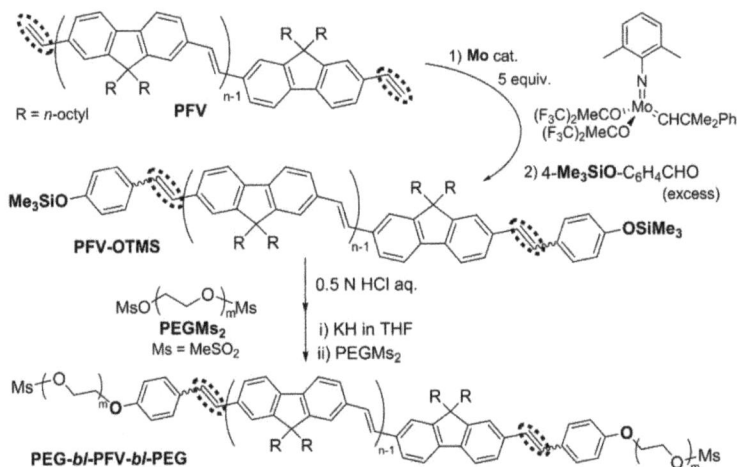

Scheme 5. Synthesis of amphiphilic triblock copolymers [25].

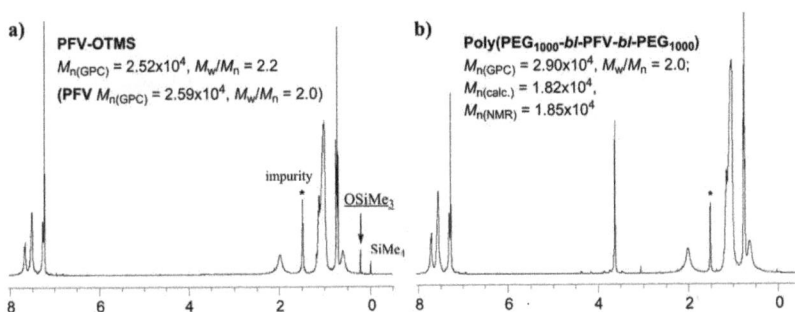

Figure 3. ^{1}H NMR spectra (in CDCl$_3$ at 25 °C) for (a) **PFV-OTMS** and (b) **PEG-*bl*-PFV-*bl*-PEG** [25].

Various block (graft) copolymers have been prepared by combination of the ADMET polymerization of 9,9-dialkyl-2,7-divinyl-fluorene with Cu-catalyzed atom transfer radical polymerization (ATRP) of styrene using macroinitiators prepared by an introduction of initiating functionalities into PFVs chain ends (Scheme 6) [27]. The

M_n value in the macroinitoator, PFV(C$_6$H$_4$OCOCMe$_2$Br)$_2$, estimated by ^{1}H NMR spectra, was relatively close to that estimated based on the exact M_n value of PFV (corrected from GPC data), strongly suggesting that both polymer chain ends could be exclusively modified by adopting the present approach. The resultant macroinitiator was added in styrene in the presence of CuBr, dNbipy (4,4-dinonyl-2,2-dipyridyl) at 90 °C for conducting subsequent ATRP. A precise synthesis of the amphiphilic ABCBA-type block copolymers could be then attained by subsequent combination with click reaction after modification of the chain end with NaN$_3$. ^{1}H NMR spectra for the resultant polymers possessed protons ascribed to PEG units, suggesting incorporation of PEG segment. Note that the M_n values estimated by ^{1}H NMR spectra (on the basis of methylene protons in the PEG segment) were very close to those calculated (based on M_n value of PFV and integration ratio of PFV and styrene). The formation of regular one-dimensional conjugated structures on the nanoscale should be thus expected by exploiting the specific assembling properties of rod-coil block copolymers, and the precise control of the amphiphilic nature as well as of the block lengths via synthesis shall open the way to fine-tuning the lateral dimensions of these nanostructures.

3.1. Precise One-Pot Synthesis of End-Functionalized Conjugated Multi-Block Copolymers via Combined Olefin Metathesis and Wittig-Type Coupling

Emerging applications of conjugated polymers require the patterning of materials on the nm length scale, and block copolymers made of covalently linked polymers represent an ideal route to control the self-assembly of these nano-sized morphologies. Formation of regular one-dimensional conjugated structures on the nano-scale should be thus expected by exploiting the specific assembling properties of rod-coil block copolymers, and a precise control of the block lengths via synthesis shall open the way to fine-tuning the lateral dimensions of these nano-structures. However, the vast majority of a significant number of studies for the generation of nano-scale morphologies from block copolymers have involved one or more of the blocks on the basis of random copolymers or random copolymers with conjugated units attached as side chains. Therefore, more relevant systems would be envisaged based on fully conjugated block copolymers from an application and scientific viewpoints.

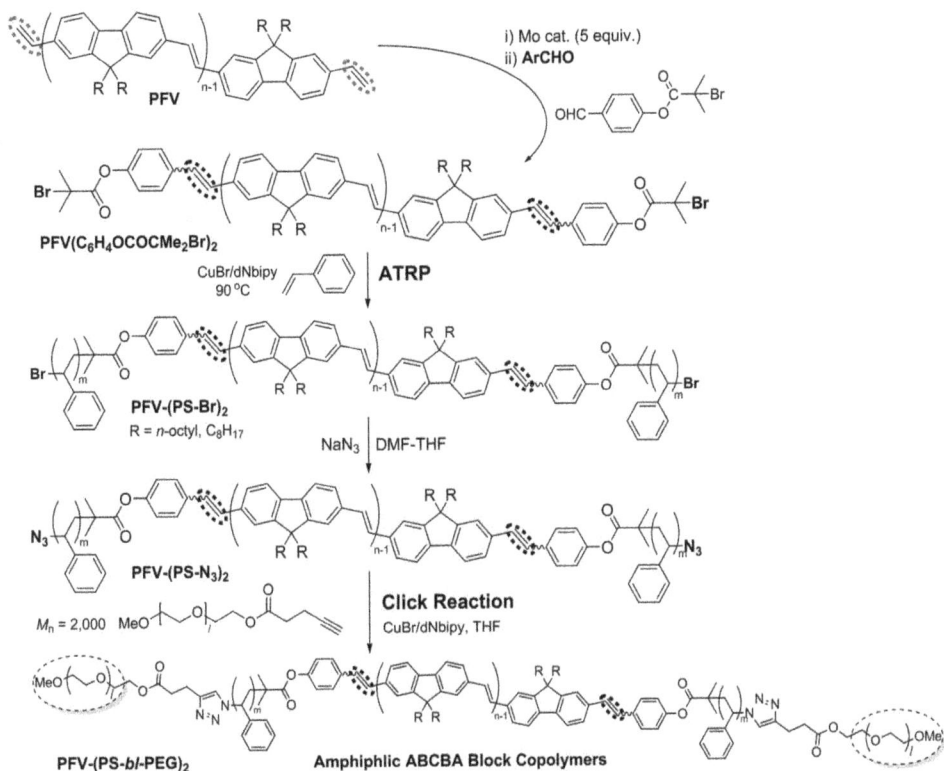

Scheme 6. Syntheses of macroinitiator, block copolymers by combination of ADMET polymerization with Cu catalyzed atom transfer radical polymerization (ATRP). Synthesis of amphiphilic ABCBA block copolymers, **PFV-(PS-bl-PEG)₂**, by combination of ATRP with click coupling [27].

A precise, one-pot synthesis of end-functionalized block copolymers consisting of PFVs and oligo(2,5-dialkoxy-1,4-phenylene vinylene) or terthiophene units as the middle segment could be prepared by olefin metathesis of the vinyl group in the PFV chain ends followed by subsequent Wittig-type coupling [29]. As shown in Scheme 7, two approaches were considered for this purpose. In method A, the vinyl groups in the PFVs' chain ends were treated with $Mo(CHCMe_2Ph)(N-2,6-Me_2C_6H_3)[OCMe(CF_3)_2]_2$ (**Mo cat.**, 5 equiv) to afford the "bis-alkylidene" species *in situ*, which were isolated by washing with cold *n*-hexane. Then, 0.5 equiv of 7PVCHO [7mer of (2,5-dialkoxy-1,4-phenylene vinylene) with CHO as the chain end] [34] or $3T(CHO)_2$ was added into the toluene solution containing the "bis-alkylidene" species. The subsequent addition of excess amount of aldehyde (ArCHO) afforded end-functionalized triblock copolymers, expressed as **[(PFV)₂-7PV]Ar₂** or **[(PFV)₂-3T]Ar₂** (Ar = C_6H_5, C_6F_5, 4-Me_3SiO-C_6H_4, terthiophene

133

(3T), ferrocene (Fc), bipyridyl (bpy)), from moderate to high yields (38-89% yields on the basis of PFVs, Table 2), which were isolated simply by pouring the reaction mixture into methanol. The M_n values in **[(PFV)$_2$-7PV]Ar$_2$**, **[(PFV)$_2$-3T]Ar$_2$** were analogous to the estimated values with unimodal molecular weight distributions (M_w/M_n = 1.14–1.36, Table 2) without significant increases after these modification procedures. Resonances ascribed to protons of the vinyl groups in the starting PFVs were not observed in the ^{1}H NMR spectra, whereas resonances ascribed to protons in the middle segment, especially **7PV**, were clearly observed. Moreover, the estimated M_n values on the basis of integration (of the middle segments, expressed as $M_{n(NMR)}$ in Table 2) were highly analogous to the calculated values [expressed as $M_{n(calcd)}$ in Table 2] on the basis of molar ratios.

As a more facile "one-pot" procedure for the synthesis (expressed as method B), the vinyl groups in the PFVs' chain ends were treated with **Mo** cat. (1.8 equiv to PFV) to afford the "bis-alkylidene" species containing "mono-alkylidene" species partially *in situ*. The mixture was then added a dichloromethane solution containing 0.5 equiv of **OPVCHO** (**3PVCHO**, **7PVCHO**) or **3T(CHO)$_2$**. The resultant reaction mixture was added into a toluene solution containing **Mo** cat (2.5 equiv) to complete the olefin metathesis (with the vinyl group in the PFV chain ends), and the subsequent addition of aldehyde (**ArCHO**) in excess amount afforded end-functionalized triblock copolymers, expressed as **[(PFV)$_2$-7PV]Ar$_2$**, **[(PFV)$_2$-3PV]Ar$_2$** or **[(PFV)$_2$-3T]Ar$_2$** (Ar = C$_6$H$_5$, C$_6$F$_5$, terthiophene (3T), ferrocene (Fc)), in high yields (70%–88% yields on the basis of PFVs, Table 2), which were isolated simply by pouring the reaction mixture into methanol. The M_n values in **[(PFV)$_2$-7PV]Ar$_2$**, **[(PFV)$_2$-3T]Ar$_2$** were analogous to the estimated values with uniform molecular weight distributions (M_w/M_n = 1.33–1.63) without significant increases after these modification procedures, and were close to those adopted in the above approach (method A). Moreover, resonances ascribed to protons of the vinyl groups in the starting PFVs were not observed in the ^{1}H NMR spectra, whereas resonances ascribed to protons in the middle segment, especially **7PV** or **3PV**, were clearly observed (Figure 4). Moreover, the estimated M_n values on the basis of integration [of the middle segments, expressed as $M_{n(NMR)}$] were highly analogous to the calculated values [expressed as $M_{n(calcd)}$] on the basis of molar ratios. These results also strongly suggest that exclusive formations of **[(PFV)$_2$-7PV]Ar$_2$**, **[(PFV)$_2$-3PV]Ar$_2$**, **[(PFV)$_2$-3T]Ar$_2$** have been achieved by adopting this methodology. The latter method should be more promising and useful, because the target end-functionalized fully conjugated block copolymers can be easily prepared in one-pot in a precise manner.

134

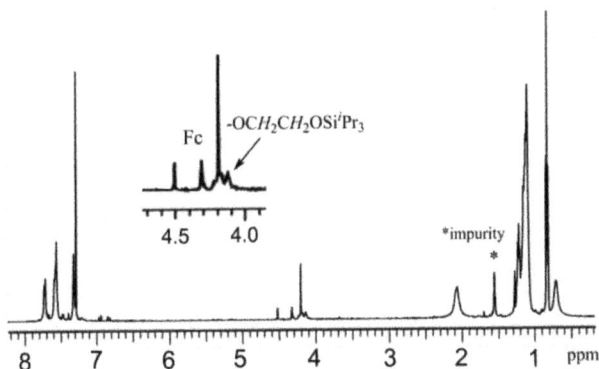

Figure 4. ^{1}H NMR spectrum (in CDCl$_3$ at 25 °C) for **[(PFV)$_2$-3PV](Fc)$_2$** ($M_{n(\text{NMR})}$ = 9000) [29].

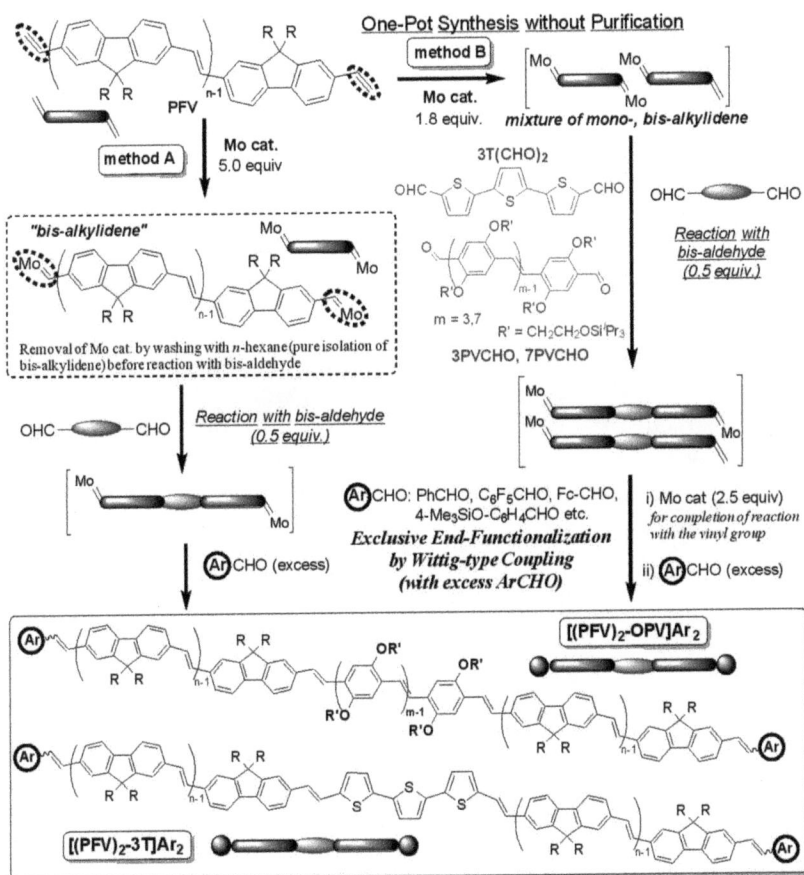

Scheme 7. Synthesis of end-functionalized triblock conjugated copolymers [29].

The resultant block copolymers containing phenolic moiety protected as SiMe$_3$ group in the chain ends, **[(PFV)$_2$-7PV](4-OSiMe$_3$-C$_6$H$_4$)$_2$, [(PFV)$_2$-3T](4-OSiMe$_3$-C$_6$H$_4$)$_2$**, were treated with HCl aq. to afford **[(PFV)$_2$-7PV](4-OH-C$_6$H$_4$)$_2$, [(PFV)$_2$-3T](4-OH-C$_6$H$_4$)$_2$** in high yields. The hydroxyl (OH) groups in the PFV chain ends were then treated with KH in THF and the subsequent reaction with mesylated poly(ethylene glycol) [PEGMs$_2$, MsO(CH$_2$CH$_2$O)$_n$Ms, Ms = MeSO$_2$: M_n = 4600] gave ABCBA type amphiphilic block copolymers in moderate yields (60, 75%, Scheme 8). The resultant copolymers possessed uniform molecular weight distributions; the copolymers were identified by ^{1}H NMR spectra and confirmed that no residual PEG was seen in GPC traces for the isolated polymer(s). The M_n values in the resultant block copolymers estimated based on the integration ratios with methylene protons of the PEG segment were very close to the calculated values (Table 3). The facts clearly indicate that facile, efficient attachments of a pseudo phenol terminus on the PFV to PEGMs$_2$, could be achieved in a precise manner by adopting the present "grafting to" approach. Moreover, importantly, the results strongly demonstrate that *the end-functionalization of PFV chain ends, reaction with the middle segments, and subsequent grafting of PEG took place with exclusively* in all cases.

Table 2. Synthesis of triblock copolymers with well-defined end functional groups [29].

Polymers	PFV[a] $M_{n(GPC)}$[b]	$M_{n(calcd)}$	Method A or B	Polymer $M_{n(GPC)}$[b]	$M_{n(calcd)}$[c]	$M_{n(NMR)}$[d]	M_w/M_n[b]	Yield e/%
[(PFV)$_2$-7PV](C$_6$H$_5$)$_2$	7,300[f]	4,560	A	19,400	13,000	12,700	1.15	38
[(PFV)$_2$-7PV](C$_6$F$_5$)$_2$	7,300[f]	4,560	A	20,300	13,200	12,800	1.16	54
[(PFV)$_2$-7PV](4-OSiMe$_3$-C$_6$H$_4$)$_2$	7,300[f]	4,560	A	19,200	13,200	12,800	1.14	62
[(PFV)$_2$-7PV](3T)$_2$	7,300[f]	4,560	A	20,400	13,300	13,100	1.16	42
[(PFV)$_2$-7PV](Fc)$_2$	7,300[f]	4,560	A	20,600	13,200	12,900	1.17	42
[(PFV)$_2$-7PV](C$_6$H$_5$)$_2$	14,000[g]	8,750	A	33,400	21,300	-	1.32	75
[(PFV)$_2$-7PV](C$_6$F$_5$)$_2$	14,000[g]	8,750	A	38,500	21,500	-	1.36	85
[(PFV)$_2$-7PV](4-OSiMe$_3$-C$_6$H$_4$)$_2$	14,000[g]	8,750	A	30,800	21,500	21,700	1.31	87
[(PFV)$_2$-7PV](bpy)$_2$	14,000[g]	8,750	A	32,000	21,500	-	1.32	53
[(PFV)$_2$-7PV](C$_6$H$_5$)$_2$	8,080[h]	5,050	B	19,700	13,900	12,100	1.50	73
[(PFV)$_2$-7PV](C$_6$F$_5$)$_2$	8,080[h]	5,050	B	20,800	14,100	12,300	1.57	80
[(PFV)$_2$-7PV](3T)$_2$	8,080[h]	5,050	B	19,200	14,400	12,500	1.43	88
[(PFV)$_2$-7PV](Fc)$_2$	8,080[h]	5,050	B	20,300	14,200	12,300	1.33	75
[(PFV)$_2$-3T](C$_6$H$_5$)$_2$	7,300[f]	4,560	A	17,600	9,500	-	1.22	72
[(PFV)$_2$-3T](4-OSiMe$_3$-C$_6$H$_4$)$_2$	7,300[f]	4,560	A	19,800	9,600	9,400	1.19	64
[(PFV)$_2$-3T](C$_6$F$_5$)$_2$	7,300[f]	4,560	A	18,800	9,600	-	1.31	80
[(PFV)$_2$-3T](3T)$_2$	7,300[f]	4,560	A	16,000	9,800	-	1.29	32
[(PFV)$_2$-3T](Fc)$_2$	7,300[f]	4,560	A	18,800	9,700	9,400	1.16	60
[(PFV)$_2$-3T](3T)$_2$	8,080[h]	5,050	B	15,400	10,800	-	1.63	70
[(PFV)$_2$-3PV](C$_6$F$_5$)$_2$	6,380[i]	3,990	B	26,600	9,900	8,900	1.50	76
[(PFV)$_2$-3PV](Fc)$_2$	6,380[i]	3,990[i]	B	22,900	9,900	9,000	1.50	71
[(PFV)$_2$-3T](C$_6$H$_5$)$_2$	14,000[g]	8,750	A	26,300	17,800	-	1.19	73
[(PFV)$_2$-3T](4-OSiMe$_3$-C$_6$H$_4$)$_2$	14,000[g]	8,750	A	23,300	18,000	18,200	1.16	78
[(PFV)$_2$-3T](C$_6$F$_5$)$_2$	14,000[g]	8,750	A	23,900	18,000	-	1.15	89
[(PFV)$_2$-3T](Fc)$_2$	14,000[g]	8,750	B	26,500	18,000	18,300	1.39	82

[a] PFV (vinyl group chain end) employed for the syntheses. [b] GPC data in THF vs polystyrene standards. [c] Calculated on the basis of molar ratio. [d] Estimated by the integration ratio (middle and end groups). [e] Isolated yield. [f] $M_{n(GPC)}$ = 7300, M_w/M_n = 1.86, $M_{n(calcd)}$ = $M_{n(GPC)}$/1.6 = 4560 (by references [21,28]), $M_{n(NMR)}$ = 4400 (estimated by the integration with the vinyl group in the ^{1}H NMR spectrum). [g] PFV (vinyl group chain end): $M_{n(GPC)}$ = 14,000, M_w/M_n = 1.60, $M_{n(calcd)}$ = $M_{n(GPC)}$/1.6 = 8750, $M_{n(NMR)}$ = 8800 (estimated by the integration with the vinyl group in the ^{1}H NMR spectrum). [h] PFV (vinyl group chain end): $M_{n(GPC)}$ = 8080, M_w/M_n = 1.79, $M_{n(calcd)}$ = $M_{n(GPC)}$/1.6 = 5050, $M_{n(NMR)}$ = 4100 (estimated by the integration with the vinyl group in the ^{1}H NMR spectrum). [i] PFV (vinyl group chain end): $M_{n(GPC)}$ = 6380, M_w/M_n = 1.99, $M_{n(calcd)}$ = $M_{n(GPC)}$/1.6 = 3990, $M_{n(NMR)}$ = 3500 (estimated by the integration with the vinyl group in the ^{1}H NMR spectrum).

We have shown that a facile, efficient synthesis of ABCBA type amphiphilic block copolymers has been established in a precise manner (as a rare example) by attachment of PEG into the both chain ends of the all-*trans*, defect-free, high molecular weight PFVs. Formation of regular one-dimensional conjugated structures on the nanoscale should be thus expected by exploiting the specific assembling properties of rod-coil block copolymers, and the control of the block lengths via synthesis opens the way to fine tuning the lateral dimensions of these nanostructures. We thus believe that the present approach should offer unique, important methodology for precise synthesis of end functionalized block conjugated polymers for targeted device materials as well as synthesis of various block copolymers containing conjugated polymer fragments.

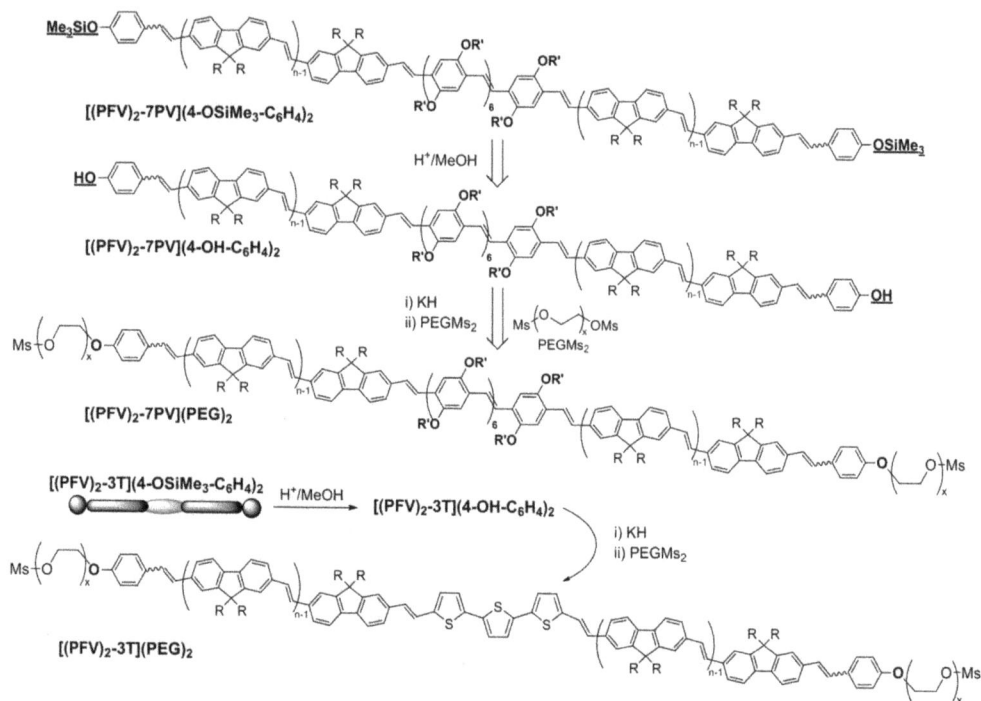

Scheme 8. Synthesis of *amphiphilic block copolymers* (by grafting PEG) [29].

Table 3. Synthesis of amphiphilic block copolymers by grafting PEG [29] [a].

Polymers	$M_{n(GPC)}$ [b]	$M_{n(calcd)}$ [c]	$M_{n(NMR)}$ [d]	M_w/M_n [b]	Yield [e]/%
[(PFV)$_2$-7PV](4-OSiMe$_3$-C$_6$H$_4$)$_2$	30800	21500	21700	1.31	87
[(PFV)$_2$-7PV](4-OH-C$_6$H$_4$)$_2$	32300	21400	-	1.33	85
[(PFV)$_2$-7PV][4-O(PEG)-C$_6$H$_4$]$_2$	30200	30500	32400	1.20	75
[(PFV)$_2$-3T](4-OSiMe$_3$-C$_6$H$_4$)$_2$	23300	18000	18200	1.16	78
[(PFV)$_2$-3T](4-OH-C$_6$H$_4$)$_2$	22900	18100	-	1.29	70
[(PFV)$_2$-3T][4-O(PEG)-C$_6$H$_4$]$_2$	22700	27100	28900	1.13	60

[a] Synthetic conditions, see Scheme 7, PEG(Ms)$_2$: M_n = 4600; [b] GPC data in THF vs polystyrene standards; [c] Calculated on the basis of molar ratio; [d] Estimated by the integration ratio; [e] Isolated yield.

3.2. Precise One-Pot Synthesis of Fully Conjugated End-Functionalized Star Polymers

On the basis of the above methodology, a facile, precise one-pot synthesis of end-functionalized star (triarm) polymers consisting of PFVs, the triblock copolymers (by incorporation of tri(2,5-dialkoxy-1,4-phenylene vinylene) or terthiophene units as the middle segment), have been prepared by olefin metathesis

138

followed by Wittig-type coupling. The key step of the synthesis is (i) olefin metathesis of the vinyl group in the PFV chain ends, (ii) treatment with core molecule [tris(4-formylphenyl)amine *etc.*], and (iii) reaction with various aldehyde (Scheme 9) [30].

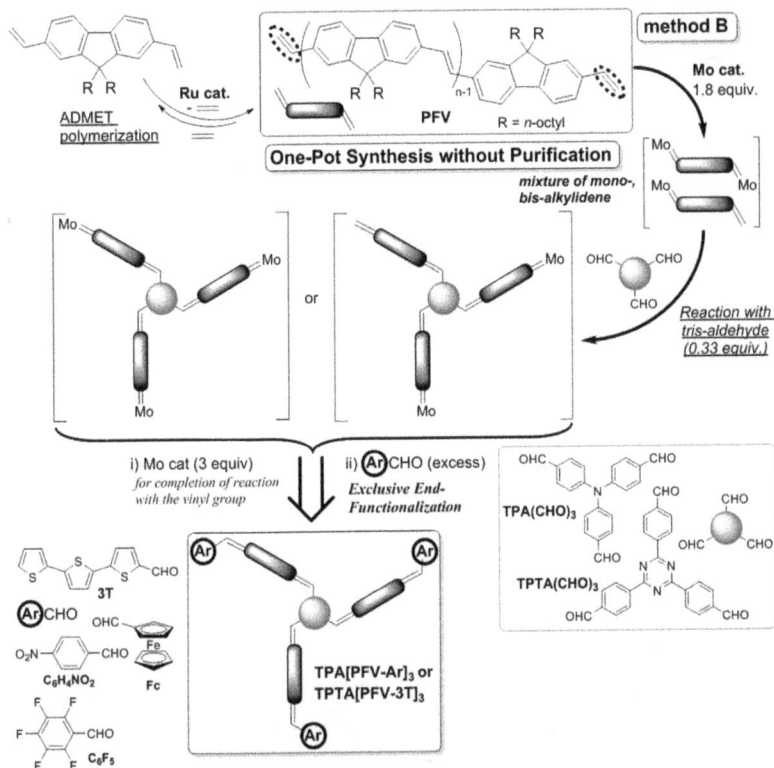

Scheme 9. Synthesis of end-functionalized triblock conjugated copolymers [30].

As shown in Scheme 9, the vinyl groups in the PFVs' chain ends were treated with **Mo** cat. (1.8 equiv to PFV), and the mixture was then added a CH_2Cl_2 solution containing **TPA(CHO)₃**. The reaction solution was added a toluene solution containing **Mo** cat (3 equiv) for completion of the olefin metathesis (with the vinyl group in the PFV chain ends), and the subsequent addition of aldehyde (**ArCHO**) in excess amount afforded the end-functionalized star (triarm) copolymers, **TPA[PFV-Ar]₃** (Ar = C_6F_5, ferrocene (Fc)). The M_n values in **TPA[PFV-Ar]₃**, estimated by GPC, were analogous to the estimated values with unimodal molecular weight distributions. Resonances ascribed to protons of the vinyl groups in the starting PFVs were not observed in the ^1H NMR spectra, whereas resonances ascribed to protons in ferrocene moiety were clearly observed. Moreover, the estimated M_n values on the basis of integration in **TPA[PFV-Fc]₃** were highly analogous to the calculated values on the basis of molar ratios, suggesting an exclusive formation of end-functionalized star conjugated polymers, **TPA[PFV-Ar]₃**. The method can also be applied for synthesis of the star polymers containing triblock copolymers as the arm segment [30].

4. Summary and Outlook

The present account introduces the synthesis of conjugated polymers, poly(arylene vinylene)s, known as promising molecular electronics, by acyclic diene metathesis (ADMET) polymerization. The method is suitable to synthesis of high molecular weight, defect-free, stereo-regular (exclusive *trans*) polymers with well-defined chain ends. The chain ends (vinyl group) in the resultant polymer prepared by ruthenium-carbene catalyst(s) can be modified by treating with molybdenum-alkylidene complex (olefin metathesis) followed by addition of various aldehyde (Wittig type cleavage), affording the end-functionalized polymers exclusively. This is the very limited method for introduction of functionality into the chain ends of the conjugated polymers. An introduction of initiating fragment, the other conjugated segment, and one-pot synthesis of end-functionalized block copolymers as can be achieved by adopting this methodology, although the ADMET polymerization is a step growth polymerization and the precise control of molecular weight (like living polymerization) cannot be achieved. Since unique emission properties have been observed probably due to an energy transfer by introduction of oligo(thiophene)s [26], and chomophores [28] into the PFV chain ends, the method thus provides a new possibility for synthesis new type of advanced optical materials on the basis of integration of functionality. Although the present approaches require molybdenum-alkylidene catalyst (reagent) for the end-functionalization, the direct chain transfer pathway [36–38] may be considered in the future. We highly believe that the facts demonstrated here would offer a new possibility for development of new advanced materials/devises for the desired purpose.

Acknowledgments: The project is partly supported by Advanced Catalytic Transformation for Carbon utilization (ACT-C), Japan Science and Technology Agency (JST).

Author Contributions: This reviewing article was prepared by Tahmina Haque and Kotohiro Nomura. T.H. prepared the initial draft and K.N. finalized the text and drawings as the corresponding author.

Conflicts of Interest: The authors declare no conflict of interest.

References

1. Müllen, K. *Organic Light Emitting Devices*; Scherf, U., Ed.; Wiley-VCH: Winheim, Germany, 2006.
2. Skotheim, T.A. *Handbook of Conducting Polymers*, 3rd ed.; Reynolds, J., Ed.; CRC Press: Boca Raton, FL, USA, 2007.

3. Jenekhe, S.A., Ed.; *Special Issue in Organic Electronics*; American Chemical Society: Washington, DC, USA, 2004; Volume 16, pp. 4381–4842.

4. Grimsdale, A.C.; Chan, K.L.; Martin, R.E.; Jokisz, P.G.; Holmes, A.B. Synthesis of light-emitting conjugated polymers for applications in electroluminescent devices. *Chem. Rev.* **2009**, *109*, 897–1091.

5. Burn, P.L.; Bradley, D.D.C.; Friend, R.H.; Halliday, D.A.; Holmes, A.B.; Jackson, R.W.; Kraft, A. Precursor route chemistry and electronic properties of poly(p-phenylenevinylene), poly[(2,5-dimethyl-p-phenylene)vinylene], and poly[(2,5-dimethoxy-p-phenylene)vinylene]. *J. Chem. Soc. Perkins Trans.* **1992**, *1*, 3225–3231.

6. Breemen, A.J.J.M.; Issaris, A.C.; de Kok, M.M.; van der Borght, M.J.A.N.; Adriaensens, P.J.; Gelan, J.M.J.V.; Vanderzande, D.J.M. Optimization of the polymerization process of sulfinyl precursor polymers toward poly(p-phenylene vinylene). *Macromolecules* **1999**, *32*, 5728–5735.

7. Lehman, S.E., Jr.; Wagener, K.B. ADMET polymerization. In *Handbook of Metathesis*; Grubbs, R.H., Ed.; Wiley-VCH: Weinheim, Germany, 2003; Volume 3, pp. 283–353.

8. Baughman, T.W.; Wagener, K.B. Recent advances in ADMET polymerization. In *Advances in Polymer Science*; Buchmeiser, M.R., Ed.; Springer: Heidelberg, Germany, 2005; Volume 176, pp. 1–42.

9. Berda, E.B.; Wagener, K.B. Advances in acyclic diene metathesis polymerization. In *Polymer Science: A Comprehensive Reference*; Matyjaszewski, K., Müllen, M., Eds.; Elsevier BV: Amsterdam, Netherlands, 2012; Volume 5, pp. 195–216.

10. Berda, E.B.; Wagener, K.B. Recent advances in ADMET polycondensation chemistry. In *Synthesis of Polymers; New Structures and Methods*; Schluter, D., Hawker, C., Sakamosto, J., Eds.; Wiley-VCH: Weinhein, Germany, 2012; pp. 587–600.

11. Atallah, P.; Wagener, K.B.; Schulz, M.D. ADMET: The future revealed. *Macromolecules* **2013**, *46*, 4735–4741.

12. Thorn-Csányi, E.; Kraxner, P. Synthesis of soluble, all-*trans* poly(2,5-diheptyl-p-phenylenevinylene) via metathesis polycondensation. *Macromol. Rapid Commun.* **1995**, *16*, 147–153.

13. Thorn-Csányi, E.; Kraxner, P. Investigations of stable molybdenum carbene complexes for the metathesis synthesis of dialkylsubstituted poly(p-phenylenevinylene)s (PPVs). *J. Mol. Catal. A* **1997**, *115*, 21–28.

14. Thorn-Csányi, E.; Kraxner, P. All-*trans* oligomers of 2,5-dialkyl-1,4-phenylenevinylenes-metathesis preparation and characterization. *Macromol. Chem. Phys.* **1997**, *198*, 3827–3843.

15. Thorn-Csányi, E.; Kraxner, P. Synthesis of soluble all-*trans* oligomers of 2,5-diheptyloxy-p-phenylenevinylene via olefin metathesis. *Macromol. Rapid Commun.* **1998**, *19*, 223–228.

16. Schlick, H.; Stelzer, F.; Tasch, S.; Leising, G. Highly luminescent poly[(m-phenylenevinylene)-co-(p-phenylenevinylene)] derivatives synthesized via metathesis condensation (ADMET). *J. Mol. Catal. A* **2000**, *160*, 71–84.

17. Thorn-Csányi, E.; Herzog, O. Synthesis of higher, *trans* configured oligomers of diisoalkyloxysubstituted divinylbenzenes (PV-oligomers) via metathesis telomerization of the corresponding lower oligomers. *J. Mol. Catal. A* **2004**, *213*, 123–128.

18. Joo, S.-H.; Jin, J.-I. All hydrocarbon main-chain thermotropic liquid crystalline polymers, poly(1,1'-biphenylene-4,4'-alkenediyl)s, prepared by the ADMET method and their hydrogenated polymers, poly(1,1'-biphenylene-4,4'-alkanediyl)s. *J. Polym. Sci., Part A* **2004**, *42*, 1335–1349.

19. Oakley, G.W.; Wagener, K. Solid-state olefin metathesis: ADMET of rigid-rod polymers and ring-closing metathesis. *Macromol. Chem. Phys.* **2005**, *206*, 15–24.

20. Pecher, J.; Mecking, S. Nanoparticles from step-growth coordination polymerization. *Macromolecules* **2007**, *40*, 7733–7735.

21. Nomura, K.; Morimoto, H.; Imanishi, Y.; Ramhani, Z.; Geerts, Y. Synthesis of high molecular weight *trans*-poly(9,9-di-*n*-octylfluorene-2,7-vinylene) by the acyclic diene metathesis polymerization using molybdenum catalysts. *J. Polym. Sci. Part A* **2001**, *39*, 2463–2470.

22. Nomura, K.; Miyamoto, Y.; Morimoto, H.; Geerts, Y. Acyclic diene metathesis polymerization of 2,5-dialkyl-1,4-divinylbenzene with molybdenum or ruthenium catalysts: Factors affecting the precise synthesis of defect-free, high-molecular-weight *trans*-poly(*p*-phenylene vinylene)s. *J. Polym. Sci., Part A: Poly. Chem.* **2005**, *43*, 6166–6177.

23. Yamamoto, N.; Ito, R.; Geerts, Y.; Nomura, K. Synthesis of all-*trans* high molecular weight poly(*N*-alkylcarbazole-2,7-vinylene)s and poly(9,9-dialkylfluorene-2,7-vinylene)s by acyclic diene metathesis (ADMET) polymerization using ruthenium-carbene complex catalysts. *Macromolecules* **2009**, *42*, 5104–5111.

24. Weychardt, H.; Plenio, H. Acyclic diene metathesis polymerization of divinylarenes and divinylferrocenes with Grubbs-type olefin metathesis catalysts. *Organometallics* **2008**, *27*, 1479–1485.

25. Nomura, K.; Yamamoto, N.; Ito, R.; Fujiki, M.; Geerts, Y. Exclusive end functionalization of all-trans-poly(fluorene vinylene)s prepared by acyclic diene metathesis polymerization: facile efficient synthesis of amphiphilic triblock copolymers by grafting poly(ethylene glycol). *Macromolecules* **2008**, *41*, 4245–4249.

26. Kuwabara, S.; Yamamoto, N.; Sharma, P.M.V.; Takamizu, K.; Fujiki, M.; Geerts, Y.; Nomura, K. Precise synthesis of poly(fluorene-2,7-vinylene)s containing oligo(thiophene)s at the chain ends: unique emission properties by the end functionalization. *Macromolecules* **2011**, *44*, 3705–3711.

27. Abdellatif, M.M.; Nomura, K. Precise synthesis of amphiphilic multiblock copolymers by combination of acyclic diene metathesis (ADMET) polymerization with atom transfer radical polymerization (ATRP) and click chemistry. *ACS Macro Lett.* **2012**, *1*, 423–427.

28. Takamizu, K.; Inagaki, A.; Nomura, K. Precise synthesis of poly(fluorene vinylene)s capped with chromophores: Efficient fluorescent polymers modified by conjugation length and end-groups. *ACS Macro Lett.* **2013**, *2*, 980–984.

29. Nomura, K.; Haque, T.; Onuma, T.; Hajjaj, F.; Asano, M.S.; Inagaki, A. Precise one-pot synthesis of end-functionalized conjugated multi-block copolymers via combined olefin metathesis and Wittig-type coupling. *Macromolecules* **2013**, *46*, 9563–9574.

30. Nomura, K.; Haque, T.; Miwata, T.; Inagaki, A.; Takamizu, K. Precise one-oot synthesis of fully conjugated end functionalized starpolymers containing poly(fluorene-2,7-vinylene) (PFV) arms. *Polymer Chem.* **2015**, *6*, 380–388.

31. Anuragudom, P.; Newaz, S.S.; Phanichphant, S.; Lee, T.R. Facile Horner–Emmons synthesis of defect-free poly(9,9-dialkylfluorenyl-2,7-vinylene). *Macromolecules* **2006**, *39*, 3494–3499.

32. Liu, Q.; Liu, W.; Yao, B.; Tian, H.; Xie, Z.; Geng, Y.; Wang, F. Synthesis and chain-length dependent properties of monodisperse oligo(9,9-di-*n*-octylfluorene-2,7-vinylene)s. *Macromolecules* **2007**, *40*, 1851–1857.

33. Nomura, K.; Abdellatif, M.M. Precise synthesis of polymers containing functional end groups by living ring-opening metathesis polymerization (ROMP): Efficient tools for synthesis of block/graft copolymers. *Polymer* **2010**, *51*, 1861–1881.

34. Abdellatif, M.M.; Nomura, K. Precise synthesis of end-functionalized oligo(2,5-dialkoxy-1,4-phenylene vinylene)s with controlled repeat units via combined olefin metathesis and Wittig-type coupling. *Org. Lett.* **2013**, *15*, 1618–1621.

35. Abdellatif, M.M.; Yorsaeng, S.; Inagaki, A.; Nomura, K. Synthesis of end functionalized oligo(2,5-dialkoxy-1,4-phenylene vinylene)s. *Macromol. Chem. Phys.* **2014**, *215*, 1973–1983.

36. Hillmyer, M.A.; Nguyen, S.B.T.; Grubbs, R.H. Utility of a ruthenium metathesis catalyst for the preparation of end-functionalized polybutadiene. *Macromolecules* **1997**, *30*, 718–721.

37. Pitet, L.M.; Hillmyer, M.A. Carboxy-telechelic polyolefins by ROMP using maleic acid as a chain transfer agent. *Macromolecules* **2011**, *44*, 2378–2381.

38. Lin, T.-W.; Chou, C.-M.; Lin, N.-T.; Lin, C.-L.; Luh, T.-Y. End group modification of polynorbornenes. *Macromol. Chem. Phys.* **2014**, *215*, 2357–2364.

Real-Time Raman Monitoring during Photocatalytic Epoxidation of Cyclohexene over V-Ti/MCM-41 Catalysts

Hsiang-Yu Chan, Van-Huy Nguyen, Jeffrey C.S. Wu, Vanesa Calvino-Casilda, Miguel A. Bañares and Hsunling Bai

Abstract: A series of V- and/or Ti-loading MCM-41 catalysts are successfully synthesized with a hydrothermal method. The photocatalytic and thermal epoxidations of cyclohexene in the presence of *tert*-butyl hydroperoxide (*t*-BuOOH) were investigated with real-time monitored by NIR-Raman spectroscopy. It suggests that both V- and Ti-loading can be responsible for the cyclohexene epoxidation. Moreover, the complementary behavior of V- and Ti-loading may be related to a similar role of activation. Interestingly, the progress of the photo-epoxidation on $V_{0.25}Ti_2$/MCM-41 photocatalyst was monitored by changes in intensity of the characteristic Raman bands without interference from the UV-light irradiation. The result, for the first time, reveals that cyclohexene was directly photo-epoxidized to 1,2-epoxycyclohexane by *t*-BuOOH during the reaction. A possible mechanism of cyclohexene photo-epoxidation is also proposed for this study.

Reprinted from *Catalysts*. Cite as: Chan, H.-Y.; Nguyen, V.-H.; Wu, J.C.S.; Calvino-Casilda, V.; Bañares, M.A.; Bai, H. Real-Time Raman Monitoring during Photocatalytic Epoxidation of Cyclohexene over V-Ti/MCM-41 Catalysts. *Catalysts* **2015**, *5*, 518–533.

1. Introduction

Nowadays, major epoxide chemicals, which are produced by epoxidation of olefins such as ethylene, propylene and cyclohexene, *etc.*, have become considerably more important, valuable and versatile intermediates in industrial organic synthesis [1–6]. Among epoxides, 1,2-epoxycyclohexane is an attractive chemical intermediates with multiple applications although it has been produced on the small scale only. For example, alicyclic molecule, which is synthesized from 1,2-epoxycyclohexane, is used in the production of pesticides, plant-protection agents, pharmaceuticals, perfumery, and dyestuffs. 1,2-epoxycyclohexane is also used as a monomer in polymerization with CO_2 to yield aliphatic polycarbonates [7]. Alternatively, it might be used for synthesis of chiral 1,2-amino alcohols and 1,2-diamines, which are known as the useful precursors of various chiral oxazolidinones, oxazinones, and phosphonamides, among others [8].

In cyclohexene epoxidation, to develop acceptable oxidative methodologies plays a major role from economic and environmental viewpoints. Among the oxidant agents for cyclohexene epoxidation, *tert*-butyl hydroperoxide (*t*-BuOOH) is the most popular. It has been extensively used due to its many advantages, such as its stability, mild oxidation, non-corrosive and non-hazardous properties. In addition, the separation process of by-products when using *t*-BuOOH is much easier than with others [9]. Another approach, which is based on utilizing clean and abundant photo-energy, instead of thermal energy, has been gained considerable interest. This concept, which directly produces 1,2-epoxycyclohexane under light radiation, is expected to be greener and less expensive than traditional methods. However, there are only a few studies of this topic which performances are satisfactory [10,11].

It is well known that the gas chromatography method is simple and preferable for quantitative analysis in cyclohexene epoxidation [12–14]. However, recently, there has been considerable discussion on the effect of temperature in injection port and column to the chromatographic selectivity values of observed products. Similar phenomena have been described for other reactions, such as the alkylation of imidazole [15]. As we noted in a previous study [16], some of new products, which could not be realized in the reaction media, can be further formed inside the injection port (at 220 °C) and the column (at 180 °C). For example, cyclohexane 1,2-diol, 2-cyclohexene-1-ol and 2-cyclohexene-1-one might be observed by gas chromatography analyses, although Raman result suggested that 1,2-epoxycyclohexane was the only product obtained in the reaction. It is important to mention that among many powerful techniques, which might be used to provide unique fingerprints for molecular analysis and monitoring of chemical reactions, Raman spectroscopy is particularly suitable to study *in situ* catalytic reaction [17–22]. Considering them together in this study, gas chromatography is used to screen the catalytic activity by determining cyclohexene conversion values while Raman technique is used to identify the reactants and selective products of epoxidation in real-time. Herein, the photo-epoxidation of cyclohexene in the presence of *t*-BuOOH was investigated for the first time over a series of V-Ti/MCM-41 photocatalysts. Additionally, the photocatalytic and thermal catalytic cyclohexene epoxidation behaviors of these catalysts were further investigated using real-time Raman spectroscopy. Most importantly, a possible mechanism for cyclohexene photo-epoxidation was also proposed based on the results of Raman technique.

2. Results and Discussion

2.1. Catalysts Characterization

Table 1 shows the nomenclature, wt.% loading and V:Ti atomic weight ratios of series V-Ti/MCM-41 catalysts. The V:Ti atomic weight ratios measured by ICP-AES analysis are corresponded to the nominal ones. As shown in Table 2, similar results from the surface area and average pore diameter has been observed. The surface areas were in the range of 815–891 m^2/g while BJH pore diameters were within 2.9–3.0 nm. These data indicate that the structure of a series MCM-41 supported in this study is very uniform.

Table 1. Nomenclature, wt.% loading and atomic weight ratios of series catalysts.

Entry	Nomenclatures	V:Ti:Si atomic weight ratio [a] (nominal)			V loading [b] (wt.%)	Ti loading [b] (wt.%)	V:Ti:Si atomic weight ratio [c] (ICP-AES)		
		V	Ti	Si			V	Ti	Si
1	V_2Ti_0/MCM-41	2	0	100	0.36	0	1.7	0	100
2	V_1Ti_1/MCM-41	1	1	100	0.18	0.29	0.9	1.3	100
3	$V_{0.25}Ti_2$/MCM-41	0.25	2	100	0.04	0.53	0.22	2.3	100
4	$V_{0.05}Ti_4$/MCM-41	0.05	4	100	0.01	0.96	0.06	4.2	100
5	V_0Ti_5/MCM-41	0	5	100	0	1.12	0	4.9	100

[a] Atomic weight ratio (V:Ti:Si), based on the assumption of preparation process by hydrothermal method; [b] Loading content (wt.%) of V and Ti, based on ICP-AES; [c] Atomic weight ratio (V:Ti:Si), calculated from ICP-AES.

Table 2. Specific surface area, average pores diameter and band gap of series catalysts.

Entry	Catalysts	Specific surface area (m^2/g)	BJH average pore diameter (nm)	Band gap (eV)
1	V_2Ti_0/MCM-41	856	3.0	3.1
2	V_1Ti_1/MCM-41	815	3.0	3.4
3	$V_{0.25}Ti_2$/MCM-41	891	2.9	4.5
4	$V_{0.05}Ti_4$/MCM-41	841	2.9	4.4
5	V_0Ti_5/MCM-41	850	2.9	4.3

Figure S1 (Supporting Information) shows a single and narrow pore size distribution of V-Ti/MCM-41 with various ratios of V-/Ti-loading, as estimated following the BJH method from the desorption isotherms. The V-Ti loadings changed very low so they did not appreciably change the average pore diameter of MCM-41. The pore diameters of the synthesized V-Ti/MCM-41 catalysts were near 3 nm and were not significantly affected by the presence of vanadia and titania. The results proved that our procedure to incorporate vanadium and titanium-containing catalysts did not deteriorate the MCM-41 structure.

The XRD patterns of V-Ti/MCM-41 catalysts prepared with different V-Ti ratios are shown in Figure 1. All catalysts had very similar patterns, indicating that all V-/Ti-loading into Si–O framework of MCM-41 had no apparent structure influence on its structure. All the V-Ti/MCM-41 catalysts exhibited a hexagonal lattice of mesoporous silica structures from a clear observation of spectra featured narrow (100) peaks. Nonetheless, the (110) and (200) peaks were not well distinguished. Additionally, there was no peak assigned to metal oxides, indicating that the metal ions were atomically dispersed on the internal and/or external surfaces of MCM-41. On the other hand, the amount of V and Ti loading was quite low (Table 1) that may also led in the absence of foreign ions peaks in XRD patterns. This observation is found to consist of Raman result of $V_{0.25}Ti_2$/MCM-41 along with that of TiO_2-anatase and V_2O_5 spectral references (Figure 2). Neither of the characteristic vibrations of titania anatase (or any other titania phase) nor those of crystalline vanadia was also apparent in the Raman spectrum of fresh $V_{0.25}Ti_2$/MCM-41.

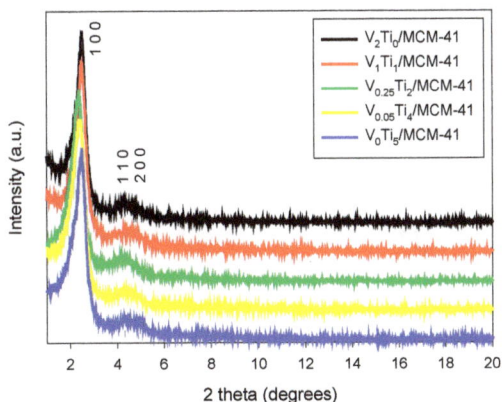

Figure 1. XRD patterns of V-Ti/MCM-41 catalysts.

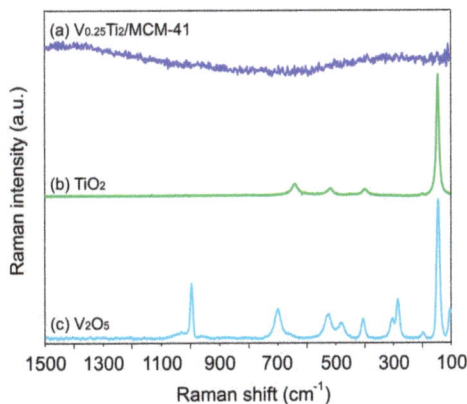

Figure 2. The 514 nm Raman spectra of (a) $V_{0.25}Ti_2$/MCM-41, (b) TiO_2 anatase and (c) V_2O_5 catalysts under ambient conditions.

The UV-vis spectra of catalysts with different Ti-/V-loading ratios in MCM-41 are displayed in Figure 3. The red-shift clearly increased with the growth amount of Ti-/V-loading. The spectra of Ti-loading exhibited a strong absorption band centered at about 220 nm, which was consistent with tetrahedral titania species on silica-based substrates [23]. Moreover, the presence of the peaks at 230 and 270 nm may attribute to the isolated Ti atoms in octa-coordinated Ti species and polymerized hexa-coordinated Ti species, respectively [24,25]. On the other hand, the absorption at 220 nm shifted towards 245 nm while a second component became increasingly apparent at 340 nm when V-loading was incremented. The absorption band near 340 nm was presented due to the charge transfer bands of highly dispersed tetra-coordinated V-oxide species present to the V^{5+} state [26]. It is noted that the V^{5+} species absorbed not only at 340 nm, but also at 260 nm [27,28]. In addition, the region band of 320–350 and 400 nm might be associated to polymeric V^{5+} species and octa-coordinated vanadium, respectively [27,29]. For further investigation, the band gap of series catalysts with different amounts of Ti and V loading in MCM-41 were calculated based upon the UV-vis results and listed in the Table 2.

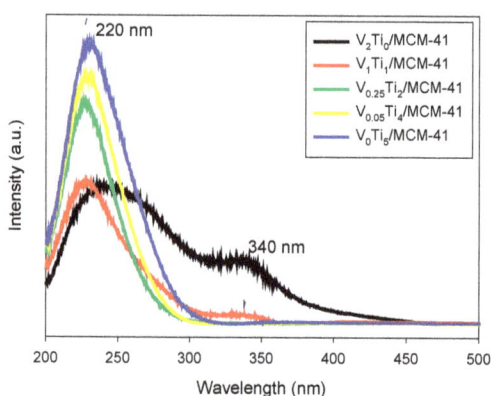

Figure 3. UV-vis spectra of the series V-Ti/MCM-41 catalysts.

X-ray photoelectron spectroscopy (XPS) has been used to determine the electronic properties of the component existed on the surface. The XPS spectra are obtained and shown in Figure 4. For various Ti-oxide and V-oxide contents, the binding energies at 459.9 and 516.7 eV have been observed from the deconvolution of spectra for the Ti $2p_{3/2}$ and V $2p_{3/2}$ band, respectively. It is noted to mention that the binding energy of Ti $2p_{3/2}$ on silica MCM-41 was higher than that of pure TiO_2 (458.2–459.0 eV) [30]. It is possible that the high binding energy reflects the tetrahedral coordinated Ti-oxide component in the strongly interaction between titania and silica Ti–O–Si, which agrees well with the value of Ti^{4+} supported on silica (458.8–459.9 eV), reported by Castillo *et al.* [31]. Importantly, all the samples showed similar binding energy of Ti, suggesting that there is no apparent interaction between Ti and V species. On the other hand, the binding energy of V $2p_{3/2}$ was in perfect agreement with V_2O_5 (516.6–517.5 eV) [30]. Due to the little amount of V on

148

MCM-41 (V loading in range of 0.01–0.36 wt.%), its XPS peaks could not be observed clearly. The XPS result is consistent with the observation from UV-vis.

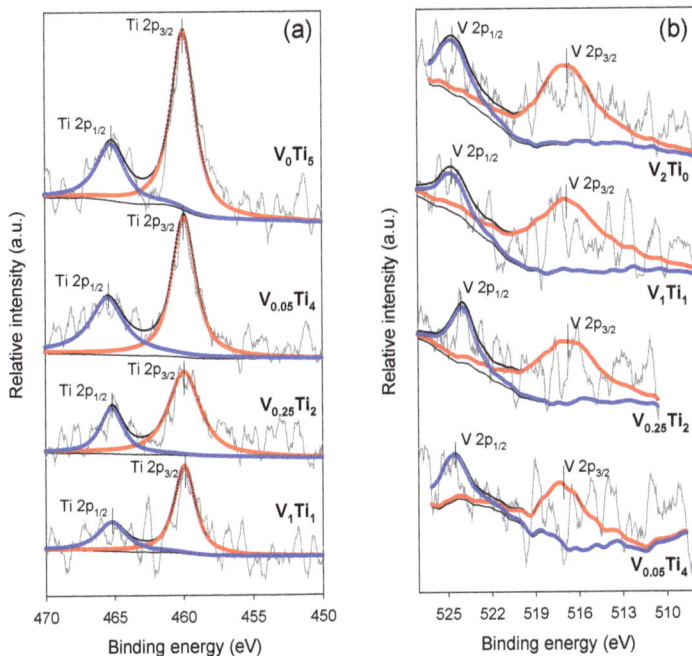

Figure 4. The XPS spectra of the series V-Ti/MCM-41 catalysts: (a) Ti 2p and (b) V 2p.

2.2. Liquid-Phase Epoxidation of Cyclohexene

Table 3 shows the blank tests, which are examined before performing the epoxidation of cyclohexene, underline that the reaction over catalysts is essentially photocatalytic. In details, negligible epoxidation was observed either when the reactor was UV illuminated in the absence of $V_{0.25}Ti_2$/MCM-41 catalyst, or if the catalyst was run at 60 °C without UV illumination. The same phenomenon was also observed in the presence of MCM-41 under light irradiation.

Table 3. Blank test results for cyclohexene epoxidation.

Entry	Conditions			Conversion of cyclohexene (%)
	Catalyst	UV irradiation	Temp. (°C)	
1	NO	NO	80	5.5
2	NO	YES	60	3.4
3	$V_{0.25}Ti_2$/MCM-41	NO	60	0.9
4	MCM-41	YES	60	0.9

The data was collected after 6 h in reaction. Cyclohexene conversion = 100% × (moles of initial cyclohexene—Moles of final cyclohexene)/moles of initial cyclohexene.

2.2.1. Effect of V-Ti/Si Weight Ratio on the Photocatalytic Epoxidation

The effect of V-Ti/Si ratio on the photocatalytic epoxidation activity is summarized in Table 4. In the presence of single active sites, V_0Ti_5/MCM-41 achieves high conversion to 1,2-epoxycyclohexane (47%) while V_2Ti_0/MCM-41 receives only 37% of conversion. We might conclude that both V- and Ti-loading can be responsible for the cyclohexene epoxidation. The complementary behavior of V- and Ti-loading may be related to a similar role of activation. We will discuss this in greater detail later in the next section. It is noted that the weight percent of V-/Ti-loading in V_2Ti_0/MCM-41 and V_0Ti_5/MCM-41 are 0.36% and 1.12%, respectively. Therefore, V_0Ti_5/MCM-41, as expected, would be more effectively than that of V_2Ti_0/MCM-41. In the presence of dual active sites, V_1Ti_1/MCM-41 promotes the best efficiency (44%), among the candidates. Moreover, decrease the V/Ti ratio will also result on the decline the photo-reaction efficiency.

Table 4. Photocatalytic epoxidation at 60 °C for 6 h using various V-Ti weight ratios.

Entry	Catalysts	V/Ti ratio	Conversion of cyclohexene (%)
1	V_2Ti_0/MCM-41	-	37
2	V_1Ti_1/MCM-41	1	44
3	$V_{0.25}Ti_2$/MCM-41	0.125	35
4	$V_{0.05}Ti_4$/MCM-41	0.0125	38
5	V_0Ti_5/MCM-41	-	47

Cyclohexene conversion = 100% × (moles of initial cyclohexene − moles of final cyclohexene)/moles of initial cyclohexene.

2.2.2. Catalytic and Photocatalytic Activity Comparison on $V_{0.25}Ti_2$/MCM-41 Catalyst

Additional epoxidation experiments were carried out to gain better insight on the role of photo/thermal driving forces. All reactions were compared in the presence of $V_{0.25}Ti_2$/MCM-41 catalyst under different temperature with/without UV irradiation, as shown in Figure 5. In more details, reaction at 80 °C without UV irradiation delivers 10.9% of conversion. The decreasing reaction temperature from 80 °C to 60 °C in the dark results in no appreciable activity within 6 h. Conversely, reaction at 60 °C under UV irradiation achieves 23.2% while reaction at 25 °C under UV irradiation delivers only 2.1% of conversion. It is worthwhile to mention that negligible epoxidation was also observed in the presence of bare MCM-41 under light irradiation (Table 3). Taking those results into consideration, the epoxidation over $V_{0.25}Ti_2$/MCM-41 catalyst takes place as a consequence of both photo-assisted and thermal reactions.

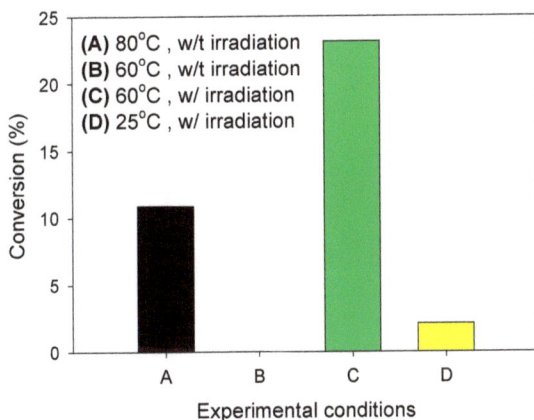

Figure 5. Catalytic and photocatalytic epoxidation of cyclohexene over $V_{0.25}Ti_2$/MCM-41 catalyst for 6 h under different conditions: (**A**) 80 °C, without irradiation; (**B**) 60 °C, without irradiation; (**C**) 60 °C, with irradiation; and (**D**) 25 °C, with irradiation.

2.3. Real-time Raman Monitoring of Cyclohexene Epoxidation

Figure 6 shows real-time Raman monitoring during epoxidation of cyclohexene with t-BuOOH over $V_{0.25}Ti_2$/MCM-41 catalyst under UV-light irradiation at 60 °C. The photocatalytic reaction was monitored by Raman spectroscopy without the interference of UV light. The full range spectra of representative Raman spectra during the reaction are illustrated in Figure 6a. The characteristic Raman bands of reactant (cyclohexene), oxidant (t-BuOOH) and product (1,2-epoxycyclohexane) correspond to those of the reference compounds (Figure S2–S4, Supporting Information). For more details, Figure 6b shows representative Raman bands of both reactants and final product. The Raman peak of 1,2-epoxycyclohexane at 784 cm^{-1} grows while those peaks belonging to cyclohexene and t-BuOOH decrease simultaneously. The most interesting result has been observed that no other Raman peaks are apparent during this reaction; thus, the Raman spectra confirms that 1,2-epoxycyclohexane is obtained as the only reaction product (100% selectivity) by photocatalytic epoxidation. Additionally, Raman spectra delivers further insight into the photo-reaction mechanism. The inset of Figure 6b shows an isosbestic point near 800 cm^{-1}, between the decreasing Raman band of cyclohexene at 824 cm^{-1} and the increasing band of the epoxide at 784 cm^{-1}; these indicate that there is no stable intermediate species formed during the photo-epoxidation of cyclohexene into 1,2-epoxycyclohexane.

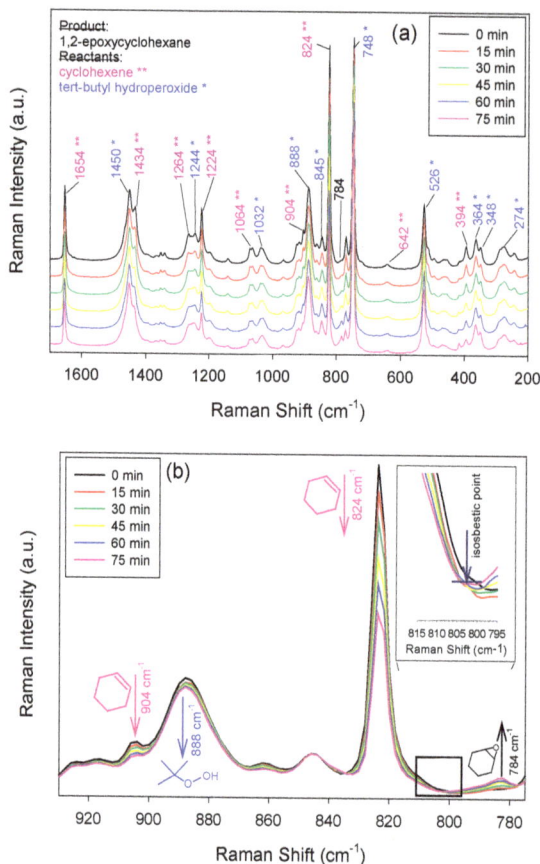

Figure 6. Representative Raman spectra during the photocatalytic cyclohexene epoxidation with t-BuOOH over $V_{0.25}Ti_2$/MCM-41 catalyst at 60 °C for 75 min: (a) full spectra; (b) partial spectra in range of 775–930 (cm^{-1}).

For comparative purposes, we performed the thermal epoxidation of cyclohexene in t-BuOOH on $V_{0.25}Ti_2$/MCM-41 catalyst at 80 °C without UV-light. Figure 7a–c show representative spectra range of 1900–860 cm^{-1}, 860–760 cm^{-1} and 800–200 cm^{-1}, respectively, that recorded every 10 min during reaction. It is clearly observed Raman bands of the reactant (cyclohexene) and the oxidant (t-BuOOH) decrease in intensity and the product (1,2-epoxycyclohexane) increases in intensity during time-on-stream. Figure 7b illustrates a zoom into the window with representative Raman bands of the participating species, showing the consumption of cyclohexene (824 cm^{-1}); those of t-BuOOH (770 cm^{-1} and 845 cm^{-1}) hardly changed in intensity during reaction since it is supplied in excess. It should be noted that the 1,2-epoxycyclohexane (784 cm^{-1}) is gradually produced. In brief summary, $V_{0.25}Ti_2$/MCM-41 catalyst is not only successful in the epoxidation of cyclohexene under UV-light irradiation but it can also work in the dark at higher temperature, 80 °C. Furthermore, Real-time Raman spectra did not find any interaction between organic compounds and the active sites of the catalyst.

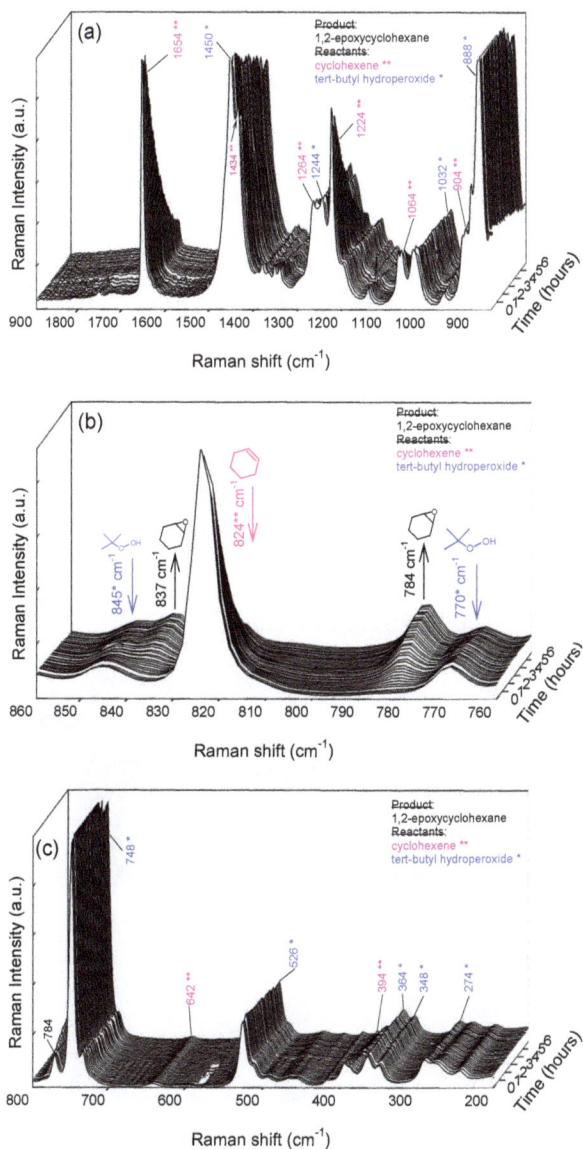

Figure 7. Representative Raman spectral range of (**a**) 1900–860 cm^{-1}; (**b**) 860–760 cm^{-1}; and (**c**) 800–200 cm^{-1} during the thermal catalytic epoxidation of cyclohexene with *t*-BuOOH over V$_{0.25}$Ti$_2$/MCM-41 catalyst at 80 °C for 6 h.

2.4. Proposed Photo-Mechanism for the Formation of Cyclohexene Epoxidation

Our UV-vis and XPS results indicate that V- and Ti-loading, which are separately well-dispersed in the mesoporous MCM-41, mainly exist in the tetra-coordinated (4-folds coordination). Thus, they can be easily excited under UV-light irradiation to form the corresponding charge-transfer excited states involving an electron transfer from O^{2-} species to V^{5+} and Ti^{4+} species, respectively (Equations (1) and (2)) [32–34].

$$[V^{5+} = O^{2-}] \rightleftharpoons [V^{4+}-O^{-}]* \tag{1}$$

$$[Ti^{4+}-O^{2-}] \rightleftharpoons [Ti^{3+}-O^{-}]* \tag{2}$$

It is in agreement that the high reactivity of these charge-transfer excited state is possible due to the electron-hole pair states are localized in close proximity [33]. On the other hand, Raman result suggests that cyclohexene is directly photo-epoxidized to 1,2-epoxycyclohexane in the presence of t-BuOOH. Taking into account these results, a possible photo-epoxidation mechanism by highly dispersed V-oxides and Ti-oxides single-site photocatalysts is proposed, as seen in Scheme 1. Firstly, either V^{5+} or Ti^{4+} species is photo-excited to $[V^{4+}-O^{-}]*$ and $[Ti^{3+}-O^{-}]*$, respectively. As mentioned above, both V- and Ti-loading can be responsible for the cyclohexene epoxidation. In addition, the complementary behavior of V- and Ti-loading may be related to a similar role of activation. Therefore, these photo-excited state species, within their lifetimes, are able to react with t-BuOOH to form (tBuO$^-$) radical, which is responsible for the formation of 1,2-epoxycyclohexane.

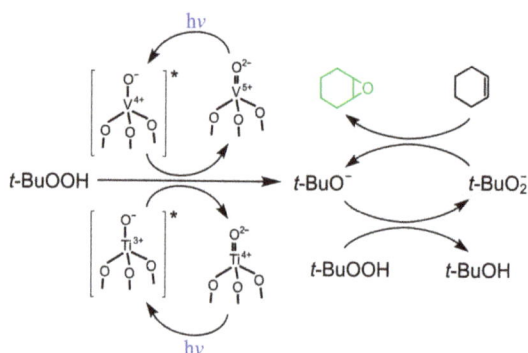

Scheme 1. Proposed photo-mechanism via vanadium and titanium active species during photocatalytic epoxidation of cyclohexene with *tert*-butyl hydroperoxide (t-BuOOH).

When both of V-oxides and Ti-oxides present, it might interact together to further enhance efficiency. In a previous study, Davydov *et al.* proposed that the charge separation occurred with a hole and an electron in Ti-O species [35]. In particular, the Cr^{5+} species of $[Cr^{5+}-O^-]^*$ can donate an electron into the surrounding Ti–O moieties while O^- could scavenge an electron from the surrounding Ti–O moieties. It is important to note that vanadium ions would similarly undergo an equivalent charge transition to $[V^{4+}-O^-]^*$ excited states because V^{5+} and Cr^{6+} ions have the same electronic structure [34]. Therefore, the charges, which might happen at or near the photocatalyst's surface, can induce the photo-epoxidation by either direct excitation via Ti–O moieties or indirect excitation via charge transition from $[V^{4+}-O^-]^*$ states. As a result, the existence of both V- and Ti-loading will further enhance the efficiency of reaction.

3. Experimental Section

3.1. Preparation of Catalysts

The mesoporous V-Ti/MCM-41 catalysts were prepared by a hydrothermal method, which was based on a procedure described by Hung *et al.* [36], using the following general gel composition (molar ratios) 1 SiO_2: 0.2 CTAB: 0.89 H_2SO_4: 120 H_2O. In our research, we named each catalyst with different metal loading weight ratios (V-Ti-Si), such as (2-0-100), (1-1-100), (0.25-2-100), (0.05-4-100), (0-5-100). The whole synthesis procedure is shown in Scheme 2.

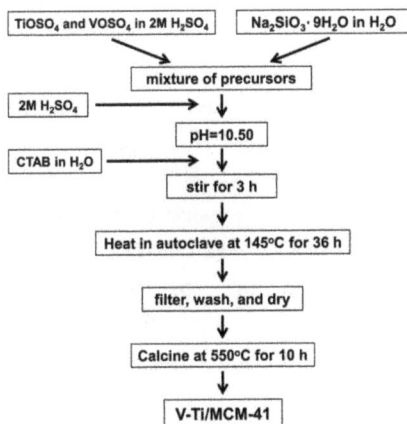

Scheme 2. Flow chart for synthesis procedure of V-Ti/MCM-41 photocatalysts.

In a typical synthesis procedure, 21.2 g of sodium metasilicate monohydrate (J.T.Baker, Avantor Performance Materials, Center Valley, PA, USA) dissolved in 100 mL of distilled water is combined with the appropriate amount of titanium oxysulfate hydrate (Sigma-Aldrich, St. Louis, MO, USA) and/or vanadyl sulfate hydrate (Acros, Geel, Belgium) metal precursors (dissolved in 20 mL of 2 M H_2SO_4 (95%–97%, Sigma-Aldrich)), respectively. The resulting mixture is stirred vigorously for 30 min. Then, 2 M H_2SO_4 is added to the above mixture to adjust the pH to 10.5 while constantly stirring to form a uniform gel. After stirring, the solution containing 7.28 g cetyltrimethylammonium bromide (CTAB, 98%, Alfa Aesar, Haverhill, MA, USA) is dissolved in 25 mL of distilled water and is added slowly into the above mixture and the combined mixture is stirred for three additional hours. The CTAB surfactant was used as the structure-directing template. The resulting gel mixture is transferred into a Teflon coated autoclave and kept in an oven at 145 °C for 36 h. After cooling to room temperature, the resulting solid is recovered, washed and dried in an oven at 80 °C for 8 h at least. Finally, the organic template is removed after calcination at 550 °C for 10 h.

3.2. Characterization of Catalysts

Powder X-ray diffraction (XRD) data was obtained on a Panalytical X'Pert Pro diffractometer (Almelo, Netherlands) to verify the crystalline structure. The diffraction patterns were checked and assigned to known crystalline phases. The light absorption of photocatalysts was fully characterized by UV-vis spectroscopy (UV-vis, Varian Cary 100, Agilent Technologies, Santa Clara, CA, USA). The Inductively Coupled Plasma—Atomic Emission Spectra (ICP-AES) Analysis was carried out by PerkinElmer Optima 2000 (PerkinElmer, Waltham, MA, USA) to determine the chemical composition of the V, Ti and Si elements in V-Ti/MCM-41 catalysts. X-ray photoelectron spectroscopy (XPS) was conducted on a Thermo Theta Probe instrument (Thermo Fisher Scientific Inc., East Grinstead, UK) to determine the electronic properties of the component existed on the surface. Additionally, nitrogen adsorption and desorption isotherms, surface area, and mesopore size distribution were measured using a Micromeritics ASAP2000 (Micromeritics, Norcross, GA, USA). Specific surface areas and pore size distributions were calculated using the Brunnauer-Emmett-Teller (BET) and Barrett-Joyner-Halenda (BJH) methods, respectively. Furthermore, Raman spectra for the characterization of the solid catalysts were obtained with a single monochromator Renishaw Raman System 1000 (Renishaw, Gloucestershire, UK) equipped with a thermoelectrically cooled CCD detector (-73 °C) and holographic Edge filter. The samples were excited with the 514 nm Ar line. The spectral resolution was 3 cm^{-1}, and the spectrum acquisition was 300 s for each sample.

3.3. Liquid-Phase Photocatalytic Epoxidation of Cyclohexene

The experiments of liquid-phase photocatalytic epoxidation of cyclohexene were performed in a three-neck round-bottom flask reactor equipped with a reflux condenser in a temperature-controlled oil bath as shown in Scheme 3. Initially, 50 mmol of *tert*-butyl hydroperoxide (*t*-BuOOH, Aldrich, 70% in water) and 5 mol% of catalysts were added to the reactor. After that, the reaction mixture was magnetically stirred and heated until the desired reaction temperature (25 °C, 60 °C and 80 °C, respectively). The reaction was started by adding 25 mmol of cyclohexene (99%, Alfa Aesar, Haverhill, MA, USA) to the mixture. The UV-light was irradiated from EXFO S1500 (EXFO Inc., Quebec, Canada) equipped with 200W Mercury-Arc lamp (0.5 mW/cm^2, filter: 320–500 nm) and guided by an optical fiber that can insert into the three-necked reactor. The light intensity was detected at the output of optical fiber by GOLDILUX Radiometer/Photometer (UV-A Probe/UV-C Probe). Reproducibility tests were run on several catalysts to confirm the accuracy of experiments that we assured the error on photocatalytic reactions were within 10%. Since the effect of Raman is a millionth to ten millionths weaker than Rayleigh scattering, any residual light at longer wavelengths from the UV light source had to be filtered off. To do so, a filter rejecting any radiation with wavelength longer than 750 nm (Thorlabs, FM201) was located between UV-light source and the sample. Its transmittance was 100% below 750 nm.

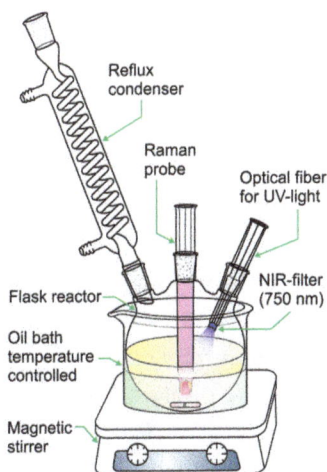

Scheme 3. Reaction system for the liquid-phase epoxidation: (1) reflux condenser; (2) three-neck flask reactor; (3) temperature-controlled oil bath; (4) Raman probe; (5) optical fiber for UV-light irradiation; (6) NIR-filter (750 nm) integrated with optical fiber; and (7) magnetic stirrer.

The liquid-phase cyclohexene epoxidation was continuously monitored by Raman spectroscopy with an in Photonics immersion probe fitted to a Perkin-Elmer Raman Station 400F system (Perkin-Elmer Inc., Shelton, CT, USA), equipped with a thermoelectrically cooled CCD detector (−73 °C) and Edge filter. The NIR filter located at the UV-light source prevented any interference from residual longer wavelengths. All the spectra were acquired every 10 min upon excitation with a near-infrared 785 nm excitation line and consisted of 6 accumulations of 10 s. The quantitative analysis of Raman spectra based on the multivariate methods analysis. The collected spectra (which include reactants and product) were analyzed by the evaluation Spectrum QUANT + V4.51 software (PerkinElmer). A partial least squares (PLS) regression was done and the calibration data, which use different calibration compositions (20 standards) of cyclohexene and 1,2-epoxycyclohexane , were stored with a correlation coefficient R^2 more than 99% and an average error of analysis less than 2%.

Additionally, reaction samples before and after 6 h in reaction were also analyzed by gas chromatography in an YL6100 GC equipped with a Porapak-N column and a flame ionization detector (FID). The cyclohexene conversion was defined in Equation (3).

$$\text{Cyclohexene conversion} = 100\% \times (\text{moles of initial cyclohexene} - \text{moles of final cyclohexene})/\text{moles of initial cyclohexene}$$

(3)

4. Conclusions

A series of V- and/or Ti-loading MCM-41 catalysts are successfully synthesized with a hydrothermal method. At low metal loading, of both vanadium and titanium ions, which mainly exists as tetrahedral coordination, are well dispersed to the Si–O framework. Both V- and Ti-loading can be responsible for the cyclohexene epoxidation. Moreover, the complementary behavior of V- and Ti-loading may be related to a similar role of activation. On the other hand, real-time Raman monitoring during thermo-photocatalytic cyclohexene epoxidation is for the first time successfully conducted without interference from UV-light irradiation. The results indicate that cyclohexene is directly photo-epoxidized to 1,2-epoxycyclohexane in the presence of t-BuOOH, thus, it runs directly with no stable intermediate. A possible mechanism of cyclohexene photo-epoxidation was proposed based on the Raman result.

Acknowledgments: We gratefully acknowledge the National Science Council of Taiwan for financial support of this research under contract number NSC 99-2923-E-002-002-MY2 (Taiwan), CTQ2011-13343-E (Spain) and bilateral NSC-CSIC project 2009TW0021.

Author Contributions: H.-Y. Chan, J.C.S. Wu, M.A. Bañares and H. Bai conceived and designed the research concept; V. Calvino-Casilda and H.-Y. Chan prepared the setup and performed the experiments; H.-Y. Chan, V.-H. Nguyen and V. Calvino-Casilda analyzed the data; V. Calvino-Casilda, M.A. Bañares, V.-H. Nguyen and J.C.S. Wu prepared the manuscript; and V.-H. Nguyen and J.C.S. Wu finished the manuscript.

Conflicts of Interest: The authors declare no conflict of interest.

References

1. Oyama, S.T. Rates, Kinetics, and Mechanisms of Epoxidation: Homogeneous, Heterogeneous, and Biological Routes. In *Mechanisms in Homogeneous and Heterogeneous Epoxidation Catalysis*; Oyama, S.T., Ed.; Elsevier: Amsterdam, The Netherlands, 2008; pp. 3–99.

2. Nguyen, V.-H.; Chan, H.-Y.; Wu, J.C.S.; Bai, H. Direct gas-phase photocatalytic epoxidation of propylene with molecular oxygen by photocatalysts. *Chem. Eng. J.* **2012**, *179*, 285–294.

3. Nguyen, V.-H.; Chan, H.-Y.; Wu, J. Synthesis, characterization and photo-epoxidation performance of Au-loaded photocatalysts. *J. Chem. Sci.* **2013**, *125*, 859–867.

4. Nguyen, V.-H.; Wu, J.C.S.; Bai, H. Temperature effect on the photo-epoxidation of propylene over V-Ti/MCM-41 photocatalyst. *Catal. Commun.* **2013**, *33*, 57–60.

5. Nguyen, V.-H.; Lin, S.D.; Wu, J.C.-S.; Bai, H. Artificial sunlight and ultraviolet light induced photo-epoxidation of propylene over V-Ti/MCM-41 photocatalyst. *Beilstein J. Nanotechnol.* **2014**, *5*, 566–576.

6. Nguyen, V.-H.; Lin, S.D.; Wu, J.C.S.; Bai, H. Influence of co-feeds additive on the photo-epoxidation of propylene over V-Ti/MCM-41 photocatalyst. *Catal. Today* **2015**, *245*, 186–191.

7. Kim, I.; Kim, S.M.; Ha, C.-S.; Park, D.-W. Synthesis and cyclohexene oxide/carbon dioxide copolymerizations of zinc acetate complexes bearing bidentate pyridine-alkoxide ligands. *Macromol. Rapid Commun.* **2004**, *25*, 888–893.

8. Anaya de Parrodi, C.; Juaristi, E. Chiral 1,2-amino alcohols and 1,2-diamines derived from cyclohexene oxide: recent applications in asymmetric synthesis. *Synlett* **2006**, *2006*, 2699–2715.

9. Gago, S.; Rodríguez-Borges, J.E.; Teixeira, C.; Santos, A.M.; Zhao, J.; Pillinger, M.; Nunes, C.D.; Petrovski, Ž.; Santos, T.M.; Kühn, F.E.; *et al.* Synthesis, characterization and catalytic studies of bis(chloro)dioxomolybdenum(VI)-chiral diimine complexes. *J. Mol. Catal. A* **2005**, *236*, 1–6.

10. Li, G.; Xu, M.; Larsen, S.C.; Grassian, V.H. Photooxidation of cyclohexane and cyclohexene in BaY. *J. Mol. Catal. A* **2003**, *194*, 169–180.

11. Larsen, R.G.; Saladino, A.C.; Hunt, T.A.; Mann, J.E.; Xu, M.; Grassian, V.H.; Larsen, S.C. A Kinetic Study of the Thermal and Photochemical Partial Oxidation of Cyclohexane with Molecular Oxygen in Zeolite Y. *J. Catal.* **2001**, *204*, 440–449.

12. Fraile, J.M.; García, J.I.; Mayoral, J.A.; Vispe, E. Optimization of cyclohexene epoxidation with dilute hydrogen peroxide and silica-supported titanium catalysts. *Appl. Catal. A* **2003**, *245*, 363–376.

13. Anand, C.; Srinivasu, P.; Mane, G.P.; Talapaneni, S.N.; Benzigar, M.R.; Vishnu Priya, S.; Al-deyab, S.S.; Sugi, Y.; Vinu, A. Direct synthesis and characterization of highly ordered cobalt substituted KIT-5 with 3D nanocages for cyclohexene epoxidation. *Micropor. Mesopor. Mat.* **2013**, *167*, 146–154.

14. Jin, F.; Chen, S.-Y.; Jang, L.-Y.; Lee, J.-F.; Cheng, S. New Ti-incorporated MCM-36 as an efficient epoxidation catalyst prepared by pillaring MCM-22 layers with titanosilicate. *J. Catal.* **2014**, *319*, 247–257.

15. Calvino Casilda, V.; Pérez-Mayoral, E.; Bañares, M.A.; Lozano Diz, E. Real-time Raman monitoring of dry media heterogeneous alkylation of imidazole with acidic and basic catalysts. *Chem. Eng. J.* **2010**, *161*, 371–376.

16. Mikolajska, E.; Calvino-Casilda, V.; Bañares, M.A. Real-time Raman monitoring of liquid-phase cyclohexene epoxidation over alumina-supported vanadium and phosphorous catalysts. *Appl. Catal. A* **2012**, *421–422*, 164–171.

17. Schmink, J.R.; Holcomb, J.L.; Leadbeater, N.E. Use of raman spectroscopy as an *in situ* tool to obtain kinetic data for organic transformations. *Chem. Eur. J.* **2008**, *14*, 9943–9950.

18. Moreno, T.; Morán López, M.A.; Huerta Illera, I.; Piqueras, C.M.; Sanz Arranz, A.; García Serna, J.; Cocero, M.J. Quantitative Raman determination of hydrogen peroxide using the solvent as internal standard: Online application in the direct synthesis of hydrogen peroxide. *Chem. Eng. J.* **2011**, *166*, 1061–1065.

19. Mozharov, S.; Nordon, A.; Littlejohn, D.; Wiles, C.; Watts, P.; Dallin, P.; Girkin, J.M. Improved Method for Kinetic Studies in Microreactors Using Flow Manipulation and Noninvasive Raman Spectrometry. *J. Am. Chem. Soc.* **2011**, *133*, 3601–3608.

20. Calvino-Casilda, V.; Banares, M.A. Recent advances in imaging and monitoring of heterogeneous catalysts with Raman spectroscopy. In *Catalysis*; The Royal Society of Chemistry: Cambridge, UK, 2012; Volume 24, pp. 1–47.

21. Bañares, M.A. Operando methodology: Combination of *in situ* spectroscopy and simultaneous activity measurements under catalytic reaction conditions. *Catal. Today* **2005**, *100*, 71–77.

22. Salkic, S.; Eckler, L.H.; Nee, M.J. Noninvasive monitoring of photocatalytic degradation of X-ray contrast media using Raman spectrometry. *J. Raman Spec.* **2013**, *44*, 1746–1752.

23. Gao, X.; Wachs, I.E. Titania–silica as catalysts: molecular structural characteristics and physico-chemical properties. *Catal. Today* **1999**, *51*, 233–254.

24. Choi, J.S.; Kim, D.J.; Chang, S.H.; Ahn, W.S. Catalytic applications of MCM-41 with different pore sizes in selected liquid phase reactions. *Appl. Catal. A* **2003**, *254*, 225–237.

25. Geobaldo, F.; Bordiga, S.; Zecchina, A.; Giamello, E.; Leofanti, G.; Petrini, G. DRS UV-Vis and EPR spectroscopy of hydroperoxo and superoxo complexes in titanium silicalite. *Catal. Lett.* **1992**, *16*, 109–115.

26. Laha, S.C.; Kumar, R. Promoter-induced synthesis of MCM-41 type mesoporous materials including Ti- and V-MCM-41 and their catalytic properties in oxidation reactions. *Micropor. Mesopor. Mat.* **2002**, *53*, 163–177.

27. Arnold, A.B.J.; Niederer, J.P.M.; Nießen, T.E.W.; Hölderich, W.F. The influence of synthesis parameters on the vanadium content and pore size of [V]-MCM-41 materials. *Micropor. Mesopor. Mat.* **1999**, *28*, 353–360.

160

28. Lewandowska, A.E.; Banares, M.A.; Tielens, F.; Che, M.; Dzwigaj, S. Different Kinds of Tetrahedral V Species in Vanadium-Containing Zeolites Evidenced by Diffuse Reflectance UV-vis, Raman, and Periodic Density Functional Theory. *J. Phys. Chem. C* **2010**, *114*, 19771–19776.

29. Peña, M.L.; Dejoz, A.; Fornés, V.; Rey, F.; Vázquez, M.I.; López Nieto, J.M. V-containing MCM-41 and MCM-48 catalysts for the selective oxidation of propane in gas phase. *Appl. Catal. A* **2001**, *209*, 155–164.

30. Wagner, C.D.; Muilenberg, G.E. *Handbook of X-ray Photoelectron Spectroscopy: A Reference Book of Standard Data for Use in X-ray Photoelectron Spectroscopy*; Perkin-Elmer Corp., Physical Electronics Division: Eden Prairie, MN, USA, 1979; pp. 68–71.

31. Castillo, R.; Koch, B.; Ruiz, P.; Delmon, B. Influence of the Amount of Titania on the Texture and Structure of Titania Supported on Silica. *J. Catal.* **1996**, *161*, 524–529.

32. Matsuoka, M.; Anpo, M. Local structures, excited states, and photocatalytic reactivities of highly dispersed catalysts constructed within zeolites. *J. Photochem. Photobiol. C* **2003**, *3*, 225–252.

33. Anpo, M.; Kim, T.-H.; Matsuoka, M. The design of Ti-, V-, Cr-oxide single-site catalysts within zeolite frameworks and their photocatalytic reactivity for the decomposition of undesirable molecules—The role of their excited states and reaction mechanisms. *Catal. Today* **2009**, *142*, 114–124.

34. Zou, J.-J.; Liu, Y.; Pan, L.; Wang, L.; Zhang, X. Photocatalytic isomerization of norbornadiene to quadricyclane over metal (V, Fe and Cr)-incorporated Ti-MCM-41. *Appl. Catal. B* **2010**, *95*, 439–445.

35. Davydov, L.; Reddy, E.P.; France, P.; Smirniotis, P.G. Transition-Metal-Substituted Titania-Loaded MCM-41 as Photocatalysts for the Degradation of Aqueous Organics in Visible Light. *J. Catal.* **2001**, *203*, 157–167.

36. Hung, C.; Bai, H.; Karthik, M. Ordered mesoporous silica particles and Si-MCM-41 for the adsorption of acetone: A comparative study. *Sep. Purif. Technol.* **2009**, *64*, 265–272.

Supported Photocatalyst for Removal of Emerging Contaminants from Wastewater in a Continuous Packed-Bed Photoreactor Configuration

Maria Emma Borges, Dulce María García, Tania Hernández,
Juan Carlos Ruiz-Morales and Pedro Esparza

Abstract: Water pollution from emerging contaminants (ECs) or emerging pollutants is an important environmental problem. Heterogeneous photocatalytic treatment, as advanced oxidation treatment of wastewater effluents, has been proposed to solve this problem. In this paper, a heterogeneous photocatalytic process was studied for emergent contaminants removal using paracetamol as a model contaminant molecule. TiO_2 photocatalytic activity was evaluated using two photocatalytic reactor configurations: Photocatalyst solid suspension in wastewater in a stirred photoreactor and TiO_2 supported on glass spheres (TGS) configuring a packed bed photoreactor. The surface morphology and texture of the TGS were monitored by scanning electron microscope (SEM). The influence of photocatalyst amount and wastewater pH were evaluated in the stirred photoreactor and the influence of wastewater flowrate was tested in the packed bed photoreactor, in order to obtain the optimal operation conditions. Moreover, results obtained were compared with those obtained from photolysis and adsorption studies, using the optimal operation conditions. Good photocatalytic activities have been observed and leads to the conclusion that the heterogeneous photocatalytic system in a packed bed is an effective method for removal of emerging pollutants.

Reprinted from *Catalysts*. Cite as: Borges, M.E.; García, D.M.; Hernández, T.; Ruiz-Morales, J.C.; Esparza, P. Supported Photocatalyst for Removal of Emerging Contaminants from Wastewater in a Continuous Packed-Bed Photoreactor Configuration. *Catalysts* **2015**, *5*, 77–87.

1. Introduction

In recent years, one of the most important aspects of environmental research has been the water pollution from emerging contaminants (ECs) such as endocrine disrupting chemicals (EDCs), pharmaceuticals and personal care products (PPCPs). Even they can be usually found in wastewater only at trace levels; their presence in aquatic environments raises the issue of their potential effects on human health and the environment [1–3]. Some adverse potential effects caused by ECs are aquatic

162

toxicity, resistance development in pathogenic bacteria, genotoxicity and endocrine disruption [4–6].

Wastewater treatment plants are designed to remove conventional pollutants, such as suspended solids and biodegradable organic compounds, but they are not designed to remove low concentrations of synthetic pollutants, such as pharmaceuticals. These synthetic pollutants are resistant to conventional wastewater treatments. Advanced oxidation processes are treatments that are considered to be effective at removing EDCs from wastewater effluents. Advanced Oxidation Technology (AOT) has been examined to solve this environmental problem [7,8].

AOT has provided innovative, highly cost-effective and catalyzed chemical oxidation processes for treatment of contaminated soil, sludge and wastewater. AOT can be broadly defined as aqueous phase oxidation methods based primarily on the intermediacy of hydroxyl radicals, $HO^•$, in the mechanism leading to the destruction of the pollutant compounds [9–11]. The hydroxyl radical can be generated photochemically and it has high effectiveness for the oxidation of organic matter [12,13].

Heterogeneous photocatalytic process is one of the most important AOTs and it is based on the oxidation of polluting compounds which can be found in air or water by means of a reaction occurring on a semiconductor catalytic surface activated by light with a specific wavelength.

TiO_2 is the most investigated semiconductor catalyst particularly because it has great potential in solving environmental pollution [14] and it is chemically stable, non-toxic and inexpensive [15–17]. However, TiO_2 has an important disadvantage because it is in powder form and, therefore, a post-treatment separation stage is needed for its use as photocatalytic material in wastewater decontamination by photocatalytic treatment [18,19]. This is the main reason why it is very useful to test the possibility of supporting the TiO_2 active phase on several materials—e.g., activated carbon, silica, glass and polymers—in order for it to be used as photocatalysts in photocatalytic reactors for wastewater decontamination [20–22].

In this study, commercial TiO_2 (Degussa P25) supported on glass spheres was evaluated as photocatalyst in a packed bed photo reactor used to remove emergent contaminants in wastewater. Paracetamol has been used as emergent contaminant model molecule, which is a common analgesic and antipyretic drug, and it is heavily used all over the world and can be found in effluent of wastewater treatment plants [23].

2. Results and Discussion

TiO_2 active phase was analyzed by XRD and its crystalline phases were obtained. Table 1 shows the percentage of anatase and rutile phase, observing that TiO_2 Degussa P25 is composed mainly by the photocatalitically active anatase phase [24,25]. In addition, Table 1 presents the textural parameters of TiO$_2$ Degussa P25, which were examined by N$_2$ adsorption–desorption porosimetry and mercury porosimetry. The textural parameters included in this table are: BET specific area (A_{BET}), total pore area (A), porosity (ε) and density (ρ).

Table 1. TiO$_2$ Degussa P25 characterization.

XRD		N$_2$ adsorption		Hg porosimetry		
% Anatase	% Rutile	A_{BET} (m^2/g))	A (m^2/g)	ε (%)	ρ_{bulk} (g/mL)	$\rho_{apparent}$ (g/mL)
81	19	51.1	63.7	92.46	0.19	2.58

Prior to the study of the photocatalytic process, experiments in the stirred photoreactor have been performed in order to quantify the photolytic effect of the UV light on the paracetamol degradation in wastewater (pH = 8) and the adsorption of the emergent contaminant molecules onto the TiO$_2$ surface. Figure 1 shows the degree of paracetamol removal from wastewater by means of both phenomenon in the photoreactor: Photolysis or adsorption. It can be concluded that the photolytic process has relatively low effect in the paracetamol removal from wastewater under UV light irradiation.

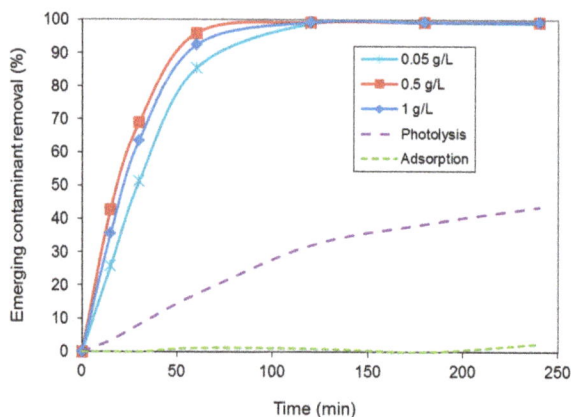

Figure 1. Paracetamol removal in the stirred batch photoreactor using TiO$_2$ suspension as photocatalyst.

The adsorption capacity of the studied photocatalytic material, TiO$_2$ Degussa P25, can be observed in Figure 1, evaluating its degree of contribution in the global process of elimination of the contaminant. As can be seen, the adsorption capacity of tested material is negligible (achieving only a 2%–3% of paracetamol removal from wastewater after 4 h of adsorption treatment) for all tested TiO$_2$ quantities.

The activity of TiO$_2$ photocatalytically active phase was studied in the stirred photoreactor configuration and the influence of the photocatalyst amount in the photoreactor has been presented in Figure 1. It can be seen that the paracetamol removal reaches an approximately constant value after 2 h of photocatalytic treatment for all the quantities of photocatalyst tested achieving values in the range 99%–100% of contaminant removal after 4 h of irradiation time. Therefore, these results from photocatalytic decontamination process can be compared with those results obtained from previous paracetamol photolysis and paracetamol adsorption studies, confirming that the paracetamol removal is mainly due to the heterogeneous photocatalytic treatment. This fact indicates that this kind of emergent contaminants in water can be degraded and mineralized by photocatalytic treatment with TiO$_2$ as photocatalyst, achieving a total elimination of contaminants from wastewater.

Moreover, the results show that the optimal photocatalyst amount is 0.50 g/L, knowing that further increases in the photocatalyst amount used do not increase the photocatalytic ability of the system. This is because when the photocatalyst amount suspended in wastewater is increased, the light path is obstructed in the water due to its turbidity increases causing an "shade" effect and, therefore, HO$^\bullet$ radical formation [26] and the effectiveness of the photocatalytic process induced by light is reduced.

Furthermore, a kinetic study of paracetamol degradation in water by photocatalysis was developed finding that the paracetamol degradation curves exhibit a mono-exponential trend, suggesting that a pseudo-first order reaction model can be applied to describe the kinetic behaviour [6,23,24,27–29] obtained from the TiO$_2$ active phase acting as photocatalyst in the heterogeneous photocatalytic treatment. Pseudo-first order kinetics with respect to paracetamol concentration (C) may be expressed as:

$$-\frac{dC}{dt} = k_{app}C \tag{1}$$

Integration of this equation leads to following equation:

$$\ln (C_0/C) = k_{app} \cdot t \tag{2}$$

where k_{app} is the apparent pseudo-first order rate constant, C_0 is the initial paracetamol concentration and t is the photocatalytic reaction time or irradiation time. The values of k_{app} were obtained directly from the regression analysis of the plot of $\ln(C_0/C)$ vs. photodegradation time for all experiments. Thus, Table 2 shows the k_{app} values obtained for all tested amounts of TiO$_2$ Degussa P25 acting as photocatalyst.

Table 2. Apparent kinetic parameters for paracetamol photodegradation.

Photocatalyst amount (g/L)	k_{app} (min^{-1})
0.05	0.036
0.10	0.040
0.25	0.038
0.50	0.045
1.00	0.042

The pH influence in the paracetamol photodegradation was examined in the stirred photoreactor using the optimal photocatalyst amount (0.50 g/L of TiO$_2$ Degussa P25) obtained from the previous kinetic study. As can be seen in the results presented in Figure 2, pH can also influence the contaminant degradation rate in the global photocatalytic process. The paracetamol removal curves show that the optimal pH is 8 and the degradation rate increases with increasing pH value. This can be attributed to enhanced formation of hydroxide ions, because at high pH more hydroxide ions available on TiO$_2$ surface can be easily oxidized and form more hydroxide ions [30,31] which consequently increases the efficiency of paracetamol degradation. On the other hand, in Figure 2, the paracetamol removal decreased at pH 10 mainly due to surface ionization of TiO$_2$. TiO$_2$ surface is positively charged in acidic media and it is negatively charged at alkaline conditions. Increasing pH gradually increases the electrostatic repulsion between TiO$_2$ surface and paracetamol [32] which is negatively charged at pH above 9.5, and the degradation rate of paracetamol is decreased at a pH higher than 9.5 [6].

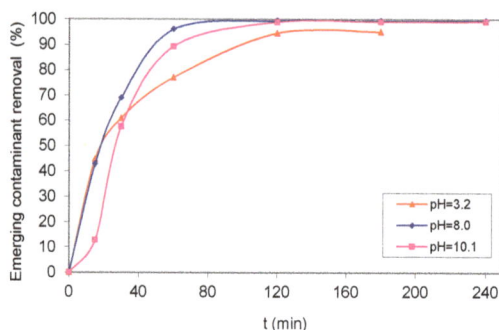

Figure 2. Influence of pH on the paracetamol removal by heterogeneous photocatalysis.

The surface morphology of the glass spheres, TiO$_2$ supported on glass spheres (TGS) and TGS, after being used for photocatalytic experiments were revealed by SEM. The SEM images (Figure 3) show that glass spheres that were TiO$_2$ supported (TGS) exhibit a similar morphology before and after photocatalytic experiments, consequently, demonstrating that the procedure's deposit of TiO_2 active phase over glass spheres is efficient, and a stable support was achieved.

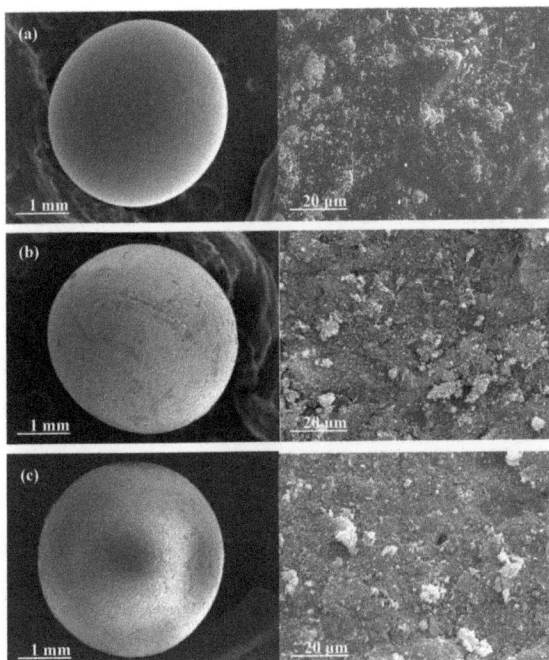

Figure 3. SEM images: (**a**) Glass sphere; (**b**) TiO_2 supported on glass spheres (TGS); (**c**) TGS after photocatalytic experiments.

The paracetamol photodegradation with TGS acting as photocatalyst in the packed-bed photoreactor was evaluated studying first the influence of the wastewater flowrate along the packed-bed photoreactor on the degree of contaminant removal (Figure 4). The photolytic effect of the irradiation light used and the photocatalytic material adsorption capacity related to the global paracetamol removal was evaluated in this reactor configuration using the optimal flowrate in order to determinate the degree of contribution of both phenomenon in the global decontamination process.

As can be seen, the optimal photocatalytic activity was achieved using the lower flowrate along the packed-bed photoreactor (0.8 mL/s), reaching a constant value of paracetamol removal (approximately 42%) after 8 h of photocatalytic treatment. Meanwhile, the global paracetamol removal decreases for higher wastewater recirculation flowrates. In addition, the paracetamol removal shows an important increase after 4 h of photocatalytic treatment when 0.8 mL/s was used as wastewater flowrate, following a pattern similar to those observed in a photocatalytic process for the removal of a mixture of pharmaceuticals from wastewater [33]. These results can be explained by an increased oxidation rate of some intermediates in the photocatalytic process.

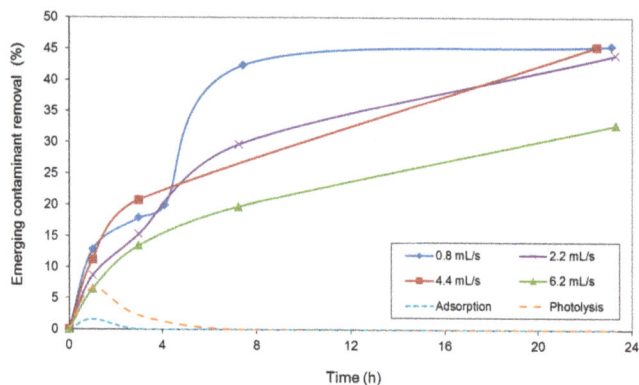

Figure 4. Emerging paracetamol removal in the fixed-bed photoreactor system.

Additionally, from photolysis and adsorption studies, carried out with the optimal flowrate (0.8 mL/s), negligible paracetamol removal was observed after 23 h of treatment. This fact confirms that the paracetamol removal is only due to the heterogeneous photocatalytic treatment in the packed bed photoreactor.

3. Experimental Section

In this paper, the activity of TiO_2 supported over glass spheres (TGS) as photocatalyst was studied evaluating the paracetamol (Acofarma, Llobregat, Spain) (as emergent contaminant model molecule) removal from wastewater. This heterogeneously photocatalyzed decontamination process, using photocatalyst particles in a packed-bed photoreactor configuration (crafted in our laboratory team), was compared with those results obtained from experiments using a TiO_2 active phase (Degussa P25, Frankfurt, Germany) suspension in wastewater in a conventional stirred photoreactor.

The photocatalytically active material, TiO_2 Degussa (Frankfurt, Germany) P25, was examined by X-ray diffraction (XRD) in order to obtain the percentage of anatase and rutile phase. Nitrogen adsorption-desorption porosimetry and mercury porosimetry techniques were used to study its specific surface area and textural properties.

Glass spheres, in the range 2–3 mm diameter, were used as support material in order to configure a packed-bed reactor with TiO_2 catalytically active phase. Before coating, the glass spheres were etched for 24 h in diluted hydrofluoric acid, washed thoroughly with deionized water and dried at 105 °C for 2 h. Titanium (IV) isopropoxide (Aldrich, Steinheim, Germany) and ethanol were mixed with HCl. A hydroxide sol was obtained by hydrolysis of titanium (IV) isopropoxide. The surface modified glass spheres were immersed for 10 min in the hydroxide sol. Finally, glass spheres were dipped into suspension for 10 min and, after coating, they were dried at 105 °C and then calcinated at 450 °C for 2 h [34,35]. The procedure was repeated until the spheres were coated with five layers obtaining TiO_2 supported over glass spheres (TGS) studied as photocatalyst.

The surface morphology and texture of the glass spheres and TGS were monitored by means of scanning electron microscope (SEM).

The paracetamol photodegradation process with TGS as photocatalyst was conducted in a packed bed photoreactor system (Figure 5). TGS were placed in the photocatalytic reactor (18 cm length, 0.6 cm internal diameter) and a solar radiation sodium vapour lamp (Philips, Amsterdam, Netherland, model 400-W G/92/2), placed at a distance of 50 cm from the packed-bed reactor, was used as light source (total radiation flux measured on the packed-bed surface was 160 mW/cm^2). Paracetamol solution in water (50 mg/L) was introduced in the wastewater tank (250 mL), photoreactor temperature was kept constant at 20 °C and the paracetamol solution was recirculated along the system using a peristaltic pump; several samples were taken during 23 h of irradiation time and they were analysed by UV-Vis spectrophotometry determining the absorbance of paracetamol at $\lambda = 244$ nm, in order to study the degradation of paracetamol as contaminant in wastewater. The optimal wastewater recirculation flowrate was studied. Moreover, paracetamol photolysis (with light source and without photocatalyst) and paracetamol adsorption onto TGS (with photocatalyst and without light source) experiments were developed in order to evaluate its contribution to the emergent contaminant degradation or global wastewater decontamination process.

The study of the photocatalytic degradation of paracetamol with TiO_2 active phase as photocatalyst suspension in wastewater was carried out in a stirred photoreactor (800 mL) using a UV radiation mercury lamp (Heraeus, Hanau, Germany, model TQ-150, 150 W) as light source. UV system is placed positioned coaxial inside reaction vessel (total radiation flux measured on the stirred photoreactor was 120 mW/cm^2). In this photoreactor the temperature was kept constant at 20 °C, the paracetamol solution (50 mg/L) and the photocatalytic active phase in powder form (TiO_2 Degussa P25) were introduced into the photoreactor and several samples were taken during 4 h of irradiation time and then analysed by UV-Vis spectrophotometry in order to follow the evolution of paracetamol concentration into the reactor. The optimal amount of photocatalyst and the pH influence were evaluated in this experimental system, pH was adjusted by using NaOH (0.1 M) or HCl (0.1 M). Additionally, paracetamol photolysis and adsorption phenomenon in this reactor configuration were evaluated.

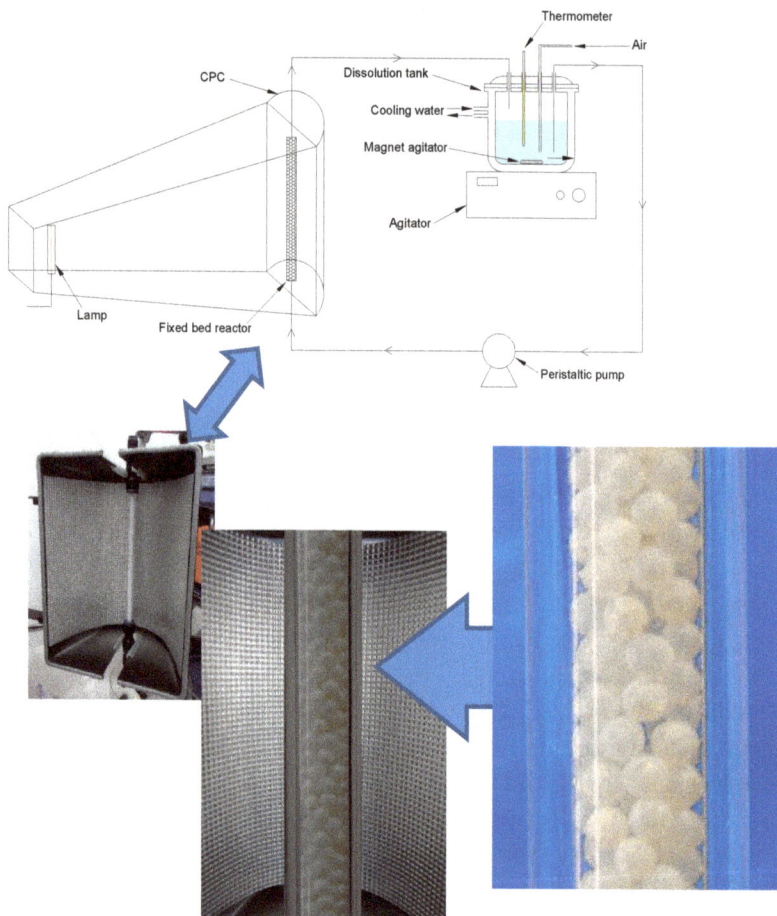

Figure 5. Packed-bed photoreactor system and glass spheres photocatalytic bed in photoreactor.

4. Conclusions

The study of the stirred photoreactor reveals that TiO_2 Degussa P25 shows a good photocatalytic activity for the paracetamol removal from wastewater, reaching a high photodegradation (between 99% and 100% of paracetamol removal after 4 h of irradiation) for all the amounts of photocatalyst tested.

It was concluded that the paracetamol degradation reaction by photocatalysis can be modeled by a pseudo-first-order kinetic equation and the optimal photocatalyst amount is 0.5 g/L obtaining a high photodegradation rate using the photocatalyst in suspension in wastewater. Photocatalytic activity is improved when water in the pH range of conventional domestic wastewater is treated.

Photocatalytic treatment in batch systems using the photocatalyst suspension in water shows disadvantages for treating high volumes of domestic or industrial wastewater, requiring a subsequent treatment after photodegradation of contaminants in order to separate the photocatalyst (TiO_2 in powder form) from the treated water effluent.

A photocatalytic decontamination process using the photocatalytic active phase supported in glass spheres configuring a packed-bed system with continuous recirculating wastewater flow has been implemented. Good photodecomposition of the emergent contaminant has been achieved—approximately 42% of paracetamol removal from wastewater after 8 h of photocatalytic treatment in the photocatalytic packed-bed continuous system. This continuous system shows clear advantages from an industrial treatment of wastewater point of view. Photocatalytic materials remain confined to the packed-bed, and no separation step is necessary for obtaining decontaminated water effluent.

Acknowledgments: This research was partially supported by the project FOTOCAT of "Fundación Cajacanarias" and by Spanish Ministry of Economy and Competitiveness (Project ENE2013-47826-C4-1R). Authors thanks the kind collaboration of Jorge Méndez-Ramos (Physics Department, University of La Laguna).

Author Contributions: All authors were involved in the conception and design of the experiments, as well as the collection, analysis and interpretation of data and writing or editing of this manuscript. All authors approved the final version of the manuscript.

Conflicts of Interest: The authors declare no conflict of interest.

References

1. Hernando, M.D.; Mezcua, M.; Fernández-Alba, A.R.; Barceló, D. Environmental risk assessment of pharmaceutical residues in wastewater effluents, surface waters and sediments. *Talanta* **2006**, *69*, 334–342.
2. Miranda-García, N.; Suárez, S.; Sánchez, B.; Coronado, J.M.; Malato, S.; Maldonado, M.I. Photocatalytic degradation of emerging contaminants in municipal wastewater treatment plant effluents using immobilized TiO_2 in a solar pilot plant. *Appl. Catal. B* **2011**, *103*, 294–301.
3. Zhang, Z.; Hibberd, A.; Zhou, J.L. Analysis of emerging contaminants in sewage effluent and river water: Comparison between spot and passive sampling. *Anal. Chimi. Acta* **2008**, *607*, 37–44.
4. Miranda-García, N.; Maldonado, M.I.; Coronado, J.M.; Malato, S. Degradation study of 15 emerging contaminants at low concentration by immobilized TiO_2 in a pilot plant. *Catal. Today* **2010**, *151*, 107–113.
5. Schriks, M.; Heringa, M.B.; van der Kooi, M.M.E.; de Voogt, P.; van Wezel, A.P. Toxicological relevance of emerging contaminants for drinking water quality. *Water Res.* **2010**, *44*, 461–476.

6. Yang, L.; Yu, L.E.; Madhumita, M.B. Degradation of paracetamol in aqueous solutions by TiO$_2$ photocatalysis. *Water Res.* **2008**, *42*, 3480–3488.

7. Pelaez, M.; Nolan, N.T.; Pillai, S.C.; Seery, M.K.; Falaras, P.; Kontos, A.G.; Dunlop, P.S.M.; Hamilton, J.W.J.; Byrne, J.A.; O'Shea, K.; *et al.* A review on the visible light active titanium dioxide photocatalysts for environmental applications. *Appl. Catal. B* **2012**, *125*, 331–349.

8. Sousa, M.A.; Gonçalves, C.; Vilar, V.J.P.; Boaventura, R.A.R.; Alpendurada, M.F. Suspended TiO$_2$-assisted photocatalytic degradation of emerging contaminants in a municipal WWTP effluent using a solar pilot plant with CPCs. *Chem. Eng. J.* **2012**, *198–199*, 301–309.

9. Litter, M.I.; Quici, N. Photochemical advanced oxidation processes for water and wastewater treatment. *Recent Pat. Eng.* **2010**, *4*, 217–241.

10. Lydakis-Simantiris, A.; Riga, D.; Katsivela, E.; Mantzavinos, D.; Xekoukoulotakis, N.P. Disinfection of spring water and secondary treated municipal wastewater by TiO$_2$ photocatalysis. *Desalination* **2010**, *250*, 351–355.

11. Robert, D.; Malato, S. Solar photocatalysis: A clean process for water detoxification. *Sci. Total Environ.* **2001**, *291*, 85–97.

12. Kositzi, M.; Poulios, I.; Malato, S.; Caceres, J.; Campos, A. Solar photocatalytic treatment of synthetic municipal wastewater. *Water Res.* **2004**, *38*, 1147–1154.

13. Wu, R.J.; Chen, C.C.; Lu, C.S.; Hsu, P.Y.; Chen, M.H. Phorate degradation by TiO$_2$ photocatalysis: Parameter and reaction pathway investigations. *Desalination* **2010**, *250*, 869–875.

14. Wei, Y.-L.; Chen, K.-W.; Wang, H.P. Study of Chromium Modified TiO$_2$ Nano Catalyst under Visible Light Irradiation. *J. Nanosci. Nanotechnol.* **2010**, *10*, 1–5.

15. Bahadur, N.; Jain, K.; Srivastava, A.K.; Govind; Gakhar, R.; Haranath, D.; Dulat, M.S. Effect of nominal doping of Ag and Ni on the crystalline structure and photo-catalytic properties of mesoporous titania. *Mater. Chem. Phys.* **2010**, *124*, 600–608.

16. Peng, Y.-H.; Huang, G.-F.; Huang, W.-Q. Visible-light absorption and photocatalytic activity of Cr-doped TiO$_2$ nanocrystal films. *Adv. Powder Technol.* **2012**, *23*, 8–12.

17. Soutsas, K.; Karayannis, V.; Poulios, I.; Riga, A.; Ntampegliotis, K.; Spiliotis, X.; Papapolymerou, G. Decolorization and degradation of reactive azo dyes via heterogeneous photocatalytic processes. *Desalination* **2010**, *250*, 345–350.

18. Li, Y.; Chen, J.; Liu, J.; Ma, M.; Chen, W.; Li, L. Activated carbon supported TiO$_2$-photocatalysis doped with Fe ions for continuous treatment of dye wastewater in a dynamic reactor. *J. Environ. Sci.* **2010**, *22*, 1290–1296.

19. Van Grieken, R.; Marugán, J.; Sordo, C.; Martínez, P.; Pablos, C. Photocatalytic inactivation of bacteria in water using suspended and immobilized silver-TiO$_2$. *Appl. Catal. B* **2009**, *93*, 112–118.

20. Borges, M.E.; Alvarez-Galván, M.C.; Esparza, P.; Medina, E.; Martín-Zarza, P.; Fierro, J.L.G. Ti-containing volcanic ash as photocatalyst for degradation of phenol. *Energy Environ. Sci.* **2008**, *1*, 364–369.

21. Palau, J.; Colomer, M.; Penya-Roja, J.M.; Martínez-Soria, V. Photodegradation of Toluene, *m*-Xylene, and *n*-Butyl Acetate and Their Mixtures over TiO$_2$ Catalyst on Glass Fibers. *Ind. Eng. Chem. Res.* **2012**, *51*, 5986–5994.

22. Song, M.Y.; Park, Y.-K.; Jurng, J. Direct coating of V_2O_5/TiO_2 nanoparticles onto glass beads by chemical vapor deposition. *Powder Technol.* **2012**, *231*, 135–140.

23. Yang, L.; Yu, L.E.; Madhumita, M.B. Photocatalytic Oxidation of Paracetamol: Dominant Reactants, Intermediates, and Reaction Mechanisms. *Environ. Sci. Technol.* **2009**, *43*, 460–465.

24. Janus, M.; Morawski, A.W. New method of improving photocatalytic activity of commercial Degussa P25 for azo dyes decomposition. *Appl. Catal. B* **2007**, *75*, 118–123.

25. Wong, C.L.; Tan, Y.N.; Mohamed, A.R. A review on the formation of titania nanotube photocatalysts by hydrothermal treatment. *J. Environ. Manag.* **2011**, *92*, 1669–1680.

26. Mangalampalli, V.; Sharma, P.; Sadanandam, G.; Ratnamala, A.; Kumari, V.D.; Subrahmanyam, M. An efficient and novel porous nanosilica supported TiO_2 photocatalyst for pesticide degradation using solar light. *J. Hazard. Mater.* **2009**, *171*, 626–633.

27. Akbarzadeh, R.; Umbarkar, S.B.; Sonawane, R.S.; Takle, S.; Dongare, M.K. Vanadia-titania thin films for photocatalytic degradation of formaldehyde in sunlight. *Appl. Catal. A* **2010**, *374*, 103–109.

28. Dalrymple, O.K.; Yeh, D.H.; Trotz, M.A. Removing pharmaceuticals and endocrine-disrupting compounds from wastewater by photocatalysis. *J. Chem. Technol. Biotechnol.* **2007**, *82*, 121–134.

29. Dimitroula, H.; Daskalaki, V.M.; Frontistis, Z.; Kondarides, D.I.; Panagiotopoulou, P.; Xekoukoulotakis, N.P.; Mantzavinos, D. Solar photocatalysis for the abatement of emerging micro-contaminants in wastewater: Synthesis, characterization and testing of various TiO_2 samples. *Appl. Catal. B* **2012**, *117–118*, 283–291.

30. Zheng, S.; Wang, Q.; Zhou, J.; Wang, B. A study on dye photoremoval on TiO_2 suspension solution. *J. Photochem. Photobiol. A* **1997**, *108*, 235–238.

31. Galindo, C.; Jacques, P.; Kalt, A. Photodegradation of deaminobenzene acid orange 52 by three advanced oxidation process: UV-H_2O_2, Uv-TiO_2 and Vis-TiO_2. Comparative mechanistic and kinetics investigation. *J. Photochem. Photobiol. A* **2000**, *130*, 35–47.

32. Brunner, M.; Schmiedberger, A.; Schmid, R.; Jager, D.; Piegler, E.; Eichler, H.; Muller, M. Direct assesment of peripheral pharmacokinetics in humans: Comparison between cantharides blister fluid sampling, *in vivo* microdialisys and saliva sampling. *Br. Clin. Pharmacol.* **1998**, *46*, 425–431.

33. Rizzo, L.; Meric, S.; Guida, M.; Kassinos, D.; Belgiorno, V. Heterogenous photocatalytic degradation kinetics and detoxification of an urban wastewater treatment plant effluent contaminated with pharmaceuticals. *Water Res.* **2009**, *43*, 4070–4078.

34. Hänel, A.; Morén, P.; Zaleska, A.; Hupka, J. Photocatalytic activity of TiO_2 immobilized on glass beads. *Physicochem. Probl. Miner. Process.* **2010**, *45*, 49–56.

35. Kobayakawa, K.; Sato, C.; Sato, Y.; Fujishima, A. Continuous-flow photoreactor packed with titanium dioxide immobilized on large silica gel beads to decompose oxalic acid in excess water. *J. Photochem. Photobiol. A* **1998**, *118*, 65–69.

Structural Evolution under Reaction Conditions of Supported $(NH_4)_3HPMo_{11}VO_{40}$ Catalysts for the Selective Oxidation of Isobutane

Fangli Jing, Benjamin Katryniok, Elisabeth Bordes-Richard, Franck Dumeignil and Sébastien Paul

Abstract: When using heteropolycompounds in the selective oxidation of isobutane to methacrolein and methacrylic acid, both the keeping of the primary structure (Keggin units) and the presence of acidic sites are necessary to obtain the desired products. The structural evolution of supported $(NH_4)_3HPMo_{11}VO_{40}$ (APMV) catalysts under preliminary thermal oxidizing and reducing treatments was investigated. Various techniques, such as TGA/DTG (Thermo-Gravimetric Analysis/Derivative Thermo-Gravimetry), H_2-TPR (Temperature Programed Reduction), *in situ* XRD (X-Ray Diffraction) and XPS (X-ray Photoelectron Spectroscopy), were applied. It was clearly evidenced that the thermal stability and the reducibility of the Keggin units are improved by supporting 40% APMV active phase on $Cs_3PMo_{12}O_{40}$ (CPM). The partial degradation of APMV takes place depending on temperature and reaction conditions. The decomposition of ammonium cations (releasing NH_3) leads to the formation of vacancies favoring cationic exchanges between vanadium coming from the active phase and cesium coming from the support. In addition, the vanadium expelled from the Keggin structure is further reduced to V^{4+}, species, which contributes (with Mo^{5+}) to activate isobutane. The increase in reducibility of the supported catalyst is assumed to improve the catalytic performance in comparison with those of unsupported APMV.

Reprinted from *Catalysts*. Cite as: Jing, F.; Katryniok, B.; Bordes-Richard, E.; Dumeignil, F.; Paul, S. Structural Evolution under Reaction Conditions of Supported $(NH_4)_3HPMo_{11}VO_{40}$ Catalysts for the Selective Oxidation of Isobutane. *Catalysts* **2015**, *5*, 460–477.

1. Introduction

Keggin-type heteropolyacid and their salts exhibit excellent performance in several catalytic reactions like oxidation, ammoxidation [1,2], oxidative dehydrogenation [3,4] and dehydration [5,6]. More particularly, vanado-molybdo-phosphoric acid ($H_4PMo_{11}VO_{40}$) is a promising bi-functional catalyst for the direct oxidation of isobutane (IBAN) into methacrylic acid (MAA) because of its acidic and redox chemical properties [7–10]. However, the thermal stability of this catalyst, which restructures on-stream, is poor [11]. Later, it was found that the use of large cations like cesium in Keggin-type heteropolymolybdates did not only improve the microstructure of the resulting catalyst, but also enhanced its thermal stability. Among counter-cations of polymolybdates that were studied in literature, ammonium ion NH_4^+ was less often used than H^+, K^+, Cs^+ and their combinations. Paul et al. [12,13] were the first authors to study the ammonium salt of 11-molybdo-1-vanado-phosphoric acid $(NH_4)_3HPMo_{11}VO_{40}$ (APMV) as a catalyst to synthesize methacrolein (MAC) and methacrylic acid starting from isobutane. The results showed that the catalyst was very selective to the desired products. For the same reaction, the catalytic behavior of the ammonia salt of 12-molybdophosphoric acid, $(NH_4)_3PMo_{12}O_{40}$, was examined by Cavani's team [14,15]. They showed that the partial decomposition of ammonium was responsible for the presence of few Mo^{5+} species, the presence of which was correlated with higher selectivity. One could add that, as it is the case for V- or Mo-containing oxides used in the selective oxidation of alkanes, V^{4+} like Mo^{5+} species, could also play a role in the activation of isobutane [16]. In an attempt to increase the surface specific reaction rate, APMV was dispersed on different types of SiO_2-based supports with high surface area, and on the cesium salt of 12-molybdophosphoric acid, $Cs_3PMo_{12}O_{40}$ (CPM). Only the latter was found to increase the catalytic performance as well as to stabilize the APMV active phase [17,18].

Indeed, from the catalytic point of view, it is mandatory to maintain the integrity of $[PMo_{11}VO_{40}]$ or $[PMo_{12}O_{40}]$ polyanions, which are known as constituting the Keggin units, also called the "primary structure". In the case of $PMo_{12}O_{40}{}^{3-}$, a central tetrahedral PO_4 group is spherically surrounded by twelve MoO_6 octahedra. P–O bonds are linked to four Mo_3O_{13} groups. Each Mo_3O_{13} group is formed by three edge-sharing MoO_6 and connected by corner-sharing Mo–O. One (to three) VO_6 groups may replace MoO_6 groups, as in $PMo_{11}VO_{40}{}^{3-}$. In the secondary structure, these Keggin units are maintained together by counter-cations including protons, and possibly water molecules at room temperature. During the catalytic reaction, water molecules may come in and out, the reason why steam is added in reactant feed to avoid the collapse of the resulting secondary structure. Owing to its thermal instability, ammonium NH_4^+ is partially eliminated as NH_3 during the pre-treatment step (calcination), as well as during the catalytic reaction. The restructuring of the solid to a stable catalyst depends on the operating conditions (temperature, atmosphere, flowrates, etc.). The common ex situ measurements, such as XRD, XPS and other spectroscopic methods, give useful information on the catalyst before and after catalytic reaction. However, to examine how the primary and secondary structures are modified when nitrogenated species are lost during calcination and reaction, and to get information about the chemical state of surface elements under stream, in situ techniques were carried out. Indeed, the selective oxidation of isobutane is supposed to proceed via a Mars-Van-Krevelen mechanism [12], stating that in the first step, the catalyst is partially reduced by isobutane, whereby the latter is oxidized by the surface lattice oxygen, before being re-oxidized in the second step by co-fed molecular oxygen. The influence of oxidizing (air) and reducing (hydrogen) atmosphere on the evolution of the precursor and the calcined catalyst was studied by *in situ* XRD measurements for bulk APMV and APMV supported on CPM (40 wt.% active phase loading) (further noted APMV/CPM). Furthermore, APMV/CPM was investigated by XPS after pretreatment in isobutane atmosphere. The catalytic properties will be first recalled, as well as few results already obtained about the reactivity of samples (TGA, H_2-TPR), to allow comparison with the results of the present *in situ* study.

2. Results and Discussion

2.1. Catalytic Evaluation for Isobutane Oxidation Reaction

The catalytic results in oxidation of isobutane for the production of methacrolein and methacrylic acid are just recalled here and listed in Table 1. They were quite different when CPM, APMV and APMV/CPM catalysts were used. The CPM support was catalytically inert. Unsupported APMV exhibited a quite low activity with only 2.5% conversion and the yield in the desired products (MAC + MAA) was as low as 1.4%. By supporting 40% of APMV on CPM, the conversion of isobutane increased by more than five times, from 2.5% to 15.3%, and correspondingly, the yield to MAC + MAA also increased to 8.0%. It is well known that both acid and redox properties are indispensable to catalyze the oxidation of isobutane [19]. The important role of the acidic sites has been discussed elsewhere [17,18]. The absence of acid sites at the surface of the CPM support is responsible for its inertness. The acid and redox properties can be modified by dispersing the active phase APMV on CPM. More importantly, a dynamical surface restructuration, which will be discussed in the present paper, may exist under reaction conditions for the supported sample. Furthermore, the supported catalyst APMV/CPM has proved an excellent stability under stream during 132 h [18].

Table 1. Catalytic performances for selective oxidation of isobutane over the different samples.

| Catalyst | Conversion, % [a] | | Selectivity, % | | | | $Y_{(MAC+MAA)}$, % | Carbon balance, % |
	O_2	IBAN	MAC	MAA	CO_x	Others		
CPM	-	-	-	-	-	-	-	100.0
APMV	10.7	2.5	20.4	34.3	35.8	9.5	1.4	99.7
APMV/CPM	61.6	15.3	10.0	42.0	30.7	17.3	8.0	99.3

[a] Reaction conditions: Temperature = 340 °C, atmospheric pressure, contact time = 4.8 s, IBAN/O_2/H_2O/inert = 27/13.5/10/49.5 (molar ratios). IBAN = isobutane, MAC = methacrolein, MAA = methacrylic acid; others products include acetic acid and acrylic acid.

2.2. Reactivity of the Precursors of APMV and APMV/CPM in Air

2.2.1. Thermal Decomposition

The main results of the TGA-DTG study [18] will just be recalled here, to be further correlated with new TP-XRD experiments. The physisorbed and crystal water were lost up to T_{pret} = 160 °C for unsupported APMV precursor, and up to 200 °C when APMV precursor was supported on CPM (Figure 1). This is a first demonstration of the stabilizing effect of CPM support. Upon increasing temperature, the decomposition of ammonium cations to ammonia (and water) and loss of proton (with formation of water) happened in the 250–424 °C range for APMV and 245–402 °C for APMV/CPM. These signals were immediately followed by a slight gain of weight (but strong DTG signal), which was assigned to the reoxidation of V^{4+} (generated by the decomposition of ammonium to NH_3) to V^{5+} species at 424 °C (APMV), and at 402 °C for APMV/CPM. Thus, all temperatures, except those for H^+ loss and V^{4+} reoxidation, were the lowest for supported APMV. Between 450 °C and 640 °C the weight remained constant and it corresponded to the formation of Keggin-type compound with oxygen vacancy, called $PMoV_{11}O_{38}$ [20]. Indeed in the 250–500 °C range, the weight loss during the transformation of anhydrous APMV to $PMo_{11}VO_{38}$ was 5.0% (theoretical value 4.8%). In the case of 40% APMV supported on CPM the observed weight loss was only 0.9%, instead of the theoretical 1.92% value. These experiments suggest that the thermal stability of APMV was globally improved when supported on CPM. At higher temperatures, the decomposition led to P, V, Mo oxides, followed by the beginning of sublimation of phosphorus oxide and MoO_3.

Therefore, at the temperature of 350 °C chosen for calcination of the precursor, it is expected that some ammonium ions remain in the secondary structure, and that most vanadium species are V^{5+}-type. Both $(NH_4)_3HPMo_{11}VO_{40}$ and $H_3PMo_{11}VO_{40}$ particles may coexist, and/or solid solutions of $(NH_4)_{3-x}H_xPMo_{11}VO_{40}$ type may be present in APMV, while things are more complicated for APMV/CPM, as shown further.

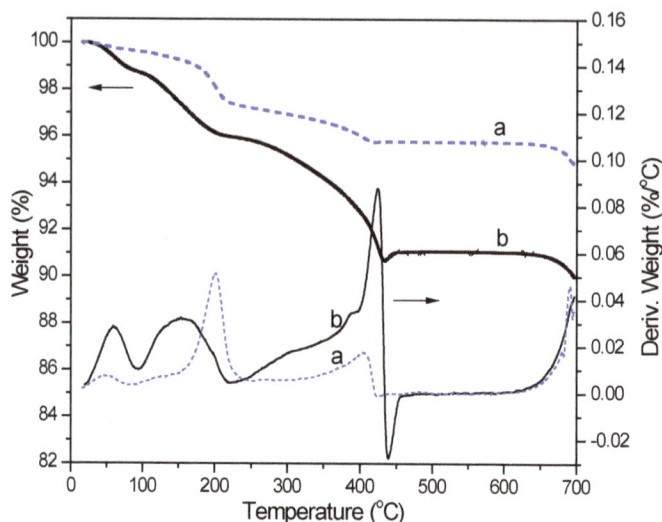

Figure 1. TGA (Thermo-Gravimetric Analysis) and DTG (Derivative Thermo-Gravimetry) curves of the fresh samples (precursors) heated in air (a: APMV/CPM and b: bulk APMV).

2.2.2. Temperature-Programmed XRD in Air

TP-XRD experiment was carried out up to 400 °C on APMV precursor (Figure 2). At room temperature (R.T.), the diffraction lines of the cubic structure typical of Keggin-type heteromolybdates were found at $2\theta = 10.8°$, $18.8°$, $24.3°$, $26.6°$, $30.9°$, $36.3°$ and $39.6°$ [21,22]. The presence of $(NH_4)_3HPMo_{11}VO_{40}$ was demonstrated by the two characteristic lines, which clearly appeared at $2\theta = 15.3°$ and $21.7°$ (triangle down in Figure 2). Some changes were detected upon heating, which consisted of shifting of diffraction lines to lower angles. An example is given for the strongest (222) diffraction line (Figure 2, right). As long as ammonium decomposed, it shifted quite regularly up to 300 °C (from $2\theta = 26.7°$ at R.T. to 26.45° at 300 °C). Then it decreased up to 26.1° at 400 °C (Figure 3), a value characteristic of the (222) line of anhydrous $H_3PMo_{12}O_{40}$. The shift to lower diffraction angles is accounted for the fact that during the transformation of $(PMo_{11}VO_{40})^{4-}$ to $[(PMo_{12}O_{40})^{3-} + V]$, the d-spacing increases (θ decreases) because the cationic radius of Mo^{6+} is bigger (0.59 Å) than that of V^{5+} (0.54 Å). Initiated by the loss of ammonium cations as ammonia, which is accompanied by the expulsion of vanadium from the primary structure, this transformation requires the reorganization of Mo atoms between Keggin units in order for the expelled V to be compensated.

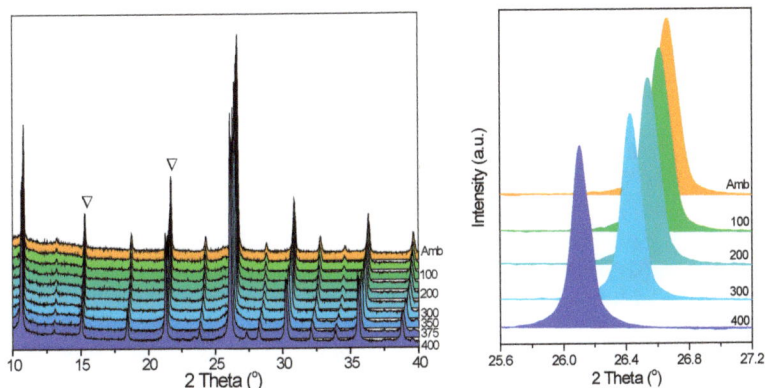

Figure 2. *In situ* XRD (X-Ray Diffraction) in air from ambient to 400 °C for the fresh APMV sample, **(left)** full range; **(right)** (222) line of APMV.

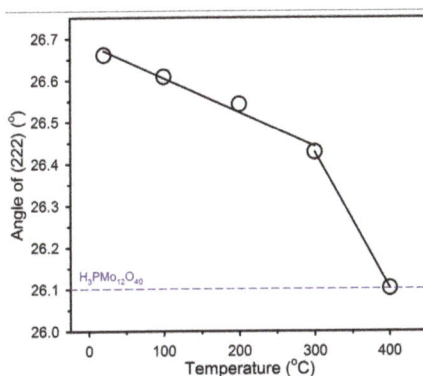

Figure 3. Variation of diffraction (222) line with an increasing temperature from ambient to 400 °C.

When the temperature increased further to 425 °C, significant changes in the crystalline phases took place (Figure 4). The characteristic lines originating from the primary structure disappeared at the expense of new diffraction lines assigned to the metal oxides or mixed-metal oxides. As an example, the pattern obtained at 425 °C exhibited diffraction peaks at $2\theta = 12.4°, 23.4°, 27.3°, 33.0°$ and $33.9°$, which were attributed to MoO_3 (diamonds in Figure 4). The lines of Mo_4O_{11} suboxide, written $(Mo_3^{6+}Mo^{4+})O_{11}$ (asterisks in Figure 4), were also found at $22.9°$ and $24.9°$, but they disappeared when the temperature reached 500 °C. The formation of this mixed valence oxide is the result of the effect of ammonia release, which, here, reduces Mo^{6+} partly to Mo^{4+}. On the contrary, the shoulder at $25.1°$, which is really ill defined (arrows in Figure 4), evolved to a distinguishable diffraction peak. Therefore, Mo_4O_{11} can be considered as an intermediate during the structural evolution. The lines of

V_2O_5 (or V_2O_4) were not observed but there are several reasons for that, beginning by the relatively small amount of V in the sample.

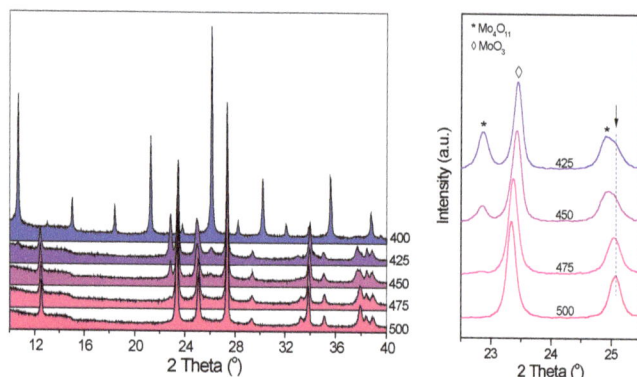

Figure 4. *In situ* XRD from 400 °C to 500 °C for fresh APMV (precursor), **(left)** full range; **(right)** diffractions from molybdenum oxides (*: Mo_4O_{11}, ◊: MoO_3).

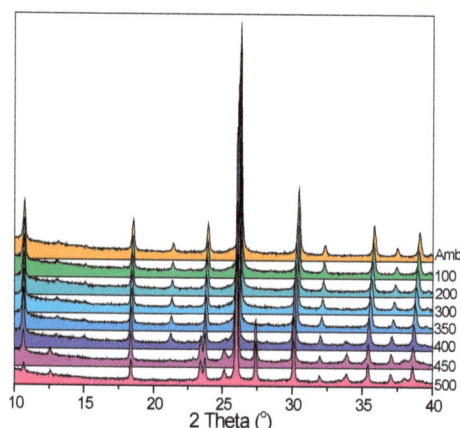

Figure 5. *In situ* XRD in air for the supported sample APMV/CPM.

In the case of CPM-supported APMV, its structural evolution followed by TP-XRD in air illustrated again the greater stability of supported APMV. At R.T. the diffraction lines were characteristic of the cubic structures of both CPM and APMV. The two characteristic lines of APMV (2θ = 15.3° and 21.7°) were more difficult to see due to 40% APMV/CPM. As temperature increased, lines shifted to low angles and accounted for the presence of $H_3PMo_{12}O_{40}$/CPM. The lines were present up to 500 °C instead of 375 °C for AMPV (Figure 5). From 400 to 500 °C, the weak diffraction peaks from MoO_3 at 27.3° and from Mo_4O_{11} at 22.9° and 24.9° were superimposed to those of the cubic structure.

2.3. Reducibility of Calcined Catalysts in Hydrogen Atmosphere

2.3.1. Temperature Programmed Reduction H_2-TPR

H_2-TPR measurements on CPM, APMV and APMV/CPM catalysts were performed to investigate the influence of the support on the reducibility of the catalytically active APMV phase (Figure 6). Almost no hydrogen consumption peak could be found for the bare CPM support but some Mo^{5+} species probably began appearing above 620 °C. Four main steps were distinguished in the reduction profile after deconvolution treatment, which were tentatively assigned in Table 2 to the successive reductions of V and Mo. Obviously the reduction of species, e.g. Mo^{6+} to Mo^{5+}, may begin before completion of the V^{4+} to V^{3+} step. Indeed it is well-known that vanadium is reduced before molybdenum when these species are present in the same oxidic compound [23]. Thus, the slight but long reduction at *ca.* 500–600 °C would be attributed to the reduction of vanadium species. The decomposition of ammonium ions, which initiated the expulsion of vanadium from Keggin units, was probably delayed in hydrogen atmosphere as compared to air, but above 550 °C it could be expected that most V species have moved in between $PMo_{12}O_{40}$ units. The reduction of Mo^{4+} to lower oxidation states might proceed above 650 and 690 °C for APMV/CPM and APMV, respectively.

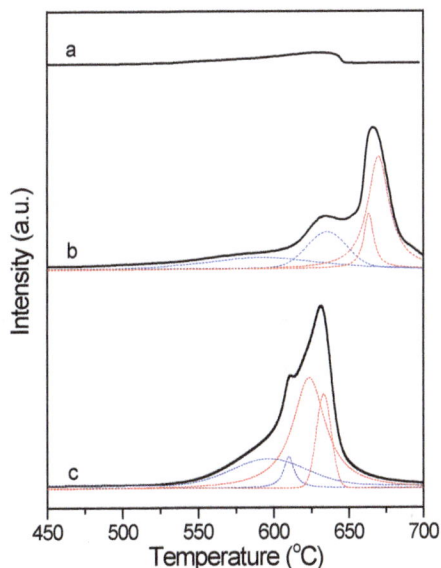

Figure 6. TPR (Temperature Programed Reduction) profiles of: (**a**) CPM; (**b**) APMV; and (**c**) APMV/CPM catalyst (Blue lines: reduction of V species, red lines: reduction of Mo species).

The main hydrogen consumption at 640–690 °C for APMV catalyst was ascribed to the reduction of Mo^{6+} to Mo^{4+} [9]. Furthermore, according to the quantitative calculation of consumed hydrogen in Table 2, the supported sample APMV/CPM consumed 1.2 times more hydrogen than APMV alone, and the reduction of Mo^{6+} to Mo^{4+} exhibited higher hydrogen consumption than reduction of V^{4+} to V^{3+} for both samples. When compared to APMV, it was observed that the reduction steps occurred at higher temperature than for that CPM-supported APMV. In other words, the reducibility of the supported catalyst was improved. On the other hand, not only the temperature of reduction was lowered, but also more metallic ions were reduced by supporting APMV.

Table 2. Tentative attribution of H_2-TPR signals for APMV and APMV/CPM.

Temperature range, °C		H_2 Consumption, mmol/g		Attribution
APMV	APMV/CPM	APMV	APMV/CPM	
500–650 (587) [a]	575–650 (595)	1.09	1.14	Reduction of V^{5+} to V^{4+}
580–670 (635)	590–625 (610)	0.84	0.27	Reduction of V^{4+} to V^{3+}
640–690 (663)	580–700 (623)	0.46	2.95	Reduction of Mo^{6+} to Mo^{5+}
600–700 (670)	610–650 (632)	1.96	0.81	Reduction of Mo^{5+} to Mo^{4+}

[a] peak maximum in parentheses.

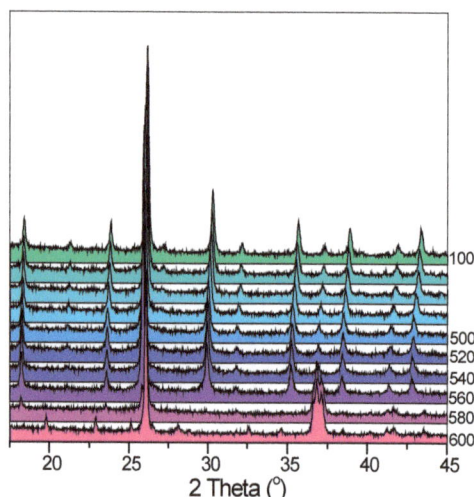

Figure 7. *In situ* XRD under hydrogen for the supported APMV/CPM catalyst.

2.3.2. TP-XRD of APMV/CPM under Hydrogen Atmosphere

To be correlated with the H_2-TPR results, the structural evolution of APMV/CPM under reductive atmosphere was studied by *in situ* H_2-TP-XRD (Figure 7). The diffraction lines assigned to the cubic structures of APMV and/or

CPM phases observed at R.T. were maintained up to 560 °C. The intensity of the weak line at $2\theta = 21.4°$, which is characteristic of APMV, decreased upon increasing temperature, but it was still visible. Though it suggests a gradual decomposition of APMV, it is worth noting that in air (TP-XRD) (Figure 5) this line became weak above 400 °C and disappeared totally at 500 °C. Therefore, when supported, the Keggin structure (whatever its composition) is significantly stabilized under a reductive atmosphere, compared to an oxidative environment. Upon increasing temperature to 580 °C, the Keggin structure collapsed, as shown by the very different and new diffractograms. No formation of $PMo_{11}VO_{38}$ could be demonstrated. The formation of metal oxides and suboxides formed by reduction was directly put in evidence. The strong diffraction line at $2\theta = 26.0°$ and the lines at $2\theta = 36.8°$ and $53.5°$ were assigned to MoO_3 and MoO_2, respectively. This finding is in good agreement with H_2-TPR results showing an overlapped reduction peak from 550 °C to 650 °C. Lines at $2\theta = 37.1°$ and $52.6°$ were assigned to V_2O_4 generated by V_2O_5 reduction. When the temperature reached 600 °C, several new diffraction peaks at low angles (16.1°, 19.8°, 23.0°, 28.1° and 28.8°) became visible, but they could not be attributed without ambiguity. They could originate from one or more mixed oxides, such as MoP_xO_y and VP_xO_y. Except the reduction of vanadium, the decomposition process was in good agreement with that suggested by TGA/DTG in air (Figure 1).

As a partial conclusion, the main findings about the reduction behavior of APMV and APMV/CPM studied by H_2-TPR and H_2-TP-XRD are: (i) the Keggin structure (in the form of APMV, $H_4PMo_{11}VO_{40}$ and/or $H_3PMo_{12}O_{40}$ + V) was stable up to *ca.* 560 °C against a deep reduction (leading to the transformation to oxides); (ii) the decomposition of the Keggin units was delayed as compared to the case of oxidative conditions where it began at *ca.* 400 °C; and (iii) the CPM support promoted the reduction of APMV.

2.4. XPS after Pretreatment under Isobutane

The composition and chemical state of the species in the first layers have already been studied for catalysts before (calcined catalysts) and after (used catalysts) the catalytic experiments by *"ex situ"* XPS technique [18]. It felt necessary to perform XPS experiments on calcined 40%APMV/CPM after pretreatement under isobutane atmosphere, to gain more information on the effects of one of the reactants on the catalyst surface (Tables 3 and 4). Values obtained in a former study by *ex situ* XPS for the same catalyst but after catalytic reaction were added for further comparison [18].

Table 3. Binding energy and surface atomic composition of 40 APMV/CPM pretreated under isobutane at various temperatures.

Temperature, T_{pret} °C	Binding energy, BE, eV			Atomic ratio [a] O1s(1)/O1s(2)	Surface atomic composition
	Cs3d$_{5/2}$	P2p	O1s		Cs/O/P/Mo/V/N
R.T. [b]	724.0	133.9	530.7/531.9	76/24	1.9/36.6/1.1/11.0/0.8/3.4
340	723.8	133.3	530.7/531.9	84/16	1.5/37.0/1.2/11.0/0.7/1.6
540	725.2	133.8	530.7/532.1	71/29	7.4/32.8/2.2/11.0/0.1/0
560	725.2	133.7	530.7/532.1	73/27	5.4/30.5/1.9/11.0/0.1/0
580	725.2	133.7	530.7/532.1	70/30	5.9/29.1/2.1/11.0/0.1/0
R.T., post-reaction [c]	724.3	133.6	-	-	2.0/-/1.0/11/0.8/1.0

[a] Ratio of photopeak area; [b] R.T. = room temperature; [c] from [18].

Table 4. Surface analysis results for V and Mo.

Temperature, T_{pret} °C	Binding energy, eV				Molar ratio
	V2p$_{3/2}$	Mo3d$_{3/2}$/Mo3d$_{5/2}$			V^{5+}/V^{4+}
	V^{5+}/V^{4+}	Mo^{6+}	Mo^{5+}	Mo^{4+}	
R.T. [a]	517.8/516.0	236.2/233.1	-	-	88/12
340	516.9/515.9	236.3/233.0	234.9/231.8	-	46/54
540	~516.3	236.1/232.9	234.9/231.4	233.3/229.9	0
560	~516.0	235.7/231.6	234.2/230.4	233.1/229.8	0
580	~515.8	236.2/232.9	234.5/231.3	233.2/229.8	0
R.T., post-reaction [b]	518.1/517.0	-/233.2	-	-	29/71

[a] R.T. = room temperature; [b] from [17].

The binding energies (BE) of N1s, Cs3d$_{5/2}$, P2p and O1s photopeaks and the surface composition expressed as atomic ratios are gathered in Table 3. Cs3d$_{5/2}$ and P2p photopeaks at 724 eV and 133.7 eV were assigned to Cs$^+$ and P^{5+} ions, respectively. The BE of P^{5+} did not change much upon temperature increase between R.T. and 560 °C, whereas *ca.* 1 eV of deviation was observed for Cs$^+$ photopeak. Typically, 724 eV stands for CsOH, meaning that the surface would be hydroxylated. The nitrogen N1s photopeak was found at *ca.* 402 eV (NH$_4^+$ ion) as a shoulder in the Mo3p$_{3/2}$ peak located itself at *ca.* 398 eV. After decomposition of the spectrum collected at room temperature, the BE of N1s could be found at 402.5 eV (Figure 8). The N1s peak became much broader at 340 °C and the surface content of N decreased by more than 50% (Table 3). At 540 °C, it disappeared while a new nitrogen species appeared at *ca.* 396.1 eV. The latter was assigned to oxynitride species [24,25]. In the case of Mo and V, the Mo3d$_{3/2}$, Mo3d$_{5/2}$ and V2p$_{3/2}$ photopeaks were decomposed to put in evidence the reduction states after pretreatment (*vide infra*) (Table 4). At room temperature, the two well-separated peaks at 236.2 eV (Mo3d$_{3/2}$) and 233.1 eV (Mo3d$_{5/2}$) were assigned to Mo^{6+} species linked to P^{5+} [26]. The wide peak at *ca.* 518 eV (V2p$_{3/2}$) was assigned to V^{5+}. This value was too high as compared to that of V$_2$O$_5$ (BE *ca.* 516.5–516.9 eV) but it was closer to those of vanadium phosphates [27,28]. In a way, the BEs for Mo and V confirmed that the Keggin structure [PMo$_{11}$VO$_{40}$] or [PMo$_{12}$O$_{40}$] was present at R.T. and at 340 °C. The O1s peak was asymmetric at any temperature. After decomposition, two species O1s(1) and O1s(2) appeared at 530.7 eV and 531.9–532.1 eV, respectively, which could be attributed to oxygen bonded to Mo and P when both are present in the same compound [26].

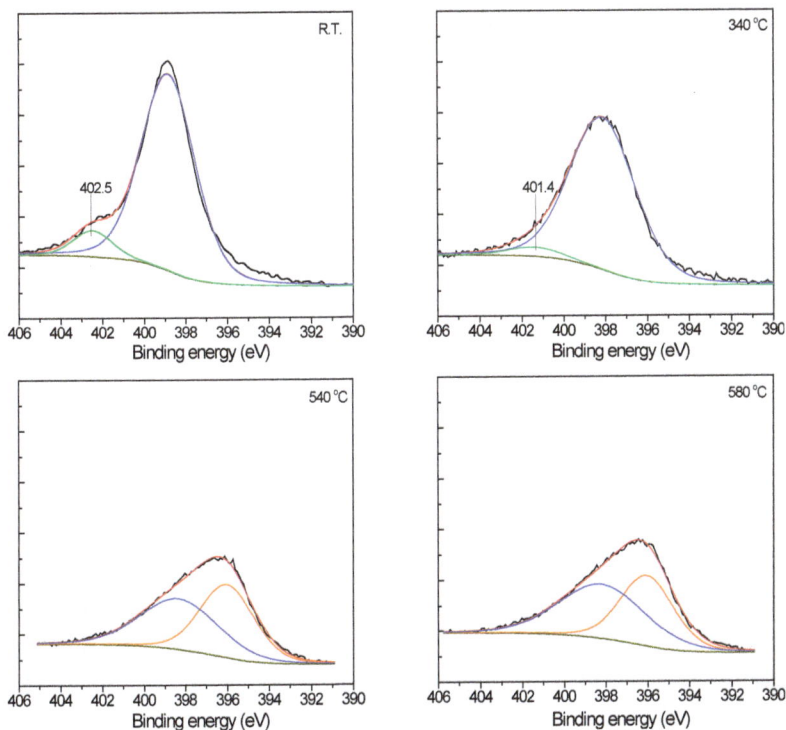

Figure 8. *In situ* XPS N1s for the APMV/CPM sample treated under isobutane.

Before examining the oxidation states of Mo and V, let us focus on the relative atomic ratios (Table 3) of N, Cs, P and their variation with T_{pret}. Significant changes could be observed upon increasing T_{pret} and particularly when it exceeded 340 °C. The surface proportions of Cs and P significantly increased, at variance of O and V proportions. The total oxygen content decreased more and more, as expected in a reduction process, while the surface P^{5+} species increased, particularly at $T_{pret} \geqslant 540$ °C. This is often observed for these labile species in mixed P, V oxides [27,29]. Together with the formation of oxynitrides, the results indicated that a complex change took place at the surface at these temperatures. In the case of the used catalyst (post-reaction) examined at R.T. [18], there was no more nitrogen, but the other values were similar to those found at R.T. in the presence of IBAN.

At 340 °C which is in the temperature range of the catalytic reaction, the surface nitrogen (as NH_4^+) decreased from 3.4 at R.T. (theoretical N/Mo = 3/11) to 1.6. This decrease confirms the partial decomposition of the ammonium cations followed by ammonia release. At $T_{pret} \geqslant 540$ °C, all NH_4^+ disappeared, in line with TP-XRD in H_2. At variance, Cs amounts first decreased from 1.9 to 1.5 at 340 °C (theoretical value Cs/Mo = 0/11), and increased up to 5.9 above 540 °C. Vanadium also underwent a change, decreasing strongly by a tenth of its value (initially V/Mo = 1/11) at $T_{pret} \geqslant 540$ °C, but below 540 °C it was approximately constant. These experiments show that the expulsion of V from the primary Keggin structure [30] may happen above 340 °C, followed by the diffusion of V from the surface to the bulk.

According to the preparation procedure for 40 wt. % APMV, the support CPM ($Cs_3PMo_{12}O_{40}$) should be covered by the active phase. Actually, the opposite trend of Cs and N when the temperature increased (up to 340 °C) was in favor of the formation of the solid solution $Cs_x(NH_4)_{3-x}HPMo_{11}VO_{40}$, which was studied in a former paper [18]. The redistribution of the surface elements especially for Cs and V with temperature suggested a cationic exchange between the CPM support and the APMV active phase. The release of ammonia resulted in a protonated intermediate, which was unstable. In air, it evolved towards a lacunary Keggin-type compound by losing constitutional water (protons plus oxygen). Hence, the cesium cations compensated the loss in ammonia (or, more precisely, the loss of protons from the resulting intermediate). Above 340 °C, when there was no more ammonium, the increase of Cs/Mo from 1.5 to 7.4/11, resulting in a Cs-enriched surface as if the CPM support—i.e., the bulk of the catalyst—was alone. However it was more probable that the solid was made of a mixture of oxides and suboxides. Therefore in situ XPS experiments confirmed that, though isobutane was a milder reducing agent than hydrogen, the Keggin structure was quite maintained in APMV/CPM up to 340 °C but that it disintegrated above 500 °C.

It is known that the redox properties of 12-phosphomolybdic acid significantly change upon replacing one Mo by V atom in the Keggin unit [31]. Hence, the oxidation state of V is of high interest. The evolution of V2p$_{3/2}$ photopeaks is shown in Figure 9 and the quantification results are reported in Table 4. The wide peak at ca. 518 eV recorded at room temperature accounted for two different oxidation states of vanadium. The decomposition of the signal was done by assigning BEs at 517.8 eV and at 516.0 eV to V^{5+} and V^{4+}, respectively. The V^{5+}/V^{4+} ratio increased with T_{pret}. Thus not only vanadium was eliminated from the primary structure, but it was also reduced at the same time. In the case of $Cs_xH_{4-x}PVMo_{11}O_{40}$, the reduction was explained by an auto-reduction process within the catalyst, the O^{2-} lattice oxygen being simultaneously oxidized to O_2 [32]. However, in our case, the explanation lies in the elimination of ammonium, which is oxidized to ammonia by means of lattice oxygen, with concomitant reduction of V and/or Mo. When the temperature was increased to 340 °C, a clear shift of the main peak (pink curves in Figure 9) was observed. At the same time, the $V_{2p3/2}$ peak became less intense and much broader compared with that observed at room temperature, indicating that the V^{5+} species were reduced to V^{4+} by isobutane (Figure 9). The calculation of V^{5+}/V^{4+} showed that the amount of V^{4+} species increased significantly from 11.9% to 53.6%. On the other hand, the decrease in the peak area suggested that the amount of the V surface species decreased. At 540 °C, the very small $V2p_{3/2}$ peak was detected at around 516 eV, which was assigned to V^{4+} species. The comparison with the same catalyst, but after reaction [17], shows that the proportion of V^{4+} was greater.

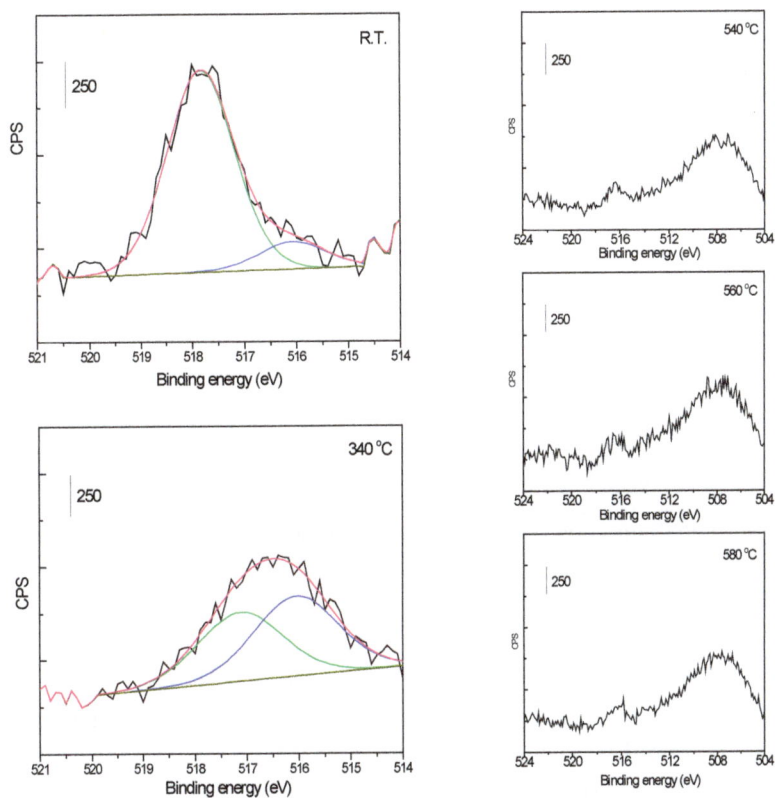

Figure 9. $V2p_{3/2}$ peaks obtained in isobutane atmosphere for the APMV/CPM sample (black: original spectrum, pink: accumulative curve, green: V^{5+} and blue: V^{4+}).

The Mo species were also reduced, as shown by the decomposition of $Mo3d_{3/2}$ and $Mo3d_{5/2}$ photopeaks at $T_{pret} \geqslant 340$ °C (Figure 10). The two well-separated peaks, observed at R.T. at 236.2 and 233.1 eV, respectively, tended to overlap and to shift to lower BEs when T_{pret} increased. Two new peaks appeared at 234.9 eV and 231.8 eV, which were attributed to Mo^{5+} species (green curves). Mo^{5+} was reduced to Mo^{4+} (orange curves) at 540 °C, so that the three oxidation states of molybdenum coexisted at the surface. A further increase in temperature did not affect the Mo signal, suggesting thereby that a "steady state" was obtained. BEs of Mo are close to those of MoO_3, like in MoO_3-P_2O_5 glasses containing high amounts of Mo [23].

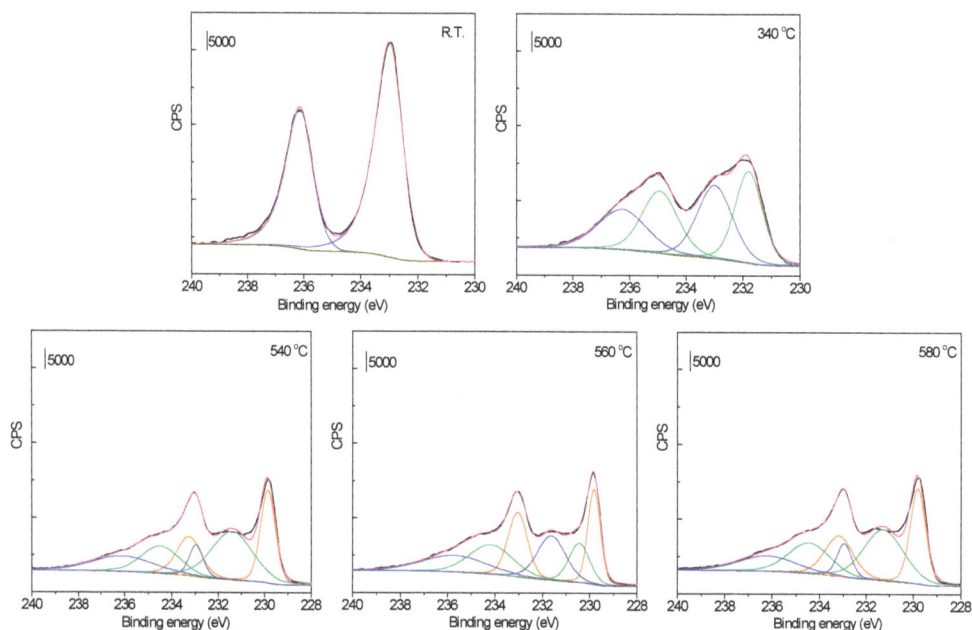

Figure 10. Reduction process of Mo species under isobutane observed by *in situ* XPS (black: original curve, pink: accumulative curve, blue: Mo^{6+}, green: Mo^{5+} and orange: Mo^{4+}).

As a conclusion, these XPS experiments after pretreatment in isobutane confirmed that not only V but also Mo species can undergo reduction and thus participate in the redox mechanism at 340 °C. They also put forth evidence of the complex restructuration of the surface. The APMV phase evolved by releasing NH_3 issued from the decomposition of ammonium cations. The resulting acidic intermediate (that could not be isolated during the experiments) rapidly loses the remaining protons, whereby the charge disequilibrium is compensated by cesium migration from the CPM support. The vanadium is expelled from the Keggin unit and then reduced by isobutane, leading to mixed-valence V species. Besides these observations, it was remarkable to find that Mo^{6+} can also be reduced to Mo^{5+}

even at temperatures as low as 340 °C. Nevertheless, the consecutive reduction to Mo^{4+}—corresponding to the decomposition of the primary structure—was not observed at 340 °C but only above 540 °C. It is noteworthy that the reduction of Mo species was not observed in our "*ex situ*" XPS experiments [18]. In the spectra of calcined and used APMV/CPM samples, the BE of Mo photopeak was constant at 233.3–233.4 eV. The proportions of Cs, N and V had more or less decreased after reaction while P was quite constant (P/Mo = 1.0–1.1/11). For 11 Mo, the relative ratios were Cs/V/N = 2.0/0.8/1.0 before and 1.5/0.7/0.7 after reaction. With *in situ* XPS, Cs/V/N = 1.9/0.8/3.4 at R.T. and 1.5/0.7/1.6 at T_{pret} = 340 °C, with similar values for P. Thus the main difference lies in the proportion of nitrogen, which is higher in reducing conditions than after reaction. As stated above after H_2-TPXRD, the stability of the Keggin structure is increased in reducing conditions. The absence of Mo^{5+} in samples after reaction may be caused by (I) the reoxidation of the surface as the used catalyst was exposed to air, and/or (II) the rapid reoxidation of Mo in the operating conditions of IBAN selective oxidation.

3. Experimental Section

3.1. Material Syntheses

The hydrated 11-molybdo-1-vanadophosphoric heteropolyacid ($H_4PMo_{11}VO_{40} \cdot xH_2O$) was synthesized according to a method described previously in the literature [2]. The number of crystal water molecules was calculated from TGA results (*vide infra*), which suggested that eight molecules of water were incorporated in the secondary structure ($H_4PMo_{11}VO_{40} \cdot 8H_2O$). The anhydrous acid will further be noted HPMV.

($NH_4)_3HPMo_{11}VO_{40}$ (APMV) was prepared by co-precipitation at 45 °C, using stoichiometric amounts of NH_4NO_3 (Sigma-Aldrich, Lyon, France) and $H_4PMo_{11}VO_{40} \cdot 8H_2O$ as precursors [17]. The resulting precipitate was dried at 70 °C to obtain the APMV precursor. The APMV catalyst was obtained by calcination in static air at 350 °C (heating rate 2 °C/min) for 3 h.

The cesium salt of 12-molybdophosphoric acid, employed as a support in the following, was prepared by controlled precipitation method [17]. An appropriate amount of Cs_2CO_3 (Sigma-Aldrich) was dissolved in water (0.1 M) and pumped into a 0.1 M aqueous solution of $H_3PMo_{12}O_{40}$ (Sigma-Aldrich) under vigorous stirring at 45 °C, generating a yellow suspension. After 2 h, the solid was recovered by removing the water under reduced pressure at 70 °C, and then dried at the same temperature for 24 h. It was finally calcined in static air at 350 °C (2 °C/min) for 3 h to obtain the solid $Cs_3PMo_{12}O_{40}$, further used as the support of APMV.

The as-prepared support was then impregnated by APMV using a deposition-precipitation method [18] as follows: CPM particles were suspended in 35 mL deionized water at 45 °C under vigorous stirring for 1 h. Then, aqueous solutions containing stoichiometric amounts of NH_4NO_3 (Sigma-Aldrich) and $H_4PMo_{11}VO_{40}$ (*i.e.*, $(NH_4NO_3):(H_4PMo_{11}VO_{40})$ = 3:1 (molar ratio)) were simultaneously pumped into the suspension to provoke APM precipitation so as to get 40 wt.% of loaded APMV. The solid noted APMV/CPM was recovered by filtration and dryness after reacting for 2 h. The conditions of calcination remained the same as for the CPM preparation.

3.2. Characterization Methods

Thermogravimetric analysis (TGA/DTG) was performed using a TA instruments (New Castle, PA, USA) 2960 SDT to study the thermal decomposition of the fresh catalyst. The samples were heated in an air flow from room temperature to 700 °C with a ramp of 3 °C/min [18].

The crystalline phases were detected by X-ray diffraction (XRD) technique on Bruker D8 Advance diffractometer (Bruker AXS, Billerica, MA, USA), using Cu Kα radiation (λ = 1.5418 Å) as an X-ray source. The reactivity of samples was studied by temperature-programmed XRD (TP-XRD) in air or hydrogen. The patterns were collected from 2θ = 10° to 60° with steps of 0.02° and an acquisition time of 0.5 s. When the measurements were carried out in air, the samples were heated in an airflow from room temperature (R.T.) to 500 °C at 1.5 °C/min heating rate, or in a reductive atmosphere (3 mol% H_2 in N_2) from R.T. to 600 °C at 10 °C/min heating rate.

The "*in situ*" surface analysis by X-ray photoelectron spectroscopy (XPS) was performed on a Kratos Axis Ultra DLD (Manchester, UK) apparatus equipped with a hemispherical analyzer and a delay-line detector. The experiments were carried out using an Al mono-chromated X-ray source (10 kV, 15 mV) with a pass energy of 40 eV (0.1 eV/step) for high-resolution spectra and a pass energy of 160 eV (1 eV/step) for the survey spectrum in the hybrid mode and slot lens mode, respectively. The measurements were performed as follows: the sample was first heated up to the required temperature of pretreatment (T_{pret} = 340, 540, 560 and 580 °C) in diluted isobutane (5 mol% isobutane in He) at 5 °C/min heating rate, then it was transferred into the analytic chamber by an interior mechanical arm to avoid any contact with ambient air. The spectra were collected after pretreatments. A blank experiment consisted of analysis of the calcined sample (without pretreatment).

The reducibility of the catalysts was evaluated by temperature-programmed reduction under hydrogen (H_2-TPR) at atmospheric pressure. One hundred milligrams of sample were loaded in a quartz reactor and pre-treated in He flow (30 mL/min) at 100 °C for 2 h. Then, helium was replaced by the reducing gas (5 mol% H_2 in He) at the same flowrate. The temperature of the reactor was increased linearly from 100 to 800 °C at a rate of 5 °C/min. The effluent gas was analyzed by a thermal conductivity detector (TCD).

3.3. Catalytic Oxidation of Isobutane

The catalytic selective oxidation of isobutane to methacrolein and methacrylic acid was performed in a fixed-bed reactor at atmospheric pressure [21]. The catalyst (0.8 g), diluted with an equal volume of SiC (0.21 mm particle size), was loaded in a stainless-steel reactor, which was then heated up to the reaction temperature (340 °C) with a ramp of 5 °C /min under airflow (10 mL/min). Then, the airflow was replaced by the reactants (IBAN/O_2/H_2O/He = 27/13.5/10/49.5 molar ratio) fed at a total flowrate of 10 mL/min. The light products, such as CO, CO_2, O_2, as well as IBAN were analyzed by a GC-TCD equipped with packed columns (Porapak Q, 3 m length; molecular sieve 13×, 3 m length). Other products, like MAC, MAA, acetic acid (AA) and acrylic acid (ACA), were analyzed by a GC-FID equipped with an Alltech (Deerfield, IL, USA) EC1000 semi-capillary column (i.d. 0.53 mm, 30 length). The reaction data were obtained after 24 h under stream, which corresponded to steady state conditions.

4. Conclusions

The catalytic performance in the oxidation of IBAN to methacrolein and methacrylic acid being significantly improved by supporting APMV on CPM (IBAN maximum conversion of 15.3% and yield to MAA + MAC of 8.0%) the structural evolution of the catalyst was studied using various *in situ* techniques. Experiments were carried out under different atmospheres to explore the changes of the catalyst under reaction conditions. Under oxidative conditions (air), the cubic system of unsupported APMV was destroyed at temperatures higher than 400 °C. When supported on CPM, the thermal stability of APMV was significantly improved, since the Keggin structure remained unchanged until 500 °C. Under reductive atmosphere, the active phase was further stabilized up to 560 °C, and its reducibility increased. It is believed that the easier formation of V^{4+} and of Mo^{5+} species increase the catalytic activity of APMV/CPM because these species are responsible for activation of isobutane [16]. Similar observation was made by Cavani *et al.* [15], who used the ammonium salt of heteropolymolybdate, $(NH_4)_3PMo_{12}O_{40}$, in the same reaction. They found that operating at isobutane-rich (IBAN/O_2 = 26/13) instead of isobutane-lean (IBAN/O_2 = 1/13) conditions was more favorable to increase the selectivity to MAA, which reached 40% instead of 15% for similar IBAN conversion (7%–8%) after 25 h on-stream.

Concerning the decomposition mechanism of APMV, two main phenomena were evidenced by our *in situ* experiments: (I) As a consequence of the decomposition of ammonium, vanadium was eliminated from the primary structure (Keggin units) during the calcination of APMV precursor as well as during the catalytic reaction which was carried out in isobutane-lean conditions. The V species remained in the solid as mixed- valence species, V^{4+} and V^{5+}, which activated IBAN and oxidized

191

the resulting adsorbed intermediate, respectively. The V^{5+} species could be reduced by isobutane under reaction conditions (340 °C), which is in agreement with the Mars and van-Krevelen mechanism. (II) As the product of oxidation of ammonium, ammonia was released from APMV, followed (or not, depending on temperature) by the elimination of the remaining protons as constitutional water. The addition of steam in the feed is supposed to slow down this process and maintain the specific acidity necessary for the reaction. The resulting charge disequilibrium was compensated by the migration of cesium ions from the support, forming Cs^+-NH_4^+-H^+ mixed salts. Furthermore, it was also found that the Mo species were also partially reduced from Mo^{6+} to Mo^{5+} under reaction conditions, thus increasing the active sites since they also participate in the redox reaction. These results show that the catalyst restructures significantly under reaction conditions.

Acknowledgments: The Fonds Européen de Développement Régional (FEDER), CNRS, Région Nord Pas-de-Calais and Ministère de l'Education Nationale de l'Enseignement Supérieur et de la Recherche are acknowledged for the funding of X-ray diffractometers. Fangli Jing thanks the China Scholarship Council (CSC) for providing the financial support during his stay in France.

Author Contributions: Fangli Jing and Sébastien Paul designed the experiments. Fangli Jing performed them and wrote the paper. Benjamin Katryniok contributed to the data analysis and writing. Sébastien Paul, Franck Dumeignil and Elisabeth Bordes-Richard contributed greatly to the interpretation of results, the preparation and improvement of the manuscript.

Conflicts of Interest: The authors declare no conflict of interest.

References

1. Rao, K.N.; Srilakshmi, C.; Reddy, K.M.; Babu, B.H.; Lingaiah, N.; Prasad, P.S.S. Heteropoly Compounds as Ammoxidation Catalysts. In *Environmentally Benign Catalysts*; Patel, A., Ed.; Springer: Dordrecht, The Netherlands, 2013; pp. 11–55.
2. Ressler, T.; Dorn, U.; Walter, A.; Schwarz, S.; Hahn, A.H.P. Structure and properties of $PVMo_{11}O_{40}$ heteropolyoxomolybdate supported on silica SBA-15 as selective oxidation catalyst. *J. Catal.* **2010**, *275*, 1–10.
3. Sun, M.; Zhang, J.; Putaj, P.; Caps, V.; Lefebvre, F.; Pelletier, J.; Basset, J.M. Catalytic Oxidation of Light Alkanes (C1–C4) by Heteropoly Compounds. *Chem. Rev.* **2013**, *114*, 981–1019.
4. Enache, D.I.; Bordes, E.; Ensuque, A.; Bozon-Verduraz, F. Vanadium oxide catalysts supported on titania and zirconia: II. Selective oxidation of ethane to acetic acid and ethylene. *Appl. Catal. A* **2004**, *278*, 103–110.
5. Katryniok, B.; Paul, S.; Capron, M.; Royer, S.; Lancelot, C.; Jalowiecki-Duhamel, L.; Belliere-Baca, V.; Rey, P.; Dumeignil, F. Synthesis and characterization of zirconia-grafted SBA-15 nanocomposites. *J. Mater. Chem.* **2011**, *21*, 8159–8168.
6. Katryniok, B.; Paul, S.; Dumeignil, F. Recent Developments in the Field of Catalytic Dehydration of Glycerol to Acrolein. *ACS Catal.* **2013**, *3*, 1819–1834.

7. Brückner, A.; Scholz, G.; Heidemann, D.; Schneider, M.; Herein, D.; Bentrup, U.; Kant, M. Structural evolution of $H_4PVMo_{11}O_{40} \cdot xH_2O$ during calcination and isobutane oxidation: New insights into vanadium sites by a comprehensive *in situ* approach. *J. Catal.* **2007**, *245*, 369–380.

8. Song, I.; Barteau, M. Redox properties of Keggin-type heteropolyacid (HPA) catalysts: Effect of counter-cation, heteroatom, and polyatom substitution. *J. Mol. Catal. A* **2004**, *212*, 229–236.

9. Sun, M.; Zhang, J.Z.; Cao, C.J.; Zhang, Q.H.; Wang, Y.; Wan, H.L. Significant effect of acidity on catalytic behaviors of Cs-substituted polyoxometalates for oxidative dehydrogenation of propane. *Appl. Catal. A* **2008**, *349*, 212–221.

10. Paul, S.; Chu, W.; Sultan, M.; Bordes-Richard, E. Keggin-type $H_4PVMo_{11}O_{40}$-based catalysts for the isobutane selective oxidation. *Sci. Chin. Ser. B* **2010**, *53*, 2039–2046.

11. Kozhevnikov, I. Heterogeneous acid catalysis by heteropoly acids: Approaches to catalyst deactivation. *J. Mol. Catal. A* **2009**, *305*, 104–111.

12. Paul, S.; LeCourtois, V.; Vanhove, D. Kinetic investigation of isobutane selective oxidation over a heteropolyanion catalyst. *Ind. Eng. Chem. Res.* **1997**, *36*, 3391–3399.

13. Sultan, M.; Paul, S.; Fournier, M.; Vanhove, D. Evaluation and design of heteropolycompound catalysts for the selective oxidation of isobutane into methacrylic acid. *Appl. Catal. A* **2004**, *259*, 141–152.

14. Cavani, F.; Etienne, E.; Mezzogori, R.; Pigamo, A.; Trifiro, F. Improvement of catalytic performance in isobutane oxidation to methacrylic acid of Keggin-type phosphomolybdates by preparation via lacunary precursors: Nature of the active sites. *Catal. Lett.* **2001**, *75*, 99–105.

15. Cavani, F.; Mezzogori, R.; Pigamo, A.; Trifirò, F. Synthesis of methacrylic acid by selective oxidation of isobutane, catalysed by Keggin-type polyoxometalates: relationship between catalytic performance, reaction conditions and chemical–physical features of the catalyst. *C.R. Acad. Sci. Ser. IIc Chim.* **2000**, *3*, 523–531.

16. Bordes, E. The role of structural chemistry of selective catalysts in heterogeneous mild oxidation catalysis of hydrocarbons. *C.R. Acad. Sci. Ser. IIc Chim.* **2000**, *3*, 725–733.

17. Jing, F.; Katryniok, B.; Bordes-Richard, E.; Paul, S. Improvement of the catalytic performance of supported $(NH_4)_3HPMo_{11}VO_{40}$ catalysts in isobutane selective oxidation. *Catal. Today* **2013**, *203*, 32–39.

18. Jing, F.; Katryniok, B.; Dumeignil, F.; Bordes-Richard, E.; Paul, S. Catalytic selective oxidation of isobutane to methacrylic acid on supported $(NH_4)_3HPMo_{11}VO_{40}$ catalysts. *J. Catal.* **2014**, *309*, 121–135.

19. Mizuno, N.; Tateishi, M.; Iwamoto, M. Oxidation of isobutane catalyzed by $Cs_xH_{3-x}PMo_{12}O_{40}$-based heteropoly compounds. *J. Catal.* **1996**, *163*, 87–94.

20. Kozhevnikov, I.V. Sustainable heterogeneous acid catalysis by heteropoly acids. *J. Mol. Catal. A* **2007**, *262*, 86–92.

21. Jing, F.; Katryniok, B.; Dumeignil, F.; Bordes-Richard, E.; Paul, S. Catalytic selective oxidation of isobutane over $Cs_x(NH_4)_{3-x}HPMo_{11}VO_{40}$ mixed salts. *Catal. Sci. Technol.* **2014**, *4*, 2938–2945.

22. Marchal-Roch, C.; Laronze, N.; Guillou, N.; Tézé, A.; Hervé, G. Study of ammonium, mixed ammonium–cesium and cesium salts derived from $(NH_4)_5[PMo_{11}V^{IV}O_{40}]$ as isobutyric acid oxidation catalysts: Part I: Syntheses, structural characterizations and catalytic activity of the ammonium salts. *Appl. Catal. A* **2000**, *199*, 33–44.

23. Courtine, P.; Bordes, E. Mode of arrangement of components in mixed vanadia catalyst and its bearing for oxidation catalysis. *Appl. Catal. A* **1997**, *157*, 45–65.

24. Hoffman, A.; Maniv, T.; Folman, M. AES and XPS studies of no adsorption on Al(100) single crystal. *Surf. Sci.* **1987**, *183*, 484–502.

25. Lee, W.J.; Lee, Y.S.; Rha, S.K.; Lee, Y.J.; Lim, K.Y.; Chung, Y.D.; Whang, C.N. Adhesion and interface chemical reactions of Cu/polyimide and Cu/TiN by XPS. *Appl. Surf. Sci.* **2003**, *205*, 128–136.

26. Khattak, G.D.; Salim, M.A.; Al-Harthi, A.S.; Thompson, D.J.; Wenger, L.E. Structure of molybdenum-phosphate glasses by X-ray photoelectron spectroscopy (XPS). *J. Non -Cryst. Solids* **1997**, *212*, 180–191.

27. Cavani, F.; de Santi, D.; Luciani, S.; Löfberg, A.; Bordes-Richard, E.; Cortelli, C.; Leanza, R. Transient reactivity of vanadyl pyrophosphate, the catalyst for *n*-butane oxidation to maleic anhydride, in response to in-situ treatments. *Appl. Catal. A* **2010**, *376*, 66–75.

28. Khattak, G.D.; Mekki, A.; Wenger, L.E. X-ray photoelectron spectroscopy (XPS) and magnetic susceptibility studies of vanadium phosphate glasses. *J. Non-Cryst. Solids* **2009**, *355*, 2148–2155.

29. Coulston, G.W.; Thompson, E.A.; Herron, N. Characterization of VPO Catalysts by X-ray Photoelectron Spectroscopy. *J. Catal.* **1996**, *163*, 122–129.

30. Cavani, F.; Mezzogori, R.; Pigamo, A.; Trifiro, F.; Etienne, E. Main aspects of the selective oxidation of isobutane to methacrylic acid catalyzed by Keggin-type polyoxometalates. *Catal. Today* **2001**, *71*, 97–110.

31. Busca, G.; Cavani, F.; Etienne, E.; Finocchio, E.; Galli, A.; Selleri, G.; Trifiro, F. Reactivity of Keggin-type heteropolycompounds in the oxidation of isobutane to methacrolein and methacrylic acid: Reaction mechanism. *J. Mol. Catal. A* **1996**, *114*, 343–359.

32. Lee, J.K.; Russo, V.; Melsheimer, J.; Kohler, K.; Schlogl, R. Genesis of V^{4+} in heteropoly compounds $Cs_xH_{4-x}PVMo_{11}O_{40}$ during thermal treatment, rehydration and oxidation of methanol studied by EPR spectroscopy. *Phys. Chem. Chem. Phys.* **2000**, *2*, 2977–2983.

194

Activated Carbon, Carbon Nanofiber and Carbon Nanotube Supported Molybdenum Carbide Catalysts for the Hydrodeoxygenation of Guaiacol

Eduardo Santillan-Jimenez, Maxime Perdu, Robert Pace, Tonya Morgan and Mark Crocker

Abstract: Molybdenum carbide was supported on three types of carbon support—activated carbon; multi-walled carbon nanotubes; and carbon nanofibers—using ammonium molybdate and molybdic acid as Mo precursors. The use of activated carbon as support afforded an X-ray amorphous Mo phase, whereas crystalline molybdenum carbide phases were obtained on carbon nanofibers and, in some cases, on carbon nanotubes. When the resulting catalysts were tested in the hydrodeoxygenation (HDO) of guaiacol in dodecane, catechol and phenol were obtained as the main products, although in some instances significant amounts of cyclohexane were produced. The observation of catechol in all reaction mixtures suggests that guaiacol was converted into phenol via sequential demethylation and HDO, although the simultaneous occurrence of a direct demethoxylation pathway cannot be discounted. Catalysts based on carbon nanofibers generally afforded the highest yields of phenol; notably, the only crystalline phase detected in these samples was Mo_2C or Mo_2C-ζ, suggesting that crystalline Mo_2C is particularly selective to phenol. At 350 °C, carbon nanofiber supported Mo_2C afforded near quantitative guaiacol conversion, the selectivity to phenol approaching 50%. When guaiacol HDO was performed in the presence of acetic acid and furfural, guaiacol conversion decreased, although the selectivity to both catechol and phenol was increased.

Reprinted from *Catalysts*. Cite as: Santillan-Jimenez, E.; Perdu, M.; Pace, R.; Morgan, T.; Crocker, M. Activated Carbon, Carbon Nanofiber and Carbon Nanotube Supported Molybdenum Carbide Catalysts for the Hydrodeoxygenation of Guaiacol. *Catalysts* **2015**, *5*, 424–441.

1. Introduction

Lignin is a complex biopolymer which is produced as a waste product by the pulp and paper industry and is typically burned on-site to operate factory processes [1]. In lieu of simply combusting this valuable source of aromatics and phenolics, various depolymerization [2] and densification [3] strategies are under development with the goal of producing value-added liquid fuels and chemicals from the lignin polymer. However, many of the products obtained from these processes

195

require further upgrading in the form of deoxygenation before they can be utilized as fuels. In order to simplify laboratory deoxygenation studies, model compounds are typically used as surrogates for the products obtained from lignin depolymerization, guaiacol being one of the most widely used models. Guaiacol possesses both phenolic and methoxy moieties which are present throughout lignin, and affords the opportunity to examine the selectivity toward hydrogenation/hydrogenolysis of the -OH and -OCH$_3$ functionalities *versus* the aromatic ring [4–6]. Guaiacol is also more representative of actual lignin product streams than other simple model compounds due to its greater propensity for coking and its tendency to be more refractory towards hydrodeoxygenation (HDO) than other common lignin models [7].

A variety of catalytic approaches have been examined in an effort to produce fuel-like compounds from oxygenates derived from lignocellulosic sources, HDO being the most widely applied. Typically, sulfided Co-Mo and Ni-Mo [8,9] catalysts, as well as supported precious metals [6], are employed for this purpose. While effective, sulfided catalysts suffer from deactivation in the absence of additional sulfur and can lead to product contamination through sulfur leaching. Given these drawbacks, systems utilizing supported precious metals have been examined as an alternative. However, these catalysts show strong selectivity for ring hydrogenation processes in addition to hydrodeoxygenation. Ring hydrogenation is somewhat undesirable in that it destroys the aromatic character of the compounds produced. Coupled with the often prohibitive cost of precious metals and increased hydrogen consumption resulting from ring hydrogenation, current catalytic strategies leave considerable room for improvement.

Given the stability and platinum-like catalytic behavior of transition metal carbides [10], these materials seem to be a logical alternative for catalyzing deoxygenation reactions. In fact, various unsupported molybdenum carbides have been shown to possess high catalytic activity in hydrocarbon conversion reactions, this being a consequence of metal-carbon bond interactions that yield a noble metal like d-state density around the Fermi level [11]. However, bulk carbides possess intrinsically low surface areas, in addition to requiring the use of highly flammable carburization gases such as methane in their synthesis. Consequently, metal carbides supported on high surface area substrates, including carbon nanomaterials, have been examined in order to enhance the catalytic properties of these formulations. Carbon-based supports are particularly attractive given that the substrate is capable of acting as the carbon source in the synthesis of the catalyst, eliminating the need for carburization gases. Indeed, metal carbides supported on activated carbon (AC) [12,13], ordered mesoporous carbon (OMC) [14], carbon nanofibers (CNF) [2,15–17] and carbon nanotubes (CNT) [18] have already shown promising activity and stability in the deoxygenation of biomass-derived molecules. The performance of these supported carbides is in most cases superior to that of the bulk molybdenum carbide as a result of improved active site accessibility through greater dispersion of carbide phases on the carbon supports [12]. Herein, we report a study of guaiacol hydrodeoxygenation using molybdenum carbides supported on three types of carbon carrier, namely activated carbon, multi-walled carbon nanotubes, and carbon nanofibers. The effectiveness and material properties of each catalyst were assessed in an attempt to gain insights into structure-activity relationships with the specific goal of identifying the carbide phases and/or supports which provide optimal HDO activity.

2. Results and Discussion

2.1. Catalyst Characterization

Synthesized catalysts—the names of which in this article are composed of a Mo loading (7.5 or 20 wt.%), a Mo precursor (Am or Ac representing ammonium molybdate and molybdic acid, respectively), and a carbon support (AC, CNF or CNT)—were analyzed via XRD in an effort to confirm the formation of crystalline molybdenum-containing phases (including molybdenum carbide phases) and measure the corresponding particle sizes. Figure 1 shows the X-ray diffractograms of representative catalysts showing distinct Mo-containing phases, while Table 1 summarizes the Mo-containing phases detected in each catalyst and provides the average particle size for each crystalline phase detected. In contrast to previous reports [12], molybdenum carbides supported on activated carbon demonstrated no crystalline phases under XRD analysis even when prepared at high temperature (1000 °C). This suggests the formation of a highly dispersed Mo-containing phase on the activated carbon support employed.

Table 1. Type and average particle size of the Mo-containing phases in each catalyst as determined by XRD.

Catalyst	Crystalline phases detected	Average particle size (nm)
7.5% Ac/AC	None	-
20% Ac/AC	None	-
7.5% Am/AC	None	-
20% Am/AC	None	-
7.5% Ac/CNF	Mo_2C & MoC	Mo_2C: 13.8 & MoC: 13.0
20% Ac/CNF	Mo_2C-ζ (zeta)	16.1
7.5% Am/CNF	Mo_2C	12.6
20% Am/CNF	Mo_2C	11.3
7.5% Ac/CNT	MoC & Mo_3O	MoC: 13.3 & Mo_3O: 6.3
20% Ac/CNT	MoO_2 & Mo_9O_{26}	MoO_2: 8.8 & Mo_9O_{26}: 9.2
7.5% Am/CNT	None	-
20% Am/CNT	Mo_2C-ζ (zeta)	16.2

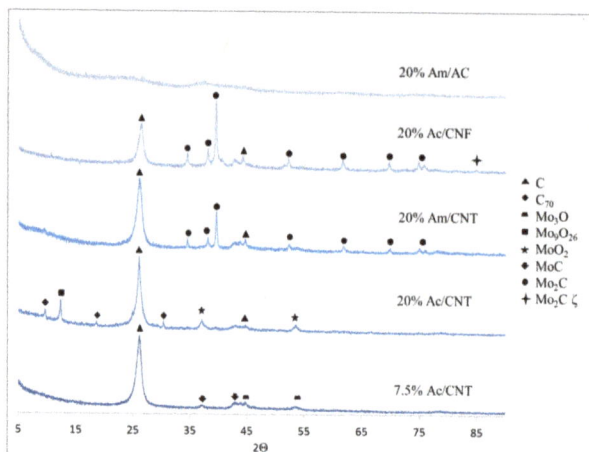

Figure 1. X-ray diffractograms of representative catalysts showing different crystalline phases.

The textural properties of the catalysts were measured by means of nitrogen physisorption, the results of these measurements being summarized in Table 2, which also includes the corresponding data for the carbon supports employed in this study. Interestingly, little change in pore size distribution occurs after carburization of the activated carbon and carbon nanofiber supported catalysts, while a significant increase in macroporosity occurs for the carbon nanotube supported catalysts. The accompanying increase in pore volume and decrease in surface area is likely a result of the agglomeration of large molybdenum particles (>2 μm) which were found to be present only in the carbon nanotube supported catalysts as demonstrated in Figure 2. Indeed, large (~2 μm) molybdenum carbide particles were observed in all the micrographs of the CNT supported catalysts containing Mo_2C or MoC, these microparticles being absent from the CNF supported formulations in which molybdenum carbide was exclusively present in the form of nanoparticles (see Figure 2). Given that the *average* particle size of molybdenum carbide is relatively similar in both CNT and CNF supported catalysts (see Table 1), it can be concluded that the use of CNT as support results in a bimodal particle size distribution containing both micro and nanoparticles, whereas the use of CNF affords solely well dispersed nanoparticles. These conclusions are also supported by SEM-EDX mapping, which reveals homogenous molybdenum dispersion throughout the CNF, as well as a uniform structural morphology even at higher magnification. In contrast, the CNT supported carbides demonstrate both a dispersed molybdenum phase on the nanotubes as well as a bulk phase which forms as an agglomeration around the nanotubes. The difference in Mo_2C morphology between the CNF and CNT supports can be rationalized on the basis of the much lower surface area and mesopore

198

volume of the CNT support, *i.e.*, at the 20 wt.% Mo loading the surface area of the CNT support is evidently insufficient to stabilize all of the formed Mo_2C in a nanodisperse state.

Table 2. Textural properties of the synthesized catalysts and of the corresponding carbon supports.

Carbon support or catalyst	BET surface area $(m^2\,g^{-1})$	Pore volume $(cm^3\,g^{-1})$	Avg. pore diameter (nm)	Pore size distribution (%)		
				Micro	Meso	Macro
AC	1330	1.00	3.4	37	61	2
7.5% Ac/AC	781	0.60	3.6	36	63	1
20% Ac/AC	624	0.49	3.7	34	65	1
7.5% Am/AC	798	0.60	3.5	36	63	1
20% Am/AC	755	0.56	3.6	35	64	1
CNF	198	0.45	10.8	5	90	5
7.5% Ac/CNF	191	0.42	9.7	5	90	5
20% Ac/CNF	163	0.38	9.2	6	86	8
7.5% Am/CNF	165	0.38	10.7	4	94	2
20% Am/CNF	144	0.33	10	5	92	3
CNT	98	0.18	8.8	4	58	38
7.5% Ac/CNT	66	0.28	9.6	3	51	46
20% Ac/CNT	71	0.26	11.2	2	49	49
7.5% Am/CNT	69	0.36	12.5	2	43	55
20% Am/CNT	55	0.22	10.5	2	49	49

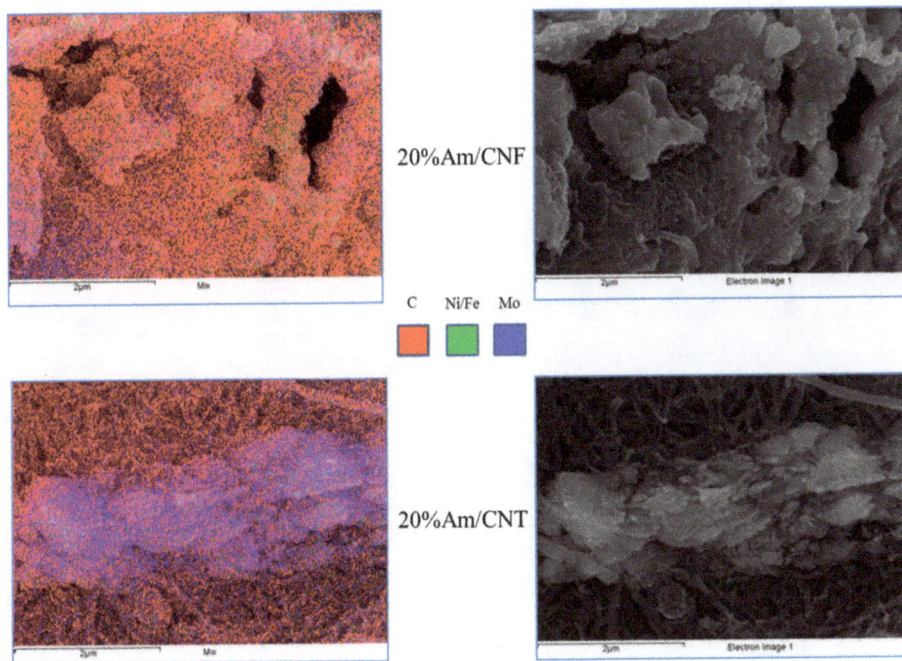

Figure 2. Scanning electron micrographs of 20% Am/CNF and 20% Am/CNT: EDX analysis (**left**) and backscattered electron images (**right**).

2.2. Catalyst Evaluation

2.2.1. Catalyst Evaluation in Water

The synthesized catalysts—as well as the bare supports—were first tested for activity in the upgrading of guaiacol in an aqueous environment. However, under these conditions the extent of guaiacol deoxygenation was low, catechol being the main reaction product (see Table S1 of the Supplementary Materials accompanying this article). Moreover, the results obtained did not reveal any significant trends, making it impossible to elucidate structure-activity relationships. Indeed, in some cases the bare carbon supports afforded better results in terms of conversion and/or selectivity than the corresponding molybdenum-containing catalysts. Therefore, the catalysts were tested in an organic environment in an effort to obtain more informative results.

2.2.2. Catalyst Evaluation in Dodecane

Table 3 summarizes the results of experiments performed to evaluate the synthesized catalysts and the bare supports in the upgrading of guaiacol in dodecane.

Table 3. Guaiacol upgrading in dodecane over different catalysts [a].

Catalyst	Guaiacol conversion (%)	Selectivity to [Yield of] catechol (%)	Selectivity to [Yield of] phenol (%)	Gravimetric mass balance [b] (%)
None	7	15 [1]	0 [0]	98
AC	12	10 [1.2]	3 [0.4]	99
7.5% Ac/AC	26	21 [5.5]	12 [3.1]	98
20% Ac/AC	61	7 [4]	9 [6]	70
7.5% Am/AC	37	20 [7.4]	1 [0.4]	98
20% Am/AC	53	31 [16]	3 [2]	95
CNF	6	19 [1]	3 [0.2]	97
7.5% Ac/CNF	40	8 [0.3]	12 [4.8]	85
20% Ac/CNF	56	9 [5]	35 [20]	96
7.5% Am/CNF	33	9 [3]	38 [12]	98
20% Am/CNF	53	11 [5.8]	37 [20]	96
CNT	16	11 [1.8]	2 [0.3]	97
7.5% Ac/CNT	72	36 [26]	3 [2]	95
20% Ac/CNT	91	24 [22]	5 [4]	98
7.5% Am/CNT	33	33 [11]	3 [1]	97
20% Am/CNT	24	9 [2]	36 [8.6]	96
GNT	13	17 [2.2]	4 [0.5]	97

[a] Reaction conditions: 0.125 g catalyst, 1.25 g guaiacol, 11.25 g dodecane, 300 °C, 580 psi of H_2, 4 h. [b] Mass balances were calculated considering only liquids and solids at the start and the end of the reaction, *i.e.*, mass balance = 100 × (mass of starting reaction mixture)/(mass of recovered liquids and solids). Gaseous reactants and products were omitted from the mass balance calculation for the reasons enumerated in Section 3.5.3.

Albeit the conversion and selectivity values observed in the experiment performed without catalyst can be attributed to thermal effects, it is interesting to note that some of the bare carbon supports display additional activity. Given that carbon nanofibers and carbon nanotubes are respectively grown using Ni and Fe catalysts and that the hydrodeoxygenation of guaiacol has been reported to proceed over both Ni [19–21] and Fe [22–24] catalysts, the activity of the bare supports could conceivably be caused by the residual presence of these metals. Indeed, the Ni and Fe content of the carbon supports was measured by means of Inductively Coupled Plasma-Optical Emission Spectrometry (ICP-OES) and determined to be non-negligible, as shown in Table 4.

Table 4. Results of ICP-OES analysis of the carbon supports used in this study.

Carbon support	Fe content	Ni content
AC	5022 ppm	666 ppm
CNF	334 ppm	1.9%
CNT	2.35%	40 ppm
GNT	837 ppm	18 ppm

However, the activity of Ni present in the CNF can be deemed to be minimal, since the conversion and selectivity values obtained using CNF are almost identical to those obtained in the blank (sans catalyst) run. Carbon nanotubes afforded higher guaiacol conversion than the blank run, albeit this cannot be entirely attributed to the presence of iron since a sample of graphitized nanotubes (GNT)—in which Fe content is reduced to 837 ppm by means of a thermal treatment (in helium) exceeding 2700 °C [25]—showed almost identical results (see Table 3). Similarly, the additional activity displayed by the bare supports relative to the blank run cannot be entirely attributed to the presence of acid sites on the surface of these materials since the graphitized nanotubes—in which these sites have been removed by the high temperatures employed during the graphitization process [26]—shows similar results (see Table 3). In a recent contribution by Jongerius *et al.*, the surface of the carbon supports in carbide catalysts were deemed inert since the high temperature employed in the carburization process effectively removes acidic functional groups from the supports [2]. However, the conversion and selectivity values obtained over graphitized nanotubes suggest that bare supports display some activity which must be taken into account during the interpretation of the results obtained using the Mo-containing catalysts.

Notably, in contrast with the results obtained in aqueous medium, higher guaiacol conversions were invariably obtained over the Mo-containing catalysts relative to the bare supports in an organic environment (using dodecane as solvent). The most abundant reaction products were typically catechol and phenol, albeit in some instances significant amounts of cyclohexane were produced (a more comprehensive account of the results of product analysis is given in Table S2 of the Supplementary Materials accompanying this article). The fact that catechol was observed in all reaction mixtures suggests that guaiacol is converted into phenol via sequential demethylation and HDO [7], which contrasts with the direct demethoxylation pathway proposed by Jongerius *et al.* [2], albeit the possibility for these two pathways to be operating in parallel cannot be discounted.

Among the aforementioned products—namely catechol, phenol, and cyclohexane—phenol represents the most desirable product, as it offers a good balance between depth of deoxygenation and value (an upgraded bio-oil or lignin depolymerization stream rich in catechol would still be unacceptably unstable and corrosive while a cyclohexane-rich product would be less valuable that a product rich in phenol). Therefore, based on the yield of phenol the best catalysts are 20% Ac/CNF, 7.5% Am/CNF, 20% Am/CNF and 20% Am/CNT (see Table 3). Tellingly, the only crystalline phase detected in all these formulations is Mo_2C (or Mo_2C-ζ, which does not seem to behave inherently differently to Mo_2C). This suggests that crystalline Mo_2C is particularly selective to phenol, which is consistent with the fact that catalysts in which Mo_2C is accompanied by MoC—as well as catalysts comprising MoC and/or different molybdenum oxides and catalysts with no crystalline Mo-containing phases—all show lower selectivity to phenol than the catalysts containing Mo_2C or Mo_2C-ζ as their only crystalline phase. Indeed, while the samples containing Mo oxides (*i.e.*, 7.5% Ac/CNT and 20% Ac/CNT) showed high guaiacol conversions, the selectivity to phenol was extremely low, the preferred products being catechol and coke (see Table S2).

Of the samples containing Mo_2C or Mo_2C-ζ as their only crystalline phase, a comparison between the results obtained using 7.5% Am/CNF and 20% Am/CNF is particularly informative, since the only difference between these two catalysts is their metal loading (Mo precursor, carbon support, Mo_2C phase and Mo_2C particle size being all the same). Interestingly, the selectivity to phenol of these two catalysts is virtually identical, the only difference in their behavior being the lower guaiacol conversion shown by 7.5% Am/CNF, which is consistent with the fact that a lower metal loading (at similar dispersion) should translate into a reduced number of active sites.

It is also instructive to compare the results obtained with the 20% Am/CNF and 20% Am/CNT catalysts, these affording very different conversion values (at similar selectivity). In an effort to assess the intrinsic activity of these catalysts, CO chemisorption was applied in order to quantify the concentration of adsorption sites present (see Table 5). Based on their very similar CO uptake, both catalysts have a similar quantity of active sites. Taking this into account and the fact that the guaiacol conversion obtained over 20% Am/CNF was over twice that obtained over 20% Am/CNT, the former formulation appears to be more intrinsically active, although this ignores possible differences in catalyst deactivation due to coke formation.

Table 5. Results of CO chemisorption and intrinsic activity of 20% Am/CNF and 20% Am/CNT.

Catalyst	CO uptake (μmol g^{-1})	Average rate (s^{-1}) [a]
20% Am/CNF	49	0.06
20% Am/CNT	53	0.03

[a] Conversion rate normalized to the number of active sites as calculated from CO uptake.

As noted above, the use of CNT as support results in a bimodal particle size distribution containing both micro and nanoparticles in catalysts where carbides are formed, whereas the use of CNF affords solely well-dispersed carbide nanoparticles. In turn, given that these catalysts show a very similar amount of active sites (see Table 5), it follows that the difference in the activity of the 20% Am/CNF and 20% Am/CNT catalysts can be attributed to the effect of particle morphology. Tellingly, spent 20% Am/CNT displayed higher amounts of coke on its surface relative to spent CNF-supported catalysts including 20% Am/CNF (see Table S2), which suggests that this morphology effect can be explained in terms of the catalyst resistance to coking—coke can cover more active sites (and arguably form more effectively) on a microparticle showing a high concentration of adjacent Mo_2C sites. Therefore, the optimization of particle morphology seems to be a promising way to further improve the performance of molybdenum carbide catalysts.

Finally, a 5% Ru/C reference catalyst was tested under identical reaction conditions for comparison purposes. Albeit this catalyst afforded quantitative guaiacol conversion (the mass balance being 94%), the catalyst displayed 100% selectivity to cyclohexane. Given that (as mentioned above) a cyclohexane-rich product would be less valuable that a product rich in phenol, the CNF-supported Mo_2C catalysts employed in this study represent a promising alternative to the costly Ru-based catalyst commonly used to catalyze this reaction.

2.2.3. Catalyst Testing in an Organic Environment at Different Temperatures

In order to study the effect of temperature on the upgrading of guaiacol in dodecane over Mo_2C/CNF—the formulation showing the best performance (in terms of yield of phenol) at 300 °C—representative CNF-supported catalysts were tested at 350 °C. The catalysts employed in these tests, namely 7.5% Am/CNF and 20% Am/CNF, were chosen in order to study the effect of temperature on two similar catalysts (their only difference being the metal loading) showing noticeably different conversion. In turn, the reaction temperature of 350 °C was chosen based on a recent report by Jongerius et al. [2] in which the best results in terms of both guaiacol conversion and phenol selectivity were obtained at this temperature. The results of these experiments are shown in Table 6 and a more comprehensive account of the results of product analysis is given in Table S3 of the Supplementary Materials accompanying this article.

Table 6. Guaiacol upgrading in an organic environment at different temperatures [a].

Catalyst	Temperature (°C)	Guaiacol conversion (%)	Selectivity to [Yield of] catechol (%)	Selectivity to [Yield of] phenol (%)	Gravimetric mass balance [b] (%)
None	300	7	15 [1]	0 [0]	98
None	350	79	5 [4]	2 [2]	59
7.5% Am/CNF	300	33	9 [3]	38 [12]	98
7.5% Am/CNF	350	99	0 [0]	49 [48]	99
20% Am/CNF	300	53	11 [5.8]	37 [20]	96
20% Am/CNF	350	98	0 [0]	48 [47]	98

[a] Reaction conditions: 0.125 g catalyst, 1.25 g guaiacol, 11.25 g dodecane, 580 psi of H_2, 4 h.
[b] Mass balances were calculated considering only liquids and solids at the start and the end of the reaction, *i.e.*, mass balance = 100 × (mass of starting reaction mixture)/(mass of recovered liquids and solids). Gaseous reactants and products were omitted from the mass balance calculation for the reasons enumerated in Section 3.5.3.

As mentioned above, the conversion and selectivity values observed in the experiments performed without catalyst can be attributed to thermal effects. Although the guaiacol conversion is considerably higher at 350 °C than at 300 °C in the absence of a catalyst, the selectivity to catechol and phenol remains minimal. Given that very small amounts of products were detected in the liquid product mixture (see Table S3) and a considerably lower mass balance was obtained from the blank experiment performed at 350 °C, it appears that most of the converted guaiacol afforded products in the gas phase which were not counted in the mass balance calculation. Other authors have explained low mass balances invoking the formation of high molecular weight condensation products not detectable through GC-based methods [2]; however, the fact that in our case mass balances were calculated gravimetrically can be used to rule out this possibility. These results contrast with those obtained in the presence of a catalyst, in which conversions of ≥98% and selectivities to phenol approaching 50% were achieved at 350 °C. Tellingly, the mass balances of these reactions were also ≥98%, which suggests that approximately half of the guaiacol was converted to the coke observed on the spent catalyst surface (see Table S3), to products that were detectable but unidentifiable via GC—which cannot be quantified using GC since their response factors are unknown—and/or to soluble higher molecular weight products undetectable via GC [2], the amount of these products being non-negligible according to a molar mass balance (see Table S4 of the Supplementary Materials). The fact that the performance of both CNF-supported catalysts was almost identical at 350 °C indicates that at this temperature the reaction is fast enough as to be driven to completion by the lower number of active sites present in the 7.5% Am/CNF catalyst. Similarly, the fact that no catechol was detected in the products of the reactions catalyzed at 350 °C suggests that if the conversion of guaiacol into phenol is intermediated by catechol, this reaction was also driven to completion (since phenol presents the most difficult C-O bond to cleave, it is particularly resistant to HDO [7]). Notably, the conversion and selectivity to phenol values obtained at 350 °C over the CNF-supported Mo_2C catalysts are among the best in the literature for carbide catalysts, being comparable to those reported by Jongerius *et al.* [2].

2.3. Catalyst Recycling

The recyclability of the most promising catalyst identified in this study, namely 20% Am/CNF, was assessed by retesting the spent catalyst (after washing and drying) in two additional guaiacol upgrading experiments, which were performed adjusting the amount of solvent and guaiacol to the amount of catalysts recovered (albeit only 8 mg of catalyst was lost during the 2nd run). The spent catalysts were tested without a reactivation step based on the results of Jongerius *et al.* [2], who determined that the recarburization of the spent Mo_2C/CNF catalysts did not afford improved results. The results of these recycling experiments are summarized in Table 7 and a more comprehensive account of the results of product analysis is given in Table S5 of the Supplementary Materials.

Table 7. Sequential guaiacol upgrading runs in an organic environment over 20% Am/CNF [a].

Run	Guaiacol conversion (%)	Selectivity to [Yield of] catechol (%)	Selectivity to [Yield of] phenol (%)	Gravimetric mass balance [b] (%)
1	76	2 [2]	50 [38]	98
2	63	3 [2]	36 [23]	97
3	53	5 [3]	36 [19]	98

[a] Reaction conditions: 300 °C, 580 psi of H_2, 4 h. Run 1: 0.125 g catalyst, 1.25 g guaiacol, 11.25 g dodecane; Run 2: 0.125 g catalyst, 1.25 g guaiacol, 11.25 g dodecane; Run 3: 0.117 g catalyst, 1.17 g guaiacol, 10.53 g dodecane. [b] Mass balances were calculated considering only liquids and solids at the start and the end of the reaction, *i.e.*, mass balance = 100 × (mass of starting reaction mixture)/(mass of recovered liquids and solids). Gaseous reactants and products were omitted from the mass balance calculation for the reasons enumerated in Section 3.5.3.

It should first be noted that the conversion and selectivity values of fresh 20% Am/CNF in Tables 3 and 7 are somewhat different. Notably, Jongerius *et al.* reported that freshly prepared batches of CNF-supported metal carbide catalysts show some variation in terms of performance, albeit these authors observed different batches of W_2C/CNF to vary considerably and different batches of Mo_2C/CNF to show little variation [2]. However, these authors made these observations at 350 °C, under conditions at which conversion values over Mo_2C/CNF were close to quantitative. Similarly, differences in temperature can be invoked to explain the disparity between the results of the recycling study shown in Table 7 and those obtained by Jongerius *et al.* [2]; in the latter case, no significant loss in guaiacol conversion or selectivity to phenol was observed at 350 °C, while Table 7 shows a ~30% drop in both guaiacol conversion and selectivity to phenol at 300 °C. However, Jongerius *et al.* did observe a 24% drop in conversion and a 20% drop in selectivity to phenol during three sequential (shorter) reactions not showing quantitative conversion [2].

In addition, the results shown in Table 7 and S5 offer some valuable insights regarding catalyst deactivation. Interestingly, although the guaiacol conversion values in Table 7 monotonically decrease in each sequential run, the amount of coke on the spent catalyst surface is also observed to decrease (see Table S5). This suggests that not all of the coke formation taking place during reaction is irreversible, as coke does not appear to accumulate with each subsequent test. This also indicates that there must be another deactivation route in addition to coking, since progressively lower conversion values are obtained in spite of the fact that the amount of coke on the catalyst surface decreases along the experiment sequence. This is in agreement with the conclusions of Jongerius *et al.* who suggested that in addition to irreversible coke formation, the encapsulation of carbide particles in the carbon support represents another potential route of catalyst deactivation [2]. However, more work is necessary to more fully understand the mechanisms of deactivation and to improve catalyst stability and recyclability.

2.4. Guaiacol Upgrading over Mo_2C/CNF in the Presence of Acetic Acid and Furfural

As mentioned in Section 1, guaiacol represents a frequently used model compound for the large number of mono- and dimethoxyphenols resulting from the densification of lignin [3]. However, bio-oil also contains other families of oxygenated compounds stemming from the holocellulosic fraction of biomass, which must be taken into account in bio-oil upgrading studies due to the fact that the reactivity of individual compounds and families of compounds can change when mixed due to synergistic [7] and/or inhibitory [27] effects. With this in mind, a guaiacol upgrading experiment over 20% Am/CNF was performed in the presence of acetic acid and furfural, which are used as model compounds to represent the fractions within bio-oil stemming from hemicellulose and cellulose, respectively [3]. The results of this experiment are summarized in Table 8, and a more comprehensive account of the results of product analysis is given in Table S6 of the Supplementary Materials.

Table 8. Guaiacol upgrading over 20% Am/CNF in the presence of acetic acid and furfural [a].

Model compound	Conversion (%)	Selectivity to [Yield of] catechol (%)	Selectivity to [Yield of] phenol (%)	Gravimetric mass balance [b] (%)
Guaiacol	6	22 [1]	63 [4]	-
Acetic acid	68	-	-	-
Furfural	80	-	-	-
TOTAL	48	-	-	97

[a] Reaction conditions: 0.125 g catalyst, 0.43 g guaiacol, 0.43 g acetic acid, 0.44 g furfural, 11.25 g dodecane, 300 °C, 580 psi of H_2, 4 h. [b] Mass balances were calculated considering only liquids and solids at the start and the end of the reaction, *i.e.*, mass balance = 100 × (mass of starting reaction mixture)/(mass of recovered liquids and solids). Gaseous reactants and products were omitted from the mass balance calculation for the reasons enumerated in Section 3.5.3.

A comparison of the guaiacol conversion values observed over 20% Am/CNF in Tables 3 and 7 clearly illustrates that the presence of acetic acid and furfural greatly inhibits guaiacol conversion. However, it is equally important to note that the presence of acetic acid and furfural greatly augment the selectivity of the catalyst to both catechol and phenol, which may be of interest from an industrial standpoint given that selectivity is commonly prioritized over conversion in industrial processes where any unreacted feed can be recirculated. In stark contrast with the limited conversion of guaiacol, the majority of both acetic acid and furfural were converted, albeit only a small amount of the products observed during the hydrotreatment of these compounds—such as ethyl acetate, THF-MeOH and γ-butyrolactone (see Table S6)—were identified in the liquid product mixture via GC analysis. It is likely that some portion of the acetic acid was converted to ethanol. However, the amount of ethanol produced could not be quantified (due to the interference of ethanol present as a stabilizer in the chloroform used to work up the reaction products). Indeed, the mass balance of this reaction (gravimetric basis) was 97%, which suggests that most of the acetic acid and furfural were converted to solid deposits on the catalyst surface or to liquid organic compounds (as opposed to being converted to gaseous products). Given the relatively low quantities of coke formed during the course of this reaction (see Table S6) it can be inferred that some portion of these reagents were converted to products that were detectable but unidentifiable via GC, and/or to soluble higher molecular weight products undetectable via GC [2]. Using a procedure similar to that described in Table S4, the quantity of unidentified products was found to be approximately 471 mg for this reaction, representing 36% of the feed.

3. Experimental Section

3.1. Catalyst Synthesis

Carbon nanofibers were prepared using a floating catalyst chemical vapor deposition (CVD) method, similar to that described by Martin-Gullon [28] and Weisenberger [29]. Multi-walled carbon nanotubes were produced via the CVD of a xylene/ferrocene feedstock between 750 and 850 °C on quartz substrates [30,31]. The AC employed was Darco KB-G obtained from Sigma Aldrich (St. Louis, MO, USA). Graphitization of the carbon nanotubes was performed at 2700 °C under helium as described elsewhere [25]. Prior to use, the surface area of CNFs and CNTs was increased by an acid treatment in which the support (2 g) was reacted with 6 M HNO_3 (60 ml) at 80 °C for 3 h under vigorous stirring, after which the solids were filtered, washed with deionized water until the pH of the washings was ~7 and dried at 120 °C. AC did not require pretreatment. Each carbon support was then impregnated with either ammonium molybdate ((NH_4)$_6Mo_7O_{24}$, Alfa Aesar, Ward Hill, MA, USA) (Am) or molybdic acid (MoO_3 ≥85%, Alfa Aesar, Ward Hill, MA, USA) (Ac) to afford the desired Mo loading (7.5 wt.% or 20 wt.%). The metal loadings were chosen based on previous work by Jongerius et al. [2] and Qin et al. [15]. The impregnation was carried out by sonicating the support (1 g), deionized water (15 mL) and the prescribed amount of precursor (Am: 0.122 g to afford 7.5 wt.% loading or 0.325 g to afford 20 wt.% loading; Ac: 0.119 g to afford 7.5 wt.% loading or 0.318 g to afford 20 wt.% loading) for 1.5 h. Excess water was removed via rotary evaporation prior to drying in a vacuum oven at 120° C for 14 h. The carbide catalysts employed in this study were prepared via the carbothermal hydrogen reduction (CHR) method [32]. In contrast to the work of Qin and co-workers [15], a carburization temperature of 700 °C proved insufficient to obtain a crystalline carbide phase. Carburization was therefore performed under a flow of 10% H_2/Ar using a ramp rate of 5 °C/min to 1000 °C followed by an isothermal step at this temperature lasting 3 h. After cooling to room temperature, the catalyst was passivated under a flow of 1% O_2/N_2 for 8 h.

3.2. Catalyst Characterization

Powder X-ray diffraction (XRD) measurements were performed on a Phillips X'Pert diffractomer using Cu Kα radiation (λ = 1.5406 Å) and a step size of 0.02°. Average crystallite sizes were calculated using the Scherrer equation. The surface area, pore volume and average pore diameter of the synthesized catalysts were determined by means of N_2 physisorption using a Micromeritics Tristar System at 77 K. N_2 at −196 °C was the sorbate. In all cases samples were degassed overnight at 250 °C prior to the measurements. Total pore volume and pore size distribution were determined using the Density Functional Theory (DFT) method. ICP-OES measurements were performed using a 100 mg sample dissolved in 10 mL of nitric acid. Heating was used to ensure that the sample was completely dissolved. Once cooled, the sample was further diluted to 25 mL with doubly distilled water. Measurements were acquired on a Varian 720-ES spectrometer equipped with a seaspray nebulizer and cyclonic class spray chamber. Parameters include a sample intake of 1 ml/min, Ar plasma flow rate of 15 L/min and an auxillary gas (Ar) flow rate of 1.5 L/min. The instrument was calibrated using a CRMS manufactured by VHG. CO pulse chemisorption experiments were performed using a Micromeritics AutoChem II 2920 Chemisorption Analyzer. 10% H_2 in Ar (50 mL/min) was flowed through the sample as the temperature was increased to 350 °C at a rate of 5 °C/min followed by a purge under flowing Ar (50 mL/min) while cooling to 35 °C. Ultra high purity CO (99.995% from Matheson-Trigas, Palm, FL, USA) was then pulsed into the system at a volume of 0.5 mL every 2 min until the intensity of the peak was constant. SEM-EDX experiments were conducted on a Hitachi S4800 scanning electron microscope with an Oxford INCA X-MAX EDS/EDX attachment. Samples were placed on carbon tape and inserted into the vacuum chamber uncoated. An emission current of 20 μA and accelerator voltage of 15 KV was used for these observations. EDX maps were collected at 0–20 KeV.

3.3. Catalyst Testing in Water

Experiments were performed in a mechanically stirred 25 mL stainless steel autoclave. Guaiacol (1.25 g), water (11.25 g), and the catalyst (0.125 g) were simultaneously added to the reactor prior to sealing, after which the reactor was purged with Ar (60 psi), charged to 1000 psi with H_2, mechanically stirred at 500 rpm and heated to the desired temperature. The autoclave temperature was measured by a type-K Omega thermocouple placed inside the reactor body. As soon as the reaction temperature was reached, the pressure was adjusted to 2000 psi. At the completion of the experiment (the reaction time being 4 h), forced air was used to facilitate cooling. Once the reactor reached room temperature, a gas sample was taken for analysis. After opening the reactor, sec-butanol (1.2 g) was added to the product mixture as an internal GC standard. The reaction mixture was then removed by a tared pipette, weighed, and separated by filtration. The catalyst was extracted with acetone to yield additional products. After being allowed to dry under ambient conditions, the filter and recovered solids were weighed again to determine the mass of recovered solids.

3.4. Catalyst Testing in Dodecane

Experiments were performed in a mechanically stirred 25 mL stainless steel autoclave using guaiacol (1.25 g), dodecane (11.25 g), and the catalyst (0.125 g) according to the procedure described in Section 3.3. The initial H_2 charge was 300 psi. As soon as the reaction temperature was reached, the pressure was adjusted to 580 psi.

At the end of the reaction, the liquid and solids were separated and the catalyst was extracted with CHCl₃ to yield additional products. Subsequent reactions using a blended feed (1.25 g) of acetic acid, furfural and guaiacol in equal proportions were run in a similar fashion.

3.5. Product Analysis

3.5.1. Liquid Analysis

Cyclohexane (HPLC Grade), ethyl acetate (HPLC Grade, manufactory), furfural (Reagent Grade) and chloroform (HPLC Grade) were purchased from Fisher Scientific. Ethanol (99.5%), cyclopentanone (99+%), cyclopentanol (99%), cyclohexanol (98%), guaiacol (99+%), levulinic acid (98+%) and sec-butanol (99%) were manufactured by Acros Organics (Geel, Belgium). Cyclohexanone (99+%), acetic acid (99.7+%), tetrahydrofurfuryl alcohol (98%), 2-methoxycyclohexanol (99%) and catechol (99%) were purchased from Alfa Aesar (Heysham, UK). γ-Butyrolactone and 1,2-pentanediol were purchased from Sigma-Aldrich (St. Louis, USA). Cyclohexanediol (mixture of *cis*- and *trans*- isomers) was purchased from TCI America (Tokyo, Japan) and phenol (99%) was purchased from City Chemical (West Haven, CT, USA).

Liquid reaction products were analyzed using an Agilent 7890A GC equipped with an Agilent Multimode inlet, a deactivated open ended helix liner and a flame ionization detector (FID). A 0.2 μL injection was employed and helium was used as the carrier gas. The FID was set to 250 °C with the following gas flow rates: H_2 = 50 mL/min; air = 450 mL/min; makeup = 19 mL/min. The inlet was isothermally maintained at 240 °C in splitless mode. An Agilent J&W DB-Wax column (30 m × 530 μm × 0.5 μm) rated to 240 °C was employed, maintaining a constant flow of 26 mL/min. The oven parameters were programmed to start at 35 °C; followed by a ramp of 4 °C/min to 60 °C; followed by a ramp of 10 °C/min to 200 °C; followed by a ramp of 40 °C/min to 240 °C and a 240 °C isotherm lasting 3.75 min. The total run time was 25 min. Chromatographic programming was performed using Agilent Chemstation software. Quantitative calibrations were conducted with solutions prepared according to Table S7 using sec-butanol as the internal standard. Conversion, selectivity and yield were calculated using the following formulas: Guaiacol conversion (%) = 100 × (mass of guaiacol loaded − mass of unconverted guaiacol)/guaiacol loaded; Selectivity to specific product (%) = 100 × mass of the specific product/(mass of guaiacol loaded − mass of unconverted guaiacol); Yield of specific product (%) = guaiacol conversion × selectivity to specific product.

3.5.2. Solid Analysis

Attempts were made to study the nature and the amount of the carbonaceous deposits on the surface of spent catalysts by means of thermogravimetric analysis (TGA)—performed under flowing air (50 mL/min) on a TA instruments Discovery Series thermogravimetric analyzer using a temperature ramp of 10 °C/minute from room temperature to 800 °C. However, these efforts proved unfruitful due to the fact that the weight loss associated with coke combustion could not be deconvoluted from the weight loss stemming from the combustion of the carbon supports (which were studied separately by subjecting fresh catalyst to TGA).

3.5.3. Gas Analysis

An Agilent 3000 Micro-GC with a configuration previously reported [33] was calibrated for possible gaseous products, including straight chain C1-C6 alkanes and alkenes. The analysis of gaseous products formed during representative guaiacol experiments proved unremarkable, no products being detected.

4. Conclusions

In this work the hydrodeoxygenation of guaiacol over molybdenum carbides supported on three types of carbon support—activated carbon, multi-walled carbon nanotubes, and carbon nanofibers—was studied. The use of activated carbon as support afforded an X-ray amorphous Mo phase, whereas crystalline carbide phases were obtained on carbon nanofibers and, in some cases, on carbon nanotubes. Scanning electron micrographs revealed the presence of large molybdenum carbide particles (>2 μm) in the CNT-supported catalysts containing molybdenum carbide phases, whereas in the CNF-supported samples molybdenum carbide was exclusively present in the form of nanoparticles. In the HDO of guaiacol in dodecane, catechol and phenol were obtained as the main products, although in some instances cyclohexane was also formed. The fact that catechol was observed in all reaction mixtures suggests that guaiacol is converted into phenol via sequential demethylation and HDO, albeit the simultaneous occurrence of a direct demethoxylation pathway cannot be discounted. Based on the yield of phenol obtained, 20% Ac/CNF, 7.5% Am/CNF, 20% Am/CNF and 20% Am/CNT were identified as the best catalysts; notably, the only crystalline phase detected in all these formulations was Mo_2C (or Mo_2C-ζ, which does not seem to behave inherently differently to Mo_2C), indicating that crystalline Mo_2C is particularly selective to phenol. At 350 °C, CNF-supported Mo_2C afforded near quantitative guaiacol conversion (≥98%), with a selectivity to phenol approaching 50%. When guaiacol HDO was performed using 20%Am/CNF in the presence of acetic acid and furfural, the conversion of guaiacol was greatly inhibited, although the selectivity of the catalyst to both catechol and phenol was increased.

Acknowledgments: This work was supported by a Seed Grant of the University of Kentucky Center for Applied Energy Research (UK CAER). Renan Sales is thanked for technical assistance. Ashley Morris, John Craddock and Matthew Weisenberger of the UK CAER Carbon Materials group are thanked for providing the nanostructured carbons employed in this work.

Author Contributions: E.S.-J., R.P. and M.C. conceived and designed the experiments; M.P., R.P. and T.M. performed the experiments; E.S.-J., M.P., R.P., T.M. and M.C. analyzed the data; E.S.-J., R.P., T.M. and M.C. wrote the paper.

Conflicts of Interest: The authors declare no conflict of interest. The founding sponsors had no role in the design of the study; in the collection, analyses, or interpretation of data; in the writing of the manuscript, and in the decision to publish the results.

References

1. Zakzeski, J.; Bruijnincx, P.C.A.; Jongerius, A.L.; Weckhuysen, B.M. The Catalytic Valorization of Lignin for the Production of Renewable Chemicals. *Chem. Rev.* **2010**, *110*, 3552–3599.
2. Jongerius, A.L.; Gosselink, R.W.; Dijkstra, J.; Bitter, J.H.; Bruijnincx, P.C.A.; Weckhuysen, B.M. Carbon nanofiber supported transition-metal carbide catalysts for the hydrodeoxygenation of guaiacol. *ChemCatChem* **2013**, *5*, 2964–2972.
3. Elliott, D.C.; Hart, T.R. Catalytic Hydroprocessing of Chemical Models for Bio-oil. *Energy Fuels* **2009**, *23*, 631–637.
4. Lin, Y.-C.; Li, C.-L.; Wan, H.-P.; Lee, H.-T.; Liu, C.-F. Catalytic Hydrodeoxygenation of Guaiacol on Rh-Based and Sulfided CoMo and NiMo Catalysts. *Energy Fuels* **2011**, *25*, 890–896.
5. Jongerius, A.L.; Jastrzebski, R.; Bruijnincx, P.C.A.; Weckhuysen, B.M. CoMo sulfide-catalyzed hydrodeoxygenation of lignin model compounds: An extended reaction network for the conversion of monomeric and dimeric substrates. *J. Catal.* **2012**, *285*, 315–323.
6. Gutierrez, A.; Kaila, R.K.; Honkela, M.L.; Slioor, R.; Krause, A.O.I. Hydrodeoxygenation of guaiacol on noble metal catalysts. *Catal. Today* **2009**, *147*, 239–246.
7. Graça, I.; Lopes, J.M.; Cerqueira, H.S.; Ribeiro, M.F. Bio-oils Upgrading for Second Generation Biofuels. *Ind. Eng. Chem. Res.* **2013**, *52*, 275–287.
8. Furimsky, E. Catalytic hydrodeoxygenation. *Appl. Catal. A* **2000**, *199*, 147–190.
9. Wang, H.; Male, J.; Wang, Y. Recent Advances in Hydrotreating of Pyrolysis Bio-Oil and Its Oxygen-Containing Model Compounds. *ACS Catal.* **2013**, *3*, 1047–1070.
10. Levy, R.B.; Boudart, M. Platinum-like behavior of tungsten carbide in surface catalysis. *Science* **1973**, *181*, 547–549.

11. Gao, Q.; Zhang, C.; Xie, S.; Hua, W.; Zhang, Y.; Ren, N.; Xu, H.; Tang, Y. Synthesis of Nanoporous Molybdenum Carbide Nanowires Based on Organic–Inorganic Hybrid Nanocomposites with Sub-Nanometer Periodic Structures. *Chem. Mater.* **2009**, *21*, 5560–5562.

12. Han, J.; Duan, J.; Chen, P.; Lou, H.; Zheng, X. Molybdenum Carbide-Catalyzed Conversion of Renewable Oils into Diesel-like Hydrocarbons. *Adv. Synth. Catal.* **2011**, *353*, 2577–2583.

13. Chang, J.; Danuthai, T.; Dewiyanti, S.; Wang, C.; Borgna, A. Hydrodeoxygenation of Guaiacol over Carbon-Supported Metal Catalysts. *ChemCatChem* **2013**, *5*, 3041–3049.

14. Han, J.; Duan, J.; Chen, P.; Lou, H.; Zheng, X.; Hong, H. Carbon-Supported Molybdenum Carbide Catalysts for the Conversion of Vegetable Oils. *ChemSusChem* **2012**, *5*, 727–733.

15. Qin, Y.; Chen, P.; Duan, J.; Han, J.; Lou, H.; Zheng, X.; Hong, H. Carbon nanofibers supported molybdenum carbide catalysts for hydrodeoxygenation of vegetable oils. *RSC Adv.* **2013**, *3*, 17485–17491.

16. Hollak, S.A.W.; Gosselink, R.W.; van Es, D.S.; Bitter, J.H. Comparison of Tungsten and Molybdenum Carbide Catalysts for the Hydrodeoxygenation of Oleic Acid. *ACS Catal.* **2013**, *3*, 2837–2844.

17. Stellwagen, D.R.; Bitter, J.H. Structure-performance relations of molybdenum- and tungsten carbide catalysts for deoxygenation. *Green Chem.* **2015**, *17*, 582–593.

18. Han, J.; Duan, J.; Chen, P.; Lou, H.; Zheng, X.; Hong, H. Nanostructured molybdenum carbides supported on carbon nanotubes as efficient catalysts for one-step hydrodeoxygenation and isomerization of vegetable oils. *Green Chem.* **2011**, *13*, 2561–2568.

19. Escalona, N.; Aranzaez, W.; Leiva, K.; Martínez, N.; Pecchi, G. Ni nanoparticles prepared from Ce substituted LaNiO₃ for the guaiacol conversion. *Appl. Catal. A* **2014**, *481*, 1–10.

20. Mortensen, P.M.; Gardini, D.; de Carvalho, H.W.P.; Damsgaard, C.D.; Grunwaldt, J.-D.; Jensen, P.A.; Wagner, J.B.; Jensen, A.D. Stability and resistance of nickel catalysts for hydrodeoxygenation: carbon deposition and effects of sulfur, potassium, and chlorine in the feed. *Catal. Sci. Technol.* **2014**, *4*, 3672–3686.

21. Zhang, X.; Long, J.; Kong, W.; Zhang, Q.; Chen, L.; Wang, T.; Ma, L.; Li, Y. Catalytic Upgrading of Bio-oil over Ni-Based Catalysts Supported on Mixed Oxides. *Energy Fuels* **2014**, *28*, 2562–2570.

22. Olcese, R.N.; Bettahar, M.; Petitjean, D.; Malaman, B.; Giovanella, F.; Dufour, A. Gas-phase hydrodeoxygenation of guaiacol over Fe/SiO₂ catalyst. *Appl. Catal. B* **2012**, *115–116*, 63–73.

23. Olcese, R.; Bettahar, M.M.; Malaman, B.; Ghanbaja, J.; Tibavizco, L.; Petitjean, D.; Dufour, A. Gas-phase hydrodeoxygenation of guaiacol over iron-based catalysts. Effect of gases composition, iron load and supports (silica and activated carbon). *Appl. Catal. B* **2013**, *129*, 528–538.

24. Olcese, R.N.; Francois, J.; Bettahar, M.M.; Petitjean, D.; Dufour, A. Hydrodeoxygenation of Guaiacol, A Surrogate of Lignin Pyrolysis Vapors, Over Iron Based Catalysts: Kinetics and Modeling of the Lignin to Aromatics Integrated Process. *Energy Fuels* **2013**, *27*, 975–984.

25. Andrews, R.; Jacques, D.; Qian, D.; Dickey, E.C. Purification and structural annealing of multiwalled carbon nanotubes at graphitization temperatures. *Carbon* **2001**, *39*, 1681–1687.

26. Gosselink, R.W.; Van Den Berg, R.; Xia, W.; Muhler, M.; De Jong, K.P.; Bitter, J.H. Gas phase oxidation as a tool to introduce oxygen containing groups on metal-loaded carbon nanofibers. *Carbon* **2012**, *50*, 4424–4431.

27. Jackson, M.A. Ketonization of model pyrolysis bio-oil solutions in a plug-flow reactor over a mixed oxide of Fe, Ce, and Al. *Energy Fuels* **2013**, *27*, 3936–3943.

28. Martin-Gullon, I.; Vera, J.; Conesa, J.A.; González, J.L.; Merino, C. Differences between carbon nanofibers produced using Fe and Ni catalysts in a floating catalyst reactor. *Carbon* **2006**, *44*, 1572–1580.

29. Weisenberger, M.; Martin-Gullon, I.; Vera-Agullo, J.; Varela-Rizo, H.; Merino, C.; Andrews, R.; Qian, D.; Rantell, T. The effect of graphitization temperature on the structure of helical-ribbon carbon nanofibers. *Carbon* **2009**, *47*, 2211–2218.

30. Craddock, J.D.; Weisenberger, M.C. Harvesting of large, substrate-free sheets of vertically aligned multiwall carbon nanotube arrays. *Carbon* **2015**, *81*, 839–841.

31. Andrews, R.; Jacques, D.; Rao, A.M.; Derbyshire, F.; Qian, D.; Fan, X.; Dickey, E.C.; Chen, J. Continuous production of aligned carbon nanotubes: a step closer to commercial realization. *Chem. Phys. Lett.* **1999**, *303*, 467–474.

32. Liang, C.; Ying, P.; Li, C. Nanostructured β-Mo_2C Prepared by Carbothermal Hydrogen Reduction on Ultrahigh Surface Area Carbon Material. *Chem. Mater.* **2002**, *14*, 3148–3151.

33. Morgan, T.; Santillan-Jimenez, E.; Harman-Ware, A.E.; Ji, Y.; Grubb, D.; Crocker, M. Catalytic deoxygenation of triglycerides to hydrocarbons over supported nickel catalysts. *Chem. Eng. J.* **2012**, *189–190*, 346–355.

Structural Evolution of Molybdenum Carbides in Hot Aqueous Environments and Impact on Low-Temperature Hydroprocessing of Acetic Acid

Jae-Soon Choi, Viviane Schwartz, Eduardo Santillan-Jimenez, Mark Crocker, Samuel A. Lewis Sr., Michael J. Lance, Harry M. Meyer III and Karren L. More

Abstract: We investigated the structural evolution of molybdenum carbides subjected to hot aqueous environments and their catalytic performance in low-temperature hydroprocessing of acetic acid. While bulk structures of Mo carbides were maintained after aging in hot liquid water, a portion of carbidic Mo sites were converted to oxidic sites. Water aging also induced changes to the non-carbidic carbon deposited during carbide synthesis and increased surface roughness, which in turn affected carbide pore volume and surface area. The extent of these structural changes was sensitive to the initial carbide structure and was lower under actual hydroprocessing conditions indicating the possibility of further improving the hydrothermal stability of Mo carbides by optimizing catalyst structure and operating conditions. Mo carbides were active in acetic acid conversion in the presence of liquid water, their activity being comparable to that of Ru/C. The results suggest that effective and inexpensive bio-oil hydroprocessing catalysts could be designed based on Mo carbides, although a more detailed understanding of the structure-performance relationships is needed, especially in upgrading of more complex reaction mixtures or real bio-oils.

Reprinted from *Catalysts*. Cite as: Choi, J.-S.; Schwartz, V.; Santillan-Jimenez, E.; Crocker, M.; Lewis, S.A., Sr.; Lance, M.J.; Meyer, H.M., III; More, K.L. Structural Evolution of Molybdenum Carbides in Hot Aqueous Environments and Impact on Low-Temperature Hydroprocessing of Acetic Acid. *Catalysts* **2015**, *5*, 406–423.

1. Introduction

Fast pyrolysis is an efficient and inexpensive method to produce liquids from lignocellulosic biomass. However, the liquid products (so-called bio-oils) are not suitable for direct application as fuels in internal combustion engines due to their high content of oxygen (35–40 wt.%, dry basis) and water (15–30 wt.%) [1]. For instance, various oxygenated hydrocarbon species present in raw bio-oils make these liquids unstable for long-term storage, corrosive, low in heating value, and poorly miscible with conventional hydrocarbon fuels. In particular, bio-oils can contain a large amount of formic and acetic acids which must be converted to less

215

corrosive species to improve the overall compatibility of the liquid with the current petroleum-based transportation fuel infrastructure [2].

The quality of bio-oils can be improved by eliminating oxygenated functionalities, hydroprocessing being one of the most promising upgrading strategies. Hydroprocessing is a well-established and widely employed operation in current petroleum refineries [3]. A major advantage of applying hydroprocessing to bio-oils is the relatively high yield of liquid fuels and a high degree of carbon atom retention [4,5]. Much of the research done thus far has involved well-established petroleum refinery catalysts, such as alumina-supported CoMo or NiMo sulfides [6]. However, extrapolating the use of sulfides to bio-oils has proven challenging due to the instability of sulfides under bio-oil upgrading conditions. Indeed, given the low sulfur content of bio-oils, the addition of sulfiding agents to the bio-oil feed is necessary to maintain active sulfide phases, a practice that risks contaminating the products with sulfur. Hydrothermal aging – which can cause phase changes in catalyst supports [7]—is another potential deactivation route as water represents a major byproduct of hydrodeoxygenation and raw bio-oils contain a significant amount of water (~25%).

Due to highly reactive oxygenated species present in raw bio-oils—such as olefins, aldehydes and ketones—extensive polymerization occurs at typical hydroprocessing temperatures (370–400 °C) leading to rapid coking and fouling of catalyst beds [8]. To circumvent this problem, the current state of the art (at various R&D stages) generally employs multiple sequential hydroprocessing steps. A common practice is to stabilize bio-oils at low temperatures (150–200 °C) over an efficient hydrogenation catalyst (e.g., Ru/C) followed by high temperature (370–400 °C) deep deoxygenation and hydrocracking over sulfided CoMo or NiMo catalysts [9,10]. Despite the relatively good activity of Ru/C in the low-temperature hydrogenation (stabilization) of bio-oils, Ru catalysts tend to generate fully hydrogenated compounds, as well as gasification products consuming excessive amounts of H_2 [11]. High cost of Ru metal utilization is another issue. Hence, despite the progress made to date, it is desirable to design novel catalysts tailored specifically to bio-oils, with an emphasis on avoiding the use of precious metals and improving durability *vis-à-vis* state-of-the-art catalysts.

Transition-metal carbides (TMCs) have been investigated extensively as potential substitutes for precious metal catalysts, since early studies found that TMCs manifest catalytic behaviors typically observed with Pt-group metal catalysts in several hydrocarbon conversion reactions [12–14]. For example, Mo_2C is often compared with Ru in hydrogenation reactions [14–16]. As Mo carbides can be prepared with high-surface area [17,18], they can be deployed as "bulk" catalysts without using support materials. This simple catalyst design could in turn be advantageous when one tries to further enhance the structural stability of catalysts.

By contrast, for supported catalysts such as Ru/C, the stability of the various components (active phases, supports and interfaces) needs to be considered.

Based on these considerations, Mo_2C appears to be a good candidate as a low-cost alternative to Ru/C catalysts for the low-temperature hydroprocessing of bio-oils. While there are recent studies focused on evaluating Mo carbides as catalysts for biomass upgrading [19–22], little has been done to investigate Mo_2C in the context of bio-oil low-temperature hydroprocessing, particularly in terms of their hydrothermal stability. In this work, the main focus was placed on evaluating the hydrothermal stability of the carbide catalysts. To this end, we prepared and compared two Mo_2C samples exposed to hot aqueous environments, probing if and how differences in initial structure affected their hydrothermal stability and intrinsic catalyst activity in the aqueous-phase hydroprocessing of acetic acid. Initial results are promising, suggesting that it is possible to design robust and active catalysts for low-temperature bio-oil hydroprocessing using inexpensive Mo carbides.

2. Results and Discussion

2.1. Synthesis and Characterization of Mo Carbides

Ammonium heptamolybdate is a precursor widely used for Mo carbide synthesis [22,23]. Effluent gas analysis performed with a mass spectrometer (Figure S1A) afforded gas concentration profiles typical of carburization processes involving this type of Mo precursor: reduction to MoO_2 mainly by H_2, followed by simultaneous O removal and C insertion at higher temperatures [16,18,23]. As expected, the carburization product, which we denote here as Mo_2C-A, presented a hexagonal close packed (hcp) Mo_2C structure (Figure 1).

The Mo_2C-B was prepared under the same conditions as the Mo_2C-A, but using an oxide precursor synthesized with a block copolymer templating method. The reason for using this unconventional precursor was to obtain a Mo_2C with a structure distinct from that of Mo_2C-A and thus help understand if structural differences could affect the catalyst performance in hot aqueous environments. This precursor was unconventional in that even though the ratio of O/Mo (2.8) was close to the theoretical value for MoO_3 (see XPS results in Table S1), the precursor contained a large amount of carbon and exhibited a crystallographically amorphous structure (Figure S2). The reduction-carburization behavior of this precursor was unique as well, with considerable C involvement (likely from the precursor) during the low temperature reduction period (see CO, CO_2 formation in Figure S1B). The resulting crystallographic structure of Mo_2C-B differed somewhat from that of Mo_2C-A even though the overall structure was hcp for both samples (Figure 1). Indeed, when crystallite sizes were estimated from peak broadening (using the Scherrer equation), the values for the two major planes (101) and (110) of Mo_2C-A were close to one another (11 nm, Table 1), while for Mo$_2$C-B, values of 13 and 19 nm were obtained, respectively, for (101) and (110) planes. It appears, therefore, that the crystallite shape of Mo$_2$C-B differed from what is commonly observed for "conventional" Mo$_2$C such as Mo$_2$C-A [18].

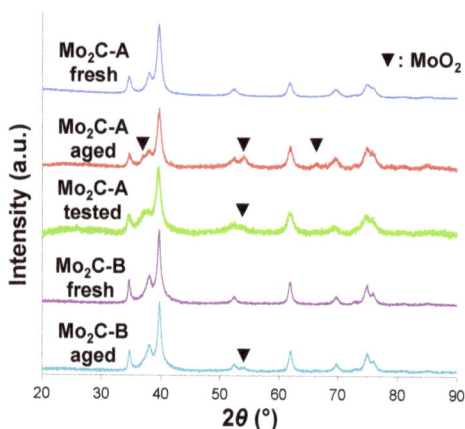

Figure 1. XRD patterns of two Mo$_2$C samples (A and B) before and after aging for 48 h in hot liquid water. The pattern for Mo$_2$C-A obtained after a 48 h aqueous-phase hydroprocessing run (10% guaiacol in water as feed) is also presented for comparison (Mo$_2$C-A tested).

Previous studies have shown that the primary particles of Mo$_2$C prepared under carburization conditions similar to those employed in the present work, are single crystals [16,18]. Consequently, crystallite sizes determined using XRD data are close to the particle sizes estimated from BET surface area as long as access to the primary particle surface is not blocked by bulk carbon deposited during the carburization. As both Mo$_2$C-A and Mo$_2$C-B in fresh state presented larger particle sizes than the calculated crystallite sizes in the present work (Table 1), it can be concluded that there was significant bulk carbon deposition during the carbide synthesis which prevented the passage of the N$_2$ probe molecules through the micropores during BET measurements. The presence of non-carbidic C was confirmed by XPS analysis (see C$_{cont.}$ in Table 2 and Figure S3), but the impact of the bulk carbon deposition was especially important for Mo$_2$C-B, which presented significantly smaller BET surface (*i.e.*, larger estimated particle sizes) and pore volume than Mo$_2$C-A, in spite of the comparable crystallite sizes.

Structural differences between the two Mo$_2$C samples can be found at the grain level as well. The SEM images in Figure 2 reveal that each carbide inherited a distinct grain morphology from its oxide precursor. For example, Mo$_2$C-B grains had sharper edges and corners and were significantly flatter than those of Mo$_2$C-A.

Higher magnification imaging showed similar morphological differences for smaller particles (Figure 3), which seems to be consistent with the anisotropic XRD peak broadening discussed above.

Another catalytically relevant carbide property worth discussing is the density of active sites determined from CO uptake and BET surface area measurements (Table 1). The site density determined in this way is generally higher when residual oxygen is minimized and carbidic carbon insertion is more complete while bulk carbon deposition is prevented [16,18]. The fact that Mo_2C-B had an inferior site density despite the higher extent of carburization (inferred from C/Mo of carbide phases in Table 2) indicates that a larger portion of active sites were likely blocked by non-carbidic C. Despite the relatively comparable $C_{cont.}$ content (Table 2), greater blockage of both pores and active sites apparently occurred over Mo_2C-B than Mo_2C-A. One possible explanation is differences in type and shape of the non-carbidic C species deposited on the two carbides.

Table 1. BET surface area, pore volume, pore size, CO uptake, density of sites, particle size and crystallite size of the catalysts studied.

Catalyst	BET surface area, S_g/m^2 g^{-1}	Pore volume /cm^3 g^{-1}	Pore size /nm	CO uptake /μmol g^{-1}	Site density [a] /$\times 10^{15}$ cm^{-2}	Particle size, D_p [b] /nm	Crystallite size, D_c [c] /nm
Mo_2C-A fresh	36	0.08	8	186	0.31	18	11/11
Mo_2C-A aged	64	0.50	36	26	0.02	10	10/11
Mo_2C-A tested	22	0.02	4	186	0.51	30	8/10
Mo_2C-B fresh	6	0.03	23	5	0.05	110	13/19
Mo_2C-B aged	15	0.02	4	12	0.05	44	12/16
Ru/C fresh	815	NM [d]	NM [d]	NA [e]	0.04 [f]	1.1 [g]	NA [e]
Ru/C aged	765	NM [d]	NM [d]	NA [e]	0.01 [f]	2.0 [g]	NA [e]

[a] CO uptake normalized to BET surface area; [b] Estimated from BET surface area; [c] Estimated from XRD peak broadening: (101) plane/(110) plane; [d] Not measured; [e] Not applicable; [f] Number of exposed Ru atoms (estimated from particle sizes) normalized to BET surface area; [g] Measured by TEM.

Table 2. Surface composition of Mo_2C catalysts analyzed by XPS [a].

Catalyst	Mo_{carb} [b] /at.%	Mo_{oxid} [c] /at.%	C_{carb} [b] /at.%	C_{cont} [d] /at.%	O_{oxid} [c] /at.%	O_{cont} [d] /at.%	C/Mo [e]	O/Mo [f]
Mo_2C-A fresh	31.5	6.4	9.9	16.2	25.7	10.5	0.31	4.03
Mo_2C-A aged	8.3	24.1	1.4	4.4	34.8	27.0	0.17	1.44
Mo_2C-A tested	13.2	14.6	4.4	31.8	22.3	13.8	0.33	1.53
Mo_2C-B fresh	40.0	4.8	15.7	13.5	14.5	11.5	0.39	3.02
Mo_2C-B aged	17.7	9.5	4.8	15.3	28.8	23.9	0.27	3.03

[a] Measurements done after Ar sputtering; [b] Carbides; [c] Oxides; [d] Contaminants, meaning that elements are not associated with carbide or oxide phases; [e] C/Mo ratio of surface carbide phases (theoretical value for bulk Mo_2C is 0.5); [f] O/Mo ratio of surface oxide phases.

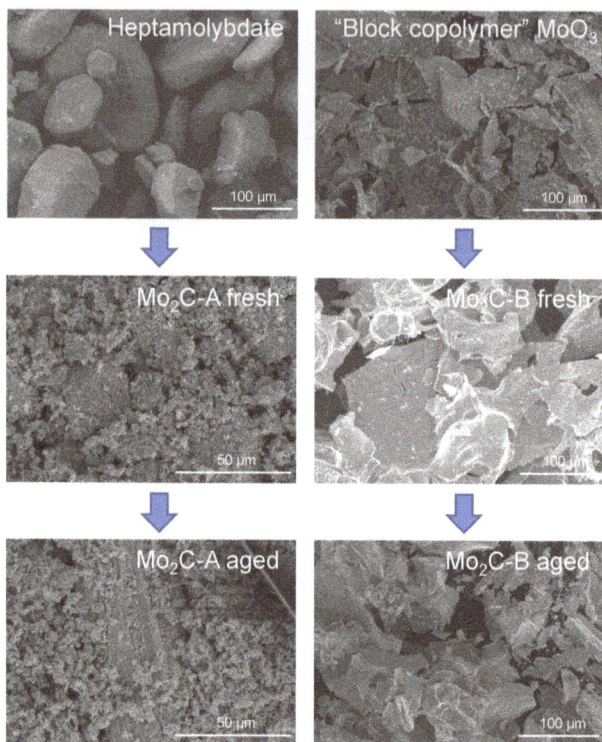

Figure 2. SEM images of the two Mo oxide precursors and resulting carbide samples before and after 48 h aging at 250 °C in liquid water.

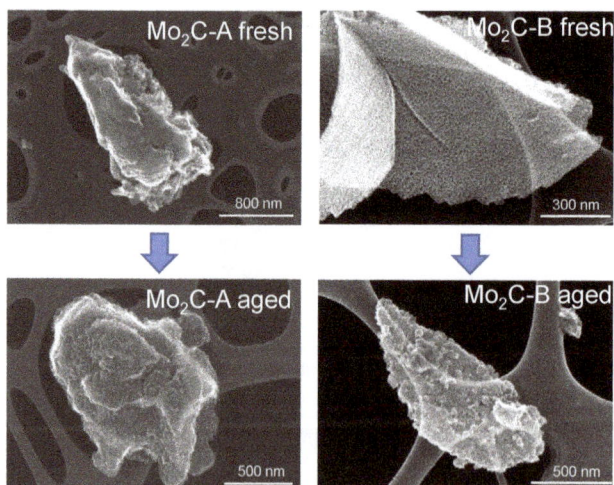

Figure 3. SEM images of two Mo_2C samples in the fresh state and after 48 h aging at 250 °C in liquid water; smaller particles taken at high magnification.

2.2. Structural Evaluation of Mo Carbides Subjected to Hot Aqueous Environments

An important technical challenge facing the development of catalysts for biomass conversion processes is the hydrothermal instability of conventional catalytic materials. For instance, γ-Al_2O_3, a popular support used in petroleum refineries for metal sulfide and precious metal catalysts, is rapidly hydrated and converted to boehmite in hot liquid water [7,24]. For the two Mo carbide samples studied in this work, despite the extended exposure to hydrothermally challenging environments (liquid water, 250 °C, 48 h), the overall bulk structure was maintained (e.g., hcp crystallographic structure and grain morphology as shown in Figures 1 and 2, respectively). This is an encouraging observation in view of the aforementioned necessity of novel catalytic materials for biomass conversion processes. Detailed characterization, however, revealed that some significant structural changes occurred on both Mo_2C-A and Mo_2C-B samples. For instance, a portion of the carbide phases were converted into MoO_2 phases as detected by XRD (Figure 1). The high oxygen affinity of Mo carbides is well documented in the literature [25], and our present results show that liquid-phase H_2O can also oxidize Mo carbides.

The fact that a considerable fraction of the carbidic (MoC_x) sites on the surface of both samples were converted into oxidic (MoO_x) sites was confirmed by XPS (Table 2). Interestingly, the number of carbidic sites (in terms of at.%) not only decreased, but the remaining carbidic sites also lost some of their C neighbors and instead gained O—as suggested by the fact that the C/Mo ratio of the carbide phases within Mo_2C-A, for instance, decreased from 0.31 to 0.17 while the content of O contaminant (*i.e.*, O not related to Mo oxides) increased from 25.7 to 34.8 at.%.

As mentioned above, it is well known that the composition of Mo sites plays a critical role in determining the surface reactivity of Mo carbides as measured by CO chemisorption [16,26]. It is thus unsurprising that the surface oxidation of Mo_2C-A led to decreased CO uptake and site density (Table 2). Given that Mo_2C-B also suffered surface oxidation, one would expect to see a similar trend. Instead, CO uptake on Mo_2C-B actually increased. This discrepancy might be related to the fact that only a fraction of "active" sites were exposed (*i.e.*, able to adsorb CO) over the fresh Mo_2C-B due to non-carbidic C deposition as discussed above. As the surface morphology changed considerably due to hydrothermal aging, it seems reasonable to conjecture that the number of exposed active sites slightly increased. Despite the increase, one should note that the CO uptake by Mo_2C-B was still much lower than those typically observed for Mo carbides prepared under similar conditions, including Mo_2C-A. Moreover, the extent of surface oxidation was relatively small for the Mo_2C-B sample. Another intriguing difference is that the surface $C_{cont.}$ increased for Mo_2C-B, meaning that a portion of the C removed from MoC_x sites ($C_{carb.}$) was retained on the surface as non-carbidic C species ($C_{cont.}$). The retention of C might

221

have helped maintain surface hydrophobicity, which would in turn have mitigated oxidation by H_2O.

The BET surface area increased after water aging for both samples (Table 1). In the case of Mo_2C-A, the concomitant increase in pore size and volume and decrease in $C_{cont.}$ (Table 2) indicates that bulk (non-carbidic) C deposits blocking carbide pores were removed during hot water aging. This interpretation is supported by the fact that—after aging—the particle size estimated from BET surface area matched the crystallite size obtained from XRD data (Table 1). Again, Mo_2C-B appears to have responded differently to hydrothermal aging, as the increased surface area was not accompanied by an increase in porosity. Instead, the pore volume of this sample decreased, which is consistent with the gain in $C_{cont.}$ (Table 2), meaning that more bulk C ($C_{cont.}$) was involved in pore blocking. The FIB-SEM images in Figure 4 seem to agree with this interpretation. The intra-grain images confirm that the aging increased pore openings for Mo_2C-A, while it decreased those for Mo_2C-B (in terms of both number and size).

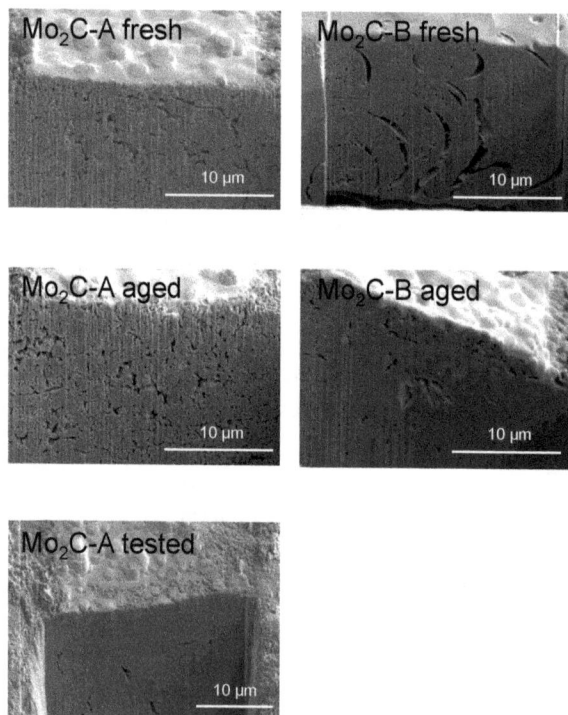

Figure 4. FIB-SEM images of the two Mo_2C samples (A and B) before and after aging for 48 h in hot liquid water. The image for Mo_2C-A obtained after a 48 h aqueous-phase hydroprocessing run (10% guaiacol in water as feed) is also presented for comparison (Mo_2C-A tested).

When Mo_2C-A was tested for 48 h at the same temperature (250 °C) but under actual hydroprocessing conditions (10% guaiacol in water + H_2), both pore opening and BET surface area decreased, likely due to increased $C_{cont.}$ as in the case of Mo_2C-B (see Figure 4 and Tables 1 and 2). Despite the BET surface area loss due to pore blocking, Mo_2C-A maintained the same degree of CO uptake after testing (Table 1). As mentioned above, the number of metallic sites measured for Mo_2C-B also did not degrade with aging and actually increased. The C from guaiacol molecules and non-carbidic C thus appear to have had similar effects with respect to mitigating H_2O-induced bulk oxidation of Mo_2C-A and Mo_2C-B, respectively (see lower bulk MoO_2 detected by XRD in Figure 1). Similar beneficial effects, *i.e.*, preventing hydration and metal particle agglomeration, have been reported for carbonaceous surface species derived from biomass compounds dissolved in liquid water over alumina catalysts [27].

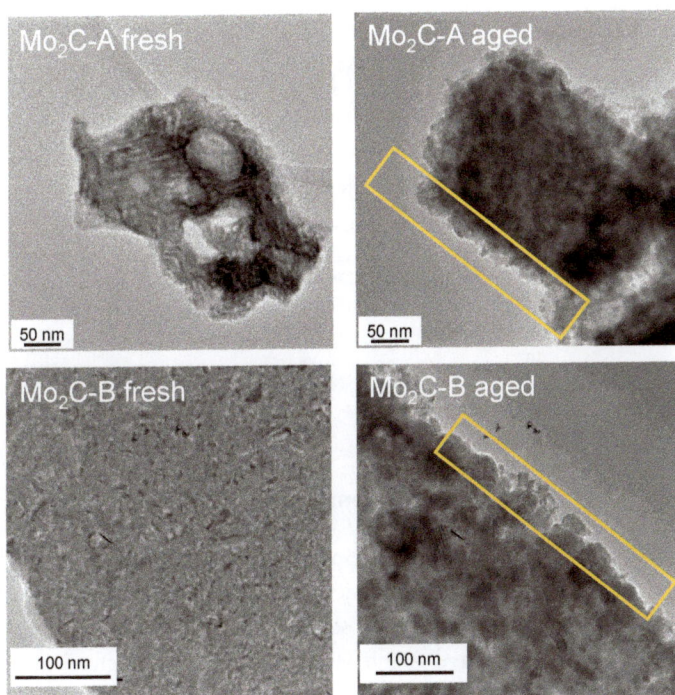

Figure 5. TEM images of Mo carbides in the fresh state and after 48 h aging at 250 °C in liquid water; the yellow boxes highlight increased surface roughness due to the formation of elliptically shaped particles.

The surface area increase of Mo_2C-B despite the pore blocking could be explained by increased surface roughness. Liquid water aging developed nanoparticles of elliptical shape on the surface of both carbide samples (Figure 5 and S4). These small particles were mainly Mo oxides in agreement with the XPS data (Table 2). Since Mo_2C-B sample initially had a lower surface area than Mo_2C-A sample (36 *vs.* 6 m^2/g) and experienced pore blocking during aging, the impact of the increased roughness on BET surface area might be discernible. In contrast, the pore opening due to water aging led to a dramatic increase in surface area for Mo_2C-A (36 to 64 m^2/g), masking the relatively small contribution of the roughness-induced surface area increase.

In summary, Mo carbides maintained overall bulk structures under hydrothermally harsh environments, but can undergo H_2O-induced oxidation. An important lesson that can be learned from these results is that the extent of oxidation can be minimized by controlling Mo carbide structure. The superior oxidation resistance of the Mo_2C-B structure compared to Mo_2C-A is further confirmed by TGA performed in air (Figure 6), in which Mo_2C-B exhibited oxidation onset temperatures (*i.e.*, temperatures at which significant weight gain started) at least 100 °C higher than Mo_2C-A.

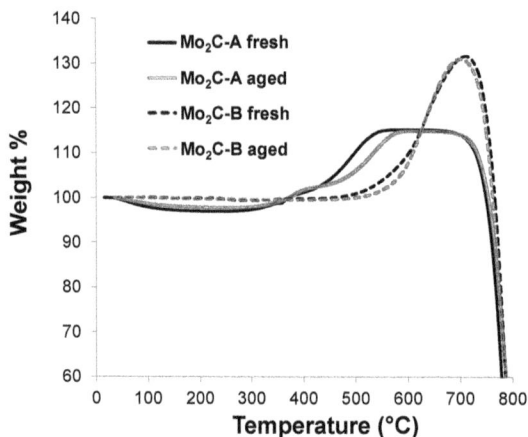

Figure 6. TGA profiles of two Mo_2C samples in the fresh state and after 48 h aging at 250 °C in liquid water. Temperature increased from room temperature to 800 °C at 10 °C/min in air (25 mL/min). The solid and dashed lines correspond to the relative weight change of Mo_2C-A and Mo_2C-B samples, respectively.

2.3. Catalytic Performance in Aqueous-Phase Hydroprocessing of Acetic Acid

To assess the catalytic effectiveness of Mo carbides in converting oxygenates in liquid water, Mo_2C-A and Mo_2C-B were evaluated using a 10 wt.% solution of acetic acid in water. This model reaction was chosen considering the fact that pyrolysis oils can contain a large amount of organic acids including acetic acid [2], and this type of model compound testing has been used previously to study the hydroprocessing performance of Ru/C and Pd/C (catalysts commonly used to stabilize bio-oils via low temperature hydroprocessing) [11]. A commercial 5% Ru/C was also evaluated before and after a 48 h water aging for comparison purposes. Carbon supports are widely studied in the context of water-based catalytic processes due to their relative hydrothermal stability [28] and Ru is one of the best catalysts known for acetic acid hydrogenation [29,30] and low-temperature processing (stabilization) of bio-oil [6,9]. Some catalytically relevant properties of the 5% Ru/C reference catalyst are reported in Table 1 and Table S2, and Figure S5. Briefly, the surface area of Ru/C decreased slightly during hydrothermal aging, while the average particle size doubled. Based on XPS, the surface Ru content decreased from 4.2 to 1.9 at.% likely due to metal leaching. This result further highlights the challenging nature of catalyst development for biomass conversion. Conversion values as well as average rates obtained in the present study are compiled in Table 3. Note that the product analysis was done only at the end of each 4 h run, therefore the conversion rate represents the apparent average activity of a catalyst. It is possible that some other deactivation mechanisms (e.g., coking during 4 h hydroprocessing) contributed to the average conversion rates. Further research using steady state flow reactor experiments could help obtain more accurate activity data and comparison among different catalysts. Another parameter which could have influenced the activity measurements is the possible effect of internal mass transfer, as the catalysts studied in this work presented significantly different particle sizes (Table 1). Olcay et $al.$ calculated the Weisz Modulus for an aqueous-phase hydrogenation of acetic acid over a Ru/C catalyst; the obtained 10^{-15} was well below the limit value of 0.15 (that is, the kinetic measurements were done in the absence of mass transport limitation) [29]. Considering that our reaction conditions were quite similar to those used in [29], we could estimate that our Ru/C catalyst was evaluated also without mass transport limitation. More detailed and rigorous kinetic experiments and catalyst morphology characterization are needed to address this question for Mo carbides.

All of the catalysts were active in acetic acid conversion at 200 and 250 °C (Table 3). The Ru/C catalyst converted more molecules of acetic acid per catalyst mass than Mo_2C. However, if the comparison is done on the intrinsic activity basis ($i.e.$, number of converted acetic acid molecules per surface area), the activity of Mo_2C-B was comparable to that of Ru/C. Considering that the Mo_2C-B sample had a relatively low BET surface area for a carbide prepared via a temperature programmed method, one could envision that the performance (mass basis) of Mo_2C could be significantly improved by increasing total surface area while maintaining the properties of the catalytic surface. In comparison, the metal dispersion of the Ru/C (over 80%) was already quite close to the theoretical maximum. Between the two Mo carbides, the intrinsic activity of fresh Mo_2C-B was 3–4 times higher than that of fresh Mo_2C-A. The Mo/C of carbide phases (determined by XPS: degree of carburization) could have resulted in such performance differences between the two carbide catalysts. It is well known that the precious-metal-like character (e.g., high hydrogenation activity) of Mo_2C catalysts increases with more complete carburization [16,26,31].

Despite the significant structural changes described earlier, the negative impact of hydrothermal aging on acetic acid conversion was minor. In fact, in the case of Mo_2C-B, the conversion at 250 °C actually doubled from 26 to 50% (Table 3). The intrinsic activity of Mo_2C-B degraded by 20%, but given that the BET surface area of Mo_2C-B increased upon aging, the net effect was that the rate per unit mass increased. For Ru/C the degradation in performance with aging seems to be mainly due to the loss in metallic surface area (*i.e.*, 50% decrease in dispersion + leaching), which attenuated the impact of the increased intrinsic rate.

Since the main objective of this reactor study was to determine if Mo carbides are catalytically active in the presence of liquid water and whether this activity are affected by hydrothermal aging, our discussion here is mainly focused on comparing the acetic acid conversion activity among different catalyst samples. We could, nonetheless, observe that ethanol was the major liquid product and that ethanol yield decreased with conversion and temperature. Consistent with the literature data [27,29], the formation of ethyl acetate (the product of an esterification reaction between ethanol and acetic acid) was low over Ru/C. In contrast, the formation of ethyl acetate was substantial over Mo_2C-B and did not change when conversion increased from 26 to 50%. The reason for this uniquely high selectivity to ethyl acetate over Mo_2C-B is not clear at this time, but could be explained in part by the surface bifunctionality of this catalyst (*i.e.*, the fact that MoC_x and MoO_x are both present). Indeed, it has been recently shown that combining Ru and acidic supports can significantly influence the carboxylic acid conversion chemistry [32], and metal oxide species created on carbide surfaces can act as acidic sites [33–37].

A large fraction of the acetic acid feed was apparently converted to gas products and CH_4 was qualitatively observed as a major product especially at high conversion (quantification was precluded by the instrumental limitations described in the Experimental section). High CH_4 formation over Ru/C catalysts at similar temperatures is well known and indicates C-C bond hydrogenolysis [28–30]. The aged Ru/C produced little liquid-phase product, suggesting the sensitivity of these acetic acid conversion reactions to Ru particle size. To better assess the potential of Mo carbides in hydroprocessing bio-oils, the catalysts will need to be compared with Ru/C in converting more complex feedstock such as multi-compound models or real bio-oils. In addition, more complete product identification and quantification is necessary.

Table 3. Catalyst performance in aqueous-phase hydroprocessing of acetic acid; catalyst loading: 0.25 g, reaction mixture: 25 g of 10% (w/w) solution of acetic acid in water, reaction pressure: 2000 psig, stirring speed: 1000 rpm, reaction duration: 4 h.

Catalyst	Reaction temperature/°C	Acetic acid conversion/%	Selectivity to ethanol/%	Selectivity to ethyl acetate/%	Average rate of acetic acid conversion (areal) [a] /μmol m^{-2}
Mo$_2$C-A fresh	200	37	NM [b]	NM [b]	0.12
Mo$_2$C-A fresh	250	44	NM [b]	NM [b]	0.14
Mo$_2$C-A aged	250	40	NM [b]	NM [b]	0.07
Mo$_2$C-B fresh	250	26	70	12	0.50
Mo$_2$C-B aged	250	50	18	14	0.39
Ru/C fresh	200	61	28	4	0.39
Ru/C fresh	250	77	10	2	0.49
Ru/C aged	250	68	1	0	1.58

[a] Rates normalized to BET surface area for Mo$_2$C and to metallic surface area estimated from TEM-measured particle size for Ru/C (for the aged Ru/C, *ca.* 55% loss of Ru due to leaching was taken into account based on XPS data shown in Table S2); [b] Not measured due to catalyst dissolution during the storage period between reactor evaluation and catalyst filtration.

3. Experimental Section

3.1. Catalysts

A standard temperature-programmed carburization method [18,37] was applied for the synthesis of Mo carbides. Ammonium heptamolybdate (Alfa Aesar, Heysham, UK) and a MoO$_3$ synthesized via a block copolymer templating method were used as precursors (the respective carburization products being denoted as Mo$_2$C-A and Mo$_2$C-B), while 15% CH$_4$/H$_2$ was used as the carburizing gas. After the synthesis (temperature ramping rate: 1 °C/min, final temperature: 700 °C, soak time: 1 h), the samples were passivated at room temperature in a 0.3% O$_2$/He flow. The block copolymer templating procedure used to prepare the MoO$_3$ precursor was adapted from a synthesis reported by Stucky and co-workers [38]. Briefly, 1 g of structure-directing copolymer Pluronic® P-123 (Aldrich, Steinheim, Germany) was dissolved in 10 g of anhydrous ethanol (99.5%, Acros Organics, Geel, Belgium). 0.01 moles of anhydrous MoCl$_5$ (99.6%, Alfa Aesar, Ward Hill, MA, USA) were then added to the resulting solution while the latter was vigorously stirred. The resulting mixture was subsequently allowed to age in a Petri dish at 60 °C for 15 days. Finally, the solids produced were calcined under static air at 300 °C for 2 h. 5 wt.% Ru/C (Alfa Aesar, Heysham, UK) was purchased and used as received for comparison purposes.

3.2. Hydrothermal Aging

Mo$_2$C and Ru/C samples were hydrothermally aged in a mechanically stirred 300 mL alloy C-276 batch reactor (Parr Instrument Co.). 2 g of sample was combined with 100 g of distilled water in the reactor. The air in the head space, as well as that dissolved in the water feed, was removed by evacuation and purging with N$_2$.

The reactor was then filled with 95 psi of N_2 and heated to 250 °C under constant stirring at 1200 rpm. The final total pressure was approximately 690 psi at the aging temperature of 250 °C. After aging for 48 h, the solids were filtered and dried in air first at room temperature, then at 110 °C overnight.

3.3. Characterization

X-ray diffraction patterns were recorded on a powder X-ray diffractometer (X'Pert PRO, PANalytical, Almelo, The Netherlands) operated at 45 kV and 40 mA using CuK_α radiation (K_α = 0.154178 nm). The crystallite sizes (D_c) were estimated from the Scherrer equation using commercial software (HighScore Plus). The full width at half maximum (FWHM) and position of the peaks were calculated by profile fitting. Before the fitting, background and $K_\alpha 2$ of XRD peaks were removed and the peak broadening effect was adjusted using LaB6 as a standard sample. FWHM values of (101) and (110) planes and a shape factor of 0.94 were used.

Microscopic imaging of Mo samples was performed using a Hitachi HF3300 high-resolution transmission-scanning transmission electron microscope (TEM/STEM), which was operated at 300 kV. This instrument was also equipped with a secondary electron detector (SE) and a Bruker X-Flash silicon drift detector (SDD) for energy dispersive spectrometry (EDS). A Hitachi S-4800 Field Emission-Scanning Electron Microscope (FE-SEM) was used to collect micrographs of the Mo_2C samples at grain level. The internal powder microstructure was visualized using a Hitachi NB-5000 FIB-SEM (Focused Ion Beam-Scanning Electron Microscope). A high-milling-rate Ga ion beam was used to cut microscopic samples out of the powder grain revealing the interior porosity. Prior to milling, the region was locally coated with tungsten in order to protect the sample from the milling process. For the Ru/C catalyst, a JEOL 2200FS-AC Aberration-corrected STEM/TEM was used to measure the distribution of Ru particle sizes (more than 100 particles analyzed) before and after the hydrothermal aging. The calculated average particle sizes were then used to estimate Ru dispersion using a simple formula, dispersion = $0.9/D_p$ (D_p in nm) [39].

The specific surface area and porosity of catalyst samples were determined by measuring N_2 adsorption isotherms using an automatic volumetric adsorption apparatus (Quantachrome, Autosorb-1-C). Average particle sizes of Mo_2C catalysts were estimated from the equation $D_p = f/(\rho S_g)$ [39], where f is a shape factor and ρ is the density of Mo_2C (9.098 g/cc). A shape factor of 6 was applied assuming spherical or cubic Mo_2C particles [18].

Pulsed CO chemisorption was performed in a tubular fixed bed reactor (AMI-200, Altamira Instruments, Inc.) to measure the amount of irreversibly chemisorbed CO on Mo_2C. Before the measurements, passivated carbide samples were reduced in a flow of H_2 at 500 °C for 1 h, then cooled down to the chemisorption temperature (room temperature) in flowing Ar. A stoichiometry of Mo:CO = 1:1 (*i.e.*,

1 molecule of CO per active site) was used to estimate the total number of active sites [18,40,41].

X-ray photoelectron spectroscopy was performed using a Thermo Scientific K-Alpha XPS instrument. The K-Alpha uses Al-Ka x-rays focused to a spot 400 microns in diameter. Emitted photoelectrons were energy analyzed using a 180° double focusing hemispherical analyzer with a 128-channel detector. Survey data were collected at a pass energy of 200 eV and an energy resolution of 1 eV/step, while core level data were collected at 50 eV pass energy and 0.1 eV/step energy resolution. Sample charging was eliminated by using the K-Alpha's dual-beam charge compensation source, which uses both low energy Ar-ion and low energy electrons. An Ar-ion gun (operated at 2 kv) was used for *in-situ* cleaning of the sample surface. Data were collected and analyzed using the Avantage data system (v.4.61, Thermo Fisher Scientific, Waltham, MA, USA).

Thermogravimetric analysis (TGA) was carried out with a TA Instruments Q5000. The temperature was increased from room temperature to 800 °C at 10 °C/min in air (25 mL/min).

3.4. Reactor Evaluation

Catalyst performance was evaluated in a mechanically stirred 50 mL stainless-steel batch reactor. The catalyst (0.25 g) was reduced *in situ* at 250 °C under 50 psi of flowing H_2 for 2 h. After allowing the system to cool to room temperature, a vacuum was drawn on the reactor and 25 g of a 10% (*w/w*) solution of acetic acid in water was pulled into the reactor vessel. After purging with argon, the system was pressurized with H_2 to 1000 psig and heated to the reaction temperature while stirring the reaction mixture at 1000 rpm. Upon reaching the reaction temperature, which typically took ~15 min, the pressure was adjusted to 2000 psig and kept at this value throughout the experiment in order to replenish any hydrogen consumed by the reaction. After 4 h at the reaction temperature, the hydrogen line feeding the reactor was closed and the system was cooled to 35 °C prior to taking a gas sample for analysis. The system was then opened and its solid and liquid contents were collected. Only qualitative information could be obtained from gas analysis due to the fact that hydrogen (which was present in overwhelming excess) swamped the detector and precluded the quantification of gaseous products. The solids and liquids recovered were separated via gravimetric filtration prior to mixing a 1 g aliquot of the recovered liquid with 0.2 g of 2-butanol, which served as the internal standard. 0.2 μL of the resulting solution were then analyzed on an Agilent 7890 GC equipped with a multimode inlet, a DB-Wax column (30 m × 530 μm × 0.5 μm) and a FID detector. Both the inlet and the FID were kept at a constant temperature of 240 °C, while the oven housing the column was programmed with a temperature ramp (10 °C/min) from 35 to 240 °C followed by an isotherm at the latter temperature lasting 9.5 min.

The column was operated with a constant flow of 26 mL/min. Quantitative data were acquired using a 3-point calibration curve created using standards of ethanol and ethyl acetate. The linear calibration curve for each individual compound had an R^2 value of $\geqslant 0.99$. Acetic acid was quantified using CE (capillary electrophoresis, HP 1600).

4. Conclusions

We studied the structure of molybdenum carbides subjected to hot aqueous environments and their catalytic performance in aqueous-phase hydroprocessing of acetic acid. The results suggest that Mo carbides have potential to be developed as hydrothermally stable catalysts, especially suitable for the low-temperature hydroprocessing of bio-oils, which is generally carried out using Ru/C-type catalysts. Key findings of the present study were:

- Bulk structures of Mo_2C were maintained after 48 h aging at 250 °C in liquid water;
- Carbide surfaces underwent significant structural changes during hydrothermal aging, particularly oxidation of carbidic Mo sites to oxidic sites, transformation of non-carbidic carbon species deposited during the carbide synthesis, and increased surface roughness;
- The removal of pore-blocking non-carbidic carbon and increased surface roughness during aging led to increased BET surface area;
- The extent of the structural changes observed during aging was sensitive to the initial carbide structure and was less severe under actual hydroprocessing conditions;
- Mo carbides showed excellent performance in the aqueous-phase hydroprocessing of acetic acid showing an intrinsic conversion activity comparable to that of Ru/C;
- The intrinsic activity of Mo carbides was shown to be sensitive to the carbide structure;
- Further investigation of carbide structures and performance in complex model or real bio-oil feedstock is needed to design robust, active, and inexpensive Mo carbide catalysts for bio-oil hydroprocessing.

Acknowledgments: This research was financially supported by the U.S. Department of Energy Bioenergy Technologies Office, the Laboratory Directed Research and Development Program of the Oak Ridge National Laboratory, and the University of Kentucky (Office of the Vice President for Research). A portion of this research was conducted at ORNL's Center for Nanophase Materials Sciences, which is a DOE Office of Science User Facility. The authors thank D. C. Elliott and T. R Hart at the Pacific Northwest National Laboratory for the useful discussion on the aqueous-phase aging and bio-oil evaluation approaches used in this study.

Kevin Perrin, Jaime Shoup and Tonya Morgan at the University of Kentucky are thanked for their technical assistance.

Author Contributions: J.-S.C., V.S., E.S.-J. and M.C. conceived the research program and designed the experiments; V.S., E.S.-J., and J.-S.C. performed the catalyst synthesis and characterization; E.S.-J., S.A.L., and J.-S.C. carried out the reactor study of the catalysts including aging, acetic acid hydroprocessing and product analysis; M.J.L. and K.L.M. performed the microscopy; HMM performed the XPS; All contributed to the writing of the paper.

Conflicts of Interest: The authors declare no conflict of interest.

References

1. Czernik, S.; Bridgwater, A.V. Overview of applications of biomass fast pyrolysis oil. *Energy Fuels* **2004**, *18*, 590–598.
2. Connatser, R.M.; Lewis, S.A.L., Sr.; Keiser, J.R.; Choi, J.-S. Measuring bio-oil upgrade intermediates and corrosive species with polarity-matched analytical approaches. *Biomass Bioenergy* **2014**, *70*, 557–563.
3. Bartholomew, C.H.; Farrauto, R.J. *Fundamentals of Industrial Catalytic Processes*, 2nd ed.; John Wiley and Sons, Inc.: Hoboken, NJ, USA, 2006.
4. Bridgwater, A.V. Catalysis in thermal biomass conversion. *Appl. Catal. A* **1994**, *116*, 5–47.
5. Bridgwater, A.V. Production of high grade fuels and chemicals from catalytic pyrolysis of biomass. *Catal. Today* **1996**, *29*, 285–295.
6. Elliott, D.C. Historical developments in hydroprocessing bio-oils. *Energy Fuels* **2007**, *21*, 1792–1815.
7. Elliott, D.C.; Sealock, L.J., Jr.; Baker, E.G. Chemical processing in high-pressure aqueous environments. 2. Development of catalysts for gasification. *Ind. Eng. Chem. Res.* **1993**, *32*, 1542–1548.
8. Baker, E.G.; Elliott, D.C. Catalytic upgrading of biomass pyrolysis oils. In *Research in Thermochemical Biomass Conversion*; Bridgwater, A.V., Kuester, J.L., Eds.; Elsevier Science Publishers Ltd.: Barking, UK, 1988; pp. 883–895.
9. Zacher, A.H.; Olarte, M.V.; Santosa, D.M.; Elliott, D.C.; Jones, S.B. A review and perspective of recent bio-oil hydrotreating research. *Green Chem.* **2014**, *16*, 491–515.
10. Elliott, D.C.; Wang, H.; French, R.; Deutch, S.; Iisa, K. Hydrocarbon liquid production from biomass via hot-vapor-filtered fast pyrolysis and catalytic hydroprocessing of the bio-oil. *Energy Fuels* **2014**, *28*, 5909–5917.
11. Elliott, D.C.; Hart, T.R. Catalytic hydroprocessing of chemical models for bio-oil. *Energy Fuels* **2009**, *23*, 631–637.
12. Sinfelt, J.H.; Yates, D.J.C. Effect of carbiding on hydrogenolysis activity of molybdenum. *Nature Phys. Sci.* **1971**, *229*, 27–28.
13. Levy, R.B.; Boudart, M. Platinum-like behavior of tungsten carbide in surface catalysis. *Science* **1973**, *181*, 547–549.
14. Oyama, S.T. Preparation and catalytic properties of transition metal carbides and nitrides. *Catal. Today* **1992**, *15*, 179–200.

15. Lee, J.S.; Locatelli, S.; Oyama, S.T.; Boudart, M. Molybdenum carbide catalysts. 3. Turnover rates for the hydrogenolysis of n-butane. *J. Catal.* **1990**, *125*, 157–170.

16. Choi, J.-S.; Bugli, G.; Djéga-Mariadassou, G. Influence of the degree of carburization on the density of sites and hydrogenating activity of molybdenum carbides. *J. Catal.* **2000**, *193*, 238–247.

17. Volpe, L.; Boudart, M. Compounds of molybdenum and tungsten with high specific surface area. II. Carbides. *J. Solid State Chem.* **1985**, *59*, 348–356.

18. Lee, J.S.; Oyama, S.T.; Boudart, M. Molybdenum carbide catalysts. 1. Synthesis of unsupported powders. *J. Catal.* **1987**, *106*, 125–133.

19. Zhang, W.; Zhang, Y.; Zhao, L.; Wei, W. Catalytic activities of NiMo carbide supported on SiO_2 for the hydrodeoxygenation of ethyl benzoate, acetone, and acetaldehyde. *Energy Fuels* **2010**, *24*, 2052–2059.

20. Jongerius, A.L.; Gosselink, R.W.; Dijkstra, J.; Bitter, J.H.; Bruijnincx, P.C.A.; Weckhuysen, B.M. Carbon nanofiber supported transition-metal carbide catalysts for the hydrodeoxygenation of guaiacol. *ChemCatChem* **2013**, *5*, 2964–2972.

21. Yu, W.; Salciccioli, M.; Xiong, K.; Barteau, M.A.; Vlachos, D.G.; Chen, J.G. Theoretical and experimental studies of C–C versus C–O bond scission of ethylene glycol reaction pathways via metal-modified molybdenum carbides. *ACS Catal.* **2014**, *4*, 1409–1418.

22. Xiong, K.; Lee, W.-S.; Bhan, A.; Chen, J.G. Molybdenum carbide as a highly selective deoxygenation catalyst for converting furfural to 2-methylfuran. *ChemSusChem* **2014**, *7*, 2146–2151.

23. Rerez-Romo, P.; Potvin, C.; Manoli, J.-M.; Chehimi, M.M.; Djéga-Mariadassou, G. Phosphorus-doped molybdenum oxynitrides and oxygen-modified molybdenum carbides: synthesis, characterization, and determination of turnover rates for propene hydrogenation. *J. Catal.* **2002**, *208*, 187–196.

24. Ravenelle, R.M.; Copeland, J.R.; Kim, W.-G.; Crittenden, J.C.; Sievers, C. Structural changes of γ-Al_2O_3-supported catalysts in hot liquid water. *ACS Catal.* **2011**, *1*, 552–561.

25. Wu, W.; Wu, Z.; Liang, C.; Ying, P.; Feng, Z.; Li, C. An IR study on the surface passivation of Mo_2C/Al_2O_3 catalyst with O_2, H_2O and CO_2. *Phys. Chem. Chem. Phys.* **2004**, *6*, 5603–5608.

26. Choi, J.-S.; Krafft, J.-M.; Krzton, A.; Djéga-Mariadassou, G. Study of residual oxygen species over molybdenum carbide prepared during in situ DRIFTS experiments. *Catal. Lett.* **2002**, *81*, 175–180.

27. Ravenelle, R.M.; Copeland, J.R.; van Pelt, A.H.; Crittenden, J.C.; Sievers, C. Stability of Pt/γ-Al_2O_3 catalysts in model biomass solutions. *Top. Catal.* **2012**, *55*, 162–174.

28. Elliott, D.C.; Hart, T.R.; Neuenschwander, G.G. Chemical processing in high-pressure aqueous environments. 8. Improved catalysts for hydrothermal gasification. *Ind. Eng. Chem. Res.* **2006**, *45*, 3376–3781.

29. Olcay, H.; Xu, L.; Xu, Y.; Huber, G.W. Aqueous-phase hydrogenation of acetic acid over transition metal catalysts. *ChemCatChem* **2010**, *2*, 1420–1424.

30. Wan, H.; Chaudhari, R.V.; Subramaniam, B. Aqueous phase hydrogenation of acetic acid and its promotional effect on *p*-cresol hydrodeoxygenation. *Energy Fuels* **2013**, *27*, 487–493.

31. Lee, J.S.; Yeom, M.H.; Park, K.Y.; Nam, I.-S.; Chung, J.S.; Kim, Y.G.; Moon, S.H. Preparation and benzene hydrogenation activity of supported molybdenum carbide catalysts. *J. Catal.* **1991**, *128*, 126–136.

32. Chen, L.; Zhu, Y.; Zheng, H.; Zhang, C.; Zhang, B.; Li, Y. Aqueous-phase hydrodeoxygenation of carboxylic acids to alcohols or alkanes over supported Ru catalysts. *J. Mol. Catal. A* **2011**, *351*, 217–227.

33. Pham-Huu, C.; Ledoux, M.J.; Guille, J. Reactions of 2- and 3-methylpentane, methylcyclopentane, cyclopentane, and cyclohexane on activated Mo_2C. *J. Catal.* **1993**, *143*, 249–261.

34. Blekkan, E.A.; Pham-Huu, C.; Ledoux, M.J.; Guille, J. Isomerization of *n*-heptane on an oxygen-modified molybdenum carbide catalyst. *Ind. Eng. Chem. Res.* **1994**, *33*, 1657–1664.

35. Ledoux, M.J.; del Gallo, P.; Pham-Huu, C.; York, A.P.E. Molybdenum oxycarbide isomerization catalysts for cleaner fuel production. *Catal. Today* **1996**, *27*, 145–150.

36. Iglesia, E.; Ribeiro, F.H.; Boudart, M.; Baumgartner, J.E. Synthesis, characterization, and catalytic properties of clean and oxygen-modified tungsten carbides. *Catal. Today* **1992**, *15*, 307–337.

37. Ribeiro, F.H.; Betta, R.A.D.; Guskey, G.J.; Boudart, M. Preparation and surface composition of tungsten carbide powders with high specific surface area. *Chem. Mater.* **1991**, *3*, 805–812.

38. Yang, P.; Zhao, D.; Margolese, D.I.; Chmelka, B.F.; Stucky, G.D. Block copolymer templating syntheses of mesoporous metal oxides with large ordering lengths and semicrystalline framework. *Chem. Mater.* **1999**, *11*, 2813–2826.

39. Boudart, M.; Djéga-Mariadassou, G. *Kinetics of Heterogeneous Catalytic Reactions*; Princeton University Press: Princeton, NJ, USA, 1984; pp. 25–26.

40. Ramanathan, S.; Oyama, S.T. New catalysts for hydroprocessing: transition metal carbides and nitrides. *J. Phys. Chem.* **1995**, *99*, 16365–16372.

41. St. Clair, T.P.; Dhandapani, B.; Oyama, S.T. Cumene hydrogenation turnover rates on Mo_2C: CO and O_2 as probes of the active site. *Catal. Lett.* **1999**, *58*, 169–171.

B-Site Metal (Pd, Pt, Ag, Cu, Zn, Ni) Promoted $La_{1-x}Sr_xCo_{1-y}Fe_yO_{3-\delta}$ Perovskite Oxides as Cathodes for IT-SOFCs

Shaoli Guo, Hongjing Wu, Fabrizio Puleo and Leonarda F. Liotta

Abstract: Perovskite oxides $La_{1-x}Sr_xCo_{1-y}Fe_yO_{3-\delta}$ (LSCF) have been extensively investigated and developed as cathode materials for intermediate temperature solid oxide fuel cells (IT-SOFCs) due to mixed ionic–electronic conductivity and high electrooxygen reduction activity for oxygen reduction. Recent literature investigations show that cathode performances can be improved by metal surface modification or B-site substitution on LSCF. Although the specific reaction mechanism needs to be further investigated, the promoting effect of metal species in enhancing oxygen surface exchange and oxygen bulk diffusion is well recognized. To our knowledge, no previous reviews dealing with the effect of metal promotion on the cathodic performances of LSCF materials have been reported. In the present review, recent progresses on metal (Pd, Pt, Ag, Cu, Zn, Ni) promotion of LSCF are discussed focusing on two main aspects, the different synthesis approaches used (infiltration, deposition, solid state reaction, one pot citrate method) and the effects of metal promotion on structural properties, oxygen vacancies content and cathodic performances. The novelty of the work lies in the fact that the metal promotion at the B-site is discussed in detail, pointing at the effects produced by two different approaches, the LSCF surface modification by the metal or the metal ion substitution at the B-site of the perovskite. Moreover, for the first time in a review article, the importance of the combined effects of oxygen dissociation rate and interfacial oxygen transfer rate between the metal phase and the cathode phase is addressed for metal-promoted LSCF and compared with the un-promoted oxides. Perspectives on new research directions are shortly given in the conclusion.

Reprinted from *Catalysts*. Cite as: Guo, S.; Wu, H.; Puleo, F.; Liotta, L.F. B-Site Metal (Pd, Pt, Ag, Cu, Zn, Ni) Promoted $La_{1-x}Sr_xCo_{1-y}Fe_yO_{3-\delta}$ Perovskite Oxides as Cathodes for IT-SOFCs. *Catalysts* **2015**, *5*, 366–391.

1. Introduction

In the last 20 years, solid oxide fuel cells (SOFCs), whose typical schematic diagram is shown in Figure 1, have been world-wide investigated since they can utilize various fuels to generate energy effectively in an environmental friendly way. For the moment, the commercialization of conventional SOFCs using the state-of-art strontium-doped lanthanum manganite $La_{1-x}Sr_xMnO_{3-\delta}$ (LSM) perovskite cathodes is restricted by the high operating temperature, around 1000 °C [1–3]. Lowering the operating temperature to about 600 °C–800 °C, the use of a wide range of materials for cell components would be allowed favoring an extensive commercialization of SOFC.

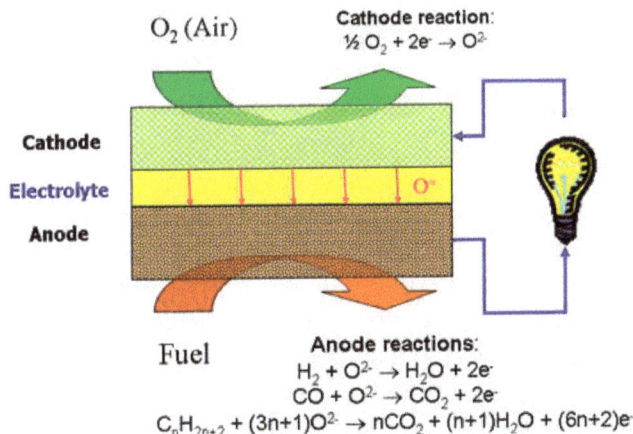

Figure 1. Schematic diagram of a solid oxide fuel cell (SOFC). (Reprinted with permission from [3], 2008, Springer Science+Business Media, LLC).

However, poor cell performance resulting mainly from the decrease in cathode kinetics becomes the biggest problem for further applications of the so-called intermediate temperature solid oxide fuel cells (IT-SOFCs). Thus, development of novel cathode materials with high electrooxygen reduction activity and stability becomes one of the major challenges in IT-SOFCs technology [4–6].

Mixed ionic and electronic conductive (MIEC) perovskite oxides, $La_{1-x}Sr_xCo_{1-y}Fe_yO_{3-\delta}$ (LSCF) [6], have attracted considerable attentions for several applications including oxygen sensors, oxygen separation membranes, cathode current-collecting materials and so on [7–10]. LSCF oxides are also promising candidates as IT-SOFCs cathodes due to excellent activity in the oxygen reduction at relative low temperature and due to the excellent ion/electron transportation properties.

The $La_{0.6}Sr_{0.4}Co_{0.2}Fe_{0.8}O_{3-\delta}$ is the most widely studied composition for its proper chemical stability, electrical conductivity, oxygen reduction activity and thermal expansion coefficient (TEC) [11–18]. Oxygen reduction performances higher more than five orders of magnitude than that of LSM materials at 900 °C are reported [19]. A comparison among $La_{0.8}Sr_{0.2}MnO_{3-\delta}$ (LSM), $Ba_{0.5}Sr_{0.5}Co_{0.8}Fe_{0.2}O_{3-\delta}$ (BSCF), $Sm_{0.5}Sr_{0.5}CoO_{3-\delta}$ (SmSC) and $La_{0.6}Sr_{0.4}Co_{0.2}Fe_{0.8}O_{3-\delta}$ (LSCF) regarding their chemical stability, electrical conductivity, oxygen reduction activity and polarization resistance showed that

LSCF was the most promising cathode material and the maximum power density (MPD) of the cell using LSCF as cathode was twice than that using LSM [20].

The applicability of LSCF as IT-SOFC cathodes is strongly dependent on the chemical composition. Usually, the oxide ion conductivity can be increased by increasing the oxygen vacancies content. The highest theoretical conductivity expected by substituting La by Sr would be at $x = 0.5$. For this composition the maximum oxygen vacancy content takes place to charge compensate the material. Moreover, at the B-site the oxidation to some $Co^{3+}(Fe^{3+})$ species to $Co^{4+}(Fe^{4+})$ may also occur [6]. However, typically, the vacancy formation is the lowest energy process and, therefore, the favorite.

While the increase of ionic conductivities is more influenced by Sr concentration at the A-site of LSCF, the increase of the electronic conductivities is affected by Fe and Co concentration at the B-site.

Moreover it has been reported that a small A-site deficiency, such as for $La_{0.58}Sr_{0.4}Co_{0.2}Fe_{0.8}O_{3-\delta}$ (L58SCF) had a particularly positive effect on the cell performance [16]. The measured current densities of cells with this A-site deficient cathode L58SCF, were as high as 1.76 A cm^{-2} at 800 °C and 0.7 V, which is about twice the current density of cells with $(La,Sr)MnO_3$/yttrium-stabilized zirconia (LSM/YSZ) composite cathodes [16].

Some efforts have been focused on the application of porous LSCF in order to supply high active surface area which is favorable for charge-transfer in oxygen reduction process [21–24]. However, the control of microstructure is supposed to be limited by the high temperature sintering step applied in the conventional fabrication of SOFCs. One possible strategy is to mix LSCF with the electrolyte gadolinia-doped ceria (GDC). It is reported that after infiltration with ionic conductive GDC, the microstructure stability, the ionic conductivity, the oxygen reduction activity as well as the adhesion to electrolyte of LSCF cathode could be improved [25–31]. For instance, the polarization resistance of $La_{0.8}Sr_{0.2}Co_{0.5}Fe_{0.5}O_{3-\delta}$-GDC composite decreased to 1.6 Ω cm^2 at 650 °C [31], and that of $La_{0.6}Sr_{0.4}Co_{0.2}Fe_{0.8}O_{3-\delta}$-GDC, fabricated by impregnation, reached values as low as 0.24 Ω cm^2 at 600 °C. Such improvement was mostly due to the formation of nano-scale (~50 nm) LSCF networks [25].

In the last few years, there have been huge interests in the cathode performance of LSCF modified at the B-site with additional metal species. Experiments have been carried out on noble metals, such as Pd [31–33], Ag [34–39], Pt [38–41] and as well on low-cost transition metals, Cu [38,39,42] Zn and Ni [43]. Although the specific mechanism of metal promotion is still under investigation, previous studies have shown that metal promotion promotes the oxygen electro-catalytic reduction and influences the TEC value, implying great prospective for such materials as

IT-SOFC cathodes. To our knowledge, no previous review focused on the B-site metal promotion effect of LSCF cathodes.

The aim of this work is to provide an overview of present research progress in this field focusing on the synthesis approaches and the promotion effects on the oxygen reduction activity of the metal (Pd, Pt, Ag, Cu, Zn, Ni) promoted LSCF cathodes for oxygen reduction. The results are compared with noble metals promoted LSM cathodes. Some perspectives on new research trends are given in the conclusion.

2. Metal Promotion: Approaches and Effects

2.1. Cathodic Reactions over LSCF Material

In order to understand the mechanism of metal promotion, it is necessary to introduce the elementary cathodic reactions over LSCF material. The mainly electrochemical reaction on IT-SOFC cathodes is the oxygen reduction reaction (ORR), as described below:

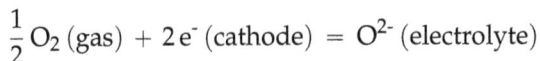

$$\frac{1}{2} O_2 \text{ (gas)} + 2 e^- \text{ (cathode)} = O^{2-} \text{ (electrolyte)}$$

Several steps are involved, such as oxygen adsorption-dissociation, oxygen surface diffusion and oxygen bulk diffusion, as it is shown in Figure 2. The former two stages are referred to as oxygen surface exchange process. It is now well established that at low operating temperature, the polarization resistance is the main reason that contributes to the power loss in IT-SOFCs. For LSCF cathodes, both surface and bulk diffusion processes play important roles, as it has been demonstrated experimentally by Jiang who investigated O_2 reduction reaction over LSCF cathode at temperatures from 700 to 900 °C in air with and without the presence of gaseous Cr species [19]. However, by using dense thin films of $La_{0.6}Sr_{0.4}Co_{0.8}Fe_{0.2}O_{3-\delta}$ prepared by pulsed laser deposition and standard photolithographic techniques, it has been revealed that the electrochemical resistance, measured at low-temperature (425–550 °C) and/or at zero or small dc (direct current) bias, is dominated by the oxygen exchange reaction at the surface of the electrode, with minor contributions from the electrode/electrolyte interface and the ohmic resistance of the electrolyte [44,45].

Three phase boundary (TPB), that is the region of contact between gaseous oxygen, cathode and electrolyte, plays a significant role in determining cathode performances by affecting possible ORR pathways. More than 20 years ago the importance of oxygen exchange kinetics in determining the magnitude of cathode resistivity values has been emphasized. More recently, a detailed quantitative interpretation of the behavior of porous mixed electrodes has been provided with

integrates information about the various macroscopic pathways for the cathodic reduction of oxygen, such as surface exchange and bulk diffusion [46].

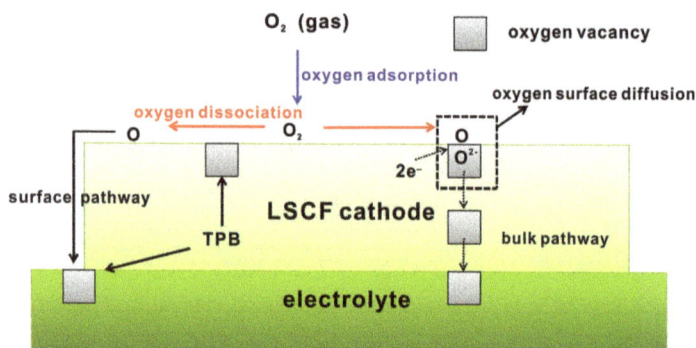

Figure 2. Schematic illustration of the two possible oxygen incorporation paths in $La_{0.6}Sr_{0.4}Co_{0.2}Fe_{0.8}O_{3-\delta}$ (LSCF) cathode.

Two possible pathways for ORR over LSCF cathodes are illustrated in Figure 2, surface pathway and bulk pathway. In the surface path, oxygen molecules dissociatively adsorb on the surface, get partially reduced, diffuse to the TPB, become fully reduced and get incorporated into the electrolyte. In the bulk pathway, oxygen molecules dissociatively adsorb on the surface, are fully reduced and incorporated as O^{2-} into the LSCF, diffuse through the bulk and get transferred into the electrolyte.

2.2. Metal Promotion

2.2.1. LSCF Surface Modification

Noble metal nanoparticles are well known to activate oxygen and to promote oxidation reactions. The dispersion of metal particles on the surface of LSCF enhances the cathode properties, combining the advantages of both noble metals and perovskite materials. In fact, most research for metal promotion is focused on this direction, herein called surface modification. It should be pointed out that since LSCF-GDC composites are one of the main subjects in IT-SOFC cathode field, surface modification is often carried out on LSCF-GDC. The influence on cathode properties coming from GDC and metal species interactions is rarely mentioned in the literature. Only few examples are reported. Wang et al. [35] studied the properties of a $La_{0.6}Sr_{0.4}Co_{0.8}Fe_{0.2}O_3$-$Ce_{0.8}Gd_{0.2}O_{1.9}$-Ag cathode on a $Ce_{0.8}Sm_{0.2}O_{1.9}$ electrolyte. The electrode consisted of a porous layer of $La_{0.6}Sr_{0.4}Co_{0.8}Fe_{0.2}O_3$-$Ce_{0.8}Gd_{0.2}O_{1.9}$ (30 wt.%) and a layer of Ag particles coated on the surface. The Ag coating was used to improve the oxygen exchange reaction activity.

The synthesis approach has great effect on the catalytic performance of metal particles by dominating the microstructure and crystallinity effects. Hence, it becomes one of the main factors that must be considered in discussions of surface modification. LSM modification by noble metals is also reported in the literature and hereinafter an example is reported for comparison with LSCF systems.

The addition of Pd, Ag, or Pt to a LSM cathode was described by using four different routes [47]. Infiltration of cathodes with a Pd solution, deposition of Pt on the electrolyte surface, mixing of $La_{0.65}Sr_{0.30}MnO_3$ (LSM) and YSZ cathode powders with different metal precursors (Pt and Pd black, Pd on activated carbon, Ag powder, Ag_2O, Ag acetate, Ag citrate, Ag_2CO_3, colloidal Ag, $AgNO_3$), and synthesis of LSM powder with the addition of $AgNO_3$.

Similarly to LSM, a widely used approach in LSCF surface modification is the impregnation or infiltration method, as recently reviewed [2]. Typically, metal precursors, usually metal nitrates aqueous solutions, are infiltrated into LSCF or LSCF/GDC backbones and decomposed into metal or metal oxide dispersed on the surface of LSCF after calcination treatment at around 800 °C. Figure 3 shows the SEM images after fuel cell testing of LSCF before and after Pd impregnation. It can be clearly seen that after impregnation, very fine metal particles can be formed and homogeneously dispersed on the surface of LSCF and even after fuel test no remarkable damage was found in the microporous structure. Experiments on perovskites modified by Ag, Pt and Cu confirmed that the impregnation method is effective for surface modification. Nevertheless, the dispersion of metal is hard to control. Usually, ethylene glycol and citric acid are used, the so called complex-polymerization method [34] and a multitude of repeated infiltration step is required in order to fill enough metal into LSCF backbone. However, due to the low specific surface area typical of perovskites and due to the sintering effects during the calcination process, metal particles tend to agglomerate, affecting the catalytic property and also leading to LSCF pore clogging [7,34].

The infiltration of porous $La_{0.6}Sr_{0.4}Co_{0.2}Fe_{0.8}O_3$ (LSCF) electrodes with $AgNO_3$ solutions in citric acid and ethylene glycol in order to deposit about 18 wt.% Ag fine particles into LSCF resulted in the enhancement of the power density of about 50% [36].

Most recently, Jun et al. [37] reported the application of plasma method technique to deposit nano-sized Ag on $La_{0.6}Sr_{0.4}Co_{0.3}Fe_{0.7}O_{3-\delta}$. Experimental results showed that Ag nanoparticles were well attached onto LSCF and strong physical linkage between Ag and LSCF was observed by FT-IR spectroscopy which can prevent agglomeration of Ag nanoparticles during heat treatment.

Huang and Chou [39] added by impregnation Pt, Ag or Cu (2 wt.% with respect to LSCF) to $La_{0.58}Sr_{0.4}Co_{0.2}Fe_{0.8}O_{3-\delta}$ (LSCF)-GDC used as cathode and investigated the effect of metal on O_2 dissociation and interfacial O transfer rates. LSCF and GDC powders were mixed at a ratio of LSCF/GDC of 100:50 in weight and designated as LSCF-50GDC. $H_2Pt(OH)_6$, $AgNO_3$ and $Cu(NO_3)_2 \cdot 3H_2O$ solution, were used as precursors, respectively. The metal-added LSCF-GDC composite was designated as LSCF-50GDC-2Cu and so on. After calcinations at 800 °C, Pt and Ag were in metallic state, while Cu was as oxide. It was found that Cu addition led to the best SOFC performance [39].

In order to develop a cathode for reduced-temperature solid oxide fuel cells, cermetting effect of LSCF with Pt was investigated [41]. However, the obtained cathode activity was only an order of 0.5 S/cm^2 at 973 K. Electrochemical results suggested that Pt has both enhancing and obstructing effects. Based on SEM observation, the main reason of these controversial results was due to poor porosity and narrow surface area determined by the high sintering temperature.

Recently, Shin *et al.* [48] found that after current conditioning, the cathode performance of LSM could be enhanced. As confirmed by SEM and TEM images, the deposition of vaporized Pt species from a current collector at the TPB resulting from the oxygen partial pressure difference was responsible for the enhancement. Similar effect was reported on LSCF, suggesting a possible approach for metal surface modification.

Figure 3. SEM pictures of fractured cross sections of (**a**) pure LSCF; (**b**) PdO-impregnated LSCF electrodes after fuel cell testing. (Reprinted with permission from [31], 2009, Elsevier).

2.2.2. Metal Ion Substitution at the B-Site of the Perovskite

It is well known for LSCF oxides that the increase of the electronic conductivities is strongly influenced by Fe and Co concentration at the B-site [6]. Moreover, LSCF are reported to be more reactive for oxygen reduction than $La_{1-x}Sr_xCoO_{3-\delta}$ (LSC) or $La_{1-x}Sr_xFeO_{3-\delta}$ (LSF) [49–52].

Crystal structure, thermal expansion, and electrical conductivity were studied for the system $La_{0.8}Sr_{0.2}Co_{1-y}Fe_yO_3$ with $0 \leqslant y \leqslant 1$ as function of Co/Fe ratio and temperature, in air [51]. The electrical conduction mechanism was attributed to the adiabatic-hopping of p-type small polarons. At high temperatures, oxygen deficiency caused lattice expansion and diminished the electrical conductivity. The observed temperature dependence of the Seebeck coefficient was attributed to changes in carrier concentration caused by a thermally excited charge disproportionation of Co^{3+} ions and by the ionic compensation of induced oxygen vacancies. The measured electrical conductivity and Seebeck coefficient as a function of the Co/Fe ratio was interpreted using a two-site hopping and the site-percolation model. It was suggested that a preferential electronic compensation of Fe ions over Co ions may occur in this system [51].

On this basis partial substitution of Co or Fe component at LSCF B-site with a third metal, with redox properties, is supposed to be effective for further cathode promotion.

The solid state reaction method has been used in the synthesis of perovskites substituted at the B-site by Cu, Zn and Ni [42,43]. In such method, high purity metal oxide powders in stoichiometric amounts are used as the starting materials and solid state reaction takes place after long term ball milling and treatment at high temperature. Solid state reaction process is simple and low cost. It provides great homogeneity and leads to direct interaction between the metal and the lattice of LSCF crystalline phase, since the metal species participate in LSCF formation process. However this approach has been found to be time and energy consuming. Furthermore, the crystalline phase and physical properties of the resulting LSCF perovskites depend critically on the experimental parameters used, thus reproducibility problems may occur and it results difficult to compare materials prepared in different laboratories.

Wang et al. [42] found that during the synthesis of LSCF-Cu by solid state reaction, Cu content exceeding 30% of B-site ions will trigger the formation of the second phase of CuO, while with lower content, no second phase can be found by X-Ray Diffraction (XRD) analysis. In Figure 4 is shown the XRD pattern of $(LaSr)(CoFeCu)O_{3-\delta}$ synthesized by solid state reaction after calcination at 1000 °C for 4 h. A rhombohedrally distorted structure was formed without second phase, except for slightly increase of c/a ratio, implying Cu atoms substitute into LSCF B-site without impacting its structure [42]. The successful synthesis of $La_{0.6}Sr_{0.4}(Co_{0.18}Fe_{0.72}X_{0.1})O_{3-\delta}$ ($x = $ Cu, Zn and Ni) by solid oxide reaction was also reported elsewhere [43].

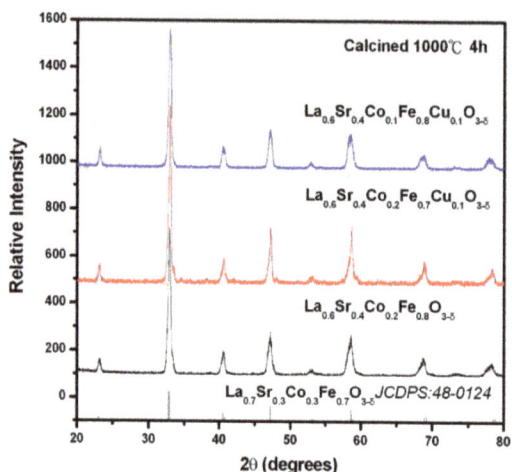

Figure 4. XRD patterns of $La_{0.6}Sr_{0.4}Co_{0.2}Fe_{0.8}O_{3-\delta}$, $La_{0.6}Sr_{0.4}Co_{0.2}Fe_{0.7}Cu_{0.1}O_{3-\delta}$ and $La_{0.6}Sr_{0.4}Co_{0.1}Fe_{0.8}Cu_{0.1}O_{3-\delta}$ powders calcined at 1000 °C for 4 h. (Reprinted with permission from [42], 2011, Elsevier.).

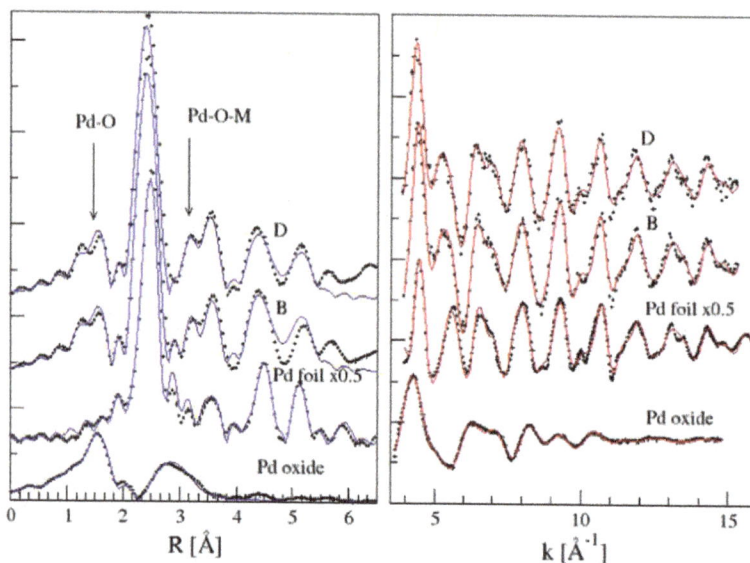

Figure 5. The fitting extended X-ray absorption fine structure (EXAFS) data of the samples LSCF0.2-Pd (B), LSCF0.8-Pd (D) and Pd oxide. (Reprinted with permission from [53], 2014, Royal Society of Chemistry).

Some of us have recently reported the first example of synthesis by one pot citrate method of $La_{0.6}Sr_{0.4}Co_{0.8}Fe_{0.17}Pd_{0.03}O_{3-\delta}$ (LSCF02-Pd) and $La_{0.6}Sr_{0.4}Co_{0.2}Fe_{0.77}Pd_{0.03}O_{3-\delta}$ (LSCF08-Pd) perovskites as possible cathode materials [53]. The local environment of Pd was investigated in details. Extended X-ray absorption fine structure (EXAFS) spectroscopy showed a very short B-site Pd-O distance of about 1.9 Å which is definitely different from the shortest Pd-O distance in palladium oxide (2.02 Å), pointing to a high Pd oxidation state in LSCF-Pd (Figure 5). Further studies showed that a Pd atomic fraction of 45% in LSCF02-Pd and of 62% in LSCF08-Pd, respectively, is arranged at the B- site of the perovskite as Pd^{4+} and the rest Pd metal clusters of about 2 nm were embedded into the matrix and strongly interact with the bulk [53]. The results indicated that even with low metal content (such as 0.03 molar fraction) only partially insertion into the lattice can be achieved and the extent of the substitution may have great relationship with the composition of B-site or the dopant metal radius [38,39,43]. The main effect of Pd insertion into the lattice of the perovskite was the formation of bulk oxygen vacancies that increased oxygen chemisorption properties of such materials. Such properties are crucial for ORR.

2.3. Cathode Performances

Hwang *et al.* investigated symmetrical electrochemical cells with various electrodes, $La_{0.6}Sr_{0.4}Co_{0.2}Fe_{0.8}O_{3-\delta}$, (LSCF), LSCF-$(Gd_{0.2}Ce_{0.8}O_2)$ GDC, LSCF-platinum (Pt) and LSCF-GDC-Pt [40]. Both the LSCF-GDC and LSCF-Pt composite electrodes performed better than the LSCF electrode for oxygen reduction. The polarization resistance was significantly reduced in the case of the composite electrodes. The incorporation of Pt particles in the LSCF electrode was found to be effective over the entire temperature range for which measurements were taken. It seems that fine platinum particles, incorporated into LSCF via chemical precipitation technique, might accelerate the oxygen adsorption at high temperature. In Figure 6, the impedance spectra, measured at 700 and 500 °C, in 20% oxygen atmosphere, taken for the LSCF and LSCF-Pt composite electrodes are shown. The addition of Pt was found to be very effective at 500 °C, as well as at 700 °C. An interesting feature is that the polarization resistance decreased with decreasing Pt content in the case of the LSCF-Pt composite electrode. At 500 °C, the polarization resistance of the LSCF-0.5 vol% Pt composite electrode was nearly one fifth of that of the LSCF electrode. In conclusion, a small amount (0.5 vol%) of Pt was found to be sufficient to reduce the polarization resistance of the LSCF.

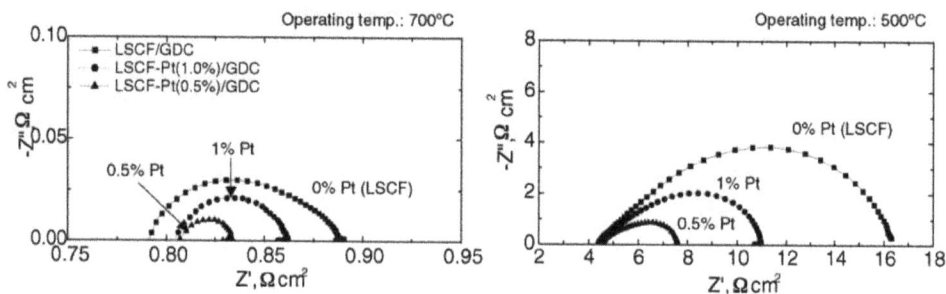

Figure 6. Impedance spectra of LSCF and LSCF-Pt (0.5 and 1 vol %) composite electrodes at 700 °C (**left**) and 500 °C (**right**). (Reprinted with permission from [40], 2005, Elsevier).

Cu^{2+} ions doped $La_{0.6}Sr_{0.4}Co_{0.2}Fe_{0.8}O_{3-}\delta$ cathodes were prepared by Wang and coworkers using solid state reaction method [42]. The electrochemical performances of the cells fabricated with such cathodes are shown in Figure 7. The maximum electrical conductivities of Cu-doped LSCF discs are higher than that of LSCF. It is evident that the doping of Cu^{2+} ions in LSCF is beneficial to the electrochemical performance of the cells.

Some representative electrochemical results in terms of MPD and AC impedance values evaluated from the area specific resistance (ASR) are summarized in Tables 1 and 2 for metal substituted LSCF perovskites, focusing on the synthesis method, metal type and content.

In general, it can be concluded that both surface modification and B-site substitution can improve cell performances by decreasing cathode resistance, especially the polarization resistance at low temperature. As above mentioned, B-site substitution leads to higher interaction between metal and LSCF lattice, hence may induce better promotion effect than surface modification.

Although Cu, Zn, Ni and Pd substituted LSCF have been successfully synthesized, only few reports investigating cell performances of substituted LSCF have been reported. Most researches dealing with Cu surface modification focus the attention on the elementary properties of the modified cathodes and, as far as we know, no systematic comparison on different synthesis approaches has been reported. Interestingly, Huang et al. [38,39] investigated the cathodic property of 2 wt.% Cu, Pt and Ag impregnated LSCF and LSCF-GDC composites respectively, despite of better catalysis activity of Pt and Ag, Cu infiltrated samples showed the best property as a function of Cu content, negative effect is also reported. In order to clarify the effect of synthesis approaches, the metal species as well as the different metal content we focus on the conclusion of promotion mechanism in this work.

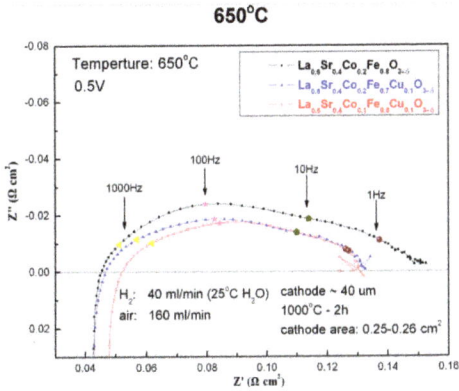

(a)

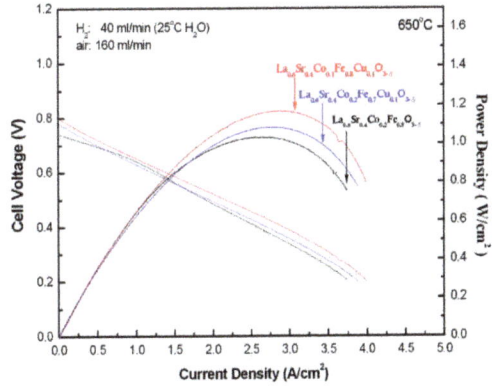

(b)

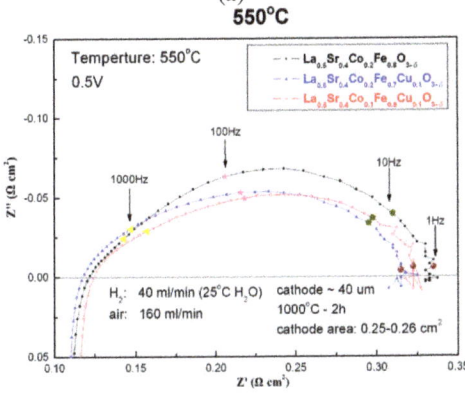

(c)

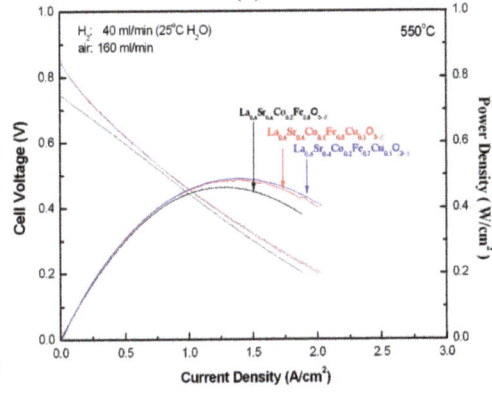

(d)

Figure 7. Nyquist plots of electrochemical impedance spectra of the single cells containing $La_{0.6}Sr_{0.4}Co_{0.2}Fe_{0.8}O_{3-\delta}$, $La_{0.6}Sr_{0.4}Co_{0.2}Fe_{0.7}Cu_{0.1}O_{3-\delta}$ and $La_{0.6}Sr_{0.4}Co_{0.1}Fe_{0.8}Cu_{0.1}O_{3-\delta}$ cathodes measured at (**a**) 650 °C and (**c**) 550 °C and I–V curves and the corresponding power densities of the anode supported single cells with various cathodes measured at (**b**) 650 °C and (**d**) 550 °C. (Reprinted with permission from [42], 2011, Elsevier).

Table 1. Summary of recent results on metal-substituted LSCF perovskites: synthesis approach, metal content and AC impedance values measured at different temperatures.

Entry	Composites	Synthesis approach	Metal content	EIS Measurement T (°C)	AC Impedance (Ω cm²)	Reference
1	$La_{0.6}Sr_{0.4}Co_{0.2}Fe_{0.8}O_{3-\delta}$	-	-	530	0.63	[36]
2	$La_{0.6}Sr_{0.4}Co_{0.2}Fe_{0.8}O_{3-\delta}$	-	-	700	<0.15	[33]
3	$La_{0.58}Sr_{0.4}Co_{0.2}Fe_{0.8}O_{3-\delta}$	-	-	800	3.86	[38]
4	$La_{0.58}Sr_{0.4}Co_{0.2}Fe_{0.8}O_{3-\delta}$-50GDC	-	-	800	5.54	[39]
5	$La_{0.6}Sr_{0.4}Co_{0.2}Fe_{0.8}O_{3-\delta}$-Ag	IMP [a]	18 wt.%	530	0.46	[36]
6	$La_{0.58}Sr_{0.4}Co_{0.2}Fe_{0.8}O_{3-\delta}$-Ag	IMP	2 wt.%	800	1.9	[38]
7	$La_{0.6}Sr_{0.4}Co_{0.8}Fe_{0.2}O_{3-\delta}$ GDC-Ag	IMP	10-15 mg cm^{-2}	600	0.9	[35]
				700	0.23	
8	$La_{0.58}Sr_{0.4}Co_{0.2}Fe_{0.8}O_{3-\delta}$-50GDC-2Ag	IMP	2 wt.%	800	6.66	[39]
9	$La_{0.58}Sr_{0.4}Co_{0.2}Fe_{0.8}O_{3-\delta}$-Cu	IMP	2 wt.%	800	1.22	[38]
10	$La_{0.58}Sr_{0.4}Co_{0.2}Fe_{0.8}O_{3-\delta}$-50GDC-2Cu	IMP	2 wt.%	800	5.85	[39]
11	$La_{0.6}Sr_{0.4}Co_{0.2}Fe_{0.7}Cu_{0.1}O_{3-\delta}$	ST [b]	2.86 wt.% [c]	700	<0.1	[42]
12	$La_{0.6}Sr_{0.4}Co_{0.1}Fe_{0.8}Cu_{0.1}O_{3-\delta}$	ST	2.87 wt.% [c]	550	~0.2	[42]
13	$La_{0.58}Sr_{0.4}Co_{0.2}Fe_{0.8}O_{3-\delta}$-Pt	IMP	2 wt.%	800	3.41	[38]
14	$La_{0.6}Sr_{0.4}Co_{0.2}Fe_{0.8}O_{3-\delta}$-Pt	IMP	0.5 vol%	700	~0.025	[40]
				500	~3	
15	$La_{0.58}Sr_{0.4}Co_{0.2}Fe_{0.8}O_{3-\delta}$-50GDC-2Pt	IMP	2 wt.%	800	12.1	[39]
16	$La_{0.6}Sr_{0.4}Co_{0.2}Fe_{0.8}$-Pd	IMP	7.5 mg cm^{-2}	700	<0.05	[33]
17	$La_{0.58}Sr_{0.4}Co_{0.2}Fe_{0.8}O_{3-\delta}$-Pd	IMP	1 wt.%	800	0.155 [d]	[32]
				750	0.193 [d]	
				700	0.247 [d]	
				650	0.468 [d]	
18	$La_{0.8}Sr_{0.2}Co_{0.5}Fe_{0.5}O_{3-\delta}$-Pd	IMP	1.2 mg cm^{-2}	750	0.22	[31]
				800	0.169 [d]	
19	$La_{0.58}Sr_{0.4}Co_{0.2}Fe_{0.8}O_{3-\delta}$-Rh	IMP	1 wt.%	750	0.204 [d]	[32]
				700	0.261 [d]	
				650	0.468 [d]	

[a] Samples prepared by the impregnation (IMP) method; [b] Samples prepared by the solid state (ST) reaction. Metal content is with respect to LSCF; [c] Metal content calculated from the given composition; [d] AC impedance evaluated from ASR.

Table 2. Summary of fuel cell performance, cell components and maximum power density (MPD) values at different temperatures.

Entry	Cell anode/Electrolyte(interlayer)/cathode	T (°C)	MPD (Wcm^{-2})	Reference
1	Ni-YSZ [a]/YSZ/GDC/La$_{0.6}$Sr$_{0.4}$Co$_{0.2}$Fe$_{0.8}$O$_{3-\delta}$	630	0.16	[36]
2	Ni-GDC/GDC/La$_{0.6}$Sr$_{0.4}$Co$_{0.2}$Fe$_{0.8}$O$_{3-\delta}$	530	0.32	[36]
		580	0.51	
3	Ni-GDC/YSZ/La$_{0.58}$Sr$_{0.4}$Co$_{0.2}$Fe$_{0.8}$O$_{3-\delta}$-50GDC	800	0.053	[39]
4	Ni-GDC/GDC/La$_{0.6}$Sr$_{0.4}$Co$_{0.2}$Fe$_{0.8}$O$_{3}$-GDC/La$_{0.58}$Sr$_{0.4}$Co$_{0.2}$Fe$_{0.8}$O$_{3-\delta}$	800	0.021	[39]
5	Ni-SDC/SDC[b]/La$_{0.6}$Sr$_{0.4}$Co$_{0.2}$Fe$_{0.8}$O$_{3-\delta}$	650	(1.03)	[42]
		550	(0.46)	
6	Ni-YSZ/YSZ/GDC/La$_{0.6}$Sr$_{0.4}$Co$_{0.2}$Fe$_{0.8}$O$_{3-\delta}$-Ag	630	0.25	[36]
		530	0.05	
		730	0.74	
7	Ni-GDC/GDC/La$_{0.6}$Sr$_{0.4}$Co$_{0.2}$Fe$_{0.8}$O$_{3-\delta}$-Ag	530	0.42	[36]
		570	0.73	
		680	~1.2	
8	Ni-GDC/GDC/La$_{0.6}$Sr$_{0.4}$Co$_{0.2}$Fe$_{0.8}$O$_{3-\delta}$-GDC/La$_{0.58}$Sr$_{0.4}$Co$_{0.2}$Fe$_{0.8}$O$_{3-\delta}$-Ag	800	0.045	[38]
9	Ni-GDC/YSZ/La$_{0.58}$Sr$_{0.4}$Co$_{0.2}$Fe$_{0.8}$O$_{3-\delta}$-50GDC-2Ag	800	0.050	[39]
10	Ni-GDC/GDC/La$_{0.6}$Sr$_{0.4}$Co$_{0.2}$Fe$_{0.8}$O$_{3-\delta}$-GDC/La$_{0.58}$Sr$_{0.4}$Co$_{0.2}$Fe$_{0.8}$O$_{3-\delta}$-Cu	800	0.052	[38]
11	Ni-SDC/SDC/La$_{0.6}$Sr$_{0.4}$Co$_{0.2}$Fe$_{0.8}$O$_{3-\delta}$	700	1.07	[42]
12	Ni-SDC/SDC/La$_{0.6}$Sr$_{0.4}$Co$_{0.2}$Fe$_{0.7}$Cu$_{0.1}$O$_{3-\delta}$	700	1.15	[42]
13	Ni-SDC/SDC/La$_{0.6}$Sr$_{0.4}$Co$_{0.1}$Fe$_{0.8}$Cu$_{0.1}$O$_{3-\delta}$	700	1.24	[42]
14	Ni-GDC/YSZ/La$_{0.58}$Sr$_{0.4}$Co$_{0.2}$Fe$_{0.8}$O$_{3-\delta}$-2Cu/La$_{0.58}$Sr$_{0.4}$Co$_{0.2}$Fe$_{0.8}$O$_{3-\delta}$-50GDC-2Cu	800	0.066	[39]
15	Ni-GDC/YSZ/La$_{0.58}$Sr$_{0.4}$Co$_{0.2}$Fe$_{0.8}$O$_{3-\delta}$-50GDC-2Cu	800	0.055	[39]
16	Ni-GDC/YSZ/GDC/La$_{0.58}$Sr$_{0.4}$Co$_{0.2}$Fe$_{0.8}$O$_{3-\delta}$-50GDC-2Cu	800	0.016	[39]
17	Ni-GDC/YSZ/La$_{0.58}$Sr$_{0.4}$Co$_{0.2}$Fe$_{0.8}$O$_{3-\delta}$-50GDC-2Cu	800	0.055	[39]
18	Ni-GDC/GDC/La$_{0.6}$Sr$_{0.4}$Co$_{0.2}$Fe$_{0.8}$O$_{3-\delta}$-GDC/La$_{0.58}$Sr$_{0.4}$Co$_{0.2}$Fe$_{0.8}$O$_{3-\delta}$-Pt	800	0.023	[38]
19	Ni-GDC/YSZ/La$_{0.58}$Sr$_{0.4}$Co$_{0.2}$Fe$_{0.8}$O$_{3-\delta}$-50GDC-2Pt	800	0.026	[39]
20	Ni-YSZ/YSZ/GDC/La$_{0.6}$Sr$_{0.4}$Co$_{0.2}$Fe$_{0.8}$O$_{3-\delta}$-Pd	800	1.99	[33]
		750	1.47	
		700	1.05	
		650	0.59	
21	Ni-YSZ/GDC/La$_{0.6}$Sr$_{0.4}$Co$_{0.2}$Fe$_{0.8}$O$_{3-\delta}$-Pd	650c	-0.15	[33]
		550	-0.04	
		800	1.83	
22	Ni-YSZ/YSZ/GDC/La$_{0.58}$Sr$_{0.4}$Co$_{0.2}$Fe$_{0.8}$O$_{3-\delta}$-Rh	750	1.46	[32]
		700	1.03	
		650	0.6	

[a] Yttrium-stabilized zircon; [b] Sm$_{0.2}$Ce$_{0.8}$O$_{2-\delta}$.

3. Metal Promotion Effects

The effect of metal promotion on oxygen transport over LSM and LSCF cathodes is still not clear, sometime controversial and detailed knowledge of the oxygen reduction mechanism is still lacking. Watanabe and coworkers [54] employed samaria-doped ceria (SDC) promoted by highly dispersed noble metals as anode material and LSM particles catalyzed with microcrystalline Pt as cathode. It was found that the anodic polarization resistance and its activation energy were greatly decreased by loading only a small amount of the catalysts (such as Ru, Rh, and Pt) onto the SDC particles. Moreover, an effective promotion of Pt species on the dissociation of O_2 molecules and/or the exchange reaction between oxygen atoms and oxygen vacancies in LSM was reported.

Contradictory effects were described for Pd, Ag and Pt promoted LSM cathodes [47]. Between 750 and 900 °C no electrocatalytic effect occurred with respect to the presence of Pt, independently on the preparation method, deposition on the electrolyte or mixing with cathode powders. Infiltration of the cathode with a Pd solution or mixing with Pd black did not result in a positive effect either. A catalytic effect was only found with Pd on activated carbon and in particular at lower temperatures. Cells prepared with Ag powder and Ag_2O showed an improved electrochemical performance compared to Ag-free cells sintered at the same temperature (920 °C). However, in comparison to Ag-free cells sintered at the standard temperature (1100 °C) lower current densities were measured [47].

Similar effects are proposed in the case of LSCF cathodes, as below discussed. Nevertheless, further fundamental studies on mechanism of ORR are highly desirable.

3.1. Effects on LSCF Surface Promoted

It is generally accepted that for LSCF surface promoted by metal particles, reactions take place between metal particles and the most external planes of LSCF crystallite. Serra [32] confirmed by XPS that in the case of Pd impregnation, around 20% of the Pd is present as metal and the rest exists in an oxidation state, which is presumably incorporated into LSCF B-site. The coexistence of surface and metal substituted into the lattice was also reported by other groups and reversible behavior under different conditions (oxygen pressure, temperature) was pointed out [32,33]. However, only few detailed investigations were done.

The presence of metal clusters at the LSCF surface will enlarge TPB by enlarging the electrode surface area directly [34,37,55]. However, the main contribution to ORR is the catalytic effect on oxygen adsorption-dissociation. Huang et al. [38,39] proposed that the promotion trend of Cu > Ag > Pt has great relationship with the oxygen affinities of the metals. Hence, it can be speculated that the higher the oxygen chemisorption over the metal with respect to the LSCF, the higher is the oxygen concentration difference leading to oxygen migration towards LSCF [35,38,39]. This mechanism suggests that a balance between oxygen adsorption-dissociation, oxygen surface and bulk diffusion is crucial for ensuring an improvement in ORR process, and can explain the influence of synthesis approach and the effect of metal content.

3.2. Effects on LSCF Promoted by Metal-in-the B-Site of the Perovskite

In order to develop new promising LSCF substituted materials it is helpful to classify the already known perovskites according to their properties. The stability of perovskites relative to their crystalline structures is frequently defined in terms of the Goldschmidt tolerance factor t [56]. Stable perovskite structures are predicted for $0.77 \leqslant t \leqslant 1.00$ [29]. The perfect cubic structure ($t = 1$) is achieved when the A-site cation has the same size as the oxygen ion (1.40 Å). The evolution of tolerance factor t as a function of A-site substitution with strontium has been calculated using the Shannon's ionic radii.

By increasing x the symmetry of the structure of $La_{1-x}Sr_xFeO_{3-\delta}$ changes from orthorhombic ($0 \leqslant x \leqslant 0.2$) via rhombohedral ($0.4 \leqslant x \leqslant 0.7$) to cubic ($0.8 \leqslant x \leqslant 1.0$). The trend of the tolerance factors of $La_{1-x}Sr_xCoO_{3-\delta}$ and $La_{1-x}Sr_xCo_{0.5}Fe_{0.5}O_{3-\delta}$ is similar to that of LSF but the values are larger and exceed unity, because of the smaller radii of Co^{3+}/Co^{4+} compared with Fe^{3+}/Fe^{4+}. A change of the symmetry from the rhombohedrally distorted ($0 \leqslant x \leqslant 0.5$) to the cubic ($0.55 \leqslant x \leqslant 0.7$) perovskite phase has been observed. Therefore with a detailed knowledge of the radii it is possible to design different materials with very similar tolerance factors, *i.e.*, crystal structures. It is well known that a larger ionic radius at the A site (e.g., substituting La^{3+} by Sr^{2+}) reduces the critical radius. The critical radius should be as large as possible in order to enhance oxygen ionic conductivity. On the one hand, substitution of La^{3+} by Sr^{2+} introduces oxygen vacancies which enhance the ionic conductivity. Indeed, as described for $La_{1-x}Sr_xCo_{0.8}Fe_{0.2}O_{3-\delta}$ the ionic conductivity increased monotonically with increasing x [56]. This indicates vacancy formation to be more important than the critical radius.

The bulk structure was investigated by Kuhn and Ozkan for Fe-based perovskite-type oxides with the formula $La_{0.6}Sr_{0.4}Co_yFe_{1-y}O_{3-\delta}$ ($y = 0.1, 0.2,$ and 0.3) [57]. They found that transition from rhombohedral to cubic structure occurred at lower temperatures for higher Co content in the perovskites. Some years later, the same group reported that for $La_{0.6}Sr_{0.4}Co_{0.2}Fe_{0.8}O_{3-\delta}$ (LSCF-6428) after substitution with Zn (10 mol% at the B-site) (LSCF-Zn), the transition temperature from rhombohedral to cubic decreased from 550 °C for LSCF-6428) to around 400–500 °C for LSCF-Zn [43]. The transition was reversible and the sample reverts back to a rhombohedral phase upon cooling. Moreover, it was reported that by doping the $La_{0.6}Sr_{0.4}Co_{0.2}Fe_{0.8}O_{3-\delta}$ at the B-site with Zn, Ni and Cu oxygen activation and vacancy generation properties increased, being the Zn-doped catalyst the best performing [43].

The electrical conductivity is also affected by metal-in-the B-site of perovskite. The maximum electrical conductivities of the $La_{0.6}Sr_{0.4}Co_{0.2}Fe_{0.7}Cu_{0.1}O_{3-\delta}$ and $La_{0.6}Sr_{0.4}Co_{0.1}Fe_{0.8}Cu_{0.1}O_{3-\delta}$ sintered at 1100 °C is 438 S cm^{-1} and 340 S cm^{-1}, which is much higher than that of $La_{0.6}Sr_{0.4}Co_{0.2}Fe_{0.8}O_{3-\delta}$ sintered at the same temperature, 81 S cm^{-1} [42]. Besides the change in material density and the formation of conductive network induced by metal species insertion into the lattice [58], the improved electronic conductivity behavior can be greatly attributed to the change in the structure after substitution, such as the Jahn-Teller distortion of the MO_6 octahedron [38,39,42].

Ozkan and co-workers confirmed the improvement in oxygen vacancies generation after B-site substitution by using thermogravimetric analyses (TGA) coupled with temperature programmed reduction (TPR) experiments [43,57]. As is shown in Figure 8, the oxygen vacancy evolution in LSCF-Zn by solid oxide reaction is better than LSCF. The evolution of oxygen vacancies can be well explained by the charge imbalance caused by low valence substitution. In fact, substitution with the same valence metal can also influence the generation of vacancies, which can be attributed to different binding capabilities of metal species with extra oxide ions [49,52,59,60]. For the same reason the diffusion coefficient of oxide ions in $LaFeO_3$ is much lower than that of $LaCoO_3$ [61,62].

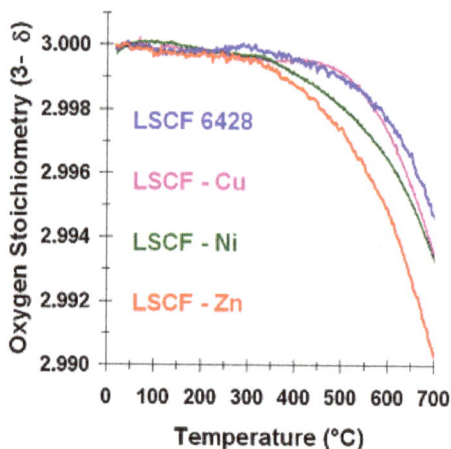

Figure 8. Oxygen vacancy generation of $La_{0.6}Sr_{0.4}Co_{0.18}Fe_{0.72}X_{0.1})O_{3-\delta}$ (where X = Zn, Ni or Cu) synthesized by solid state reaction in air as a function of temperature measured by TGA. $La_{0.6}Sr_{0.4}Co_{0.2}Fe_{0.8}O_{3-\delta}$ catalyst served as the baseline for comparison. (Reprinted with permission from [43], 2011, Elsevier).

Similarly, some of us found an increase of oxygen vacancy content in presence of Pd as suggested by carrying out TGA and Oxygen storage capacity (OSC) experiments [53]. Among the LSCF and LSCF-Pd samples prepared with two different Fe content, $La_{0.6}Sr_{0.4}Co_{0.8}Fe_{0.17}Pd_{0.03}O_{3-\delta}$ (LSCF0.2-Pd) and $La_{0.6}Sr_{0.4}Co_{0.2}Fe_{0.77}Pd_{0.03}O_{3-\delta}$ (LSCF0.8-Pd) showed higher oxygen chemisorption capacity than the corresponding undoped LSCF perovskites, $La_{0.6}Sr_{0.4}Co_{0.8}Fe_{0.17}O_{3-\delta}$ (LSCF0.2) and $La_{0.6}Sr_{0.4}Co_{0.2}Fe_{0.77}O_{3-\delta}$ (LSCF0.8), giving the higher weight loss (see Table 3). The increase of oxygen vacancy content in presence of Pd was also suggested by carrying out TGA experiments. It was confirmed that LSCF0.2 and LSCF0.2-Pd showed the highest oxygen chemisorption capacity giving the highest weight loss (%). Once again, metal doping had a beneficial effect. Moreover, such Pd doped perovskites have very similar lattice parameters than the LSCF0.2 and LSCF0.8 samples and the XRD patterns have been refined in the rhombohedral R3CH space group [53]. Then, the insertion of Pd^{4+} cations into the perovskite lattice did not produce any appreciable modification of the structure. This is not surprising because the ionic radius of Pd^{4+} matches that of the Fe^{2+} cation.

Table 3. Results of oxygen storage capacity (OSC) and thermogravimetric analysis (TGA) analyses on LSCF and LSCF-Pd perovskites [53].

Sample	O_2 Chemisorbed values (mL g^{-1}) at 600 °C	Weight loss (%) due to O_2 release
LSCF0.2	2.2	1.90
LSCF0.2-Pd	2.5	2.05
LSCF0.8	1.1	1.21
LSCF0.8-Pd	1.3	1.35

In general, the improvement in MIEC can supply more TPBs for ORR and at the same time promotes the electrochemical process by accelerating the electronic transportation [32].

In the 1982, Kilner and Brook determined by experimental methods and theoretical calculations the effects of structure and host cation type on the migration enthalpies and the effect of dopant cation size on the association enthalpies, for doped non-stoichiometric oxides with oxygen ion conductivity [63]. It was concluded that dopant cation size in the association enthalpies is the most important factor in the determination of the magnitude of oxygen ion conduction. Some years later, in the 1993, oxygen permeability measurements were carried out on $La_{0.6}Sr_{0.4}CoO_{3-\delta}$ [64]. The occurrence of order-disorder transitions in the range of temperatures 790–940 °C has been confirmed by combined thermogravimetric analysis (TGA) and differential scanning calorimetry (DSC) measurements of the sample slowly cooled or quenched from high temperature after annealing in different atmosphere. The oxygen permeability found upon exposing the material to a stream of air and of He increased sharply at the onset of the transition from the low-temperature vacancy-ordered state to the defective perovskite structure.

Definitely in a defective perovskite higher vacancy content can be related to higher ionic conductivity as demonstrated from theoretical calculations and experimental studies as well.

However, the specific effect of metal at the B-site on the SOFC performances of promoted perovskites is difficult to explain. It has been suggested by Serra *et al.* [32] that Pd substitution may accelerate the redox cycles of the charge carriers, Co^{3+}/Co^{4+} and Fe^{3+}/Fe^{4+} at B-site, hence may improve the reduction of oxygen atoms. Other metals which have been studied include Pt [40,41,47], Ag [35,47] and Cu [38,42,43]. The reactivity for O_2 dissociation over Pt is well known to be much better than Cu while that over Ag is worse than Cu. However, the SOFC performance with Pt-added LSCF as the SOFC cathode has been shown to be much worse than that with Cu or Ag-added LSCF [38].

Notably, the oxygen-ion migration in the LSCF lattice can become the rate-limiting step in the overall process of oxygen transport from the cathode to the anode via the oxygen-ion conducting materials. When the ionic conductivity increases, the oxygen ion migration becomes faster and thus the interfacial O transfer can become the rate limiting step. When metal is added to the perovskite and it is not well interacting with the perovskite lattice, an interfacial resistance for O transfer occurs.

Thus, the variation of the SOFC performance after metal addition can be due to either the effect of the oxygen dissociation reactivity of the metal or that of the interfacial oxygen transfer rate between the metal phase and the oxygen-ion conducting cathode phase. The combined effect of oxygen dissociation and interfacial oxygen transfer rates seems the balance factor (see paragraph 3.3).

Finally, we want to mention the effect of metal substitution on TEC values of LSCF. It has been reported that TEC values can be adjusted by B-site substitution on $LaCoO_3$ [65], $LaFeO_3$ [66], $La_{1-x}Sr_xFeO_{3-\delta}$ [58] and $La_{1-x}Sr_xCoO_{3-\delta}$ [67]. Previous studies suggested that B-site substitution also affects the TEC of LSCF, which may have relationship with the formation of oxygen vacancies [7]. Taken Cu substitution for example, the calculated TEC of $La_{0.6}Sr_{0.4}Co_{0.2}Fe_{0.7}Cu_{0.1}O_{3-\delta}$ and $La_{0.6}Sr_{0.4}Co_{0.1}Fe_{0.8}Cu_{0.1}O_{3-\delta}$ is 17.6×10^{-6} K^{-1} and 12.2×10^{-6} K^{-1}, respectively, which is different from that of undoped $La_{0.6}Sr_{0.4}Co_{0.2}Fe_{0.8}O_{3-\delta}$, 15.9×10^{-6} K^{-1}. The decreased TEC for $La_{0.6}Sr_{0.4}Co_{0.1}Fe_{0.8}Cu_{0.1}O_{3-\delta}$ is considerably beneficial to cathodic use in IT-SOFC [42].

3.3. A Balance between Two Effects

The influence of metal species and metal content can be well explained by the matching between oxygen adsorption-dissociation and oxygen interfacial diffusion, *i.e.*, the balance between metal clusters and metal in the perovskite lattice. In order to transfer the as dissociated O into LSCF bulk, surface oxygen vacancy as well as high fractional lattice-site interaction between metal and LSCF is needed.

In Figure 9(a–c) the schematic diagram of the Cu content effect on the electrochemical reaction of oxygen over a composite cathode $La_{0.58}Sr_{0.4}Co_{0.2}Fe_{0.8}O_{3-\delta}$ (LSCF)-50GDC is shown [39]. The a, b and c diagrams simulate the behavior of LSCF-GDC doped with Cu, 1 wt.%, 2wt.% and 5wt.%, respectively. The electrochemical reaction rate is related to the amount of TPB. Without any metal enhancement, the electrochemical reaction takes place at the TPB by adsorbing and dissociating O_2 (Figure 9a). The adsorbed O species (O_{ads}) is transformed to an oxygen ion (O^{2-}) via the charge transfer reaction, the formed O^{2-} then migrates via the oxygen-ion conducting material. The presence of the Cu species, Cu 1wt % as an example, can enhance the dissociation of O_2 and thus accelerates the electrochemical reaction (Figure 9a). On the other hand, although Cu may enhance O_2 dissociation, it also forms a separate phase from that of the oxygen-ion conducting materials. The interfacial transfer of the formed O species from the metal to the oxygen vacancy can be retarded due to a poor interaction of the metal with the lattice of the oxygen-ion conducting materials; as a consequence, the electrochemical reaction rate decreases. Therefore, the addition of metal can be detrimental to the electrochemical reaction rate if the interaction of the metal with the oxygen vacancy is not good enough. In Figures 9b and 9c the blockage of the TPB by the Cu particles is displayed. Such Cu particles should obstruct the path for interfacial O transfer and decrease its rate. If the overall rate of electrochemical reaction cannot be compensated by the higher rate of O_2 dissociation, the SOFC performance should become worse [39].

In conclusion, the authors pointed out that an optimum Cu content exists. When the Cu content increases from 1 to 2 wt.%, the O_2 dissociation reactivity increases and thus the SOFC performance can become better. However, as the Cu content increases further to 5wt %, the SOFC performance drops dramatically.

It is worth noting that the extent of interaction of the metal with the lattice of the oxygen-ion conducting LSCF perovskite strongly depends on the synthesis approach, calcination temperature and may be associated with the size of the metal cation in comparison with that of the A-site or B-site cation in the ABO_3 perovskite. As reported by Huang and Chou [39], the cation radius of Cu^{2+} (0.73 Å) is close to that of LSCF B-site Co^{2+} (0.75 Å); this results in a good interaction of Cu with LSCF B-site cations due to the insertion of Cu^{2+} cation in the LSCF B-site lattice. On the other hand, the cation radius of Ag^+ (1.26 Å) is close to that of the LSCF A-site La^{3+} (1.16 Å), resulting in a good interaction of Ag with LSCF A-site cations due to the insertion of Ag^+ cation in the LSCF A-site lattice. On the contrary, the size of Pt cation, such as Pt^{4+} (0.63 Å), is much smaller than that of LSCF B-site cations. This would result in a poor interaction of Pt with LSCF B-site cations and in the stabilization of metallic Pt. Consequently, although Pt has excellent reactivity for O_2 dissociation, the interfacial O transfer should be slow and thus the electrochemical reaction rate can become low. In conclusion, the rate of interfacial O transfer can be associated with the extent of interaction of the metal with the lattice of the oxygen-ion conducting materials. A balance between the reactivity of oxygen dissociation and the rate of interfacial O transfer is needed for the best SOFC performance.

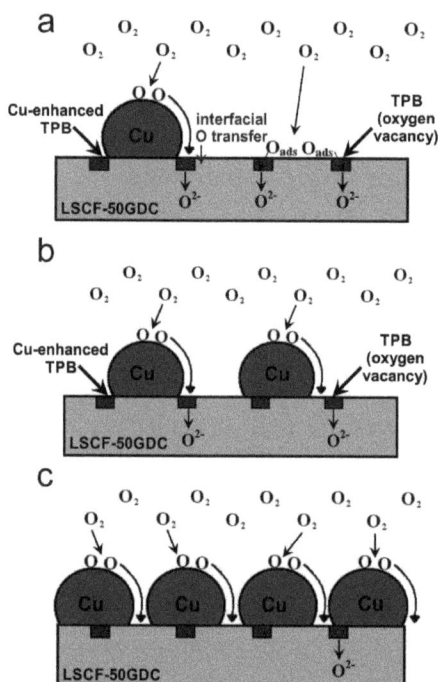

Figure 9. Schematic diagram of the Cu content on the electrochemical reaction of oxygen over the cathode TPB. The diagrams of a, b and c simulate LSCF-GDC doped with Cu, 1 wt.%, 2wt.% and 5wt.%, respectively. (Reprinted with permission from [39], 2010, WILEY-VCH Verlag GmbH & Co. KGaA, Weinheim).

Ab initio modeling of multicomponent BSCF/LSCF perovskite solid solutions has proven to be extremely helpful in understanding oxygen ion conduction mechanism. Quite a complex system of $(La,Sr)CoO_{3-\delta}$ has recently been investigated by *ab initio* calculations. One study interprets the experimentally observed increased oxygen exchange rate at such interfaces [68] as caused mainly by a redistribution of Sr from the perovskite to the interface region coupled with a respective strong local increase of the oxygen vacancy concentration. Another investigation focuses more on the strain caused at the hetero interface leading to increased oxygen vacancy mobility [69]. There is also an idea that the transition state structure of the perovskite may play an important role in the course of the vacancy diffusion [70–73]. Variations in structural disorder also were pointed out as a factor.

4. Conclusions and Perspectives

Recent results for metal (Pd, Pt, Ag, Cu, Zn, Ni) promoted LSCF cathodes for IT-SOFCs have been reviewed in this paper, focusing on the different synthesis approaches used (infiltration, deposition, solid state reaction, one pot citrate method) and as well on the effects of metal promotion on structural properties, oxygen vacancies content and cathodic performances.

It has been discussed that both noble and transition metals can improve LSCF cathodic properties by affecting electrochemical reactions, oxygen bulk transportation and TEC. A good match between oxygen adsorption-dissociation, oxygen surface and bulk diffusion is necessary to achieve optimum ORR performances. Besides metal species and content, the synthesis approach is also a key factor for the final performance. As reported by Huang and Chou [39], the addition of proper metal (Pt, Ag, Cu) loading (2 wt.%) can promote the electrochemical reaction rate of LSCF-GDC composite cathode, but can be detrimental when the amount of TPB dramatically decreases. Indeed, metal addition increases the O_2 dissociation reactivity but results in an interfacial resistance for O transfer. A balance between the rates of O_2 dissociation and interfacial O transfer is needed for the best SOFC performance.

However, investigations on metal-promoted LSCF cathodes are still challenging. When noble metals are used in order to promote LSCF, the melting point of such metals must be considered and related to the operating temperature. For instance, the melting point of silver is 961 °C; therefore, metallic silver is a potential component for the cathode in SOFCs operated at less than 800 °C because of its good oxygen reduction activity, high electrical conductivity, and its relatively low cost [6].

Moreover, it should be pointed out here that the addition of noble metals, in particular Pd and Pt, by impregnation method on the surface of the perovskite does not seem a good solution to improve the performances of cathodes due to agglomeration tendency of metal particles occurring after high-temperature operation. Moreover, high loading of noble metals should be avoided since one target to commercialization of SOFC is to decrease the price. To this respect, the addition of very small amount of noble metals and the use of new synthesis approaches, such as the one pot citrate method is highly recommended since it allows the stabilization of the doping metal (Pd) in two forms in dynamic equilibrium, partially as nanoparticles and partially as cations into the lattice. Such a method may open new perspectives for the design of new high-performing B-site doped perovskite.

Among other methods used for the synthesis of perovskites metal substituted at the B-site, the solid state reaction must be mentioned. Such a process is simple and low cost. It provides greater homogeneity and leads to deep interaction between the metal and the LSCF crystalline phase, since the metal species participate in LSCF formation process. However, the physical properties of the resulting LSCF perovskites depend critically on the experimental parameters used; thus, reproducibility problems may occur.

Deeper investigations need to be carried out in order to get fundamental understanding in the kinetics of ORR over metal-promoted LSCF cathodes. Theoretical investigation, such as density functional theory (DFT) calculations will be a powerful tool in this field, since detailed information is not obtained by experimental studies. The main use of DFT studies is to provide specific micro-scale mechanisms of chemical processes that are not attainable from experiment alone and that help to interpret experimental measurements.

As previously mentioned, DFT investigations have been successfully applied to investigate processes determining cathode performance and oxygen vacancy formation and migration in LSM, BSCF and LSCF [70–73]. Since such investigation has been not yet reported on metal (Pd, Pt, Ag, Cu, Zn, Ni) substituted LSCF, it will be meaningful to carry out such studies on perovskite metal substituted at the B-site. Meanwhile, in order to get more insights into the structural modifications occurring under operating conditions typical of IT-SOFCs in metal promoted LSCF cathodes, such as the reversible behavior between metal nanoclusters and metal ions into the lattice of the perovskite, "in situ" EXAFS and XRD experiments are highly recommended.

The controlling of metal dispersion, particle microstructure and the level of substitution remains an outstanding challenge. The application of new synthesis approaches and new metal species is still the main direction. Most recently the application of plasma and electrical depositions in surface modification and sol-gel in B-site substitution has been tried and instead of noble metal, more transition metals are used, especially on B-site substitution.

Moreover, even in the case of excellent properties of fresh LSCF cathodes, the chemical and thermal stability under operating condition is an important issue to be addressed for practical applications [51,74,75]. A study of Ni-YSZ/YSZ/GDC/LSCF-Ag cell showed that the good initial performances of Ag infiltrated cathodes are maintained after 150 h operation at 730 °C (Figure 10 [36]). Moreover, no significant degradation of out-put power was observed for 150 h at 0.7 V and 700 °C (Figure 11).

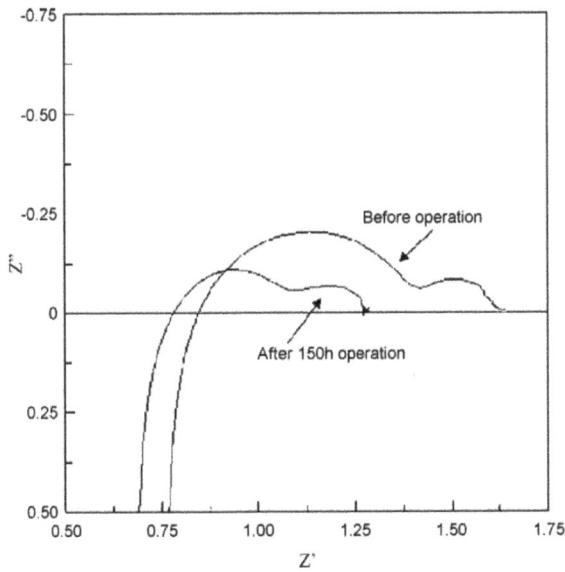

Figure 10. Impedance spectra of Ni–YSZ/YSZ/GDC/LSCF–Ag before operation and after 150 h operation at 730 °C. (Reprinted with permission from [36], 2008, Elsevier).

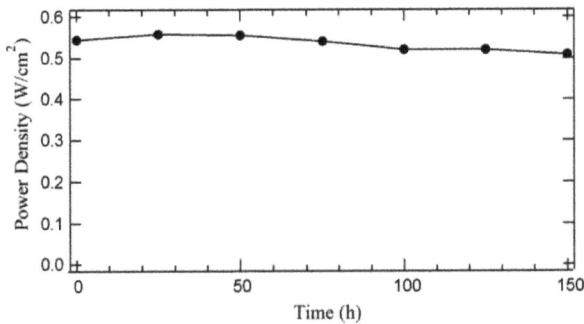

Figure 11. Time dependence of the power density at 700 °C and 0.7 V. (Reprinted with permission from [36], 2008, Elsevier).

Future improvements of LSCF cathodes should claim Sr-low content perovskites obtained by insertion into the LSCF A-site lattice of metal cation with radius similar to that of Sr^{2+}. Such materials can be advantageous, helping to decrease long-time degradation processes caused by cation segregation and poisoning by gaseous chromium oxide or SO_x species, forming $SrCrO_4$ or $SrSO_4$ precipitates that may block the active surface [70].

However, it is worth noting that about Sr segregation there are some contradictory results. Indeed, it has been recently reported [76] that surface-decoration of perovskites can strongly affect the oxygen reduction activity, and therefore is a new and promising approach to improve SOFC cathode materials. In this study, it has been demonstrated that a small amount of secondary phase (La-, Co-, and Sr-(hydr)oxides/carbonates) on a (001) $La_{0.8}Sr_{0.2}CoO_{3-\delta}$ surface can either significantly activate or passivate the electrode. Although the physical origin for the enhancement is not fully understood, results from atomic force microscopy, X-ray diffraction, and X-ray photoelectron spectroscopy suggest that the observed enhancement for "Sr"-decorated surfaces can be attributed largely to catalytically active interface regions between surface Sr-enriched particles and the LSC surface.

Finally, IT-SOFC properties are also limited to a large extent by solid state reactions at the interface with the electrolyte. Despite of excellent initial cathode performances, the solid-state reactions occurring at high temperature between LSCF and the electrolyte, YSZ or YSZ-GDC, are still big problem to be addressed [77]. For commercialization, the chemical and physical match between LSCF cathode and GDC electrolyte is still challenging.

Acknowledgments: The authors acknowledge the financial support provided by the China Scholarship Council for supporting Guo's scholarship. S. Guo thanks COST Action CM 1104 for supporting her STSM (COST-STSM-ECOST-STSM-CM1104-010414-042095) in the group of Eugene Kotomin to carry out DFT calculations.

Author Contributions: Guo Shaoli wrote the first draft of the article which was then extensively improved by the comments and suggestions of Hongjing Wu and Fabrizio Puleo; Leonarda Francesca Liotta coordinated the entire work and performed the revision.

Conflicts of Interest: The authors declare no conflict of interest.

References

1. Stambouli, A.B.; Traversa, E. Solid oxide fuel cells (SOFCs): A review of an environmentally clean and efficient source of energy. *Renew. Sustain. Energy Rev.* **2002**, *6*, 433–455.
2. Vohs, J.M.; Gorte, R.J. High-Performance SOFC Cathodes Prepared by Infiltration. *Adv. Mater.* **2009**, *21*, 943–956.
3. Jiang, S.P. Development of lanthanum strontium manganite perovskite cathode materials of solid oxide fuel cells: A review. *J. Mater. Sci.* **2008**, *43*, 6799–6833.
4. Orera, A.; Slater, P.R. New Chemical Systems for Solid Oxide Fuel Cells. *Chem. Mater.* **2010**, *22*, 675–690.
5. Brett, D.L.J.; Atkinson, A.; Brandon, N.P.; Skinner, S.J. Intermediate temperature solid oxide fuel cells. *Chem. Soc. Rev.* **2008**, *37*, 1568–1578.
6. Sun, C.W.; Hui, R.; Roller, J. Cathode materials for solid oxide fuel cells: A review. *J. Solid State Electrochem.* **2010**, *14*, 1125–1144.
7. Mori, M.; Liu, Y.; Itoh, T. $La_{0.6}Sr_{0.4}Co_{0.2}Fe_{0.8}O_{3-\delta}$ Current Collectors via Ag Infiltration for Microtubular Solid Oxide Fuel Cells with Intermediate Temperature Operation. *J. Electrochem. Soc.* **2009**, *156*, B1182–B1187.

8. Kim, J.H.; Song, R.H.; Chung, D.Y.; Hyun, S.H.; Shin, D.R. Degradation of cathode current-collecting materials for anode-supported flat-tube solid oxide fuel cell. *J. Power Sources* **2009**, *188*, 447–452.

9. Teraoka, Y.; Honbe, Y.; Ishii, J.; Furukawa, H.; Moriguchi, I. Catalytic effects in oxygen permeation through mixed-conductive LSCF perovskite membranes. *Solid State Ionics* **2002**, *152–153*, 681–687.

10. Kusaba, H.; Shibata, Y.; Sasaki, K.; Teraoka, Y. Surface effect on oxygen permeation through dense membrane of mixed-conductive LSCF perovskite-type oxide. *Solid State Ionics* **2006**, *177*, 2249–2253.

11. Itoh, T.; Nakayama, M. Using *in situ* X-ray absorption spectroscopy to study the local structure and oxygen ion conduction mechanism in $(La_{0.6}Sr_{0.4})(Co_{0.2}Fe_{0.8})O_{3-\delta}$. *J. Solid State Chem.* **2012**, *192*, 38–46.

12. Itoh, T.; Shirasaki, S.; Ofuchi, H.; Hirayama, S.; Honma, T.; Nakayama, M. Oxygen partial pressure dependence of *in situ* X-ray absorption spectroscopy at the Co and Fe K edges for $(La_{0.6}Sr_{0.4})(Co_{0.2}Fe_{0.8})O_{3-\delta}$. *Solid State Commun.* **2012**, *152*, 278–283.

13. Marinha, D.; Dessemond, L.; Djurado, E. Electrochemical investigation of oxygen reduction reaction on $La_{0.6}Sr_{0.4}Co_{0.2}Fe_{0.8}O_{3-\delta}$ cathodes deposited by Electrostatic Spray Deposition. *J. Power Sources* **2012**, *197*, 80–87.

14. Simrick, N.J.; Bieberle-Hütte, A.; Ryll, T.M.; Kilner, J.A.; Atkinson, A.; Ruppal, J.L.M. An investigation of the oxygen reduction reaction mechanism of $La_{0.6}Sr_{0.4}Co_{0.2}Fe_{0.8}O_3$ using patterned thin films. *Solid State Ionics* **2012**, *206*, 7–16.

15. Raj, I.A.; Nesaraj, A.S.; Kumar, M.; Tietz, F.; Buchkremer, H.P.; Stoever, D. On the Suitability of $La_{0.60}Sr_{0.40}Co_{0.20}Fe_{0.80}O_3$ Cathode for the Intermediate Temperature Solid Oxide Fuel Cell (ITSOFC). *J. New. Mat. Electrochem. Syst.* **2004**, *7*, 145–151.

16. Mai, A.; Haanappel, V.C.; Uhlenbruck, S.; Tietz, F.; Stöver, D. Ferrite-based perovskites as cathode materials for anode-supported solid oxide fuel cells. *Solid State Ionics* **2005**, *176*, 1341–1350.

17. Wang, W.G.; Mogensen, M. High-performance lanthanum-ferrite-based cathode for SOFC. *Solid State Ionics* **2005**, *176*, 457–462.

18. Wailer, D.; Lane, J.A.; Kilner, J.A.; Steele, B.C.H. The effect of thermal treatment on the resistance of LSCF electrodes on gadolinia doped ceria electrolytes. *Solid State Ionics* **1996**, *86–88*, 767–772.

19. Jiang, S.P. A comparison of O_2 reduction reactions on porous $(La,Sr)MnO_3$ and $(La,Sr)(Co,Fe)O_3$ electrodes. *Solid State Ionics* **2002**, *146*, 1–22.

20. Rembelski, D.; Viricelle, J.P.; Combemale, L.; Rieu, M. Characterization and Comparison of Different Cathode Materials for SCSOFC: LSM, BSCF, SSC, and LSCF. *Fuel Cells* **2012**, *12*, 256–264.

21. Zhao, E.Q.; Zhen, J.; Zhao, L.; Xiong, Y.P.; Sun, C.W.; Brito, M.E. One dimensional $La_{0.8}Sr_{0.2}Co_{0.2}Fe_{0.8}O_{3-\delta}/Ce_{0.8}Gd_{0.2}O_{1.9}$ nanocomposite cathodes for intermediate temperature solid oxide fuel cells. *J. Power Sources* **2012**, *219*, 133–139.

22. Baqué, L.; Caneiro, A.; Moreno, M.S.; Serquis, A. High performance nanostructured IT-SOFC cathodes prepared by novel chemical method. *Electrochem. Commun.* **2008**, *10*, 1905–1908.

23. Yoon, J.; Araujo, R.; Grunbaum, N.; Baqué, L.; Serquis, A.; Caneiro, A.; Zhang, X.H.; Wang, H.Y. Nanostructured cathode thin films with vertically-aligned nanopores for thin film SOFC and their characteristics. *Appl. Surf. Sci.* **2007**, *254*, 266–269.

24. Gandavarapu, S.R.; Sabolsky, K.; Gerdes, K.; Sabolsky, E.M. Direct foamed and nano-catalyst impregnated solid-oxide fuel cell (SOFC) cathodes. *Mater. Lett.* **2013**, *95*, 131–134.

25. Shah, M.; Barnett, S.A. Solid oxide fuel cell cathodes by infiltration of $La_{0.6}Sr_{0.4}Co_{0.2}Fe_{0.8}O_{3-\delta}$ into Gd Doped Ceria. *Solid State Ionics* **2008**, *179*, 2059–2064.

26. Harris, J.; Metcalfe, C.; Marr, M.; Kuhn, J.; Kesler, O. Fabrication and characterization of solid oxide fuel cell cathodes made from nano-structured LSCF-SDC composite feedstock. *J. Power Sources* **2013**, *239*, 234–243.

27. Qiang, F.; Sun, K.N.; Zhang, N.Q.; Zhu, X.D.; Le, S.R.; Zhou, D.R. Characterization of electrical properties of GDC doped A-site deficient LSCF based composite cathode using impedance spectroscopy. *J. Power Sources* **2007**, *168*, 338–345.

28. Rahman, H.A.; Muchtar, A.; Muhamad, N.; Abdullah, H. Structure and thermal properties of $La_{0.6}Sr_{0.4}Co_{0.2}Fe_{0.8}O_{3-\delta}$-SDC carbonate composite cathodes for intermediate- to low-temperature solid oxide fuel cells. *Ceram. Int.* **2012**, *38*, 1571–1576.

29. Murray, E.P.; Sever, M.J.; Barnett, S.A. Electrochemical performance of $(La,Sr)(Co,Fe)O_3$-$(Ce,Gd)O_3$ composite cathodes. *Solid State Ionics* **2002**, *148*, 27–34.

30. Yang, M.; Yan, A.Y.; Zhang, M.; Hou, Z.F.; Dong, Y.L.; Cheng, M.J. Effects of GDC interlayer on performance of low-temperature SOFCs. *J. Power Sources* **2008**, *175*, 345–352.

31. Chen, J.; Liang, F.L.; Chi, B.; Pu, J.; Jiang, S.P.; Jian, L. Palladium and ceria infiltrated $La_{0.8}Sr_{0.2}Co_{0.5}Fe_{0.5}O_{3-\delta}$ cathodes of solid oxide fuel cells. *J. Power Sources* **2009**, *194*, 275–280.

32. Serra, J.M.; Buchkremer, H.P. On the nanostructuring and catalytic promotion of intermediate temperature solid oxide fuel cell (IT-SOFC) cathodes. *J. Power Sources* **2007**, *172*, 768–774.

33. Sahibzada, M.; Benson, S.J.; Rudkin, R.A.; Kilner, J.A. Pd-promoted $La_{0.6}Sr_{0.4}Co_{0.2}Fe_{0.8}O_3$ cathodes. *Solid State Ionics* **1998**, *113–115*, 285–290.

34. Liu, Y.; Mori, M.; Funahashi, Y.; Fujishiro, Y.; Hirano, A. Development of micro-tubular SOFCs with an improved performance via nano-Ag impregnation for intermediate temperature operation. *Electrochem. Commun.* **2007**, *9*, 1918–1923.

35. Wang, S.; Kato, T.; Nagata, S.; Honda, T.; Kaneko, T.; Iwashita, N.; Dokiya, M. Performance of a $La_{0.6}Sr_{0.4}Co_{0.8}Fe_{0.2}O_3$-$Ce_{0.8}Gd_{0.2}O_{1.9}$-Ag cathode for ceria electrolyte SOFCs. *Solid State Ionics* **2002**, *146*, 203–210.

36. Sakito, Y.; Hirano, A.; Inanishi, N.; Takeda, Y.; Yamamoto, O.; Liu, Y. Silver infiltrated $La_{0.6}Sr_{0.4}Co_{0.2}Fe_{0.8}O_3$ cathodes for intermediate temperature solid oxide fuel cells. *J. Power Sources* **2008**, *182*, 476–481.

37. Jun, S.H.; Uhm, Y.R.; Song, R.H.; Rhee, C.K. Synthesis and properties of Ag nanoparticles attached $La_{0.6}Sr_{0.4}Co_{0.3}Fe_{0.7}O_{3-\delta}$. *Curr. Appl. Phys.* **2011**, *11*, S305–S308.

38. Huang, T.J.; Shen, X.D.; Chou, C.L. Characterization of Cu, Ag and Pt added $La_{0.6}Sr_{0.4}C_{o0.2}Fe_{0.8}O_{3-\delta}$ and gadolinia-doped ceria as solid oxide fuel cell electrodes by temperature-programmed techniques. *J. Power Sources* **2009**, *187*, 348–355.

39. Huang, T.J.; Chou, C.L. Oxygen Dissociation and Interfacial Transfer Rate on Performance of SOFCs with Metal-Added $(LaSr)(CoFe)O_3$-$(Ce,Gd)O_{2-\delta}$ Cathodes. *Fuel Cells* **2010**, *10*, 718–725.

40. Hwang, H.J.; Moon, J.W.; Lee, S.; Lee, E.A. Electrochemical performance of LSCF-based composite cathodes for intermediate temperature SOFCs. *J. Power Sources* **2005**, *145*, 243–248.

41. Sasaki, K.; Tamura, J.; Hosoda, H.; Lan, T.N.; Yasumoto, K.; Dokiya, M. Pt-perovskite cermet cathode for reduced-temperature SOFCs. *Solid State Ionics* **2002**, *148*, 551–555.

42. Wang, S.F.; Yeh, C.T.; Wang, Y.R.; Hsu, Y.F. Effects of $(LaSr)(CoFeCu)O_{3-\delta}$ cathodes on the characteristics of intermediate temperature solid oxide fuel cells. *J. Power Sources* **2012**, *201*, 18–25.

43. Lakshminarayanan, N.; Choi, H.; Kuhn, J.N.; Ozkan, U.S. Effect of additional B-site transition metal doping on oxygen transport and activation characteristics in $La_{0.6}Sr_{0.4}Co_{0.18}Fe_{0.72}X_{0.1})O_{3-\delta}$ (where X = Zn, Ni or Cu) perovskite oxides. *Appl. Catal. B* **2011**, *103*, 318–325.

44. Baumann, F.S.; Fleig, J.; Habermeier, H.U.; Maier, J. Impedance spectroscopic study on well-defined $(La,Sr)(Co,Fe)O_{3-\delta}$ model electrodes. *Solid State Ionics* **2006**, *177*, 1071–1081.

45. Xiong, H.; Lai, B.K.; Johnson, A.C.; Ramanathan, S. Low-temperature electrochemical characterization of dense ultra-thin lanthanum strontium cobalt ferrite $(La_{0.6}Sr_{0.4}Co_{0.8}Fe_{0.2}O_3)$ cathodes synthesized by RF-sputtering on nanoporous alumina-supported Y-doped zirconia membranes. *J. Power Sources* **2009**, *193*, 589–592.

46. Steele, B.C.H.; Bae, J.-M. Properties of $La_{0.6}Sr_{0.4}Co_{0.2}Fe_{0.8}O_{3-x}$ (LSCF) double layer cathodes on gadolinium-doped cerium oxide (CGO) electrolytes II. Role of oxygen exchange and diffusion. *Solid State Ionics* **1998**, *106*, 255–261.

47. Haanappel, V.A.C.; Rutenbeck, D.; Mai, A.; Uhlenbruck, S.; Sebold, D.; Wesemeyer, H.; Röwekamp, B.; Tropartz, C.; Tietz, F. The influence of noble-metal-containing cathodes on the electrochemical performance of anode-supported SOFCs. *J. Power Sources* **2004**, *130*, 119–128.

48. Shin, S.M.; Yoon, B.Y.; Kim, J.H.; Bae, J.M. Performance improvement by metal deposition at the cathode active site in solid oxide fuel cells. *Int. J. Hydrogen Energy* **2013**, *38*, 8954–8964.

49. Teraoka, Y.; Zhang, H.M.; Furukawa, S.; Yamazoe, N. Oxygen permeation through perovskite-type oxides. *Chem. Lett.* **1985**, *11*, 1743–1746.

50. Matsumoto, Y.; Yamada, S.; Nishida, T.; Sato, E. Oxygen Evolution on $La_{1-x}Sr_xFe_{1-y}Co_yO_3$ Series Oxides. *J. Electrochem. Soc.* **1980**, *127*, 2360–2364.

51. Tai, L.W.; Nasrallah, M.M.; Anderson, H.U.; Sparlin, D.M.; Sehlin, S.R. Structure and electrical properties of $La_{1-x}Sr_xCo_{1-y}Fe_yO_3$. Part I. The system $La_{0.8}Sr_{0.2}Co_{1-y}Fe_yO_3$. *Solid State Ionics* **1995**, *76*, 259–271.

52. Teraoka, Y.; Zhang, H.M.; Yamazoe, N. Oxygen-sorptive properties of defect perovskite-type $La_{1-x}Sr_xCo_{1-y}Fe_yO_{3-\delta}$. *Chem. Lett.* **1985**, *9*, 1367–1370.

53. Puleo, F.; Liotta, L.F.; La Parola, V.; Banerjee, D.; Martorana, A.; Longo, A. Palladium local structure of $La_{1-x}Sr_xCo_{1-y}Fe_{y-0.03}Pd_{0.03}O_{3-\delta}$ perovskites synthesized by one pot citrate method. *Phys. Chem. Chem. Phys.* **2014**, *16*, 22677–22686.

54. Watanabe, M.; Uchida, H.; Shibata, M.; Mochizuki, N.; Amikura, K. High Performance Catalyzed-Reaction Layer for Medium Temperature Operating Solid Oxide Fuel Cells. *J. Electrochem. Soc.* **1994**, *141*, 342–346.

55. Wang, J.H.; Liu, M.L.; Lin, M.C. Oxygen reduction reactions in the SOFC cathode of Ag/CeO_2. *Solid State Ionics* **2006**, *177*, 939–947.

56. Richter, J.; Holtappels, P.; Graule, T.; Nakamura, T.; Gauckler, L.J. Materials design for perovskite SOFC cathodes. *Monatsh. Chem.* **2009**, *140*, 985–999.

57. Kuhn, J.N.; Ozkan, U.S. Effect of Co content upon the bulk structure of Sr- and Co-doped $LaFeO_3$. *Catal. Lett.* **2008**, *121*, 179–188.

58. Vogt, U.F.; Holtappels, P.; Sfeir, J.; Richter, J.; Duval, S.; Wiedenmann, D.; Zuttel, A. Influence of A-site variation and B-site substitution on the physical properties of $(La,Sr)FeO_3$ based Perovskites. *Fuel Cells* **2009**, *9*, 899–906.

59. Zhang, H.M.; Yamazoe, N.; Teraoka, Y. Effects of B site partial substitutions of perovskite-type $La_{0.6}Sr_{0.4}CoO_3$ on oxygen desorption. *J. Mater. Sci. Lett.* **1989**, *8*, 995–996.

60. Teraoka, Y. Influence of constituent metal cations in substituted $LaCoO_3$ on mixed conductivity and oxygen permeability. *Solid State Ionics* **1991**, *48*, 207–212.

61. Ishigaki, T.; Yamauchi, S.; Mizusaki, J.; Fueki, K.; Tamura, H. Tracer diffusion coefficient of oxide ions in $LaCoO_3$ single crystal. *J. Solid State Chem.* **1984**, *54*, 100–107.

62. Ishigaki, T.; Yamauchi, S.; Mizusaki, J.; Mizusaki, J.; Fureki, K.; Naito, H.; Adachi, T. Diffusion of oxide Ions in $LaFeO_3$ single crystal. *J. Solid State Chem.* **1984**, *55*, 50–53.

63. Kilner, J.A. A study of oxygen ion conductivity in doped non-stoichiometric oxides. *Solid State Ionics* **1982**, *6*, 237–252.

64. Kruidhof, H.; Bouwmeester, H.J.M.; Doorn, R.H.E.; Burggraaf, A.J. Influence of order-disorder transitions on oxygen permeability through selected nonstoichiometric perovskite-type oxides. *Solid State Ionics* **1993**, *63–65*, 816–822.

65. Hrovat, M.; Katsrakis, N.; Reichmann, K.; Bernik, S.; Kuscer, D.; Holc, J. Characterisation of $LaNi_{1-x}Co_xO3$ as a possible SOFC cathode material. *Solid State Ionics* **1996**, *83*, 99–105.

66. Ohzeki, T.; Hashimoto, T.; Shozugawa, K.; Matsuo, M. Preparation of $LaNi_{1-x}Fe_xO_3$ single phase and characterization of their phase transition behaviors. *Solid State Ionics* **2010**, *181*, 1771–1782.

67. Yasumoto, K.; Inagaki, Y.; Shiono, M.; Dokiya, M. An $(La,Sr)(Co,Cu)O_{3-\delta}$ cathode for reduced temperature SOFCs. *Solid State Ionics* **2002**, *148*, 545–549.

68. Sase, M.; Hermes, F.; Yashiro, K.; Sato, K.; Mizusaki, J.; Kawada, T.; Sakai, N.; Yokokawa, H. Enhancement of Oxygen Surface Exchange at the Hetero-interface of $(La,Sr)CoO_3/(La,Sr)_2CoO_4$ with PLD-Layered Films. *J. Electrochem. Soc.* **2008**, *155*, B793–B797.

69. Han, J.W.; Yildiz, B. Mechanism for enhanced oxygen reduction kinetics at the $(La,Sr)CoO_{3-\delta}/(La,Sr)_2CoO_{4+\delta}$ hetero-interface. *Energy Environ. Sci.* **2012**, *5*, 8598–8607.

70. Kuklja, M.M.; Kotomin, E.A.; Merkle, R.; Mastrikov, Y.A.; Maier, J. Combined theoretical and experimental analysis of processes determining cathode performance in solid oxide fuel cells. *Phys. Chem. Chem. Phys.* **2013**, *15*, 5443–5471.

71. Mastrikov, Y.A.; Merkle, R.; Heifets, E.; Kotomin, E.A.; Maier, J. Pathways for Oxygen Incorporation in Mixed Conducting Perovskites: A DFT-Based Mechanistic Analysis for $(La,Sr)MnO_{3-\delta}$. *J. Phys. Chem. C* **2010**, *114*, 3017–3027.

72. Merkle, R.; Mastrikov, Y.A.; Kotomin, E.A.; Kuklja, M.M.; Maier, J. First Principles Calculations of Oxygen Vacancy Formation and Migration in $Ba_{1-x}Sr_xCo_{1-y}Fe_yO_{3-\delta}$ Perovskites. *J. Electrochem. Soc.* **2012**, *159*, B219–B226.

73. Mastrikov, Y.A.; Merkle, R.; Kotomin, E.A.; Kuklja, M.M.; Maier, J. Formation and migration of oxygen vacancies in $La_{1-x}Sr_xCo_{1-y}Fe_yO_{3-\delta}$ perovskites: Insight from *ab initio* calculations and comparison with $Ba_{1-x}Sr_xCo_{1-y}Fe_yO_{3-\delta}$. *Phys. Chem. Chem. Phys.* **2013**, *15*, 911–918.

74. Fossdal, A.; Menon, M.; Wærnhus, I.; Wiik, K.; Einarsrud, M.A.; Grande, T. Crystal Structure and Thermal Expansion of $La_{1-x}Sr_xFeO_3$ Materials. *J. Am. Ceram. Soc.* **2004**, *87*, 1952–1958.

75. Emmerlich, J.; Linke, B.M.; Music, D.; Schneider, J.M. Towards designing $La_{1-x}Sr_xCo_yFe_{1-y}O_{3-\delta}$ with enhanced phase stability: Role of the defect structure. *Solid State Ionics* **2014**, *255*, 108–112.

76. Mutoro, E.; Crumlin, E.J.; Biegalski, M.D.; Christen, H.M.; Shao-Horn, Y. Enhanced oxygen reduction activity on surface-decorated perovskite thin films for solid oxide fuel cells. *Energy Environ. Sci.* **2011**, *4*, 3689–3696.

77. Yamamoto, O.; Takeda, Y.; Kanno, R.; Noda, M. Perovskite-type oxides as oxygen electrodes for high temperature oxide fuel cells. *Solid State Ionics* **1987**, *22*, 241–246.

A Mechanistic Study of Direct Activation of Allylic Alcohols in Palladium Catalyzed Amination Reactions

Yasemin Gumrukcu, Bas de Bruin and Joost N. H. Reek

Abstract: We here report a computational approach on the mechanism of allylicamination reactions using allyl-alcohols and amines as the substrates and phosphoramidite palladium catalyst **1a**, which operates in the presence of catalytic amount of 1,3-diethylurea as a co-catalyst. DFT calculations showed a cooperative hydrogen-bonding array between the urea moiety and the hydroxyl group of the allyl alcohol, which strengthens the hydrogen bond between the O-H moiety of the coordinated allyl-alcohol and the carbonyl-moiety of the ligand. This hydrogen bond pattern facilitates the (rate-limiting) C-O oxidative addition step and leads to lower energy isomers throughout the catalytic cycle, clarifying the role of the urea-moiety.

Reprinted from *Catalysts*. Cite as: Gumrukcu, Y.; de Bruin, B.; Reek, J.N.H. A Mechanistic Study of Direct Activation of Allylic Alcohols in Palladium Catalyzed Amination Reactions. *Catalysts* **2015**, *5*, 349–365.

1. Introduction

The direct activation of allylic alcohols, without pre-activation of the alcohol moiety, in palladium catalyzed allylic substitution reactions for application in C-C and C-X bond formation, is of growing interest [1–3]. Traditionally, allylic alcohols are pre-activated by transforming the alcohol moiety into a better leaving group, such as a halide, tosylate, carboxylate or phosphate. Alternatively they can be activated by stoichiometric Lewis acid adducts, such as BEt_3 and $Ti(O-i-Pr)_4$ [4–14]. However, all these approaches lead to formation of substantial (in most cases stoichiometric) amounts of (salt) waste, which can be avoided by the direct catalytic activation of the allylic alcohol [15–19]. Ozawa and co-workers reported the first examples in which allylic alcohols were directly used as coupling partners in Pd-catalyzed allylic substitution reactions by the π-allyl palladium complexes of substituted diphosphinidenecyclobutene ligands (DPCB-Y) [15]. Mechanistic studies demonstrated that C-O bond dissociation is the rate-determining step in these reactions (proposed by Ozawa and Yoshifuji to proceed via a palladium-hydride intermediate) [20], which was later also confirmed by le Floch and co-workers on the basis of DFT calculations (invoking ammonium promoted protonation of the allyl alcohol) [21]. The theoretical study further showed that nucleophilic attack on the π-allyl complex first generates a cationic allyl amine, which assists in dissociation of

264

the hydroxyl group in the oxidative addition step. Another prominent example was reported by Oshima and co-workers, showing that a EtOAc/H_2O biphasic system allows the *in situ* activation of allylic alcohols with tppts based Pd(0) catalyst by hydration of the alcohol moiety [22]. Several water molecules assist in delocalization of the developing negative charge in the transition state, and thus lower the activation energy according to the reported DFT calculations. Similarly, the use of methanol as a solvent in allylic alkylation reactions of simple ketones with allylic alcohols was reported to be crucial for the C-O bond cleavage step, which was catalyzed by the [(dppf)Pd(allyl)] in the presence of pyrolidine as co-catalyst [23]. Solvation of the hydroxyl/hydroxy moiety in the charge-developing transition state was again proposed to lower the activation energy.

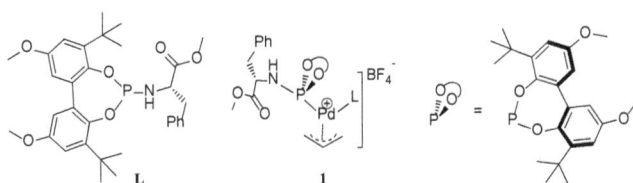

Scheme 1. The structure of phosphoramidite ligand (**L**) and schematic representation of Pd-allyl complex, **1**.

We recently disclosed a new phosphoramidite-based Pd(π-allyl)catalyst (**1**) for allylic substitution reactions, using unactivated allylic alcohols to selectively produce linear alkylated and aminated products (Scheme 1) [24]. The addition of catalytic amount of 1,3-diethylurea (3 mol%) increased the catalytic activity and improved the reproducibility of these reactions. Detailed kinetic studies revealed zero order kinetics in the nucleophilic amine and first order kinetics in the allylic alcohol and 1,3-diethylurea. On this basis, we proposed the mechanism depicted in Scheme 2. The catalyst precursor **1** was generated *in situ* by mixing [(η^3-allyl)Pd(cod)]BF_4 with ligand **L**. Nucleophilic attack of amine is expected to generate intermediate **2**, and the coordination of the allyl alcohol to palladium results in formation of intermediate **3** and the allylamine product. Experimentally, we observed product inhibition; the reaction becomes slower with increasing amounts of the product. This is most likely due to a shift in the equilibrium between intermediate **3** and **2** in favor of **2** at higher product concentrations. In line with observations reported by others, the allylic alcohol C-O bond oxidative addition step is likely the rate-determining step. Preliminary DFT calculations suggested that hydrogen-bonding interactions between the hydroxyl group of the alcohol and both the carboxyl group of the ligand and the 1,3-diethylurea moiety assist in the C-O bond activation step. Mechanistic details of the direct activation of allylic alcohols by this system, and in particular the role of the 1,3-diethylurea additive in this process, are required to facilitate further development of this type of catalyst systems. Here we report the mechanism of the Pd-catalyzed direct amination of allylic alcohols based on DFT calculations. We also found the mechanism by which the 1,3-diethylurea additive activates the catalyst as it forms H-bonds with the allylalcohol substrate.

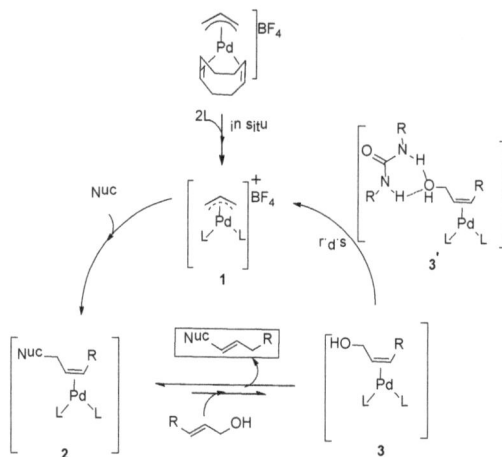

Scheme 2. Proposed mechanism for amination reactions of allylic alcohols [22].

2. Results and Discussion

The mechanism we explored computationally is based on our experimental kinetic studies (Scheme 2). The rate limiting step derived from the kinetic studies is the C-O bond oxidative addition in intermediate **3**, judging from the first order rate dependency on allyl alcohol and 1,3-diethylurea and zero order behavior in the amine (=nucleophile). We took this information as a starting point to build our computational model. For the theoretical calculations, we employed allyl alcohol and *N*-methylaniline as the substrate and nucleophile, respectively. The reaction produces the corresponding allyl-amine as the product (Scheme 2). To avoid complicated conformational searches, the ligand **L** was simplified by removing the bulky substituents of the backbone and replacing the chiral benzyl group by a methyl group (*S* chirality maintained, see **L′**, Scheme 3a). Notably, the preferred *R* chirality on bi-phenyl backbones of the phosphoramidite ligand **L** in the palladium homo-complex **1** [25] complies with the DFT calculations using the simplified ligand **L′** in **I**. Further note that the chiral element of the ligand was not used to induce enantioselectivity, neither experimentally nor in the DFT calculations. In the Figures and Schemes, ligand **L′** is drawn in a simplified manner showing only the relevant parts of the ligand for clarity (Scheme 3b). We performed all calculations in Turbomole with the BP86 functional. We employed the large def2-TZVP basis set and Grimme's version 3 (DFT-D3, disp3) dispersion corrections.

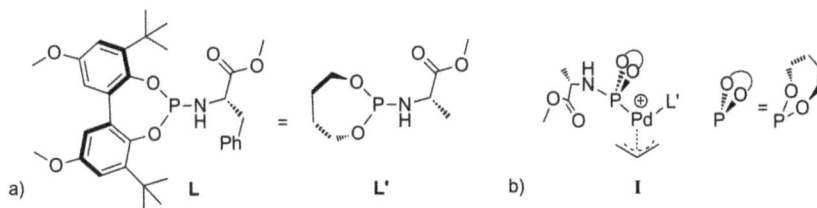

Scheme 3. (a) The ligand **L** and the simplified ligand **L′**. (b) Schematic representation of the **L′** based Pd-allyl complex **I**.

We started by optimizing the geometries of several isomers of the π-allyl complex **I** having different conformation of ligand **L′** around the metal center. In subsequent studies we took the structure with the lowest energy as the reference point ($G^0_I = 0$ kcal mol^{-1}). The coordination geometry of the palladium center in this structure is square planar, with the P-donor of the two ligands **L′** and the two donor sites of the bidentate allyl-ligand surrounding the metal. The steric interactions between the two ligands **L′** mutually and with the allyl ligand appear to be minimized in **I** (Scheme 4). This geometry of the Pd(**L′**) moiety is used for the calculation of succeeding reaction intermediates.

First, we investigated the nucleophilic attack of the amine to the allyl moiety of complex **I**. In principle, the nucleophilic attack could follow either an inner-sphere mechanism (first coordination of the nucleophile to the metal, followed by a reductive elimination step) or outer-sphere attack (direct attack of the nucleophile to the allyl moiety, without pre-coordination of the nucleophile to the metal). This strongly depends on the basicity of amines [26,27]. Strong nucleophiles (pK_a > 20) are expected to attack the Pd metal and hence likely follow an inner-sphere mechanism while weaker nucleophiles (pK_a < 20), such as N-methylaniline (pK_a~10), are believed to attacked the allyl-moiety directly and likely follow an outer-sphere mechanism. Thus, in this computational study we focused on the outer-sphere mechanism.

Outer-sphere nucleophilic attack at the allyl moiety produces the cationic Pd-allylamine complex **IIb** initially. Following this, rapid isomerization of **IIb** is expected to form the thermodynamically more stable rotamer **IIa** by C-C bond rotation (Scheme 4). The relative energies of the product **IIa** (8.2 kcal mol^{-1}) is lower than product **IIb** (11.7 kcal mol^{-1}). H-bonding between the N-H moiety of the allyl-amine and the carboxylic group of the ligand **L′** in complex **IIa** explains its relative stability (Scheme 5). Notably, the direct formation of **IIa** by the nucleophilic attack at **I** is not favorable. Steric hindrance around the palladium-allyl moiety prevents nucleophilic attack in any other direction.

267

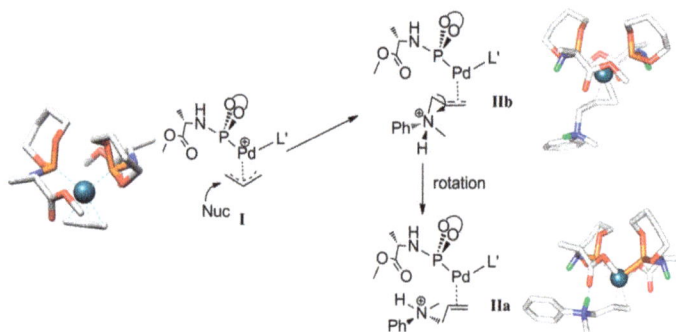

Scheme 4. Calculated structure and schematic representation of complex I, formation of complex IIb by nucleophilic attack and IIa by rotational isomerization ($\Delta G^0{}_{\text{IIa-IIb}} = -3.5$ kcal mol^{-1}).

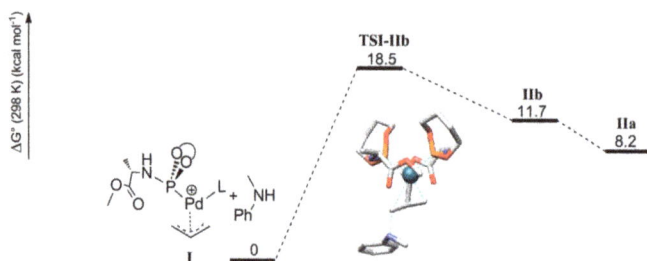

Scheme 5. Calculated pathway for nucleophilic attack of the amine at complex **I**, which forms complex **IIb**, which then converts to **IIa** via rotation (See Scheme 4 for the calculated structures and schematic representation of **IIa** and **IIb**).

In the precursor **I**, the overall positive charge is delocalized over the Pd^{II} atom, the P-donors of the ligands **L′** and the allyl moiety, while in **IIa** and **IIb** the charge is largely concentrated on the quaternary ammonium moieties. Hence, in the absence of stabilization by discrete solvent (or more likely substrate alcohol) molecules, the complexes **IIa** and **IIb** are artificially destabilized in the applied gas phase DFT calculations. However, when we applied cosmo dielectric solvent corrections, the effects of H-bonding in stabilization of some of the intermediates was partially lost due to overcompensation of dielectric forces [28], which does not represent the known effects of H-bonding in apolar solvents used experimentally. Thus, despite the abovementioned problems with charge relocation for some steps in this study, we still decided to focus mostly on a description of the mechanistic pathways on the basis of pure gas phase DFT calculations. The cosmo corrected pathways are, however, reported in the supporting information and in Scheme 8 for comparison, and we will further debate the effects of cosmo corrections in the discussion about the rate limiting step of the catalytic cycle (*vide infra*). The overestimated entropies in the gas phase as compared to solution data were corrected by the suggested 6 kcal mol^{-1} correction term for each step involving a change in the number of species (see experimental section for details).

The second step considered in the DFT computed mechanism is the displacement of the protonated allyl-amine coordinated to Pd^0 by the allyl-alcohol substrate (Scheme 6). Since gas phase DFT calculations become very unreliable when complete charge changes of the computed molecules are involved (e.g., converting a cationic species to a neutral species), we treated this step by keeping the cationic ammonium product H-bonded to the coordinated allyl alcohol. Furthermore, in previous studies, similar treatments (again to avoid charge-associated problems in gas phase calculations) were found to actually have a barrier-lowering effect on the subsequent oxidative addition step due to H-bond assisted C-O bond breaking [19]. The coordination of the allyl alcohol substrate to Pd^0 can occur in two manners, leading to two possible configurations **IVa** and **IVb** based on the position of the hydroxyl group with respect to the carboxyl group of ligand **L'**. In **IVa** the hydroxyl group of the allylalcohol is H-bonded to the carbonyl group of **L'**, whereas in **IVb** it is not. In both cases the protonated allylamine is H-bonded to the oxygen atom of the hydroxyl group of the Pd coordinated allyl-alcohol substrate (Scheme 6). The computed relative free energies of **IVa** and **IVb** (relative to precursor **I**) are –0.3 and 6.1 kcal mol^{-1}, respectively. Hence species **IVa** is the most stable intermediate. Complex **IVa** might also form via rotational isomerization of **IVb**.

To elucidate the possible role of the 1,3-diethylurea moiety, we also optimized the geometries **IIIa** and **IIIb**, which are analogues of **IVa** and **IVb** lacking the H-bonded allyl-ammonium moiety. The energies and geometries of **IIIa** and **IIIb** were compared to their analogues **IIIa'** and **IIIb'**, each containing a 1,3-diethylurea moiety double H-bonded to the hydroxyl of the coordinated allyl-alcohol (Scheme 7). Complex **IIIa** is more stable than **IIIb** (relative energies 0 and 3.5 kcal mol^{-1}, respectively), in line with the fact that **IIIb** lacks the internal hydrogen bonding between the hydroxyl moiety of the allylalcohol and the carboxyl group of **L'**. The hydrogen bonded 1,3-diethylurea moiety further stabilizes **IIIa'** substantially (ΔG^0 = –15.9 kcal mol^{-1}) whereas species **IIIb'** shows less stabilization (ΔG^0 = –7.4 kcal mol^{-1}). Hence, the H-bonding pattern in **IIIa'** based on internal H-bond donation from the hydroxyl-group to ligand **L'** and hydrogen bond acceptation from the 1,3-diethylurea H-bond donor is clearly cooperative, a feature that is commonly observed and is a consequence of polarization. It is very likely that this also plays an important role in the oxidative addition process under the experimental conditions. Unfortunately, it was not possible to compute the effect of 1,3-diethyl urea H-bonding on the barrier of the rate determining oxidative addition process. This is due to the abovementioned requirement to keep the overall charge neutral to prevent any unrealistic charge effects in the gas phase calculations. Hence, to still mimic the effect of a hydrogen bond, we used the associated allyl-ammonium salt product, which can deliver a proton after the transition state to keep an overall 1+ charge for each complex involved in that catalytic cycle.

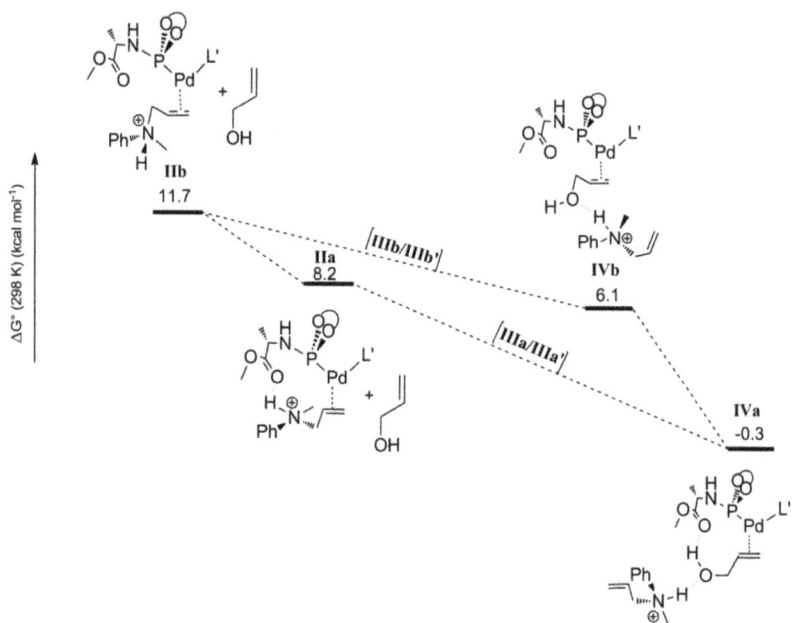

Scheme 6. Substitution of the coordinated allyl-ammonium product by the allyl alcohol substrate leading to species **IVa** and **IVb**.

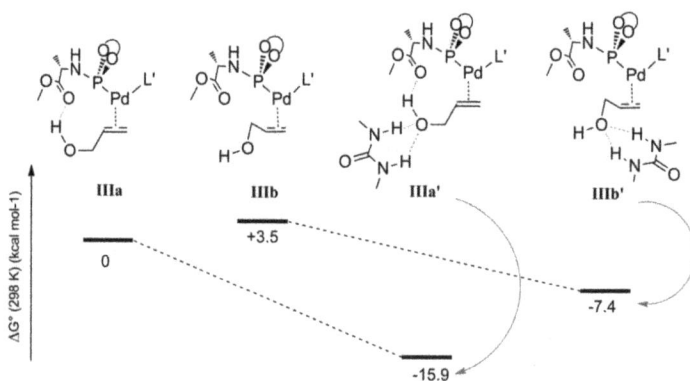

Scheme 7. The energy profile of the allyl alcohol coordinated Pd complexes in the presence and absence of urea molecule, **IIIa-IIIa′** and **IIIb-IIIb′**.

The last step of the computed catalytic cycle involves oxidative addition of the C-O bond of the allyl alcohol substrate in intermediates **IVa** and **IVb** to regenerate the palladium-allyl intermediate **I**. Since this process involves breaking the C-O bond and liberation of a hydroxide (OH$^-$) moiety being a poor leaving group, this process is expected to be facilitated by hydrogen bonding. In good agreement, the **TSIVa-I** barrier (9.5 kcal mol^{-1}) for oxidative addition of the C-O bond is lower than **TSIVb-I** (15.3 kcal mol^{-1}) reflecting the effects of cooperative H-bonding (Scheme 8). The process leads to elimination of water, because the H-bonded ammonium salt transfers its proton to the hydroxide-leaving group in a concerted manner. Hence, unfavorable uncompensated charge relocations in the gas phase calculations play a smaller role in the transition states **TSIVa-I** and **TSIVb-I** than in the precursors **IVa** and **IVb**. Besides, the relative barriers on going from **IVa** and **IVb** to **TSIVa-I** and **TSIVb-I** are likely somewhat underestimated, while the absolute barriers of **TSIVa-I** and **TSIVb-I** *versus* **I** suffer less from such uncompensated charge relocation effects.

In the gas phase calculations without cosmo dielectric solvent corrections, the transition state barrier for nucleophilic attack of the amine at the allyl moiety (**TSI-IIb** = 18.5 kcal mol^{-1}) is much higher than the transition state for the oxidative addition step (**TSIVa-I** = 9.5 kcal mol^{-1}), which would indicate that the former should be the rate determining step. However, the experimental kinetic experiments clearly indicate that the oxidative addition is rate determining. Concerning the rate determining step, the experimental data and the computational model without cosmo corrections are therefore not in agreement. This is due to the abovementioned charge concentration effects in this reaction, so that the gas phase DFT numbers do not accurately describe the solution data. Therefore we also investigated the overall pathway including cosmo dielectric solvent corrections. The results taking toluene as the medium are shown in the bottom part of Scheme 8. Indeed, this leads to a shift in the rate-limiting step from the nucleophilic attack to the oxidative addition step, in good agreement with the experimental data. To describe the effects of H-bonding, however, some of the steps seem to be better described without cosmo, which is the reason we focused on discussing those calculations first.

As reported, we have discovered by kinetic analysis that product inhibition takes place during the catalysis. The coordination of the allylamine to Pd is computed to be more favorable than coordination of the allyl alcohol. We have examined this by calculating the product coordinated Pd complexes with two possible configurations and determined the relative energies with respect to allyl alcohol coordinated Pd complexes **IIIa** and **IIIb** (Scheme 9). The formation of these complexes is expected to take place along with the formation of isomers **IVa** and **IVb** (the hydrogen bonded analogues) from **IIa** and **IIb**, which represent the product-coordinated complexes. Deprotonation of the product in complex **IIa/IIb** affords complexes **Va** and **Vb**.

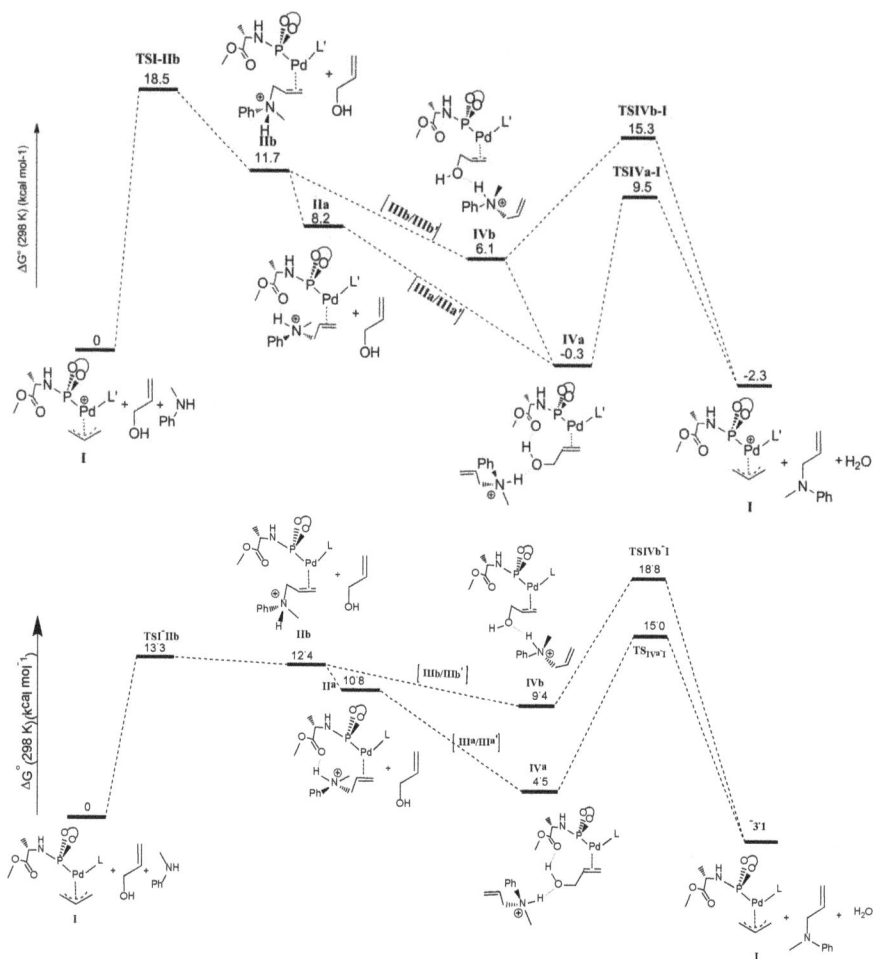

Scheme 8. Calculated catalytic pathway of the complete cycle. **Top:** Free energies without cosmo corrections; **Bottom:** Free energies with cosmo corrections ($\varepsilon = 2.38$; toluene).

Notably, both isomers **Va** and **Vb** have the same relative free energy, indicating the poor secondary interaction between the product and the complex, besides the alkene coordination energy. The relative energies of **IIIa** and **IIIb** with respect to **V** were found to be 4.9 and 8.4 kcal mol^{-1}, respectively. Note that the allyl amine adducts **Va** and **Vb** are more stable than the allyl alcohol adducts, in agreement with the experimental observations, showing product inhibition. This is likely due to sigma bond inductive effects of nitrogen *versus* oxygen. The lower energy of **IIIa** compared to **IIIb** is due to the hydrogen bond with the carbonyl of the ligand. The same trend is observed for intermediate **IVa** being lower in energy than the

IVb, and the larger difference in these complexes is again a result of cooperative hydrogen bonding.

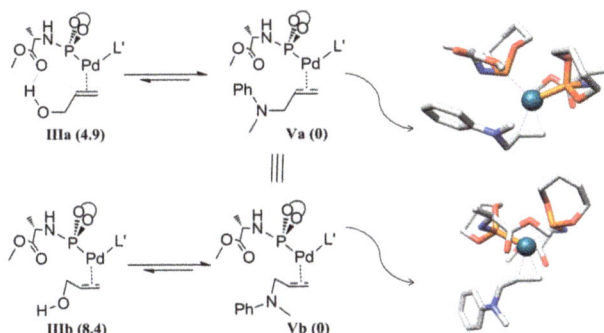

Scheme 9. Schematic representation of the product inhibition equilibriums and calculated structures of **Va** and **Vb**. The values between parentheses are the relative energies (kcal mol^{-1}). The orientation of the allylamine at Pd does not lead to a significant energy difference (complex **Va** and **Vb** have the same relative free energy).

2.1. Summary of the DFT Computed Pathway

The most interesting finding of the DFT study is the cooperative hydrogen bond between the hydroxyl and carbonyl groups (substrate-ligand) and urea (or the ammonium salt we used as a reference). We found experimentally that phosphoramidite ligands that do not have the ester functionality did not show the urea co-catalyzed activation of allyl alcohols, which is in line with the DFT calculations. The interaction between the coordinated allyl alcohol and ligand **L′** proved to be significant in the whole catalytic cycle, overall lowering the relative free energy of the Pd-intermediates (Scheme 8). The function of 1,3-diethylurea is determined to further lower the barrier for C-O oxidative addition (Scheme 10a). In addition, the calculations show a preference for the coordination of the product leading to product inhibition, however, a cooperative urea binding to **IIIa** will shift the equilibrium in favor of the substrate complex (Schemes 7 and 9). The computed catalytic cycle complies with the experimental kinetic data if we include cosmo dielectric solvent corrections. Oxidative addition of the C-O bond of the coordinated allyl alcohol in intermediate **IVa** via the rate limiting transition state **TSIVa-I** is assisted by H-bonding. This was computed using the ammonium product as the model (to reduce undesired effects of charge changes), but in practice this does explain the rate enhancing effect of the urea (Scheme 10b). This mechanism deviates from the mechanism proposed by Ozama, Samec and ourself involving allyl alcohol activation with palladium hydrides [29,30].

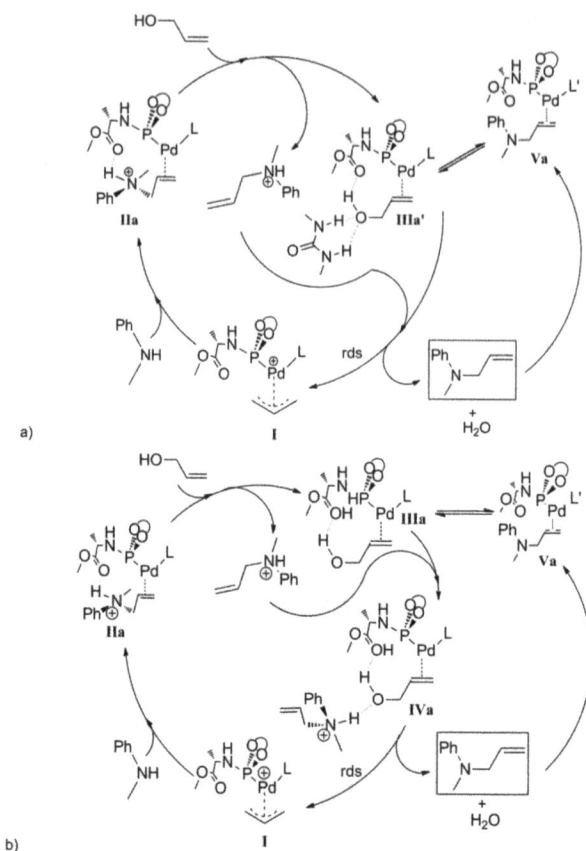

Scheme 10. (a) Proposed catalytic cycle in agreement with the experimental studies. (b) DFT calculated catalytic cycle using the allyl ammonium product as a model for urea H-bonding (intermediate IVa as a model for IIIa').

2.2. Bond Length Analysis

We analyzed the bond length changes throughout the calculated catalytic pathway in order to obtain more information about the effect of H-bond interactions on the formation of the isomeric intermediates *a* and *b* (see Table 1). First, we examined the nucleophilic attack of the *N*-methylaniline to complex **I**, which involves intermediates **IIb** and its respective transition state structure **TSI-IIb** (Figure 1). The Pd-P1 and Pd-P2 bond lengths show only small changes on going from the precursor **I** to the transition state **TSI-IIb** and subsequently formed the **IIb**, which has virtually same bond distances as **I** (Table 1).

274

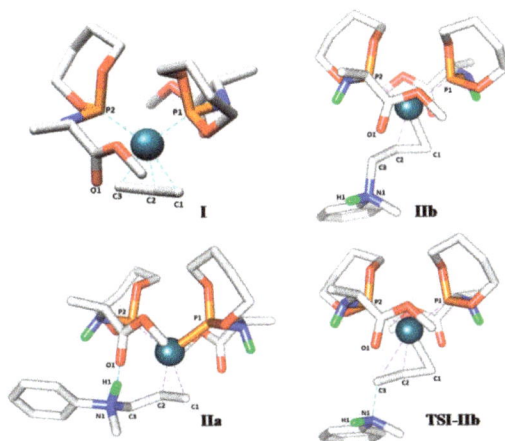

Figure 1. In all structures the related atoms are labeled such that C1 refers to the allyl carbon trans to P2, C2 is the central allyl carbon and C3 is the terminal allyl carbon trans to P1.

Table 1. Bond lengths (Å) of the calculated structures of intermediates involved in nucleophilic attack.

Bond	I	IIa	IIb	TSI-IIb
*d*Pd-P1	2.261	2.274	2.261	2.256
*d*Pd-P2	2.262	2.245	2.26	2.266
*d*Pd-C1	2.233	2.176	2.143	2.147
*d*Pd-C2	2.205	2.149	2.162	2.176
*d*Pd-C3	2.225	3.167	2.99	2.853
*d*C3-N1	-	1.521	1.639	1.841
*d*N1-H1	-	1.054	1.028	1.025
*d*H1-O1	-	1.8	-	-

The attack of the amine leads to elongation of the Pd-C3 bond (**I** Pd-C3 = 2.225 Å, **TSI-IIb** Pd-C3 = 2.853 Å) while the Pd-C1 and Pd-C2 bonds shorten, as expected. On the other hand, generated by the rotational isomerization, the coordinated ligands of **IIa** require a rearrangement of the ligands around palladium to allow the H-bond interactions between the proton of the amine (H1) and the carboxyl moiety of the ligand (O1). Thus, the bond lengths of Pd-P1 and Pd-P2 show clear changes (compare Pd-P1 = 2.261 Å, Pd-P2 = 2.260 Å in **IIb** with Pd-P1 = 2.274 Å, Pd-P2 = 2.245 Å in **IIa**) (Table 1). Besides, the newly formed C3-N1 bond is also affected by the H-bond interactions. The shorter C3-N1 bond of the lower energy isomer **IIa** must be the result of H1-O1 interaction as it directs the positioning of the amine (**IIa** C3-N1 = 1.521 Å, **IIb** C3-N1 = 1.639 Å) (Table 1).

Next, we analysed the bond length changes occurring in the oxidative addition step, which involves the simultaneous carbon-hydroxyl bond dissociation (C3-O2), hydroxyl-hydrogen bond formation (H1-O2, water), the release of the product and formation of the complex I (Table 2). The intermediate **IVa** was found lowest in energy (-0.3 kcal mol^{-1}), which also needed lower activation energy than **IVb** to generate the products (**TSIVa-I** = 9.5 and **TSIVa-I** = 15.3 kcal mol^{-1}). Judging from the comparative Pd-C3 distances of **TSIVa-I** and **TSIVb-I** (3.223 and 2.816 Å, respectively) relative to **I** (2.225 Å), the **TSIVa-I** represents an early transition state (Table 2). Moreover, we determined that the additional H-bond interaction in **TSIVa-I**, between the proton acceptor carbonyl (O1) and proton donor hydroxyl group (H2), facilitates the C3-O2, N1-H1 bond dissociation and H1-O2 bond formation processes of the transition state by altering the partial charges of the participating atoms (Figure 2). The C3-O2 and N1-H1 bonds are longer in **TSIVa-I** (1.950 and 1.760 Å) compared to **TSIVb-I** (1.844 and 1.360Å) and the H1-O2 bond distance in **TSIVa-I** (1.018 Å) is shorter than in **TSIVb-I** (1.171 Å), which might be the effect of complementary H-bonding in **TSIVa-I**.

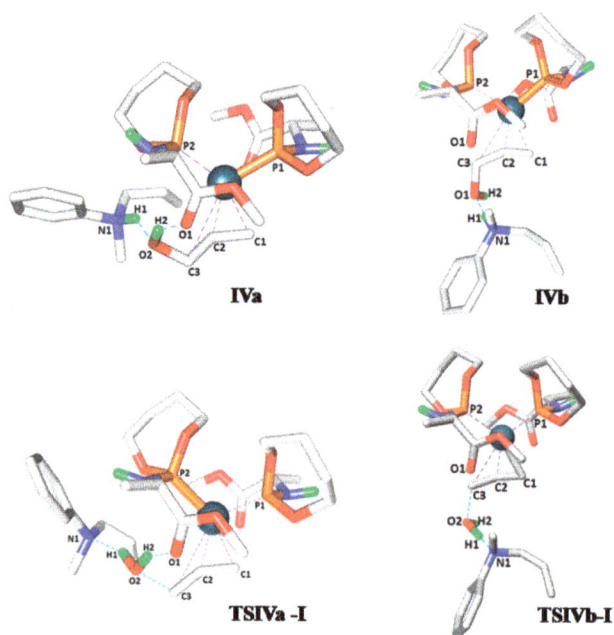

Figure 2. Transition state structures of **IVa**, **TSIVa-I**, **IVb** and **TSIVb-I**.

Table 2. Bond lengths (Å) of the calculated structures of intermediates involved in oxidative addition step.

Bond	I	IVa	IVb	TSIVa-I	TSIVb-I
dPd-P1	2.261	2.272	2.259	2.244	2.26
dPd-P2	2.262	2.235	2.247	2.284	2.271
dPd-C1	2.233	2.218	2.167	2.128	2.143
dPd-C2	2.205	2.261	2.158	2.334	2.199
dPd-C3	2.225	3.182	3.048	3.223	2.816
dC3-O2	-	1.454	1.501	1.950	1.844
dN1-H1	-	1.064	1.081	1.760	1.360
dH1-O2	-	1.793	1.575	1.018	1.171
dH2-O2	-	0.988	0.975	0.997	0.976
dH2-O1	-	1.845	-	1.720	-

3. Experimental Section

Computational Details. Geometry optimizations were carried out with the Turbomole program package [31] coupled to the PQS Baker optimizer [32] via the BOpt package [33], at the ri-DFT level using the BP86 [34,35] function and the resolution-of-identity (ri) method [36]. We optimized the geometries of all stationary points at the def2-TZVP basis set level [37], employing Grimme's dispersion corrections (DFT-D3, disp3 version) [38]. The identity of the transition states was confirmed by following the imaginary frequency in both directions (IRC). All minima (no imaginary frequencies) and transition states (one imaginary frequency) were characterized by calculating the Hessian matrix. ZPE and gas-phase thermal corrections (entropy and enthalpy, 298 K, 1 bar) from these analyses were calculated. The relative free energies obtained from these calculations are reported in the Schemes and Figures of this chapter.

By calculation of the partition function of the molecules in the gas phase, the entropy of dissociation or coordination for reactions in solution is overestimated (overestimated translational entropy terms in the gas phase compared to solutions). For reactions in "solution", we therefore corrected the Gibbs free energies for all steps involving a change in the number of species. Several methods have been proposed for corrections of gas phase to solution phase data. The minimal correction term is a correction for the condensed phase (CP) reference volume (1 L mol^{-1}) compared to the gas phase (GP) reference volume (24.5 L mol^{-1}). This leads to an entropy correction term (SCP = SGP + Rln{1/24.5} for all species, affecting relative free energies (298 K) of all associative steps of -2.5 kcal mol^{-1} [39]. Larger correction terms of -6.0 kcal mol^{-1} have been suggested based on solid arguments [40,41]. While it remains a bit debatable which entropy correction term is best to translate gas phase DFT data into free energies relevant for reactions in solution, in this chapter we adapted the suggested correction term of -6.0 kcal mol^{-1} [40,41]. Separately, additional dielectric constant corrections (comso [42]) were taken into account with single point calculations, using the dielectric constants of toluene ($\varepsilon = 2.38$), ethylacetate ($\varepsilon = 6.02$) and water ($\varepsilon = 78.54$). The cosmo corrected energy pathways are shown in the supporting information and in Scheme 8.

4. Conclusions

We studied by DFT calculations the mechanistic pathway of the amination reaction of allylic alcohols using catalyst **1**, which operates in the presence of catalytic amount of 1,3-diethylurea as a co-catalyst. The DFT calculation without cosmo dielectric solvent corrections predict the nucleophilic attack of the amine to be rate determining, while experimentally the oxidative addition step was found to be the rate-determining step. However, when including cosmo corrections, the DFT model is in good agreement with the experimental kinetic data. Yet, some steps in the catalytic cycle, which involve effects of H-bonding, seem to be better described without cosmo corrections. Notably, the operation mode of the urea additive in the overall mechanism of the catalytic reaction has become clear. The urea moiety hydrogen bonds to the hydroxyl of the allyl alcohol, which in turn hydrogen bonds to the carbonyl of the ligand. The hydrogen bonding is cooperative, most likely because of polarization effects. The hydrogen bonds are present throughout the cycle, including the transition state for oxidative addition (**TSIVa-I**). Clearly, in this transition state, the hydrogen bond leads to a lower energy compared to the non-hydrogen bonded analogue **IVb**. The oxidative addition process from **IVa** was computed to be exergonic, affording the aminated product, water and regenerating the allyl form of the catalyst (**I**). The effect of the H-bonding pattern is reflected by insignificant different bond length changes for the pathways with and without H-bonding, both in the nucleophilic attack of the amine on the allyl moiety and on the C-O oxidative addition step. In all cases, the hydrogen bond interaction leads to lower energy isomers. Concisely, H-bonding between the carboxyl group of the ligand **L′** and the protons of the coordinated allyl amine or alcohol, **IIa** or **IVa** leads to lower energy isomers with lower barriers throughout the catalytic cycle.

Acknowledgments: This research has been performed within the framework of the CatchBio program. The authors gratefully acknowledge the support of the Smart Mix Program of the Netherlands Ministry of Economic Affairs and the Netherlands Ministry of Education, Culture and Science.

Author Contributions: Yasemin Gumrukcu performed the DFT calculations with help and supervision provided by Bas de Bruin and Joost Reek. The paper was written by all three authors.

Conflicts of Interest: The authors declare no conflict of interest.

References

1. Yamamoto, T.; Akimoto, M.; Saito, O.; Yamamoto, A. Interaction of palladium(0) complexes with allylic acetates, allyl ethers, allyl phenyl chalcogenides, allylic alcohols, and allylamines. Oxidative addition, condensation, disproportionation, and π-complex formation. *Organometallics* **1986**, *5*, 1559–1567.

2. Sundararaju, B.; Achard, M.; Bruneau, C. Transition metal catalyzed nucleophilic allylic substitution: activation of allylic alcohols via π-allylic species. *Chem. Soc. Rev.* **2012**, *41*, 4467–4483.

3. Kayaki, Y.; Koda, T.; Ikariya, T. Halide-Free Dehydrative Allylation Using Allylic Alcohols Promoted by a Palladium−Triphenyl Phosphite Catalyst. *J. Org. Chem.* **2004**, *69*, 2595–2597.

4. Trost, B.M.; van Vranken, D.L. Asymmetric transition metal-catalyzedallylicalkylations. *Chem. Rev.* **1996**, *96*, 395–422.

5. Kimura, M.; Tomizawa, T.; Horino, Y.; Tanaka, S.; Tamaru, Y. Et₃B-Pd-promoted allylation of benzaldehyde with allylic alcohols. *Tetrahedron Lett.* **2000**, *41*, 3627–3629.

6. Kimura, M.; Horino, Y.; Mukai, R.; Tanaka, S.; Tamaru, Y. Strikingly Simple Direct α-Allylation of Aldehydes with Allyl Alcohols: Remarkable Advance in the Tsuji−Trost Reaction. *J. Am. Chem. Soc.* **2001**, *123*, 10401–10402.

7. Kimura, M.; Futamata, M.; Shibata, K.; Tamaru, Y. Pd-Et₃B-catalyzed alkylation of amines with allylic alcohols. *Chem. Commun.* **2003**, 234–235.

8. Kimura, M.; Futamata, M.; Mukai, R.; Tamaru, Y. Pd-Catalyzed C₃-Selective Allylation of Indoles with Allyl Alcohols Promoted by Triethylborane. *J. Am. Chem. Soc.* **2005**, *127*, 4592–4593.

9. Trost, B.M.; Quancard, J. Palladium-Catalyzed Enantioselective C-3 Allylation of 3-Substituted-1*H*-Indoles Using Trialkylboranes. *J. Am. Chem. Soc.* **2006**, *128*, 6314–6315.

10. Itoh, K.; Hamaguchi, N.; Miura, M.; Nomura, M. Palladium-catalysed reaction of aryl-substituted allylic alcohols with zinc enolates of β-dicarbonyl compounds in the presence of titanium(IV) isopropoxide. *J. Chem. Soc. Perkin Trans.* **1992**, *1*, 2833–2835.

11. Satoh, T.; Ikeda, M.; Miura, M.; Nomura, M. Palladium-Catalyzed Etherification of Allyl Alcohols Using Phenols in the Presence of Titanium(IV) Isopropoxide. *J. Org. Chem.* **1997**, *62*, 4877–4879.

12. Yang, S.C.; Hung, C.W. Palladium-Catalyzed Amination of Allylic Alcohols Using Anilines. *J. Org. Chem.* **1999**, *64*, 5000–5001.

13. Yang, S.C.; Tsai, Y.C.; Shue, Y.J. Direct Platinum-Catalyzed Allylation of Anilines Using Allylic Alcohols. *Organometallics* **2001**, *20*, 5326–5330.

14. Shue, Y.J.; Yang, S.C.; Lai, H.C. Direct palladium(0)-catalyzed amination of allylic alcohols with aminonaphthalenes. *Tetrahedron Lett.* **2003**, *44*, 1481–1485.

15. Ozawa, F.; Okamoto, H.; Kawagishi, S.; Yamamoto, S.; Minami, T.; Yoshifuji, M. (π-Allyl)palladium Complexes Bearing Diphosphinidenecyclobutene Ligands (DPCB): Highly Active Catalysts for Direct Conversion of Allylic Alcohols. *J. Am. Chem. Soc.* **2002**, *124*, 10968–10969.

16. Banerjee, D.; Jagadeesh, R.V.; Junge, K.; Junge, H.; Beller, M. An Efficient and Convenient Palladium Catalyst System for the Synthesis of Amines from Allylic Alcohols. *ChemSusChem* **2012**, *5*, 2039–2044.

17. Banerjee, D.; Jagadeesh, R.V.; Junge, K.; Junge, H.; Beller, M. Efficient and Convenient Palladium-CatalyzedAmination of Allylic Alcohols with N-Heterocycles. *Angew. Chem. Int. Ed.* **2012**, *51*, 11556–11560.

18. Hsu, Yi-C.; Gan, K.H.; Yang, S.C. Palladium-catalyzed allylation of acidic and less nucleophilic anilines using allylic alcohols directly. *Chem. Pharm. Bull.* **2005**, *53*, 1266–1269.

19. Ghosh, R.; Sarkar, A. Palladium-Catalyzed Amination of Allyl Alcohols. *J. Org. Chem.* **2011**, *76*, 8508–8512.

20. Ozawa, F.; Ishiyama, T.; Yamamoto, S.; Kawagishi, S.; Murakami, H. Catalytic C–O Bond Cleavage of Allylic Alcohols Using Diphosphinidenecyclobutene-Coordinated Palladium Complexes: A Mechanistic Study. *Organometallics* **2004**, *23*, 1698–1707.

21. Piechaczyk, O.; Thoumazet, C.; Jean, Y.; le Floch, P. DFT Study on the Palladium-Catalyzed Allylation of Primary Amines by Allylic Alcohol. *J. Am. Chem. Soc.* **2006**, *128*, 14306–14317.

22. Kinoshita, H.; Shinokubo, H.; Oshima, K. Water Enables Direct Use of Allyl Alcohol for Tsuji-Trost Reaction without Activators. *Org. Lett.* **2004**, *6*, 4085–4088.

23. Huo, X.; Yang, G.; Liu, D.; Liu, Y.; Gridnev, I.D.; Zhang, W. Palladium-Catalyzed Allylic Alkylation of Simple Ketones with Allylic Alcohols and Its Mechanistic Study. *Angew. Chem. Int. Ed.* **2014**, *53*, 6776–6780.

24. Gumrukcu, Y.; de Bruin, B.; Reek, J.N.H. Hydrogen-Bond-Assisted Activation of Allylic Alcohols for Palladium-Catalyzed Coupling Reactions. *ChemSusChem* **2014**, *7*, 890–896.

25. The bi-phenyl backbone of the phosphoramidite ligand **L** in the palladium homocomplex **1** has a prevalent *R* configuration that we determined by CD and VT-NMR spectroscopy.

26. Muzart, J. Procedures for and Possible Mechanisms of Pd-Catalyzed Allylations of Primary and Secondary Amines with Allylic Alcohols. *Eur. J. Org. Chem.* **2007**, *19*, 3077–3089.

27. Comas-Vives, A.; Stirling, A.; Lledos, A.; Ujaque, G. The Wacker Process: Inner- or Outer-Sphere Nucleophilic Addition? New Insights from *Ab Initio* Molecular Dynamics. *Chem. Eur. J.* **2010**, *16*, 8738–8747.

28. Aquino, A.J.A.; Tunega, D.; Haberhauer, G.; Gerzabek, M.H.; Lischka, H. Solvent Effects on Hydrogen Bonds. A Theoretical Study. *J. Phys. Chem. A* **2002**, *106*, 1862–1871.

29. Sawadjoon, S.; Sjöberg, P.J. R.; Orthaber, A.; Matsson, O.; Samec, J.S.M. Mechanistic Insights into the Pd-Catalyzed Direct Amination of Allyl Alcohols: Evidence for an Outer-Sphere Mechanism Involving a Palladium Hydride Intermediate. *Chem. Eur. J.* **2014**, *20*, 1520–1524.

30. Gumrukcu, Y.; de Bruin, B.; Reek, J.N.H. Dehydrative cross-coupling reactions of allylic alcohols with olefins. *Chem. Eur. J.* **2014**, *20*, 10905–10909.

31. Ahlrichs, R. *Turbomole Version 6.4*; University of Karlsruhe: Germany, 2012.

32. Baker, I. An algorithm for the location of transition states. *J. Comput. Chem.* **1986**, *7*, 385–395.

33. Budzelaar, P.H.M. Geometry optimization using generalized, chemically meaningful constraints. *J. Comput. Chem.* **2007**, *28*, 2226–2236.

34. Becke, A.D. Density-functional exchange-energy approximation with correct asymptotic behavior. *Phys. Rev. A* **1988**, *38*, 3098–3100.

35. Perdew, J.P. Density-functional approximation for the correlation energy of the inhomogeneous electron gas. *Phys. Rev. B* **1986**, *33*, 8822.

36. Sierka, M.; Hogekamp, A.; Ahlrichs, R. Fast evaluation of the Coulomb potential for electron densities using multipole accelerated resolution of identity approximation. *J. Chem. Phys.* **2003**, *118*, 9136–9148.

37. Schaefer, A.; Horn, H.; Ahlrichs, R. Fully optimized contracted Gaussian basis sets for atoms Li to Kr. *J. Chem. Phys.* **1992**, *97*, 2571–2577.

38. Grimme, S.; Antony, J.; Ehrlich, S.; Krieg, H. A consistent and accurate *ab initio* parametrization of density functional dispersion correction (DFT-D) for the 94 elements H-Pu. *J. Chem. Phys.* **2010**, *132*, 154104–154119.

39. Dzik, W.I.; Xu, X.; Zhang, X.P.; Reek, J.N. H.; de Bruin, B. 'Carbene Radicals' in Co^{II}(por)-Catalyzed Olefin Cyclopropanation. *J. Am. Chem. Soc.* **2010**, *132*, 10891–10902.

40. Wertz, D.H. Relationship between the gas-phase entropies of molecules and their entropies of solvation in water and 1-octanol. *J. Am. Chem. Soc.* **1980**, *102*, 5316–5322.

41. Schneider, N.; Finger, M.; Haferkemper, C.; Bellemin-Laponnaz, S.; Hofmann, P.; Gade, L. Metal Silylenes Generated by Double Silicon-Hydrogen Activation: Key Intermediates in the Rhodium-Catalyzed Hydrosilylation of Ketones. *Angew. Chem. Int. Ed.* **2009**, *48*, 1609–1613.

42. Klamt, A.; Schüürmann, G. COSMO: A new approach to dielectric screening in solvents with explicit expressions for the screening energy and its gradient. *J. Chem. Soc. Perkin Trans.* **1993**, *2*, 799–805.

Ni-CeO$_2$/C Catalysts with Enhanced OSC for the WGS Reaction

Laura Pastor-Pérez, Tomás Ramírez Reina, Svetlana Ivanova,
Miguel Ángel Centeno, José Antonio Odriozola and
Antonio Sepúlveda-Escribano

Abstract: In this work, the WGS performance of a conventional Ni/CeO$_2$ bulk catalyst is compared to that of a carbon-supported Ni-CeO$_2$ catalyst. The carbon-supported sample resulted to be much more active than the bulk one. The higher activity of the Ni-CeO$_2$/C catalyst is associated to its oxygen storage capacity, a parameter that strongly influences the WGS behavior. The stability of the carbon-supported catalyst under realistic operation conditions is also a subject of this paper. In summary, our study represents an approach towards a new generation of Ni-ceria based catalyst for the pure hydrogen production via WGS. The dispersion of ceria nanoparticles on an activated carbon support drives to improved catalytic skills with a considerable reduction of the amount of ceria in the catalyst formulation.

Reprinted from *Catalysts*. Cite as: Pastor-Pérez, L.; Reina, T.R.; Ivanova, S.; Centeno, M.Á.; Odriozola, J.A.; Sepúlveda-Escribano, A. Ni-CeO$_2$/C Catalysts with Enhanced OSC for the WGS Reaction. *Catalysts* **2015**, *5*, 298–309.

1. Introduction

Nowadays, synthesis gas (a mixture of hydrogen and carbon monoxide) is generally produced by steam reforming of hydrocarbons, especially if the principal objective is the generation of gas streams with high H$_2$/CO ratios. For clean energy production, pure hydrogen is required as a feed gas for electricity generation in low temperature fuel cells [1]. In this sense, some processes have been considered in order to minimize the reformate CO concentration for PEM fuel cell applications as for example water-gas shift (WGS). Conventionally, in an integrated fuel processor, the WGS reactor is the first clean-up unit, and it is the reaction that removes the most important amount of CO in the H$_2$ stream [2].

Due to its high oxygen storage capacity and reducibility (via the Ce^{4+}↔Ce^{3+} redox process), cerium oxide is widely used as catalyst support and promoter in the water-gas shift reaction [3]. The role of the ceria surface is determinant for this reaction, both in terms of reducibility but also in terms of extension (large exposed surface areas are needed). Furthermore ceria reducibility is favored by the presence of metals [4,5].

One of the key points that convert ceria in a highly suitable support for the shift reaction relies on its ability to provide oxygen to the process. Indeed the WGS is a redox reaction, thus the oxygen mobility of the catalyst is regarded as a main issue in the catalytic design [6]. The later makes obligatory an accurate study of the redox features exhibited by the WGS systems, namely, the oxygen storage complete capacity (OSCC) and oxygen storage capacity (OSC). Recently, a great increase in both the OSC and the OSCC has been proposed as one of the main reasons to explain the activity promotion when ceria is dispersed on a high surface alumina [7]. In this way, promoting redox properties of the catalysts, namely surface and bulk oxygen mobility should benefit the WGS performance. Therefore, oxygen supply from the support is a highly desired feature for an effective WGS catalyst. Dispersing ceria over a high surface support such as carbon provides higher surface/bulk ratios improving oxygen mobility. In addition, the limited supply and extensive applications of CeO_2 makes desirable to optimize its use [8].

On the other hand, ceria-supported noble metals are mostly used due to the high activity in the WGS reaction. However, their elevated prices motivate the search for cheaper and more abundant metals which can be useful in the shift reaction. Ni-based catalysts are widely employed in many industrial processes such as reforming of alcohols, hydrogenation reactions and hydrocracking or oxidation processes [9]. Previous studies have addressed the behavior of NiO/CeO_2 or Ni/CeO_2 systems as catalysts for the title reaction. Despite good activities were reported, Ni containing catalysts suffer for deactivation mainly due to the well-known metallic particles sintering process together with the accumulation of carbon deposits on the catalysts surface [10,11].

Another point to consider is the adaptation of the new catalysts formulation to the budding market demands. The emerging fuel cell technology for portable electronic devices is changing the requisites for the WGS catalysts. Apart from developing a solid with the properties mentioned above, an industrial catalyst for WGS reaction requires in addition high long term stability and good resistance toward start/stop situations [12].

In a previous work [13] Ni_xCeO_2/C catalysts with different CeO_2 contents ($x = 10, 20, 30, 40\%$wt.) were tested in the WGS reaction. Detailed characterization was performed, deepening in both physical and chemical characteristics of these materials. The results showed that there was a clear effect of the ceria loading on the catalytic activity. It could be seen that the Ni-CeO_2-carbon system yielded catalysts which showed better performance than bulk Ni/CeO_2.

In this scenario, and taking into account our previous results [13] the objective of this work is to correlate the OSC of these catalysts with their high activity. A study of the enhancement of the oxygen storage capacity when ceria is dispersed on carbon is carried out. For this study, a sample with a 20 wt.% CeO_2 loading was chosen due to the interesting results obtained before. Additionally, and differently to our previous study, water-gas shift reactions employing more demanding WGS conditions were carried out. Further, stability tests and start/stop cycles were developed in order to evaluate the viability of this catalyst in real applications.

2. Results and Discussion

2.1. Catalysts Composition and Textural Properties

The chemical composition and the main textural properties of the prepared catalysts are presented in Table 1. The enhanced textural properties of the carbon materials are evidenced, all of them exhibiting larger surface area compared to the bulk solids. As previously reported, carbon is playing a role of textural promoter in these type of catalysts [13]. Data in Table 1 show a decrease of the BET surface area for ceria-carbon based samples, attributed to the blockage of porosity by ceria crystallites and/or an effect of mass increment and the much lower porosity of ceria as compared with carbon. The addition of nickel also produces a decrease in the BET surface area of the catalysts, which is also attributed to presence of high amounts of Ni (15 wt.%).

Regarding the elemental analysis, the actual Ni and CeO_2 content of the catalysts were determined by ICP (Inductively Coupled Plasma) measurements, and they are reported in Table 1. The obtained values are very close to the targeted ones, this validating the preparation method used.

Table 1. Catalysts composition and textural properties of the prepared samples.

Sample	$S_{BET}(m^2/g)$	$V_{micro}(cm^3/g)$	$V_{meso}(cm^3/g)$	Ni (wt.%) *	CeO_2 (wt.%)*
C	1487	0.52	0.62	–	–
$20CeO_2/C$	1083	0.37	0.47	–	22.2
CeO_2	101	0.04	0.07	–	–
$Ni20CeO_2/C$	807	0.28	0.34	12.7	21.6
Ni/CeO_2	70	0.03	0.04	14.1	–

* Determined by ICP analysis.

2.2. OSCC and OSC

As mentioned above, redox properties constitute one of the main issues to be optimized for the successful design of an efficient WGS catalyst. An accurate understanding of such skills is obtained by the oxygen storage capacity measurements. In particular, OSCC measurements provide information about the maximum reducibility of the samples, while OSC informs about the most reactive and most available oxygen atoms that are involved in the redox process [14]. This study was carried out at two different temperatures (150 °C and 250 °C) both of them low enough to avoid any possible carbon combustion due to the oxygen pulses. Actually our TGA data (not shown for sake of briefness) point that carbon combustion in our catalysts under air atmosphere starts at 320 °C. The selected temperatures are relevant points in the catalytic study. Therefore, the analysis of the redox behavior of both catalysts at these temperatures provides valuable information to understand their WGS performance. Figure 1 shows the OSCC results of the prepared solids. As expected, the OSCC increases with the temperature for both supports and catalysts, indicating higher degree of ceria reduction at 250 °C. In the case of the supports (Figure 1A) it must be underlined the superior OSCC exhibited for the CeO_2/C system at the studied temperatures. These data indicate that the dispersion of ceria nanoparticles on a high surface carrier as activated carbon results in a notorious improvement of its reducibility. In other words, enhanced redox properties are achieved for the CeO_2/C solid in comparison to the bulk CeO_2. Even though carbon is not playing a chemical role in the oxygen storage process, it acts as an ideal media to support ceria nanoparticles providing high surface area (as shown in Table 1). Similar results were obtained when CeO_2 is dispersed over gamma alumina [15]. In this sense, dispersing ceria over a high-surface carrier opens up the possibility of having higher surface/bulk ratios thus improving the oxygen mobility no matter the used support. However it is worth noting that our ceria/carbon based samples present higher OSCC values than those observed for ceria/alumina systems measured under the same conditions [15]. The latter indicates the suitability of carbon as a textural promoter allowing better dispersion of ceria nanoparticles.

Nickel addition (Figure 1B) enhances the OSCC for both, bulk CeO_2 and CeO_2/C support. This result agrees with H_2-TPR data recently published for these solids evidencing a remarkable increase of the support reducibility due to the strong Ni-CeO_2 interaction [13]. Again, as observed for the bare supports, much higher OSCC values are obtained when ceria and nickel phases are dispersed on the activated carbon.

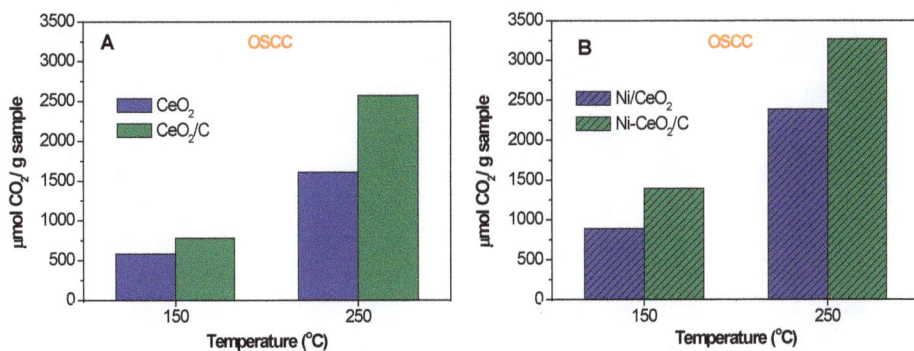

Figure 1. Oxygen storage complete capacity (OSCC) of the prepared solids. (**A**) supports (**B**) Ni based catalysts.

The OSC study is summarized in Table 2. At 150 °C the supports presented low oxygen mobility; however, the OSC improves when the temperature is raised. In a similar way to the OSCC, the OSC of the CeO_2/C sample is always higher than that of the bulk CeO_2 support. As stated above, Ni influences CeO_2 reducibility due to the intimate Ni-CeO_2 contact [16]. This effect is highlighted at low temperature (150 °C), where the metallic particles broadly boost the OSC of the parent supports. In any case, for all the studied temperatures, the Ni-CeO_2/C catalyst shows the best oxygen mobility. The promoted redox features of this catalyst are related with two inherent aspects of its composition: (i) the dispersion of ceria nanoparticles on the activated carbon, which increases the surface/bulk ratio, thus potentiating the oxygen mobility in the ceria lattice and, (ii) the facilitated CeO_2 reduction due to the presence of Ni, arising from a strong metal-support interaction.

Table 2. OSC in μmol CO_2/g sample for the studied samples at different temperatures.

Sample	$OSC_{150\,°C}$	$OSC_{250\,°C}$
CeO_2	10	180
CeO_2/C	25	255
Ni/CeO_2	410	726
Ni-CeO_2/C	620	850

2.3. WGS Behavior

The catalysts were evaluated under a surrogate post reforming stream (7 mol% CO, 30 mol% H_2O, 50 mol% H_2, and 9 mol% CO_2 balanced with He) with the aim to test their viability for a real application, for instance the use of a WGS reactor in an integrated fuel processor for pure hydrogen production. The catalytic activity of the prepared Ni-ceria based catalysts in the shift reaction is presented in Figure 2. It should be mentioned that the supports were also tested under the same conditions, showing an almost nil activity in the studied temperature range. This underlines the importance of the metallic phase to achieve good performance in the WGS when ceria-based solids are considered. As intended from the plot, the samples are not very effective in the low temperature range (<200 °C). At these conditions the solids present limited oxygen mobility (low OSC values) that may contribute to the observed low CO conversion. Nevertheless, the catalytic activity rapidly increases with the temperature. The Ni-CeO_2/C sample evidences superior WGS activity compared to the Ni/CeO_2 one in the whole temperature range. Taking into account that the amount of metallic phase (Ni) is comparable in both materials, as obtained by ICP, the higher activity in WGS correlates with the boosted OSC and OSCC demonstrated for this solid. In addition to the excellent redox skills, the Ni-Ce/C sample presents greater metallic dispersion and superior specific surface area contributing to activity enhancement. This catalyst quadruplicates the activity of the Ni/CeO_2 solid at 240 °C, and reaches equilibrium conversion at 260 °C. Compared to others Ni-ceria catalysts previously reported, our carbon-supported sample performs better even under harder WGS conditions [17,18]. This is a very promising result since this material seems to be as efficient as some well-known noble metal based catalysts for this reaction [19]. However, contrary to the noble metal based solids, Ni-ceria solids also catalyze the methanation reaction. As can be seen in Figure 2, both catalysts exceed the equilibrium conversion at high temperatures. The additional CO consumption is due to the methanation process. The high concentrations of CO used in this stream favor the methanation reaction [20]. The good point is that the methanation reaction for both catalysts starts at temperatures higher than 260 °C and thus, this does not affect our WGS operating window.

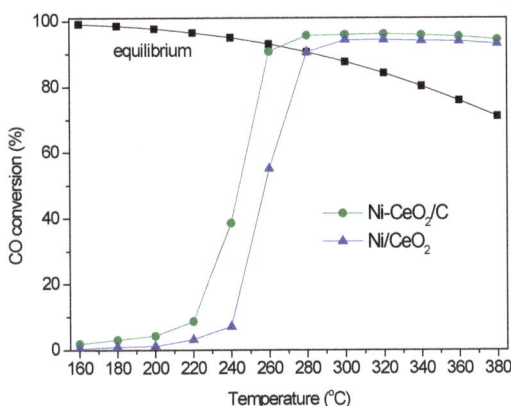

Figure 2. CO conversion *vs.* reaction temperature for catalysts reduced at 350 °C. Gas mixture: 7 mol% CO, 30 mol% H_2O, 50 mol% H_2, and 9 mol% CO_2 balanced in He.

This result is very interesting from the point of view of catalyst design since, according to our data, a much better catalytic activity is obtained when CeO_2 is dispersed on activated carbon. This approach permits to develop very active WGS catalysts reducing significantly the amount of ceria on the catalyst formulation. This outstanding behavior is linked to a strong enhancement of the redox properties when ceria nanoparticles are dispersed on a high surface carrier.

From the industrial perspective, the catalytic stability is as relevant as the activity. In this sense, a complete stability study of the selected sample was carried out. The stability test was developed at 250 °C, a reaction temperature at which the methanation reaction was not observed and good WGS activity was achieved. Figure 3 shows the long term stability test of the $Ni-CeO_2/C$ catalyst. For the aim of comparison the same test was carried out with the Ni/CeO_2 solid; however, this catalyst exhibited fast deactivation and very poor CO conversion. Therefore it will not be longer considered. Regarding our Ni-CeO$_2$/C material, at the first stages of the reaction the system exhibits a good catalytic behavior. However, the CO conversion notably drops after 8 h of continuous reaction and remains stable after more than 120 h. Afterwards, several start/stop cycles were introduced simulating likely situations in real application. It was observed that the catalyst recovers the conversion after the start/stop operation. Apparently, once the initial deactivation occurs (at the early reaction stages) the start-up/shutdown actions do not influence the activity.

For a deeper understanding of this result, a second stability test was carried out. At this time, the catalyst was directly submitted to a series of start/stop cycles to check its actual tolerance under these conditions. Indeed, this type of stability experiment is considered as the most exigent test for a WGS catalyst, since during the stop stages the system is cooled down at room temperature with the reaction flow passing through the catalytic bed [21]. The later involves liquid water may condense on the pores of the catalyst damaging the system. Figure 4 reveals that the Ni-CeO$_2$/C catalyst withstands four start/stop cycles within 6 h of reaction while preserving the initial activity. However, the activity drops after the fourth cycle, and it reaches practically the same value attributed to the steady state in the long term stability test. This result is very interesting from the catalyst deactivation point of view. Our data reveal that in a long term stability test high conversions are preserved during at least 10 h of reaction, but start-up/shutdowns cycles strongly affect the catalysts stability, in such a way that the activity drop takes place earlier (after 6 h of reaction).

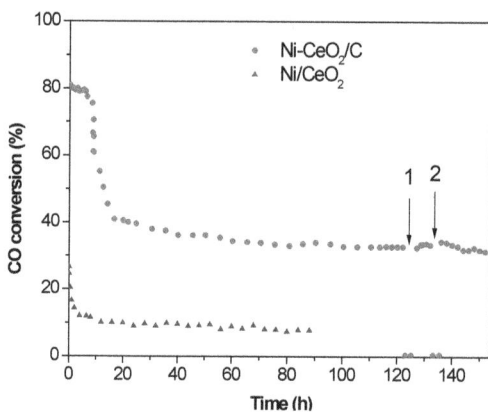

Figure 3. Long term stability test of the prepared catalysts at 250 °C. 1 and 2 start/stop cycles. Gas mixture: 7 mol% CO, 30 mol% H$_2$O, 50 mol% H$_2$, and 9 mol% CO$_2$ balanced in He.

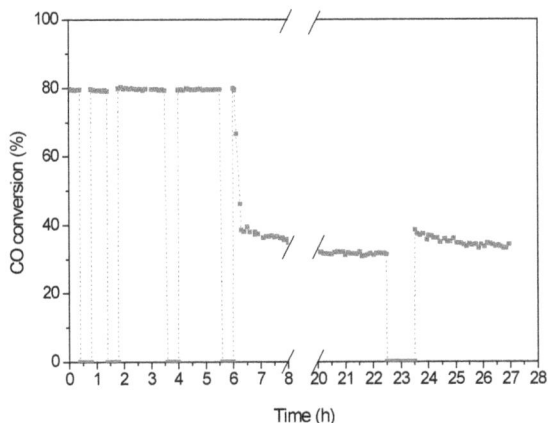

Figure 4. Stability test: start/stop cycles of the Ni-CeO$_2$/C catalyst at 250 °C. Gas mixture: 7 mol% CO, 30 mol% H$_2$O, 50 mol% H$_2$, and 9 mol% CO$_2$ balanced in He.

In both cases, long term and start-up/shut-down, the deactivation is related to the well-known phenomenon of Ni particles sintering. Actually, as previously proposed, the presence of water potentiates Ni particles agglomeration especially at high temperatures [22,23]. TEM images presented in Figure 5, corresponding to the Ni-CeO$_2$/C catalyst, evidences the Ni sintering process.

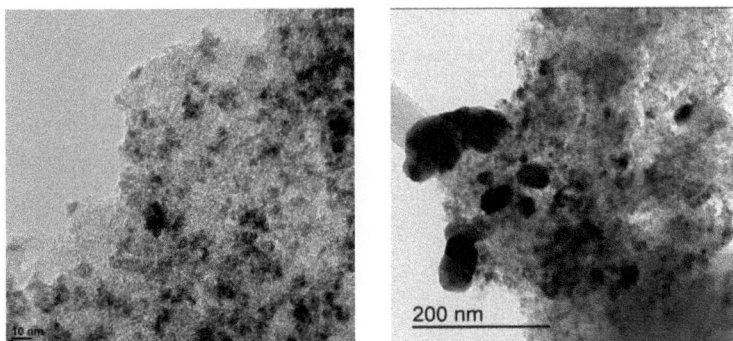

Figure 5. TEM micrographs of the Ni-CeO$_2$/C catalyst before (scale 10 nm) (**a**) and after (scale 200 nm) (**b**) the stability reaction test at 250 °C.

Our stability study indicates that water favors Ni agglomeration. This reduces the catalytic surface area and results in a decreased activity [24]. According to our data, the sintering process occurs earlier when liquid water (during the start/stop cycles) enters in contact with the catalyst.

3. Experimental Section

3.1. Catalysts Preparation

The procedure used for the catalyst synthesis was the same than in our previous work [13]. The support was an industrial activated carbon (RGC30, from Westvaco, New York, NY, USA). This carbon was grinded and meshed (300–500 μm). The corresponding amount of $Ce(NO_3)_3 \cdot 6H_2O$ (99.99%, Sigma–Aldrich, St. Louis, MO, USA) to obtain 20 wt.% of CeO_2 was dissolved in acetone. Dried carbon was added to the solution, in a proportion of 10 mL/g of support, with stirring. After 12 h, the excess of solvent was slowly removed under vacuum at 40 °C and the solid was then dried in the oven overnight. Finally, the dried solid was heat treated during 4 h at 350 °C under flowing He (50 mL/min), with a heating rate of 1 °C/min, in order to slowly decompose the cerium nitrate to form CeO_2, trying to avoid the modification of the carbon surface by the evolved nitrogen oxides [25].

Nickel addition to the CeO_2/C solid was carried out using the proper amount of $Ni(NO_3)_2 \cdot 6H_2O$ (99.9%, Sigma-Aldrich, St. Louis, MO, USA) in acetone to obtain 15 wt.% Ni, using 10 mL of solution per gram of solid. After stirring for 12 h, the solvent was removed under vacuum at 40 °C. Finally, the solid was treated at 350 °C for 4 h under flowing He (50 mL/min). The catalyst prepared was labeled Ni-CeO_2/C. For the sake of comparison, a Ni/CeO_2 catalyst was also synthesized. The ceria support was prepared by homogeneous precipitation from an aqueous solution of $Ce(NO_3)_3 \cdot 6H_2O$ (99.99%, Sigma-Aldrich, St. Louis, MO, USA) containing an excess of urea. The solution was heated at 80 °C and kept at this temperature, with slow stirring, during 12 h. The solid formed was filtered and calcined at 350 °C for 4 h. The CeO_2 support prepared in this way was impregnated with the Ni precursor as described for the carbon supported catalysts.

3.2. Textural Properties

The textural properties of the supports were characterized by nitrogen adsorption measurements at −196 °C. Gas adsorption experiments were performed in home-made fully automated manometric equipment. Prior to the adsorption experiments, samples were out-gassed under vacuum (10^{-4} Pa) at 250 °C for 4 h. The specific surface area was estimated after application of the BET equation.

3.3. Elemental Analysis

The actual metal loading of the different catalysts was determined by ICP-OES in a Perkin Elmer device (Optimal 3000). To this end, the metal was extracted from the catalysts by digestion in HNO_3/H_2O_2 (4:1) for 30 min, in a microwave oven at 200 °C.

3.4. OSCC and OSC Experiments

For the Oxygen Storage Complete Capacity (OSCC) 100 mg of catalyst was loaded into a U-shaped quartz reactor and the temperature was raised in a He flow (50 mL/min) until 350 °C. Then, the system was cooled and set to the desired temperature (150, 250 and 350 °C). For each temperature 10 O_2 pulses of 1 mL were injected every 2 min. After that, the sample was submitted to 10 CO pulses of 1 mL each (every 2 min). The OSCC was calculated from the sum of the CO_2 formed after each CO pulse. The sample was then degassed during 10 min in a He flow to perform the OSC measurements. Four alternating series of pulses ($CO-O_2-CO-O_2-CO-O_2-CO-O_2$) were applied for each temperature. The OSC was determined by the average amount of CO_2 per pulse formed after the first CO pulse of the alternated ones. This method is based on the one proposed by Duprez et al. [26]. The gas composition at the exit of the reactor was analyzed by a mass spectrometer PFEIFFER Vacuum PrismaPlus controlled by Quadera® software (version 4.0, INFICON, LI-9496 Balzers, Furstentum Liechtenstein).

Several assumptions were contemplated for the OSC calculations. Concretely, it was considered that (i) only oxygen atoms bonded to the cerium participate in the oxygen storage process; (ii) the surface is assumed homogeneous and (iii) only one of the four oxygen atoms is involved in the storage ($2CeO_2 \rightarrow Ce_2O_3 + "O"$)

3.5. TEM

TEM images were taken with a JEOL electron microscope (model JEM-2010) working at 200 kV. It was equipped with an INCA Energy TEM 100 analytical system and a SIS MegaView II camera. Copper grids with a holey-carbon film support were used for the microscopy measurements. The samples were suspended in ethanol and placed on the grids for the analysis.

3.6. WGS Catalytic Tests

The catalytic behavior of the prepared samples in the low temperature water-gas shift reaction was evaluated in a fixed bed flow reactor under atmospheric pressure in the range of temperatures from 160 °C to 380 °C. For the stability tests the temperature employed was 250 °C. Trying to simulate a more close to real outgas mixture from a reformer, experiments with a feed gas composition of 7 mol% CO, 30 mol% H_2O, 50 mol% H_2, and 9 mol% CO_2 balanced to 100 mL/min with helium was tested. Activity tests were performed using 0.150 g of catalyst diluted with SiC, to avoid thermal effects. The corresponding contact time was 0.09 g·s/mL. Prior to reaction, the catalysts were reduced under flowing H_2 (50 mL/min) during 2 h at 350 °C. The composition of the gas stream exiting the reactor was determined by mass spectrometry (Pfeiffer, OmniStar GSD 301). The stabilization time for each temperature was 1 h and the CO conversion percentage was calculated by this equation:

$$CO\,conversion\,(\%) = 100 - (x_{CO}/x_{COinitial}) \cdot 100 \tag{1}$$

where x_{CO} is the molar concentration of CO in the outlet of the reactor and $x_{COinitial}$ is the CO concentration in the initial gas mixture. The carbon balance was checked taking into account all the carbon containing products.

4. Conclusions

The impact of the oxygen storage capacity on the WGS behaviour of a Ni/CeO_2 based catalysts is underlined in this paper. The dispersion of ceria nanoparticles on a high surface activated carbon drives to a strong enhancement of the catalyst's oxygen mobility. The excellent redox skills together with the better metallic dispersion of the Ni-CeO_2/C catalyst due to the carbon textural promotion, result in a remarkable improvement of the WGS activity.

The stability of the Ni-CeO_2/C catalysts under continuous operation, as well as interrupting cycles, was tested. Although high activity was found under these conditions at early stages of the reaction, a notable activity loss was observed and attributed to the sintering of Ni particles. Despite the issue of the stability should be improved, the Ni-CeO_2/C catalyst could be considered as an interesting alternative for the low temperature WGS reaction and merits further investigations.

As final remark, it should not be disregarded that the high WGS activity observed for the developed sample was achieved with a relatively low amount of cerium oxide on its composition. This fact indicates that our preparation procedure could constitute an alternative approach towards a new generation of Ni/CeO_2 based catalysts.

Acknowledgments: Financial support of Ministerio de Ciencia e Innovacion of Spain (Project MAT2010-21147) is gratefully acknowledged. L.P.P. acknowledges her grant BES-2011-0406508. The Spanish Ministerio de Economía y Competitividad (MINECO) is also acknowledged for the funding linked to the projects ENE2012-374301-C03-01 and ENE2013-47880-C3-2-R.

Author Contributions: Laura Pastor Pérez was the main author of this paper. Tomás Ramírez Reina carried out the OSC experiments and he provided assistance in the manuscript writing. Svetlana Ivanova, Miguel Ángel Centeno, José Antonio Odriozola and Antonio Sepúlveda Escribano took part in the article revision and they contributed in the response to reviewers' comments

Conflicts of Interest: The authors declare no conflict of interest.

References

1. Ladebeck, J.R.; Wagner, J.P. Catalyst development for water-gas shift. In *Handbook of Fuel Cells-Fundamentals, Technology and Applications*; Vielstich, W., Lamm, A., Gasteiger, H.A., Eds.; Wiley, New York, NY, USA, 2003; Volume 3, Chapter 16, Part 2; pp. 190–201.

2. Ratnasamy, C.; Wagner, J.P. Water-gas shift catalysis. *Catal. Rev. Sci. Eng.* **2009**, *51*, 325–440.

3. Deng, W.; De Jesus, J.; Saltsburg, H.; Flytzani Stephanopoulos, M. Low-content gold-ceria catalysts for the water-gas shift and preferential CO oxidation reactions. *Appl. Catal. A* **2005**, *291*, 126–135.

4. Summers, J.C.; Ausen, S.A. Interaction of cerium oxide with noble metals. *J. Catal.* **1979**, *58*, 131–143.

5. Andreeva, D.; Idakiev, V.; Tabakova, T.; Ilieva, L.; Falaras, P.; Bourlinos, A.; Travlos, A. Low-temperature water-gas shift reaction over Au/CeO_2 catalysts. *Catal. Today* **2002**, *72*, 51–57.

6. Colussi, S.; Katta, L.; Amoroso, F.; Farrauto, R.J.; Trovarelli, A. Ceria-based palladium zinc catalysts as promising materials for water gas shift reaction. *Catal. Commun.* **2014**, *47*, 63–66.

7. Reina, T.R.; Ivanova, S.; Centeno, M.A.; Odriozola, J.A. Low-temperature CO oxidation on multicomponent gold based catalysts. *Front. Chem.* **2013**, *12*, 1–9.

8. Serrano-Ruiz, J.C.; Sepúlveda-Escribano, A.; Rodríguez-Reinoso, F.; Duprez, D. Pt-Sn catalysts supported on highly dispersed ceria on carbon. *J. Mol. Catal. A* **2007**, *268*, 227–234.

9. Barrio, L.; Kubacka, A.; Zhou, G.; Estrella, M.; Martínez-Arias, A.; Hanson, J.C. Unusual physical and chemical properties of Ni in $Ce_{1-x}Ni_xO_{2-y}$ oxides: structural characterization and catalytic activity for the water-gas shift reaction. *J. Phys. Chem. C* **2010**, *114*, 12689–12697.

10. Jacobs, G.; Chenu, E.; Patterson, P.M.; Williams, L.; Sparks, D.; Thomas, G.; Davis, B.H. Water-gas shift: comparative screening of metal promoters for metal/ceria systems and role of the metal. *Appl. Catal. A* **2004**, *258*, 203–214.

11. Li, Y.; Fu, Q.; Flytzani-Stephanopoulos, M. Low-temperature water-gas shift reaction over Cu- and Ni-loaded cerium oxide catalysts. *Appl. Catal. B* **2000**, *27*, 179–191.

12. Ilinich, O.; Ruettinger, W.; Liu, X.; Farrauto, R. $Cu-Al_2O_3-CuAl_2O_4$ water-gas shift catalyst for hydrogen production in fuel cell applications: Mechanism of deactivation under start-stop operating conditions. *J. Catal.* **2007**, *247*, 112–118.

13. Pastor-Pérez, L.; Buitrago-Sierra, R.; Sepúlveda-Escribano, A. CeO_2-promoted Ni/activated carbon catalysts for the water-gas shift (WGS) reaction. *Int. J. Hydrogen Energy* **2014**, *39*, 17589–17599.

14. Kacimi, S.; Barbier, J.; Taha, R.; Duprez, D. Oxygen storage capacity of promoted Rh/CeC_2 catalysts. Exceptional behaviour of $RhCu/CeO_2$. *Catal. Lett.* **1993**, *22*, 343–350.

15. Reina, T.R.; Ivanova, S.; Delgado, J.J.; Ivanov, I.; Tabakova, T.; Idakiev, V.; Centeno, M.A.; Odriozola, J.A. Viabilty of $Au/CeO_2-ZnO/Al_2O_3$ catalysts for pure hydrogen production by the Water-Gas Shift reaction. *ChemCatChem* **2014**, *6*, 1401–1409.

16. Senanayake, S.D.; Rodriguez, J.A.; Stacchiola, D. Electronic metalsupport interactions and the production of hydrogen through the water-gas shift reaction and ethanol steam reforming: Fundamental studies with well-defined model catalysts. *J. Phys. Chem. C* **2012**, *116*, 9544–9549.

17. Lin, J.H.; Guliants, V.V. Hydrogen production through water-gas shift reaction over supported Cu, Ni, and Cu-Ni nanoparticles catalysts prepared from metal colloids. *ChemCatChem* **2012**, *4*, 1611–1121.

18. Lin, J.H.; Biswas, P.; Guliants, V.V.; Misture, S. Hydrogen production by water-gas shift reaction over bimetallic Cu-Ni catalysts supported on La-doped mesoporous ceria. *Appl. Catal. A* **2010**, *387*, 87–94.

19. Burch, R. Gold catalysts for pure hydrogen production in the water-gas shift reaction: Activity, structure and reaction mechanism. *Phys. Chem.* **2006**, *47*, 5483–5500.

20. Wang, T.; Porosoff, M.D.; Chen, J.G. Effects of oxide supports on the water-gas shift reaction over Pt-Ni bimetallic catalysts: Activity and methanation inhibition. *Catal. Today* **2014**, *15*, 61–69.

21. Liu, X.; Ruettinger, W.; Xu, X.; Farrauto, R. Deactivation of Pt/CeO_2 water-gas shift catalysts due to shutdown/startup modes for fuel cell applications. *Appl. Catal. B* **2005**, *56*, 69–75.

22. Senanayake, S.D.; Evans, J.; Agnoli, S.; Barrio, L.; Chen, T.L.; Hrbek, J.; Rodriguez, J.A. Water-gas shift and CO methanation reactions over $NiCeO_2(111)$ catalysts. *Top. Catal.* **2011**, *54*, 34–41.

23. Bunluesin, T.; Gorte, R.J.; Graham, G.W. Studies of the water-gas shift reaction on ceria-supported Pt, Pd, and Rh: implications for oxygen-storage properties. *Appl. Catal. B* **1998**, *15*, 107–114.

24. Gonzalez-De la Cruz, V.M.; Holgado, J.P.; Pereñíguez, R.; Caballero, A. Morphology changes induced by strong metal-support interaction on a Ni-ceria catalytic system. *J. Catal.* **2008**, *257*, 307–314.

25. Serrano-Ruiz, J.C.; Ramos-Fernandez, E.V.; Silvestre-Albero, J.; Sepúlveda-Escribano, A.; Rodríguez-Reinoso, F. Preparation and characterization of CeO_2 highly dispersed on activated carbon. *Mater. Res. Bull.* **2008**, *43*, 1850–1857.

26. Royer, S.; Duprez, D. Catalytic oxidation of carbon monoxide over transition metal oxides. *ChemCatChem* **2011**, *3*, 24–65.

Influence of the Synthesis Method for Pt Catalysts Supported on Highly Mesoporous Carbon Xerogel and Vulcan Carbon Black on the Electro-Oxidation of Methanol

Cinthia Alegre, María Elena Gálvez, Rafael Moliner and María Jesús Lázaro

Abstract: Platinum catalysts supported on carbon xerogel and carbon black (Vulcan) were synthesized with the aim of investigating the influence of the characteristics of the support on the electrochemical performance of the catalysts. Three synthesis methods were compared: an impregnation method with two different reducing agents, sodium borohydride and formic acid, and a microemulsion method, in order to study the effect of the synthesis method on the physico-chemical properties of the catalysts. X-ray diffraction and transmission electron microscopy were applied. Cyclic voltammetry and chronoamperometry were used for studying carbon monoxide and methanol oxidation. Catalysts supported on carbon xerogel presented higher catalytic activities towards CO and CH_3OH oxidation than catalysts supported on Vulcan. The higher mesoporosity of carbon xerogel was responsible for the favored diffusion of reagents towards catalytic centers.

Reprinted from *Catalysts*. Cite as: Alegre, C.; Gálvez, M.E.; Moliner, R.; Lázaro, M.J. Influence of the Synthesis Method for Pt Catalysts Supported on Highly Mesoporous Carbon Xerogel and Vulcan Carbon Black on the Electro-Oxidation of Methanol. *Catalysts* **2015**, *5*, 392–405.

1. Introduction

Carbon materials such as charcoals, activated carbons, carbon blacks and graphite have been widely used as support in the synthesis of catalysts with important applications in the chemical industry [1]. The properties of such supports, as well as their porous structure and surface chemistry, influence the dispersion and stability of the metal employed as active phase, determining the activity of the catalyst [2,3].

In the field of fuel cells, electrocatalysts are responsible for the oxidation of the fuel (hydrogen, methanol, *etc.*) at the anode, and the reduction of oxygen at the cathode, thus generating electrical power. In PEMFC the catalysts are usually based on Pt and supported on carbon black with high surface area. In the last few decades, growing literature has appeared concerning new carbon supports such as carbon nanotubes [3], carbon nanofibers [4], ordered mesoporous carbons and carbon xerogels and aerogels [2,5,6]. These materials could lead to the production of

295

more stable and highly active catalysts, with low platinum loadings (<0.1 mg$\cdot$cm^{-2}) and therefore low cost [7].

Direct methanol fuel cells (DMFCs) are very promising for portable devices [2], given the ease of use of a liquid fuel, which is easily storable and transportable. However the commercialization of DMFCs is still hindered by some technical challenges, mainly: (1) slow kinetics of both the oxygen reduction and the methanol oxidation and (2) methanol crossover [2]. For these reasons, improved catalysts are required. Research is mainly focusing on (i) the development of new catalyst syntheses taking into account alloy composition, nanometric dimensions and uniform distribution of the catalyst and (ii) the development of new carbon supports with high electric conductivity, high surface area and high mesoporosity in the pore size range of 20–40 nm [5].

Carbon gels are materials with interesting properties to be used as electrocatalyst supports in fuel cells (FC). Carbon xerogels in particular are attracting much attention for their unique and controllable properties (high surface area and mesoporosity with narrow pore size distribution as well as high purity (>99.5% C)) which, in turn, are attributed to the mechanism reaction, being similar to the sol-gel process [8].

In a previous work [9], Pt and PtRu catalysts supported on carbon xerogels (CXGs) were synthesized. Carbon supports presented high surface area and pore volume values, 577 m$^2\cdot$g^{-1} and 1.82 cm$^3\cdot$g^{-1} respectively, with pore size equal to 7.1 nm. Pt/CXG and PtRu/CXG catalysts presented higher catalytic activities than commercial ones, namely Pt/C and PtRu/C from E-TEK. The present work involves the synthesis of a carbon xerogel with improved properties (higher surface area, 660 m$^2\cdot$g^{-1} and larger pore size, 23 nm), to be used as Pt support. In order to compare the results obtained, Pt nanoparticles were also deposited on a commercial carbon black (Vulcan-XC-72R). Also, three different synthetic methods were assessed given that the catalyst preparation procedure has a strong influence on its final properties. It is necessary to optimize each synthesis method, taking into consideration the particular characteristics of the support, in order to obtain a properly dispersed active phase, with the most appropriate crystal size.

2. Results and Discussion

2.1. Textural and Chemical Characterization

2.1.1. Characterization of the Carbon Supports

Table 1 presents the textural properties of the carbon supports, determined from their corresponding nitrogen adsorption isotherms. Carbon supports were called "CXG" for carbon xerogels and "Vulcan", for the commercial carbon black. Significantly higher values of surface area (660 m$^2\cdot$g^{-1}) and pore volume (1.79 cm$^3\cdot$g^{-1}) were obtained for the carbon xerogel, in comparison to those determined for Vulcan carbon black (224 m$^2 \cdot$g^{-1}), pointing to a more developed porous structure in the case of the CXG.

Although the average pore size measured for the carbon xerogel, 23.3 nm, is higher than the value obtained for the carbon black, 11 nm, both carbon materials are predominantly mesoporous. BJH-mesopore volumes are 1.66 cm$^3 \cdot$g^{-1} (for the CXG) and 0.41 cm$^3 \cdot$g^{-1} (for Vulcan), much more relevant than their respective micropore contributions, which become almost negligible in the case of the carbon black. In carbon aerogels and xerogels, micropores are related to the stacking arrangement of the aromatic layers that compose the primary particles of their structure, with a particle size of 3–25 nm [10]. In their three-dimensional arrangement, mesopores appear as a consequence of the void space and interconnections among these primary particles [6]. As shown in Table 1, t-plot analysis of the adsorption isotherms yielded similar values for mesopore volume in the range of 2 to 25 nm as for wider mesopores and macropores, 25–300 nm, in the case of the carbon black. However, for the synthesized carbon xerogel, narrow mesopores (in the range of 2 to 25 nm) represent a relatively higher fraction of the total mesopore volume. Abundance of mesopores in the structure of these materials favors, in principle, their application as supports in the preparation of Pt-based electrocatalysts. Pores in the range of wide mesopores ensure an adequate deposition of the active phase, avoiding pore blockage, and facilitating both the access of the reactants to the metallic sites and the diffusion of the products out of the porous structure of the catalyst [5,11,12].

2.1.2. Catalyst Characterization

Table 2 shows the textural properties obtained from their corresponding nitrogen adsorption isotherms for the several catalysts prepared, using both the carbon black Vulcan and the carbon xerogel, and following the different synthetic routes for Pt deposition. Impregnation and microemulsion synthetic methods were employed to support Pt nanoparticles on the different carbon materials. Impregnation and reduction with formic acid (FAM) or sodium borohydride (SBM) were employed as general synthetic methods, whereas a microemulsion based method (ME) was introduced as a new synthetic route not previously used for carbon xerogels. In general, surface area decreases notably after Pt-loading, as a consequence of partial blockage of the porous structure of the carbon materials. In spite of this, the prepared electrocatalysts show areas in the 177–478 m$^2 \cdot$g^{-1} range, higher in the case of the ones prepared using the carbon xerogel in comparison to Vulcan, due to the more developed porous structure of the starting carbon material.

Table 1. Carbon support textural properties.

Sample	S_{BET} (m²·g⁻¹)	Total pore volume (0.7–180 nm) (cm³·g⁻¹)	Micropore volume (cm³·g⁻¹)			Mesopore volume (cm³·g⁻¹)			Average pore size (nm)
			V_{micro} (0–0.7 nm)	V_{micro} (0.7–2 nm)	V_{micro} (0–2 nm)	V_{meso} (2–25 nm)	V_{meso} (25–300 nm)	V_{BJH} (1.5–300 nm)	
Vulcan	224	0.47	0.01	0.03	0.04	0.24	0.17	0.41	11.0
CXG	660	1.79	0.06	0.07	0.13	0.97	0.69	1.66	23.3

Table 2. Catalysts textural properties.

Sample	S_{BET} (m⁴·g⁻¹)	Total pore volume (0.7–180 nm) (cm³·g⁻¹)	Micropore volume (cm³·g⁻¹)			Mesopore volume (cm³·g⁻¹)			Average pore size (nm)
			V_{micro} (0–0.7 nm)	V_{micro} (0.7–2 nm)	V_{micro} (0–2 nm)	V_{meso} (2–25 nm)	V_{meso} (25–300 nm)	V_{BJH} (1.5–300 nm)	
Pt/Vulcan-SBM	187	0.47	0.02	0.01	0.03	0.21	0.21	0.42	14.8
Pt/Vulcan-FAM	191	0.38	0.02	0.01	0.03	0.20	0.13	0.33	11.9
Pt/Vulcan-ME	177	0.35	0.02	0.01	0.03	0.17	0.15	0.32	10.3
Pt/CXG-SBM	475	1.34	0.07	0.05	0.12	0.45	0.77	1.22	24.5
Pt/CXG-FAM	478	1.08	0.09	0.05	0.14	0.34	0.59	0.93	23.5
Pt/CXG-ME	349	1.27	0.06	0.04	0.10	0.38	0.77	1.15	25.5

Micropore volume remains almost constant upon the introduction of the metallic component, with a slight reduction in the amount of wide micropores, which become partially blocked by the deposition of Pt particles of lower size. On the other hand, metal loading affects mostly the mesopore volume of the carbon supports, depending on the initial porous structure of each material and the synthesis method employed for the preparation of the catalyst. Mesopore volume decreases as a consequence of the introduction of the active phase, but this decrease is more noticeable in the case of the catalysts prepared using formic acid as a reducing agent. Following the borohydride synthesis route, Pt-deposition using carbon black as support results in only a slight blockage of narrow mesoporosity. The use of formic acid results in a more extensive pore blockage, either in the case of employing the Vulcan carbon black or the carbon xerogel as supports. A wider range of mesopore sizes, 30–60 nm, was affected by introduction of the metallic active phase when following the formic acid synthesis route, vis-á-vis the use of sodium borohydride as reducing agent. On the other hand, pore volume and surface area decrease dramatically in the case of the catalysts prepared through the microemulsion method. In this last case, micropore volume is totally blocked upon active phase loading. This can be due to the presence of some remains of the surfactant employed within the synthesis of this catalyst, which remain bonded to the carbon surface, as evidenced in TGA oxidation of this catalyst (not shown). The presence of this material, even upon extensive washing, still blocks part of the porous structure of the support.

Figure 1 shows the X-Ray patterns for the prepared electrocatalysts. Peaks observed indicate the presence of the face-centered cubic (fcc) structure typical of Pt and represented by the (111), (200), (220) facets.

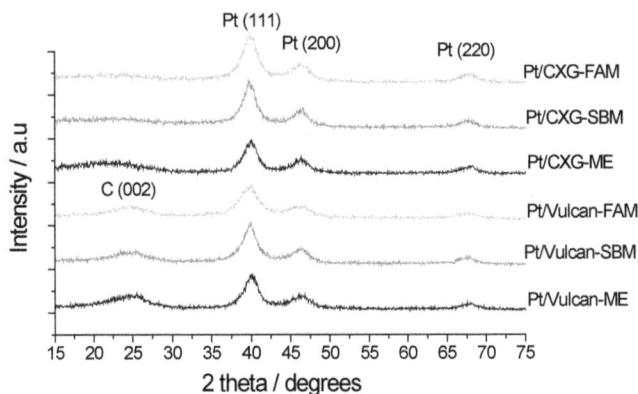

Figure 1. XRD patterns for the assessed catalysts.

Scherrer's equation was applied to (220) reflectances for the calculation of Pt particle size in each case. Crystallite sizes obtained range from 3.4 to 4.6 nm, as presented in Table 3. Broader peaks correspond, according to Scherrer's law, to small average crystal sizes. The broadening of the peaks evidences also lower crystallinity, typical of small particles having lattice strain [13]. Pt crystal size increases for the catalysts prepared using sodium borohydride as reducing agent in the case of using CXG as support, whereas the opposite effect is encountered for the carbon black support.

Table 3. Crystallite size obtained by XRD and percentage of metal (weight) deposited measured by ICP-AES.

Catalysts	Pt crystal size (nm)	% Pt (ICP-AES)
Pt/CXG-SBM	4.2	20.2
Pt/CXG-FAM	3.6	17.9
Pt/CXG-ME	3.9	17.1
Pt/Vulcan-SBM	3.4	16.8
Pt/Vulcan-FAM	4.6	16.7
Pt/Vulcan-ME	4.4	17.4

Pt catalysts were characterized by ICP-AES (Table 3) to determine the percentage of metal deposited. The metallic loading ranges from 16% to 20% (weight). As can be seen in the micrographs obtained by TEM (Figure 2), catalysts present different particle distribution. In principle, the synthetic route using formic acid provides the catalysts with a more adequate Pt-distribution, independently from the support (Figure 2b,d) with average particle sizes centered at 2.7 nm in both cases (see histograms). Synthesis by means of the borohydride route, Figure 2a,e, shows the presence of both isolated particles of lower size and agglomerates of particles in the case of the CXG, where a bimodal particle size distribution is observed: one peak is centered around 3.2 nm and another about 4.5 nm. In the case of Vulcan, there is also some presence of particle agglomerates, with particle size distribution centered around 2.8 nm. The microemulsion route leads to Pt-particles that tend to agglomerate, forming chain-like structures (Figure 2c,f) although particles appear to be individually slightly smaller than in the other cases, with an average size centered at 3.6 nm for Vulcan and at 4.3 nm for the CXG.

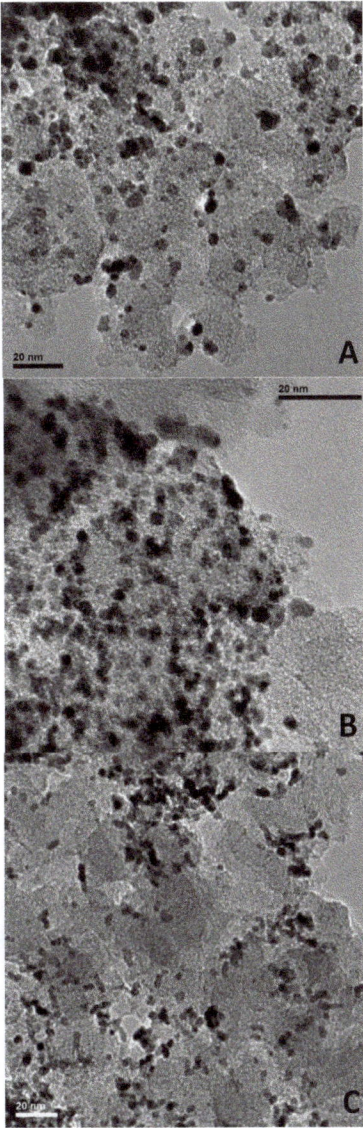

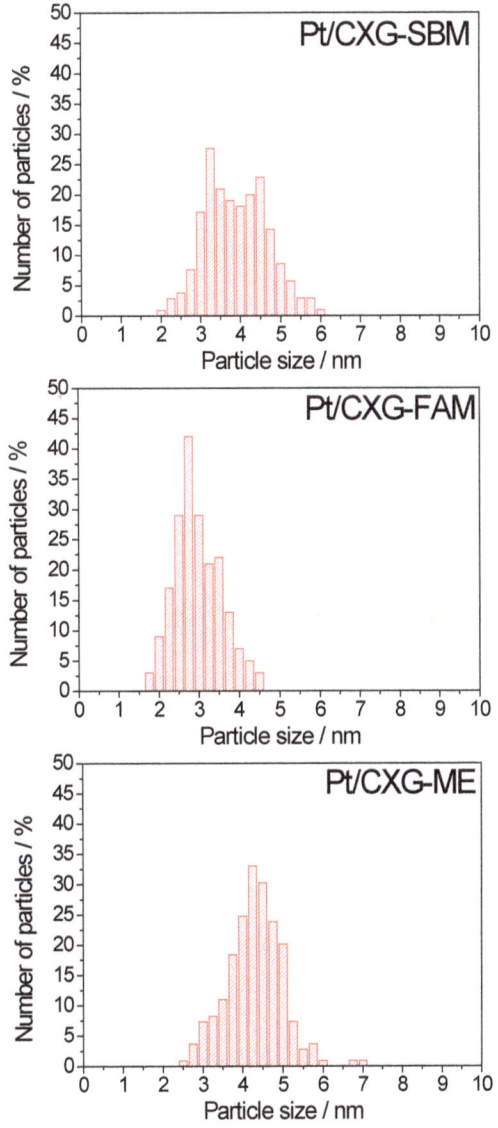

Figure 2. *Cont.*

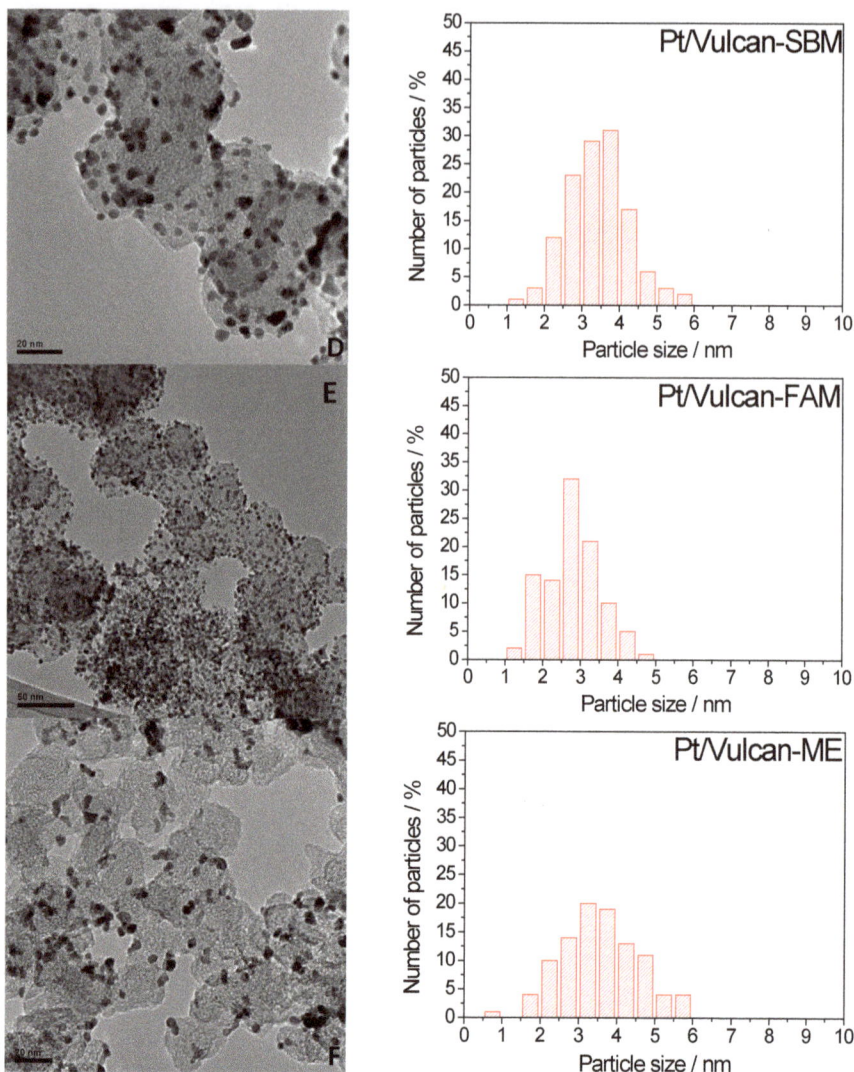

Figure 2. TEM images and corresponding histograms for (**a**) Pt/CXG-SBM; (**b**) Pt/CXG-FAM; (**c**) Pt/CXG-ME; (**d**) Pt/Vulcan-SBM; (**e**) Pt/Vulcan-FAM; and (**f**) Pt/Vulcan-ME.

2.2. Electrochemical Characterization

Pt catalysts were characterized by cyclic voltammetry, and CO stripping was performed in order to establish the influence of the support on the potential for its oxidation. Figure 3 shows the CO-stripping voltammograms obtained for the assessed catalysts. Two scans were performed: during the first one, CO was oxidized; whereas the second one is equivalent to the voltammogram in the base electrolyte.

The COads oxidation peak potential of Pt supported onto both carbon supports was obtained around 0.81–0.83 V *vs.* RHE, for the catalysts synthesized by impregnation methods. However, for the catalysts synthesized by the microemulsion method, the COads oxidation peak potential occurred at more negative potentials, (0.78 V *vs.* RHE) along with another oxidation peak at low potentials (0.70 V *vs.* RHE). Maillard and co-workers [14] investigated the performance of carbon black supported catalysts with different degrees of metallic agglomeration towards the electro-oxidation of CO. They observed that in fact CO monolayer oxidation was strongly influenced by Pt particle size. They reported remarkable activity for Pt agglomerates in comparison to isolated Pt particles or polycrystalline Pt, catalysing CO oxidation at considerably lower overpotentials.

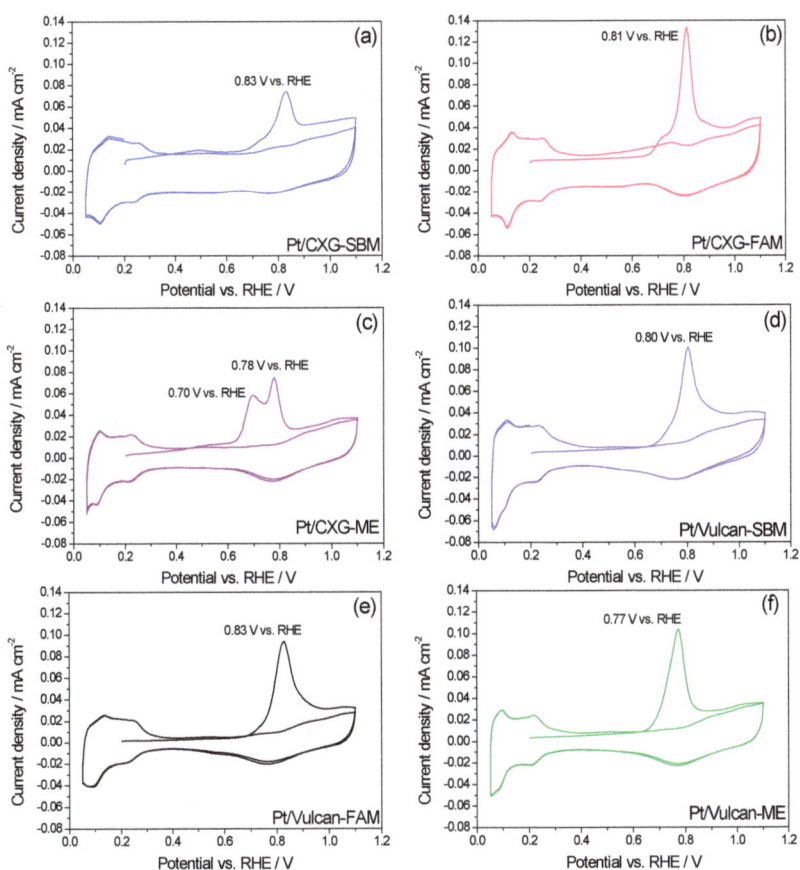

Figure 3. CO stripping voltammograms for (**a**) Pt/CXG-SBM; (**b**) Pt/CXG-FAM; (**c**) Pt/CXG-ME; (**d**) Pt/Vulcan-SBM; (**e**) Pt/Vulcan-FAM; and (**f**) Pt/Vulcan-ME.

CO stripping was also employed to determine the electrochemical Pt active areas (ECSA) as shown in Table 4. Catalysts show similar ECSA values between 36.5 and 47.9 $m^2 \cdot g^{-1}$, except for Pt/Vulcan-ME, which shows a much lower ECSA, 14.7 $m^2 \cdot g^{-1}$. As previously shown in TEM micrographs, this catalyst showed the worst particle distribution, which is responsible for this low ECSA. Contrary to expectations, catalysts with the lowest crystal size do not present the highest ECSA.

Table 4. ECSA obtained from CO stripping and peak mass activity for methanol oxidation.

Catalysts	Pt crystal size (nm)	ECSA ($m^2 \cdot g^{-1}$)	Peak mass activity ($A \cdot g^{-1}$ Pt)	Peak specific activity ($mA \cdot cm^{-2}$ Pt)
Pt/CXG-SBM	4.2	47.9	363	0.96
Pt/CXG-FAM	3.6	38.6	367	1.13
Pt/CXG-ME	3.9	36.5	470	1.29
Pt/Vulcan-SBM	3.4	39.2	330	0.84
Pt/Vulcan-FAM	4.6	41.4	300	0.47
Pt/Vulcan-ME	4.4	14.7	232	1.58

The behavior of the electrocatalysts towards the oxidation of methanol was studied in a deaerated 2 M CH_3OH + 0.5 M H_2SO_4 solution. Resulting cyclic voltammograms recorded are shown in Figure 4. Table 4 also shows the peak activities towards the oxidation of methanol, expressed both in terms of mass activity ($A \cdot g^{-1} \cdot$ Pt) and specific activity ($mA \cdot cm^{-2} \cdot$ Pt).

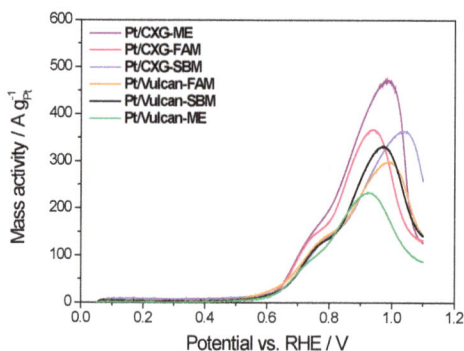

Figure 4. Cyclic voltammograms in a 2M CH_3OH + 0.5M H_2SO_4 solution for the assessed catalysts. Scan rate = 0.02 $V \cdot s^{-1}$.

The highest mass activity value is obtained with Pt/CXG-ME catalyst. This higher activity might be due to its higher activity towards the oxidation of CO, an intermediate of CH_3OH oxidation reaction. Since this catalyst presents the highest activity towards CO oxidation, CH_3OH oxidation becomes also speeded up. Catalysts supported on carbon xerogels, whatever the synthetic method employed, are more active than catalysts supported on Vulcan. The higher mesopore volume of carbon xerogels, allowing an optimal diffusion of reagents towards the catalytic centres, is responsible for the higher catalytic activity. Among the different methods, both impregnation routes (FAM and SBM) lead to catalysts with similar behaviours, particularly when using CXG as catalyst support. Pt/CXG-FAM and Pt/Vulcan-SBM present very similar crystal sizes. Pt/Vulcan-SBM presents a better metal dispersion, (due to its lower crystal size) and so a slightly higher ECSA. In spite of this, its performance is lower, due to the higher mesopore volume of the carbon xerogel in comparison to the carbon black Vulcan. In the CXG, Pt particles are better located within the three phase boundaries, where Pt is in contact with both the ionomer, nafion, and CH_3OH, favoring a higher catalytic activity. The catalyst Pt/Vulcan-ME, showed the poorest activity, probably due to its bad metallic distribution, as could be seen in TEM micrographs. Regarding the specific activity, calculated dividing mass activity by the ECSA, catalysts with the lower ECSA show the higher specific activity, as a result of dividing by a lower value.

The activity of these catalysts towards methanol oxidation was also analyzed by chronoamperometric tests at 600 mV as shown in Figure 5. As in the CV tests, Pt/CXG-ME showed the highest activity. Nevertheless, this activity did not seem very stable with time, given the certain activity decay observed during the experiment.

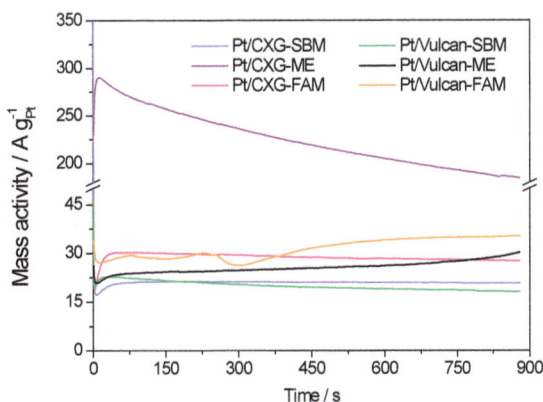

Figure 5. Chroamperometric tests performed at 0.6 V *vs.* RHE in a 2 M CH_3OH + 0.5 M H_2SO_4 solution for the assessed catalysts.

3. Experimental Section

3.1. Synthesis of Carbon Xerogels

Resorcinol (1,3-dihydroxybenzoic acid)-formaldehyde organic gels were synthesized by the sol-gel method first proposed by Pekala *et al.* [8]. The necessary amounts of resorcinol (R) (98% Sigma-Aldrich, St. Louis, MO, USA) and sodium carbonate deca-hydrated (Panreac, Castellar del Vallès (Barcelona, Spain)) (C) (molar ratio R/C = 800) were dissolved in deionised water under stirring. Subsequently, the required volume of formaldehyde (F) (37 wt.% in H_2O, contains 10%–15% Methanol as stabilizer, Sigma-Aldrich, St. Louis, MO, USA) (molar ratio resorcinol/formaldehyde R/F = 0.5, and dilution rate equal 5.7, as described elsewhere [15]) was added to the former mixture and pH was adjusted to 6 with an aqueous solution of 2 M HNO_3. The mixture was stirred for 30 min and then poured into sealed flasks, followed by curing for 24 h at room temperature, 24 h at 50 °C and 3 days at 85 °C until curing was complete as described elsewhere [16]. Afterwards gels were washed with acetone for three days to exchange the initial solvent, water. Acetone was daily replaced after vacuum filtration. This procedure allows keeping the original gel structure, as surface tension caused by evaporation of the solvent is lower for acetone than for water [6]. Finally wet gels were sub-critically dried at 110 °C for 5 h. Organic xerogels are grinded prior to the treatment with acetone and after the drying process, in an agate mortar until a fine powder is obtained. Xerogels were pyrolyzed with the following heating program steps: (1) 2 h at 150 °C; (2) 1 h at 300 °C; (3) 1 h at 600 °C; and (4) 2 h at 800 °C as described elsewhere [17].

3.2. Pt catalysts Synthesis

Pt was deposited on former carbon supports by impregnation and reduction with two different reducing agents: sodium borohydride and formic acid. The amount of metallic precursor was calculated to obtain a metal loading of 20% w/w. An aqueous solution of H_2PtCl_6 (Sigma-Aldrich, St. Louis, MO, USA) was slowly added to a dispersion of the carbon support in ultrapure water. Mixture was sonicated during the addition of the metallic precursor. pH was adjusted to 5 with a solution of NaOH. Subsequently, sodium borohydride was added in excess to reduce the metallic precursor, maintaining temperature under 18 °C. Finally, catalyst was filtered and thoroughly washed with deionized water, and dried overnight at 60 °C. The catalyst so obtained was named Pt/C-SBM, being C, CXG or Vulcan.

When using formic acid as reducing agent, a dispersion of the carbon support in a 2M solution of formic acid was prepared and heated at 80°C. Then an aqueous solution of H_2PtCl_6 (Sigma-Aldrich, St. Louis, MO, USA) was slowly added under magnetic stirring. Catalyst were filtered and thoroughly washed with ultrapure water, and dried overnight at 60 °C. The catalyst so obtained was named Pt/C-FAM.

The microemulsion route consisted on the synthesis of Pt nanoparticles by means of a water in oil microemulsion [18]. The microemulsion was composed of a commercial surfactant, n-heptane as the non-polar phase and 2-propanol as co-surfactant. A commercial surfactant (Brij30, Sigma-Aldrich, St. Louis, MO, USA) and n-heptane were mixed and stirred, before adding dropwise an aqueous solution of H_2PtCl_6 (8 mM). The co-surfactant, 2-propanol was added until an optically transparent mixture was observed, indicating the formation of the microemulsion. The whole mixture was stirred for 4 h, after which an 0.1 M aqueous solution of the reducing agent ($NaBH_4$) was slowly added to the microemulsion under continuous stirring. Finally this suspension was slowly added to a suspension of the carbon support in ethanol under sonication. The catalyst was thoroughly washed with ethanol and water and subsequently dried overnight at 60 °C [19]. The catalyst so obtained was named Pt/C-ME.

3.3. Carbon Supports and Catalysts Textural, Structural and Morphological Characterization

The nature and characteristics of the carbon supports were characterized using nitrogen physisorption. Textural properties such as specific surface area, pore volume and mesoporosity were calculated from nitrogen adsorption-desorption isotherms, measured at −196 °C using a Micromeritics (Norcross, GA, USA) ASAP 2020. Total surface area and pore volume were determined using the Brunauer-Emmet-Teller (BET) equation and the single point method, respectively. Microporosity was determined using the t-plot method and mesopore volume and medium pore size were calculated by applying the Barret-Joyner-Hallenda method. Catalysts were characterized by X-Ray Diffraction (XRD), and XRD patterns were performed using a Bruker (Billerica, MA, USA) AXS D8 Advance diffractometer, with a θ–θ configuration and using Cu Kα radiation. Particle sizes were evaluated using TEM images obtained in a JEOL (Tokyo, Japan) 2100F microscope operated with an accelerating voltage of 200 kV and equipped with a field emission electron gun providing a point resolution of 0.19 nm. The standard procedure involved dispersing 3 mg of the sample in ethanol in an ultrasonic bath for 15 min. The sample was then placed in a Cu carbon grid where the liquid phase was evaporated.

3.4. Electrochemical Characterization

Catalyst electrochemical activity towards the oxidation of carbon monoxide and methanol was studied using cyclic voltammetry and chronoamperometry. A flow cell with a three-electrode assembly at room temperature and an AUTOLAB (Utrecht, The Netherlands) potentiostat-galvanostat were used to carry out the electrochemical characterization. The counter electrode consisted of a pyrolytic graphite bar and the reference electrode was a reversible hydrogen electrode (RHE). All potentials in the text refer to the latter. The working electrode consisted of a pyrolytic graphite disk with a thin layer of the different electrocatalysts deposited onto it. For the preparation of this layer, an aqueous suspension of Pt/CXG or Pt/Vulcan catalysts was obtained by ultrasonically dispersing it in Nafion and ultrapure water (Merck-Millipore, Billerica, MA, USA). Subsequently, an aliquot of 40 μL of the dispersed suspension was deposited on top of the graphite disk (7 mm) and dried under inert atmosphere prior its use. After preparation, the electrode was immersed into deaerated 0.5 M H_2SO_4 electrolyte, prepared from high purity reagents (Merck-Millipore, Billerica, MA, USA) and ultrapure water (Merck-Millipore, Billerica, MA, USA). The electrolyte was saturated with pure N_2 or CO (99.997%, Air Liquide, Madrid, Spain), depending on the experiments. Activation of the electrode was performed by potential cycling solution between 0.05 and 1.10 V at a scan rate of 0.5 $V \cdot s^{-1}$ until a stable voltammogram in the base electrolyte (0.5 M H_2SO_4) was obtained.

CO stripping voltammograms were obtained after bubbling this gas in the cell for 10 min at 0.20 V, followed by electrolyte exchange and nitrogen purging to remove the excess of CO, and oxidation by scanning the potential up to the upper limit previously established for Pt catalysts. The admission potential was selected considering that, for this value, maximum adsorbate coverage is achieved for CO adsorption on Pt. Electrochemical Pt active areas were determined from the integration of the current involved in the oxidation of a CO monolayer taking into account that CO adsorbs on Pt and assuming 420 $\mu C \cdot cm^{-2}$. Cyclic voltammograms for the electrooxidation of methanol were carried out in a 2 M CH_3OH + 0.5 M H_2SO_4 solution, at scan rate of 0.02 $V \cdot s^{-1}$, between 0.0 and 1.10 V. Chronoamperometries were performed at 0.60 V vs. RHE in a 2 M CH_3OH + 0.5 M H_2SO_4 solution in order to evaluate the performance of the electrocatalysts for the oxidation of methanol. Every experiment was carried out at room temperature (25 °C ± 1 °C).

4. Conclusions

A highly mesoporous carbon xerogel was synthesized as support for Pt nanoparticles. Three different synthetic methodologies were employed for the synthesis of Pt-based catalysts: two impregnation methods and a microemulsion method. These methods were also applied to a commercial carbon black, Vulcan.

Carbon xerogel showed a high developed surface area and high mesoporosity, much greater than the one shown for the commercial support.

Catalysts supported on carbon xerogel provided lower crystal sizes and good metallic distribution, particularly when using formic acid as reducing agent.

Pt-based catalysts were tested for CO oxidation and the CH_3OH oxidation. Pt synthesized by the microemulsion method supported on the carbon xerogel was the highest active catalyst for the oxidation of CH_3OH. This fact was explained by the higher activity of this catalyst towards the oxidation of CO, an intermediate of the oxidation of methanol.

Catalysts supported on carbon xerogels were more active than catalysts supported on the commercial carbon black support, Vulcan, due to the higher mesoporosity of the former, favoring the diffusion of reagents to and from the catalytic centers.

Acknowledgments: The authors wish to thank the Spanish Ministry of Economy and Competitiveness (Secretaría de Estado de I+D+I) and FEDER for financial support under the project CTQ2011-28913-C02-01.

Author Contributions: C.A. performed the synthesis and characterization of both carbon xerogels and catalysts assessed. The whole work was supervised by M.E.G, R.M. and M.J.L.

Conflicts of Interest: The authors declare no conflict of interest.

References

1. Job, N.; Ribeiro Pereira, M.F.; Lambert, S.; Cabiac, A.; Delahay, G.; Colomer, J.-F.; Marien, J.; Figueiredo, J.L.; Pirard, J.-P. Highly dispersed platinum catalysts prepared by impregnation of texture-tailored carbon xerogels. *J. Catal.* **2006**, *240*, 160–171.
2. Figueiredo, J.L.; Ribeiro Pereira, M.F.; Serp, P.; Kalck, P.; Samant, P.V.; Fernandes, J.B. Development of carbon nanotube and carbon xerogel supported catalysts for the electro-oxidation of methanol in fuel cells. *Carbon* **2006**, *44*, 2516–2522.
3. Samant, P.V.; Fernandes, J.B.; Rangel, C.M.; Figueiredo, J.L. Carbon xerogel supported Pt and Pt–Ni catalysts for electro-oxidation of methanol in basic medium. *Catal. Today* **2005**, *102–103*, 173–176.
4. Sebastián, D.; Suelves, I.; Lázaro, M.J.; Moliner, R. Carbon nanofibers as electrocatalyst support for fuel cells: Effect of hydrogen on their properties in CH_4 decomposition. *J. Power Sources* **2009**, *192*, 51–56.
5. Arbizzani, C.; Beninati, S.; Manferrari, E.; Soavi, F.; Mastragostino, M. Cryo- and xerogel carbon supported PtRu for DMFC anodes. *J. Power Sources* **2007**, *172*, 578–579.
6. Al-Muhtaseb, S.A.; Ritter, J.A. Preparation and Properties of Resorcinol-Formaldehyde Organic and Carbon Gels. *Adv. Mater.* **2003**, *15*, 101–114.
7. Antolini, E. Carbon supports for low-temperature fuel cell catalysts. *Appl. Catal. B* **2009**, *88*, 1–24.
8. Pekala, R.W. Organic aerogels from the polycondensation of resorcinol with formaldehyde. *J. Mater. Sci.* **1989**, *24*, 3221–3227.
9. Alegre, C.; Calvillo, L.; Moliner, R.; González-Expósito, J.A.; Guillén-Villafuerte, O.; Martínez Huerta, M.V.; Pastor, E.; Lázaro, M.J. Pt and PtRu electrocatalysts supported on carbon xerogels for direct methanol fuel cells. *J. Power Sources* **2011**, *196*, 4226–4235.
10. Yoshizawa, N.; Hatori, H.; Soneda, Y.; Hanzawa, Y.; Kaneko, K.; Dresselhaus, M.S. Structure and electrochemical properties of carbon aerogels polymerized in the presence of Cu^{2+}. *J. Non-Cryst. Solids* **2003**, *330*, 99–105.
11. Job, N.; Marie, J.; Lambert, S.; Berthon-Fabry, S.; Achard, P. Carbon xerogels as catalyst supports for PEM fuel cell cathode. *Energy Convers. Manag.* **2008**, *49*, 2461–2470.

12. Job, N.; Lambert, S.; Chatenet, M.; Gommes, C.; Maillard, F.; Berthon-Fabry, S.; Regalbuto, J.R.; Pirard, J.P. Preparation of highly loaded Pt/carbon xerogel catalysts for Proton Exchange Membrane fuel cells by the Strong Electrostatic Adsorption method. *Catal. Today* **2009**, *150*, 119–127.

13. Maillard, F.; Savinova, E.R.; Simonov, P.A.; Zaikovskii, V.I.; Stimming, U. Infrared Spectroscopic Study of CO Adsorption and Electro-oxidation on Carbon-Supported Pt Nanoparticles: Interparticle *versus* Intraparticle Heterogeneity. *J. Phys. Chem.* **2004**, *108*, 17893–17904.

14. Maillard, F.; Schreier, S.; Savinova, E.R.; Weinkauf, S.; Stimming, U. Influence of particle agglomeration on the catalytic activity of carbon-supported Pt nanoparticles in CO monolayer oxidation. *Phys. Chem. Chem. Phys.* **2005**, *7*, 385–393.

15. Job, N.; Théry, A.; Pirard, R.; Marien, J.; Kocon, L.; Rouzaud, J.N.; Béguin, F.; Pirard, J.P. Carbon aerogels, cryogels and xerogels: Influence of the drying method on the textural properties of porous carbon materials. *Carbon* **2005**, *43*, 2481–2494.

16. Morales-Torres, S.; Maldonado-Hódar, F.J.; Pérez-Cadenas, A.F.; Carrasco-Marín, F. Structural characterization of carbon xerogels: From film to monolith. *Microporous Mesoporous Mater.* **2012**, *153*, 24–29.

17. Gorgulho, H.F.; Gonçalves, F.; Pereira, M.F.R.; Figueiredo, J.L. Synthesis and characterization of nitrogen-doped carbon xerogels. *Carbon* **2009**, *47*, 2032–2039.

18. Sebastián, D.; Alegre, C.; Calvillo, L.; Pérez, M.; Moliner, R.; Lázaro, M.J. Carbon supports for the catalytic dehydrogenation of liquid organic hydrides as hydrogen storage and delivery system. *Int. J. Hydrog. Energy* **2013**, *39*, 4109–4115.

19. Sebastián, D.; Lázaro, M.J.; Suelves, I.; Moliner, R.; Baglio, V.; Stassi, A.; Aricò, A.S. The influence of carbon nanofiber support properties on the oxygen reduction behavior in proton conducting electrolyte-based direct methanol fuel cells. *Int. J. Hydrog. Energy* **2012**, *37*, 6253–6260.

Selective Oxidation of Raw Glycerol Using Supported AuPd Nanoparticles

Carine E. Chan-Thaw, Sebastiano Campisi, Di Wang, Laura Prati and
Alberto Villa

Abstract: Bimetallic AuPd supported on different carbonaceous materials and TiO_2 was tested in the liquid phase oxidation of commercial grade and raw glycerol. The latter was directly obtained from the base-catalyzed transesterification of edible rapeseed oil using KOH. The best catalytic results were obtained using activated carbon and nitrogen-functionalized carbon nanofibers as supports. In fact, the catalysts were more active using pure glycerol instead of the one obtained from rapeseed, where strong deactivation phenomena were present. Fourier transform infrared (FT-IR) and TEM were utilized to investigate the possible reasons for the observed loss of activity.

Reprinted from *Catalysts*. Cite as: Chan-Thaw, C.E.; Campisi, S.; Wang, D.; Prati, L.; Villa, A. Selective Oxidation of Raw Glycerol Using Supported AuPd Nanoparticles. *Catalysts* **2015**, *5*, 131–144.

1. Introduction

The use of biomass for the production of renewable raw materials and their conversion to high value chemicals is still a young field, but a significant potential has been shown [1]. Only 3.5% of the existing biomass production is presently being used for human needs. Most of this is used for human food (around 62%), 33% for energy use, paper and construction needs, and the remaining 5% is used for clothing, detergents and chemicals. The other 96.5% of the biomass production is used in the planetary ecosystem. A recent EU directive (2009/28/EC) has set the target of achieving, by 2020, a 20% share of energy from renewable energy sources in the EU's overall energy consumption. In this context, special consideration is paid to the role played by the use of waste oils or non-edible vegetable oils as feedstock that do not interfere with the food chain.

Vegetable oils are composed of triglycerides, and their transesterification results in obtaining three moles of fatty acid methyl esters (FAMEs) and one mole of glycerol. The products of the hydrogenation of these FAMEs can be used as biodiesel or as non-negligible products, such as lubricants [2], surfactants, solvents [3], polymers and fine chemicals [4]. The tremendous growth of the biodiesel industry has concomitantly been accompanied by an over-production of raw glycerol [5,6]. Together with the high cost of disposal, this unused product was often released into landfills. Therefore, from an economical and environmental aspect, it is worthwhile

311

to focusing on this raw product. The valorization of a huge amount of compound that is considered as waste, hence being low cost, would significantly affect the price of biodiesel. For this reason, since the end of the 20th century, intensive research has been focused on the use of glycerol as a benign solvent [7] or as starting material for subsequent transformations [8–15]. However, nearly almost all results that can be found in the literature were obtained with pure commercial glycerol, whereas few reports used raw glycerol as a starting material [16–18].

Crude glycerol is normally obtained by a simple transesterification reaction of vegetable oils with methanol using sodium (Na) or potassium (K) hydroxide as catalysts. These ions together with the alcohol or the remaining free fatty acids represent the main impurities that could thus deposit on the active sites on the catalyst surface or modify any pre-established reaction pathway, during the subsequent transformations of glycerol. Skrzyńska et al. recently reported on the potential behavior of crude glycerol impurities at various pH in liquid phase oxidation [18]. Konaka et al. [19] intentionally prepared a potassium-supported zirconia-iron oxide catalyst for crude glycerol conversion into allyl alcohol. A proper amount of K, in their case 5 mol%, successfully produced allyl alcohol, a useful chemical for the preparation of resins, paints and plasticizers [20]. An amount of K over the mentioned value is detrimental for the reaction. Therefore, the composition of the crude oil could influence any catalytic transformation. Among the possible routes for glycerol valorization, selective oxidation is of importance, because glycerol acts as a platform molecule that is convertible through the use of inexpensive oxidizing agents (air, oxygen, H_2O_2, etc.) in a variety of value-added derivatives (glyceric acid, tartronic acid, hydroxypyruvic acid, etc.).

Since Prati et al. have demonstrated that gold supported on activated carbon (AC) in the presence of a base was catalytically active and selective for the oxidation of ethylene glycol to produce glycolate [9], gold as a catalyst for liquid phase oxidation of alcohols has been the subject of many studies [21–23]. In particular, gold catalysts were shown to be more resistant to oxygen poisoning and selective for the oxidation of primary alcohols. In order to increase the catalytic performance of gold catalyst, the support effect, as well as the introduction of a second metal has been deeply investigated. It was shown that alloying Pd to Au nanoparticles leads to a significant enhancement of the catalytic activity in the selective oxidation of glycerol, also increasing the durability of the catalyst [24–26]. On the other hand, the choice of the support has a strong influence on the catalytic activity of the Au catalyst [27,28].

Indeed, the support has the role of avoiding coalescence and agglomeration of metal nanoparticles by reducing their mobility. In many cases, the strong metal-support interactions (SMSI) can also play a non-negligible role in the reaction mechanism. Furthermore, the surface properties of the catalysts (acidity, hydrophobicity/hydrophilicity, etc.) are principally determined by the support [28].

Herein, we investigate the catalytic performance of supported AuPd catalysts, prepared by sol immobilization, in the selective oxidation of both commercial glycerol and that directly obtained from the transesterification of rapeseed oil. To study the support effect, carbonaceous materials having different textural and surface properties, such as activated carbon, carbon nanotubes (CNTs), carbon nanofibers (CNFs) and N-doped carbon nanofibers (N-CNFs), were utilized, whilst TiO$_2$ represents the oxide materials. Particular attention has been devoted to the impact of the impurities present in the raw glycerol on the activity and durability of the AuPd catalysts.

2. Results and Discussion

Raw glycerol was obtained from the transesterification of edible rapeseed vegetable oil. GC analysis revealed that the rapeseed vegetable oil was mainly composed of mono- and poly-unsaturated fatty acids; namely oleic acid (9-octadecenoic, C18:1), linoleic acid (9,12-octadecandienoic, C18:2), and the conjugated isomers thereof, and linolenic acid (9,12,15-octadecantrienoic, C18:3). Palmitic (hexadecanoic acid, C16:0) and stearic acids (octadecanoic acid, C18:0), saturated fatty acids, are in less quantity (Table 1).

Table 1. Typical fatty acid composition of rapeseed.

Oil	Fatty acids composition % by weight				
	C16:0	C18:0	C18:1	C18:2	C18:3
Rapeseed	4.39	1.94	63.93	20.67	9.07

The transesterification was carried out at 40 °C for 2 h using KOH as the base. At the end of the reaction, the glycerol solution was separated from the reaction media and used in the consecutive reactions without any pretreatment. Selective raw glycerol oxidation was performed using different AuPd bimetallic systems, supported on carbonaceous material or TiO$_2$ (glycerol 0.3 M, 50 °C, 3 atm O$_2$, glycerol/Au 1000 mol/mol, NaOH/glycerol 4 mol/mol). For understanding the importance of glycerol purity on the catalytic performance, a commercial grade glycerol was also studied under the same reaction conditions. AuPd catalysts were synthesized by impregnation of preformed metal nanoparticles using polyvinyl alcohol as a protective agent. We reported on a two-step procedure for the preparation of activated carbon-supported AuPd alloy nanoparticles (NPs) [24–26]. In this procedure, after immobilization of a preformed gold sol on activated carbon, a sol of palladium was generated in the presence of Au/AC using H$_2$ as the reducing agent. Herein, we extended this procedure to other supports (CNFs, N-CNFs, CNTs, TiO$_2$).

The resulting catalysts show a similar particle size for AuPd on AC, CNTs and N-CNFs (3.4, 3.5 and 3.7 nm for Au-Pd/AC, Au-Pd/CNTs and Au-Pd/N-CNFs, respectively) and are even distributed differently (Table 2). In the case of CNFs and TiO$_2$ as supports, bigger AuPd particles are measured (4.5 and 4.1 nm for CNFs and TiO$_2$, respectively) (Table 2). As already highlighted in the literature, these results confirmed the beneficial effect of the presence of heteroatoms on the carbon nanofibers' surface in terms of stabilization of the metal nanoparticles with a better particle size distribution.

Table 2. Statistical median and standard deviation of 1% Au$_6$Pd$_4$ catalysts. AC, activated carbon; N-CNF, N-doped carbon nanofiber.

Catalyst	Statistical median (nm)	Standard deviation σ
1% Au$_6$Pd$_4$/CNFs	4.5	1.5
1% Au$_6$Pd$_4$/N-CNFs	3.7	0.9
1% Au$_6$Pd$_4$/CNTs	3.5	1.3
1% Au$_6$Pd$_4$/TiO$_2$	4.1	1.2
1% Au$_6$Pd$_4$/AC	3.4	0.7
1% Au$_6$Pd$_4$/AC after reaction using raw glycerol	4.6	1.2
1% Au$_6$Pd$_4$/AC after reaction using purified glycerol	3.5	0.7

In the case of AuPd on activated carbon, alloyed AuPd nanoparticles with a uniform composition and homogeneity were obtained [24–26]. In the other cases, most of the nanoparticles were alloyed, but the Au/Pd ratio was inhomogeneous from cluster to cluster. On the contrary, a partial segregation of Au and Pd was detected on 1% Au$_6$Pd$_4$/CNF, as evidenced by the elemental mapping (Figure 1).

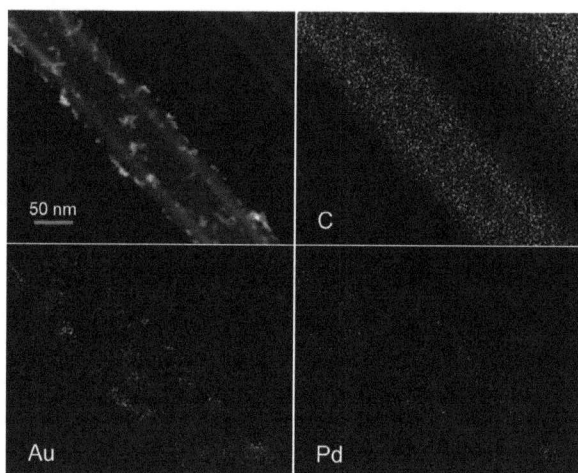

Figure 1. STEM image and element mapping on 1% Au$_6$Pd$_4$/CNF.

The catalysts were first tested in the oxidation of pure glycerol. The initial activity, expressed as mol of glycerol converted per mol of metal per hour, calculated at 15 min of reaction, as well as a selectivity at 90% conversion are reported in Table 3. One percent Au_6Pd_4/AC resulted in the most active catalysts with an initial activity (3205 mol/mol h) at least three-times more active than 1% Au_6Pd_4/N-CNFs (1076 mol/mol h) and 1% Au_6Pd_4/CNTs (815 mol/mol h), despite a similar AuPd particle size (3.4–3.7 nm). This result is in agreement with the findings that AuPd homogeneous alloys are more active than inhomogeneous phases in the alcohol oxidation due to a synergistic effect [24–26,29]. One percent Au_6Pd_4/CNFs and 1% Au_6Pd_4/TiO_2 resulted in less activity (675 and 628 mol/mol h, respectively), probably due the presence of bigger particle sizes (4.5 and 4.1 nm) (Table 2).

The selectivity to glyceric acid is slightly higher for 1% Au_6Pd_4/CNFs and 1% Au_6Pd_4/TiO_2 (78–79%) compared to the one of 1% Au_6Pd_4/CNTs (70%) (Table 3). This trend can be explained by observing the AuPd particle sizes (4.5, 4.1 and 3.5 nm, for Au_6Pd_4 supported on CNFs, TiO_2 and CNTs, respectively). Indeed, it has been reported that the metal nanoparticle size is one of the factors influencing the selectivity in glycerol oxidation, with larger particles giving higher selectivity towards glyceric acid [8–15].

Table 3. Glycerol oxidation over 1% Au_6Pd_4-based carbon and oxide supports. Glycerol was obtained from the transesterification of rapeseed oil. Reaction conditions: glycerol, 0.3 M; substrate/total metal = 1000 mol/mol; total volume, 10 mL; 4 eq NaOH; 50 °C; 3 atm O_2.

Origin of glycerol	Catalyst	Initial activity [a]	Selectivity (%) [b]		
			Glyceric acid	Tartronic acid	C1 + C2 product
	1% Au_6Pd_4/AC	3205	77	5	14
	1% Au_6Pd_4/N-CNFs	1076	66	22	8
Pure glycerol	1% Au_6Pd_4/CNTs	815	70	8	10
	1% Au_6Pd_4/CNFs	675	78	11	9
	1% Au_6Pd_4/TiO_2	628	79	8	12
	1% Au_6Pd_4/AC	1672	72	15	11
	1% Au_6Pd_4/N-CNFs	736	71	17	10
Raw glycerol	1% Au_6Pd_4/CNTs	651	72	15	12
	1% Au_6Pd_4/CNFs	269	81	8	10
	1% Au_6Pd_4/TiO_2	230	78	10	10
Purified raw glycerol	1% Au_6Pd_4/AC	3150	75	7	15
	1% Au_6Pd_4/TiO_2	598	79	3	14
Pure glycerol + fatty acids	1% Au_6Pd_4/AC	1523	73	14	13
	1% Au_6Pd_4/TiO_2	185	79	5	16

[a] Glycerol converted moles per hour per mole of metal, calculated after 15 min of reaction;
[b] selectivity calculated at 90% conversion.

On the other hand, 1% Au_6Pd_4/AC showed a comparable selectivity as 1% Au_6Pd_4/CNFs and 1% Au_6Pd_4/TiO$_2$ (77–79%) despite smaller AuPd particle sizes (3.5, 4.1 and 4.5 nm for AuPd on AC, TiO$_2$ and CNFs, respectively). This result can be justified by the better selectivity to glyceric acid evidenced in the presence of the pure AuPd alloy, instead of the inhomogeneous phases [24–26]. One percent Au_6Pd_4/N-CNFs showed the lowest selectivity to glyceric acid (66%) with the formation of a high amount of tartronic acid, which is the product of the consecutive reaction of glycerol to tartronate. Probably the presence of nitrogen groups altered the AuPd active sites by an electronic effect, promoting the overoxidation reaction.

The same catalysts have been tested using raw instead of pure glycerol. All of the catalysts resulted in less activity, but following the same order as when tested in pure glycerol: AC > N-CNFs > CNTs > CNFs > TiO$_2$. We attributed the lower activity to the presence of impurities that adsorbed onto the AuPd catalysts, partially blocking the active sites. In order to prove this hypothesis, the raw glycerol was purified from the impurities derived from the transesterification process. This process consists first in the addition of H_2SO_4 to the glycerol solution to convert the soap into free fatty acid. Therefore, free fatty acids were removed by extraction with hexane, and then, methanol was removed by evaporation. The glycerol solution was then separated and treated with activated carbon in order to remove free ions. Lastly, the activated carbon was separated by filtration.

The most active 1% Au_6Pd_4/AC and the less active 1% Au_6Pd_4/TiO$_2$ catalysts were tested with the purified glycerol. Similar results as the one obtained with commercial pure glycerol on both catalysts were obtained. Finally, the same catalysts have been tested, adding rapeseed oil (10 wt% with respect to glycerol) to pure glycerol (Table 3). The results highlighted that the addition of fatty acid to pure glycerol has a detrimental effect on the activity, with an initial activity even lower than the one obtained using raw glycerol. Figure 2 reported the conversion/time curves using 1% Au_6Pd_4/AC and 1% Au_6Pd_4/TiO$_2$, clearly showing that the reaction profile is strongly influenced by the type of glycerol. Indeed, in the presence of pure or purified glycerol, full conversion was obtained after 1 h, whereas in the presence of fatty acids, a lower activity was observed, possibly due to the blocking of some active sites by the adsorbed species.

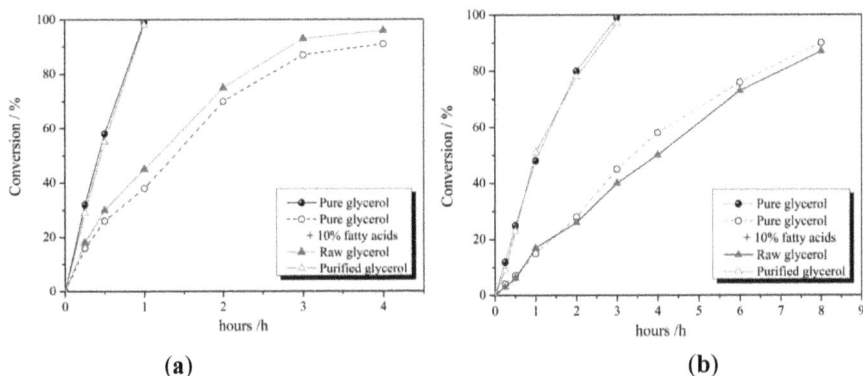

Figure 2. Reaction profile for pure glycerol, pure glycerol + 10% fatty acid, raw glycerol and purified glycerol using (**a**) 1% Au_6Pd_4/AC and (**b**) 1% Au_6Pd_4/TiO$_2$.

To investigate in more detail this latter finding, Fourier transform infrared spectroscopy studies were carried out on 1% Au_6Pd_4/TiO$_2$ after reaction with both raw and pure glycerol. Figure 3 shows the recorded IR spectra of the used catalysts after reaction using pure and raw glycerol. In the case of glycerol derived from the transesterification of rapeseed oil, the bands at 2921 and 2846 cm^{-1} were assigned to the sp^3 asymmetric CH$_2$ stretch and the symmetric CH$_2$ stretch, respectively. The band at 2959 cm^{-1} was attributed to the sp^2 C-H stretch in C=C-H, highlighting thus the presence of C=C. The assignment of these three bands are in agreement with Wu *et al.* [30]. Moreover, a conjugated v(C=C) stretch is highlighted by the bands at 1638 and 1595 cm^{-1}. The presence of these carbon double bonds could be explained by the presence of some traces of oleic, linoleic and linolenic acids in the raw glycerol. Probably, these unsaturated species were adsorbed on the catalyst during the glycerol oxidation. Indeed, these species are not present in the IR spectrum of the catalyst used for the selective pure glycerol oxidation (Figure 3). The bands in common between the pure and the raw glycerol are at 1456 and 1402 cm^{-1} and were attributed to O-H in plane [30] and to C-O-H bending vibration [31], respectively.

This finding can be extended to the carbon-based AuPd catalyst. However, the nature of the support makes investigation using infrared spectroscopy more difficult. The characteristic band of C=C at 2959 cm^{-1} and the one at 1638–1595 cm^{-1}, attributed to the conjugated C=C, are due to the presence in the raw glycerol of some fatty acids. These long unsaturated chains could easily block access to the active sites, as already reported by Gil *et al.* [16,17]. They indeed ascribed the decrease of the catalytic activity when using crude glycerol as the feedstock with the presence of impurities, namely the methyl esters. The latter produce important changes in the surfaces properties of the support and the leaching of gold into the liquid reaction solution. During the phase separation after the transesterification, despite all care

317

being taken, it becomes obvious that traces of the fatty acids were mixed together with the co-product, glycerol.

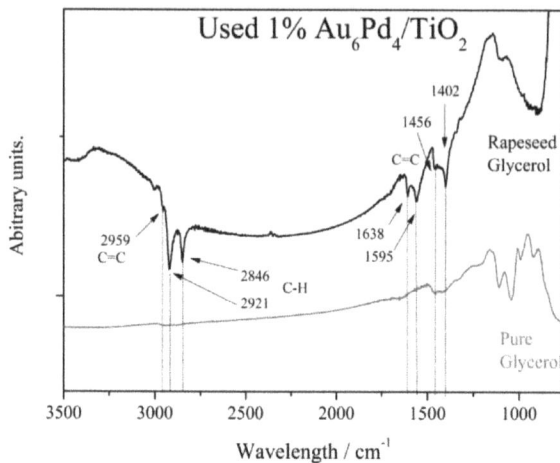

Figure 3. FT-IR spectra of the used 1% Au_6Pd_4/TiO_2 after glycerol oxidation. Reaction conditions: glycerol, 0.3 M; substrate/total metal = 1000 mol/mol; total volume, 10 mL; 4 eq NaOH; 50 °C; 3 atm O_2.

The effect of the nature of the glycerol on the durability of the catalyst was evaluated by a recycling test using the most active catalyst, 1% Au_6Pd_4/AC. Recycling experiments were carried out just by filtering the catalyst and adding a fresh solution of glycerol using purified and non-purified raw glycerol (Figure 4).

Figure 4 shows that, when the glycerol was not formerly purified, the activity of the 1% Au_6Pd_4/AC catalyst rapidly decreases, while the selectivity to glyceric acid increase. Indeed, from Run 1 to Run 8, a drop of conversion (from 89% to 6%) occurred. Along the eight successive reactions, the selectivity to glyceric acid is constantly increasing (from 77% to 87%). Such an increase in the value of the selectivity suggests an increase of the bimetallic AuPd particle size. This assumption is confirmed by measuring the mean particle size of the AuPd on the used catalysts with an increase from 3.4 to 4.7 nm (Table 2), with the presence of aggregated AuPd particles in some regions (Figure 5).

On the contrary, the conversion (about 94%), as well as the selectivity to glyceric acid (around 77%) remained constant over the eight runs using the purified glycerol (Figure 4). TEM investigation confirmed that the morphology of the AuPd nanoparticles was not significantly modified during the stability test, with AuPd mean sizes of 3.4 and 3.5 nm before and after the reaction, respectively (Table 2).

The impact of the purity of the glycerol on the stability of the catalyst was confirmed by analyzing the leaching of metal in both cases. Moreover, ICP analyses of the collected solution of the eight runs resulted in a loss of 15 wt% of total metal in the case of non-purified glycerol against only 1% in the case of the purified one.

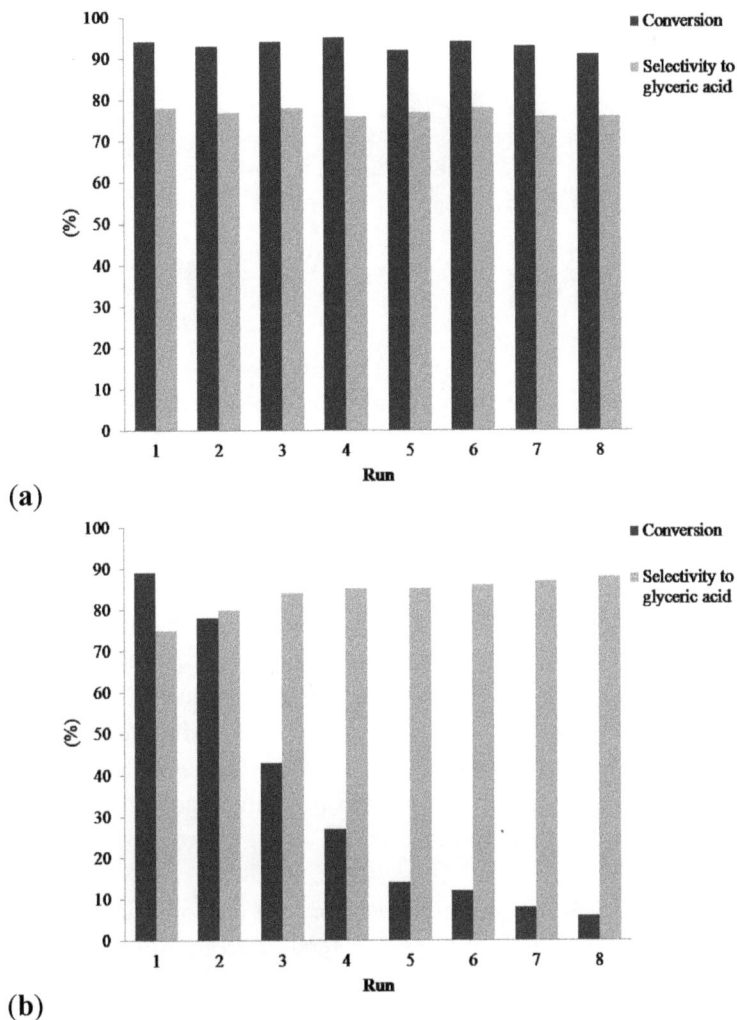

(a)

(b)

Figure 4. Evolution of the glycerol conversion over 1% Au_6Pd_4/AC and over the eight runs using (a) the purified glycerol and (b) the non-purified glycerol directly obtained from transesterification of rapeseed oil. Reaction conditions: glycerol, 0.3 M; substrate/total metal = 1000 mol/mol; total volume, 10 mL; 4 eq NaOH; 50 °C; 3 atm O_2.

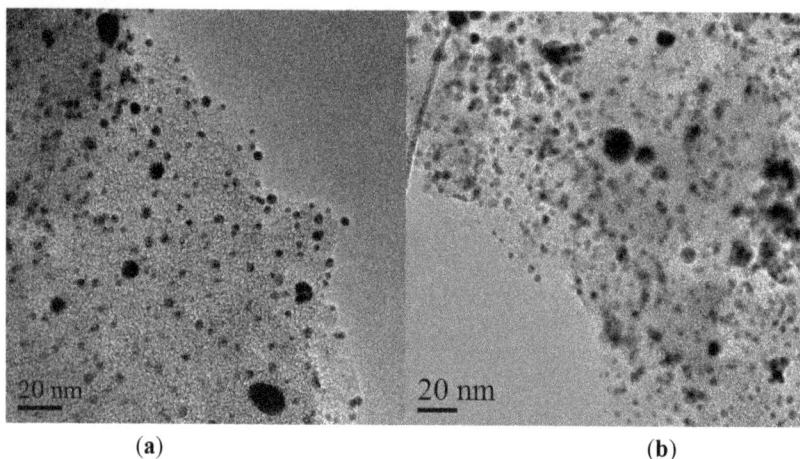

Figure 5. TEM images of AuPd/AC (**a**) before and (**b**) after the recycling tests using raw glycerol.

3. Experimental Section

3.1. Materials

Fatty acid methyl esters (FAMEs) and, therefore, the by-product, raw glycerol, were obtained by the homogeneous transesterification of rapeseed oil with methanol (MeOH:triglycerides = 6:1 mol/mol) and KOH (1% by weight) as the catalyst. This mixture was magnetically stirred for two hours at 40 °C. After the reaction, by simple decantation methyl esters are separated from the raw glycerol. The final methyl ester was obtained after removing the excess of alcohol under vacuum and a final filtration on silica powder. In the case of non-purified raw glycerol, the glycerol was used as obtained.

Glycerol (87 wt% solution), glyceric acid and tartronic acid were purchased from Fluka. $NaAuCl_4$ $2H_2O$ and Na_2PdCl_4 were from Aldrich (99.99% purity), activated carbon from Camel (X40S; SA = 900–1100 m^2/g; PV = 1.5 mL/g; pH 9–10), commercial carbon nanotubes, Baytubes (average diameter of 10 ± 2 nm and a specific surface area of 288 m^2/g), from Bayer, commercial CNFs PR24-PS from Applied Science (average diameter of 88 ± 30 nm and a specific surface area of 43 m^2/g) and the 3% N-containing CNFs from the pre-oxidized CNFs by thermal treatment (10 g) with NH_3 at 600 °C for 4 h, as reported in [32]. TiO_2 (P25) was purchased from Degussa (SA = 50 m^2/g). $NaBH_4$ of purity >96% from Fluka and polyvinyl alcohol (PVA) (M_w = 13,000–23,000 87–89% hydrolysed) from Aldrich were used. Gaseous oxygen from SIAD was 99.99% pure.

3.2. Catalyst Preparation

1% Au_6Pd_4 Bimetallic Catalysts

$NaAuCl_4$ $2H_2O$ (0.072 mmol) was dissolved in 140 mL of H_2O, and PVA (2% w/w) was added (0.706 mL). The yellow solution was stirred for 3 minutes, and 0.1 M $NaBH_4$ (2.15 mL) was added under vigorous magnetic stirring. The ruby red Au(0) sol was immediately formed. A UV-visible spectrum of the gold sol was recorded to check the complete $AuCl_4^-$ reduction and the formation of the plasmon peak. Within a few minutes of sol generation, the gold sol (acidified until pH 2 by sulfuric acid) was immobilized by adding the support under vigorous stirring. The amount of support was calculated as having a final metal loading of 0.73 wt%. After 2 h, the slurry was filtered and the catalyst washed thoroughly with distilled water (neutral mother liquors). A check of Au loading was performed directly on the catalysts, confirming the quantitative adsorption of Au NPs of the sol. The Au/support was then dispersed in 140 mL of water; Na_2PdCl_4 (10 wt% in Pd solution) (0.0386 mL) and PVA (2% w/w) (0.225 mL) were added. H_2 was bubbled (50 mL/min) under atmospheric pressure and room temperature for 2 h. The slurry was filtered and the catalyst washed thoroughly with distilled water. The total metal loading of 1 wt% and the Au-Pd molar ratio 6:4 was confirmed by ICP analyses for all of the catalysts. The catalysts were labeled as 1% Au_6Pd_4/AC, 1% Au_6Pd_4/N-CNFs, 1% Au_6Pd_4/CNTs, 1% Au_6Pd_4/CNFs and 1% Au_6Pd_4/TiO_2.

3.3. Catalytic Tests

Glycerol was obtained as a by-product by traditional transesterification starting from a commercial and edible rapeseed vegetable oil and using KOH as the base. The composition of the rapeseed oil was completed by GC using a HP 7820A gas chromatograph equipped with a non-bonded, bis-cyanopropyl polysiloxane (100 m) capillary column. The homogeneous transesterification of the rapeseed oil was carried out at 40 °C for two hours, as described in this section.

To obtain a purified glycerol, the resulted glycerol solution was diluted in a ratio of 1/10 in order to reduce the fluid viscosity and used as obtained or further purified. In the latter case, H_2SO_4 was added in order to adjust the pH to 2 and to convert the soap into free fatty acid. The free fatty acid was removed by extraction with hexane. The obtained methanol-glycerol mixture was evaporated in a rotavapor in order to remove the methanol. The glycerol solution was then separated and treated with activated carbon in order to remove the free ions. The activated carbon was separated by filtration. Finally, the glycerol solution was diluted with water in order to obtain the desired concentration (0.3 M). The presence of possible impurities was verified using high-performance liquid chromatography (HPLC) using a column (Alltech OA-10308, 300 mm × 7.8 mm) with UV and refractive index (RI) detection or GC using an HP 7820A gas chromatograph equipped with a capillary column (HP-5 30 m × 0.32 mm, 0.25 μm film, by Agilent Technologies)) and a TCD detector. GC-MS was used for the identification of the products.

Glycerol oxidation: reactions were carried out in a 30-mL glass reactor equipped with a thermostat and an electronically-controlled magnetic stirrer connected to a 5000-mL reservoir charged with oxygen (3 atm). The oxygen uptake was followed by a mass-flow controller connected to a PC through an A/D board, plotting a flow time diagram.

Glycerol 0.3 M and the catalyst (substrate/total metal = 1000 mol/mol) were mixed in distilled water (total volume, 10 mL) and 4 equivalents of NaOH. The reactor was pressurized at 3 atm of oxygen and set to 50 °C. Once this temperature was reached, the gas supply was switched to oxygen, and the monitoring of the reaction started. The reaction was initiated by stirring. Recycling tests were carried out under the same conditions (substrate/total metal = 1000 mol/mol, 50 °C, 3 atm O_2, 1250 rpm, alcohol 0.3 M, 1 h for purified and 3 h for non-purified glycerol). A test was performed adding to a 0.3 M solution of pure glycerol 10 wt% of fatty acids (rapeseed).

The catalyst was recycled in the subsequent run after filtration without any further treatment. Samples were removed periodically and analyzed by high-performance liquid chromatography (HPLC) using a column (Alltech OA- 10308, 300 mm × 7.8 mm) with UV and refractive index (RI) detection to analyze the mixture of the samples. Aqueous H_3PO_4 solution (0.1 wt%) was used as the eluent. Products were identified by comparison with the original samples.

3.4. Characterization

3.4.1. Catalyst Characterization

(1) The metal content was checked by ICP analysis of the filtrate or, alternatively, directly on the catalyst after burning off the carbon, on a Jobin Yvon JY24.

(2) The morphology and microstructures of the catalysts were characterized by transmission electron microscopy (TEM). The powder samples of the catalysts were ultrasonically dispersed in ethanol and mounted onto copper grids covered with holey carbon film. A Philips CM200 FEG electron microscope, operating at 200 kV and equipped with an EDX DX4 analyzer system and a FEI Titan 80-300 electron microscope, operating at 300 kV and equipped with an EDX SUTW detector, were used for TEM observation.

(3) Fourier transform infrared (FT-IR) spectra were recorded with a JASCO FT/IR 410 spectrometer using a mercury cadmium telluride (MCT) detector. Spectra with 4 cm^{-1} resolution, 64 scans and a scan speed of 0.20 cm s^{-1} were recorded

at room temperature using KRS5 thallium bromo-iodide windows in the range 4000–1000 cm^{-1}. The 1% Au$_6$Pd$_4$/TiO$_2$ used, which was filtered and dried after the reaction, was mixed with KBr to obtain a pellet (pressed at 10 MPa for 10 minutes).

4. Conclusions

One percent Au$_6$Pd$_4$ nanoparticles deposited on different supports were investigated in pure and raw glycerol. We addressed the best catalytic performance of 1% Au$_6$Pd$_4$/AC in the presence of uniform alloy AuPd nanoparticles. Indeed, in the other cases, inhomogeneity in the AuPd composition, together with the partial segregation of Au and Pd, was detected. A general decrease in activity was observed when raw glycerol was used as the substrate. FT-IR results revealed the presence of traces of unsaturated fatty acids on the catalyst surface. These unsaturated compounds probably partially blocked the AuPd active sites, decreasing the activity. The effect of the nature of the glycerol was investigated on the durability of 1% Au$_6$Pd$_4$/AC. A strong deactivation was observed in the presence of non-purified glycerol due to the agglomeration and leaching of the metal nanoparticles. On the contrary, the catalyst showed a good stability when purified glycerol was used.

Author Contributions: Carine E. Chan-Thaw made the catalytic tests and worked on the redaction of the article. Sebastiano Campisi performed some additional catalytic tests. Di Wang made the TEM and EDX analysis. Laura Prati contributed by overviewing the manuscript, and finally, Alberto Villa synthesized the catalysts and overviewed the article.

Conflicts of Interest: The authors declare no conflict of interest.

Acknowledgments: TEM characterization was carried out in KIT and sponsored by Karlsruhe Nano Micro Facility (KNMF).

References

1. Zhou, C.-H.; Beltramini, J.N.; Fan, Y.-X.; Lu, G.Q. Chemoselective catalytic conversion of glycerol as a biorenewable source to valuable commodity chemicals. *Chem. Soc. Rev.* **2008**, *37*, 527–549.
2. Lawate, S.S.; Lal, K. High oleic polyol esters, compositions and lubricants functional fluids and greases containing the same. European Patent 0712834A1, November 1994.
3. Heidbreder, A.; Gruetzmacher, R.; Nagorny, U.; Westfechtel, A. Use of fatty ester-based polyols for polyurethane casting resins and coating materials. US Patent 2002/0161161 A1, October 2002.
4. Gallezot, P. Conversion of biomass to selected chemical products. *Chem. Soc. Rev.* **2012**, *41*, 1538–1558.
5. Mittelbach, M.; Remschmidt, C. *Biodiesel: The Comprehensive Handbook*; Boersedruck Ges M.B.H.: Vienna, Austria, 2004.
6. McCoy, M. Glycerin surplus. *Chem. Eng. News* **2006**, *84*, 7–8.

7. Gu, Y.; Jérôme, F. Glycerol as a sustainable solvent for green chemistry. *Green Chem.* **2010**, *12*, 1127–1138.

8. Carrettin, S.; McMorn, P.; Johnston, P.; Griffin, K.; Kiely, C.J.; Hutchings, G.J. Oxidation of glycerol using supported Pt, Pd and Au catalysts. *Phys. Chem. Chem. Phys.* **2003**, *5*, 1329–1336.

9. Porta, F.; Prati, L. Selective oxidation of glycerol to sodium glycerate with gold-on-carbon catalyst: an insight into reaction selectivity. *J. Catal.* **2004**, *224*, 397–403.

10. Demirel-Gülen, S.; Lucas, M.; Claus, P. Liquid phase oxidation of glycerol over carbon supported gold catalysts. *Catal. Today* **2005**, *102*, 166–172.

11. Dimitratos, N.; Villa, A.; Bianchi, C.L.; Prati, L.; Makkee, M. Gold on Titania: Effect of preparation method in the liquid phase oxidation. *Appl. Catal. A* **2006**, *311*, 185–192.

12. Dimitratos, N.; Villa, A.; Prati, L. Liquid Phase Oxidation of glycerol using a single phase (Au–Pd) alloy supported on activated carbon: Effect of reaction conditions. *Catal. Lett.* **2009**, *133*, 334–340.

13. Villa, A.; Wang, D.; Su, D.S.; Prati, L. Gold sols as catalysts for glycerol oxidation: The role of stabilizer. *ChemCatChem* **2009**, *1*, 510–514.

14. Chan-Thaw, C.E.; Villa, A.; Katekomol, P.; Su, D.; Thomas, A.; Prati, L. Covalent triazine framework as support for liquid phase reaction. *Nano Lett.* **2010**, *10*, 537–541.

15. Villa, A.; Veith, G.M.; Prati, L. Selective Oxidation of glycerol under acidic conditions using gold Catalysts. *Angew. Chem. Int. Ed.* **2010**, *49*, 4499–4502.

16. Gil, S.; Marchena, M.; Sánchez-Silva, L.; Sánchez, P.; Romero, A.; Valverde, J.L. Effect of the operation conditions on the selective oxidation of glycerol with catalysts based on Au supported on carbonaceous materials. *Chem. Eng. J.* **2011**, *178*, 423–435.

17. Gil, S.; Marchena, M.; María Fernández, C.; Sánchez-Silva, L.; Romero, A.; Valverde, J.L. Catalytic oxidation of crude glycerol using catalysts based on Au supported on carbonaceous materials. *Appl. Catal. A* **2013**, *450*, 189–203.

18. Skrzyńska, E.; Wondołowska-Grabowska, A.; Capron, M.; Dumeignil, F. Crude glycerol as a raw material for the liquid phase oxidation reaction. *Appl. Catal. A* **2014**, *482*, 245–257.

19. Konaka, A.; Tago, T.; Yoshikawa, T.; Nakamura, A.; Masuda, T. Conversion of glycerol into allyl alcohol over potassium-supported zirconia–iron oxide catalyst. *Appl. Catal. B* **2014**, *146*, 267–273.

20. Krähling, L.; Krey, J.; Jakobson, G.; Grolig, J.; Miksche, L. Allyl Compounds. In *Ullmann's Encyclopedia of Industrial Chemistry*; Wiley-VCH Verlag GmbH & Co. KGaA: Weinheim, Germany, 2000.

21. Corma, A.; Garcia, H. Supported gold nanoparticles as catalysts for organic reactions. *Chem. Soc. Rev.* **2008**, *37*, 2096–2126.

22. Della Pina, C.; Falletta, E.; Prati, L.; Rossi, M. Selective oxidation using gold. *Chem. Soc. Rev.* **2008**, *37*, 2077–2095.

23. Besson, M.; Gallezot, P. Selective oxidation of alcohols and aldehydes on metal catalysts. *Catal. Today* **2000**, *57*, 127–141.

24. Wang, D.; Villa, A.; Porta, F.; Su, D.; Prati, L. Single-Phase Bimetallic System for the Selective Oxidation of Glycerol to Glycerate. *Chem.Comm.* **2006**, *18*, 1956–1958.
25. Villa, A.; Campione, C.; Prati, L. Bimetallic gold/palladium catalysts for the selective liquid phase oxidation of glycerol. *Catal. Lett.* **2007**, *115*, 133–136.
26. Wang, D.; Villa, A.; Porta, F.; Prati, L.; Su, D. Bimetallic Gold/Palladium Catalysts: Correlation between Nanostructure and Synergistic Effects. *J. Phys. Chem. C* **2008**, *112*, 8617–8622.
27. Prati, L.; Villa, A.; Chan-Thaw, C.E.; Arrigo, R.; Wang, D.; Su, D.S. Gold catalyzed liquid phase oxidation of alcohol: the issue of selectivity. *Faraday Discuss.* **2011**, *152*, 353–365.
28. Prati, L.; Villa, A.; Lupini, A.R.; Veith, G.M. Gold on carbon. One billion catalysts under a single label. *Phys. Chem. Chem. Phys.* **2012**, *14*, 2969–2978.
29. Prati, L.; Villa, A.; Porta, F.; Wang, D.; Su, D.S. Single-phase gold/palladium catalyst: The nature of synergistic effect. *Catal. Today* **2007**, *122*, 386–390.
30. Wu, N.; Fu, L.; Su, M.; Aslam, M.; Chun Wong, K.; Dravid, V.P. Interaction of fatty acid monolayers with cobalt nanoparticles. *Nano Lett.* **2004**, *4*, 383–386.
31. Nor Hidawati, E.; Mimi Sakinah, A.M. Treatment of glycerin pitch from biodiesel production. *Int. J. Chem. Environ. Eng.* **2011**, *2*, 309–313.
32. Arrigo, R.; Haevecker, M.; Wrabetz, S.; Blume, R.; Lerch, M.; McGregor, J.; Parrott, E.P.J.; Zeitler, J.A.; Gladden, L.F.; Knop-Gericke, A.; Schloegl, R.; Su, D. Tuning the acid/base properties of nanocarbon by functionalization via amination. *J. Am. Chem. Soc.* **2010**, *132*, 9616–9630.

Fischer-Tropsch Synthesis on Multicomponent Catalysts: What Can We Learn from Computer Simulations?

José L. C. Fajín, M. Natália D. S. Cordeiro and José R. B. Gomes

Abstract: In this concise review paper, we will address recent studies based on the generalized-gradient approximation (GGA) of the density functional theory (DFT) and on the periodic slab approach devoted to the understanding of the Fischer-Tropsch synthesis process on transition metal catalysts. As it will be seen, this computational combination arises as a very adequate strategy for the study of the reaction mechanisms on transition metal surfaces under well-controlled conditions and allows separating the influence of different parameters, e.g., catalyst surface morphology and coverage, influence of co-adsorbates, among others, in the global catalytic processes. In fact, the computational studies can now compete with research employing modern experimental techniques since very efficient parallel computer codes and powerful computers enable the investigation of more realistic molecular systems in terms of size and composition and to explore the complexity of the potential energy surfaces connecting reactants, to intermediates, to products of reaction. In the case of the Fischer-Tropsch process, the calculations were used to complement experimental work and to clarify the reaction mechanisms on different catalyst models, as well as the influence of additional components and co-adsorbate species in catalyst activity and selectivity.

Reprinted from *Catalysts*. Cite as: Fajín, J.L.C.; Cordeiro, M.N.D.S.; Gomes, J.R.B. Fischer-Tropsch Synthesis on Multicomponent Catalysts: What Can We Learn from Computer Simulations?. *Catalysts* **2015**, *5*, 3–17.

1. Introduction

The Fischer-Tropsch synthesis is a chemical process discovered by the chemists Franz Fischer and Hans Tropsch in the nineteen-twenties [1]. The main reaction of this chemical process consists in the combination of hydrogen and carbon monoxide (synthesis gas, with H_2/CO ratio of ~2 that is controlled by the water gas shift reaction) on a solid catalyst to produce liquid hydrocarbons. During the Fischer-Tropsch process many different hydrocarbons, e.g., methane, alkanes, alkenes, branched alkanes, and oxygenated compounds, e.g., alcohols, aldehydes, ketones, and fatty acids, from secondary reaction routes are also produced [2–6].

In order to produce synthetic fuels, typically from coal, natural gas or biomass, the process conditions and catalyst composition are usually chosen to favor the production of hydrocarbons with a long-straight chain ($>C_5$) and to minimize the formation of methane, olefins, branched alkanes and oxygenated species [7–11]. Thus, the route for the production of long-straight chain alkanes can be well described by the following chemical equation:

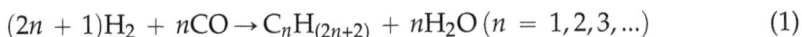

$$(2n + 1)H_2 + nCO \rightarrow C_nH_{(2n+2)} + nH_2O \,(n = 1,2,3,...) \qquad (1)$$

Usually, only relatively small quantities of non-alkane products are formed, although catalysts favoring some of these byproducts have been also developed [12,13].

The Fischer-Tropsch synthesis process was intensively used during the World War II period for the production of synthetic fuels from coal due to the difficulties felt by some countries to grant access to sufficient amounts of crude oil [14]. Since then, catalysts based on iron, cobalt and ruthenium metals have been the most widely used. Unfortunately, all are far from being perfect since other factors beyond activity loss due to catalyst aging (e.g., conversion of the active phase, loss of active specific surface area, fouling, *etc.*) are found to affect their performances or hamper their improvement. For instance, iron, despite being very cheap, forms many different phases, but the most active one is still unresolved; cobalt seems to be the most effective of these three transition metals but it is dramatically poisoned by sulfur species, which prevents its utilization when used with syngas from coal or biomass; and ruthenium is associated with interesting activities at low temperature but it is very expensive. Additionally, energy costs are quite large with any of the catalysts developed so far and the process becomes very interesting only in situations associated with high prices of crude oil.

Nowadays, oil reserves are decreasing quite fast, which is being accompanied by a concomitant raise of the petroleum price. Hence, many researchers are focusing their current scientific interests in the development of new catalysts for the Fischer-Tropsch synthesis [15]. Not only Fe, Co and Ru are being considered in this quest for novel catalysts for the Fischer-Tropsch synthesis reaction. Other transition metals (and alloys of different metals) are being considered, especially those from group VIIIA in the Periodic Table because they display some activity in the C–C coupling reaction during CO hydrogenation [13,16–29]. Yet, this is not a trivial task since small variations in the catalyst composition change dramatically the reaction yield and/or selectivity. For instance, the addition of Pt or Ru to Co enhances the activity of the catalyst for the Fischer-Tropsch reaction [30], but the addition of high quantities of Pt to Co based catalysts changes their selectivity toward methanol production [31].

The inclusion of other components into the traditional catalysts for the Fischer-Tropsch reaction can be used to improve the activity and selectivity of the catalyst, as for example with the addition of ionic additives to ruthenium nanoparticles, which increases the activity and modifies the selectivity toward long-straight hydrocarbon chains, olefins or alcohols, depending on the ionic addictive used [12]. Other elements can be added to the catalyst for preventing its poisoning by deposition of species like carbon [32,33] or sulfur [34]. High selectivities toward other byproducts can be also achieved by the use of promoters. For instance, the inclusion of Fe into $Ni/\gamma-Al_2O_3$ is accompanied by an increase in the CO methanation [35]. The latter reaction is also enhanced when Ca, La, K, Ni and/or Co are added to a Ru/TiO_2 catalyst, being especially active the $Ca-Ni-Ru/TiO_2$ one [36]. The production of olefins is favored by catalysts based on Fe-Mn-V-K [37] or based on Fe-Mn nanoparticles supported on carbon nanotubes [38]. Branched hydrocarbons can be obtained from the Fischer-Tropsch synthesis on Fe-Si-K and Co/SiO_2 based catalysts through the alkylidene mechanism [39].

The activity and selectivity of the multicomponent catalysts are determined by the catalyst composition, but also by other factors such as the temperature [40], the oxidized/reduced level of the catalyst components, which is modulated with promoters [41–43], or the size of the catalyst particles [11,28,44,45]. Furthermore, the incorporation of different species into the catalyst (i) can modulate the catalyst activity by conferring a different microstructure to the active sites of the catalyst [46]; (ii) can establish synergic effects with preexistent phases [47] or (iii) can have a separate role from the other catalyst phases [31,48].

The examples above clearly show that our understanding on the role of each catalyst phase and on the most favorable mechanistic pathways in catalysts suggested for the Fischer-Tropsch synthesis is still in its infancy. Therefore, scientific knowledge gathered from experimental and computational studies on catalyst models working under well-controlled conditions is mandatory. Herewith, it is reviewed the literature on the application of computational methods to the study of the Fischer-Tropsch synthesis on multicomponent catalysts, in particular that with emphasis in the determination of the preferential reaction mechanisms.

2. Computational Studies on Catalyst Surface Models

As referred above, the addition of extra components to the traditional Fischer-Tropsch catalysts can enhance their performance by increasing the activity [30,36] and selectivity [12,31,37–39], or by avoiding catalyst poisoning [32–34]. In this section we will show how computational methods can help in the understanding of the Fisher-Tropsch catalysis by solid catalysts. In fact, computational approaches have been widely used in the study of reaction mechanisms on simple catalysts models, which are usually based on monometallic surfaces. In the case of the Fischer-Tropsch synthesis reaction, Ojeda *et al.* [19] studied through density functional theory, *cf.*

Table 1, the formation of methane from CO and H_2 on Fe(110) and Co(0001) surfaces. Their calculations show that the reaction of direct CO bond cleavage,

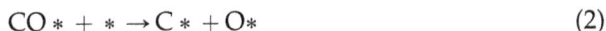

$$CO* + * \rightarrow C* + O* \qquad (2)$$

where * denotes a free site on the catalyst surface, is energetically much more expensive than the dissociation of the C–O via the hydroxymethylidyne, COH, intermediate,

$$CO* + H* \rightarrow COH* + * \qquad (3a)$$

$$COH* + * \rightarrow C* + OH* \qquad (3b)$$

and also more energetically expensive than the dissociation following the route through the formyl (HCO or oxymethylidyne) intermediate,

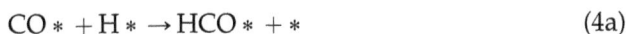

$$CO* + H* \rightarrow HCO* + * \qquad (4a)$$

$$HCO* + * \rightarrow CH* + O* \qquad (4b)$$

with energetic data summarized in Table 2. A representation of the reaction steps corresponding to the carbide, hydroxymethylidyne and formyl routes for the CO dissociation in the Fischer-Tropsch reaction can be seen in Figure 1.

On both surfaces the barriers calculated for the dissociation reactions described by Equation (2) (Fe, 1.96 eV; Co, 3.80 eV) and 3b (Fe, 1.63 eV; Co, 3.26 eV) are significantly larger than those calculated for reaction 4b (Fe, 0.79 eV; Co, 0.95 eV), *cf.* Table 2. Additionally, the barriers for formation of the HCO species (Equation (4a)) are smaller, on Fe(110), or identical, on Co(0001), to those leading to the COH intermediate [19]. These results suggest that the breakage of the C–O bond occurs only after hydrogenation of the carbon atom of the CO species [19]. Using another computational code, Inderwildi *et al.* reported that the C–O bond dissociation from the formyl species was also much more favorable than the direct C–O bond breakage on Fe(111) [49], with values 1.17 eV (Equation 4b) and 1.76 eV (Equation (2)), and on Co(0001) [50], with values 1.00 eV (Equation 4b) and 2.82 eV (Equation (2)). A similar mechanistic profile for the dissociation of the C–O bond was obtained when the unity bond index-quadratic exponential potential (UBI-QEP) method [68] was applied to the synthesis of C_1–C_2 alkanes or alkenes on Co [69].

Table 1. Computational details for the calculations devoted to the study of the Fischer-Tropsch process on transition metal surfaces that are reviewed in this article.

Code [a]	DFT [b]	Basis set [c]	Transition metal surface(s) [d]	Ref.
DACAPO	PW91	PW-USPP	Fe(110) and Co(0001)	Ojeda et al. [19]
DACAPO	RPBE	PW-USPP	Ru(0001) and Ru(10$\bar{1}$9)	Vendelbo et al. [23]
VASP	PBE	PW-PAW	Co(0001), Pt@Co(0001), and Ru@Co(0001)	Balakrishnan et al. [32]
CASTEP	PW91	PW-USPP	Pt(111), Pd(111), and Ru(0001)	Inderwildi et al. [49]
CASTEP	PW91	PW-USPP	Co(0001)	Inderwildi et al. [50]
VASP	RPBE	PW-USPP	Ru(0001)	Loveless et al. [51]
VASP	PW91	PW-USPP	Ru(10$\bar{1}$5)	Ciobica et al. [52]
VASP	PBE	PW-PAW	Ru(11$\bar{2}$1)	Shetty et al. [53]
VASP	PBE	PW-PAW	Ru(10$\bar{1}$0)B and Co(10$\bar{1}$0)B	Shetty et al. [54]
VASP	PBE	PW-PAW	Co(0001), Co(10$\bar{1}$2), and Co(11$\bar{2}$0)	Liu et al. [55]
VASP	PW91	PW-PAW	Ni(111), Ni(110), Rh@Ni(111), Rh@Ni(110), Ru@Ni(111) and Ru@Ni(110)	Fajín et al. [56]
Siesta	PBE	DZP-TMPP	Co(0001)	Cheng et al. [57]
Siesta	PBE	DZP-TMPP	Ru(001) [e], Fe(210), Rh(211), and Re(001)[e]	Cheng et al. [58]
VASP	PW91	PW-PAW	Rh(111) and Rh(211)	van Grootel et al. [59]
SeqQuest	PBE	DZP-TMPP	Ni(111)	Mueller et al. [60]
-	GGA	PW-USPP	Rh(111)	Zhang et al. [61]
VASP	PW91	PW-PAW	Rh(111), Ni(111), and Rh-Ni(111)	Lee et al. [62]

[a] For further details please refer to the websites https://wiki.fysik.dtu.dk/dacapo/, http://www.vasp.at/, http://www.uam.es/siesta/, and http://dft.sandia.gov/Quest/. [b] PW91 (Perdew-Wang 91 functional [63]), RPBE (Revised Perdew-Burke-Ernzerhof functional [64]), PBE (Perdew-Burke-Ernzerhof functional [65]) and GGA (generalized gradient approximation). [c] PW-USPP (plane-wave for valence and ultrasoft pseudopotential for core electrons), PW-PAW (plane-wave for valence and projected augmented wave method for core electrons [66,67]), DZP-TMPP (double zeta plus polarization for valence and Troullier-Martins norm-conserving scalar relativistic pseudopotential for core). [d] Periodic slab models were employed to define the different Miller index surfaces. [e] Two rows were removed in the top layer of the Ru(001) and Re(001) surfaces for obtaining a stepped model surface.

Table 2. Activation energy barriers (eV) for the CO bond break through different reaction routes on several metallic surfaces [a].

Reaction route	Fe(110)	Co(0001)	Fe(111)
CO* + * → C* + O*	1.96 [19]	3.80 [19]; 2.82 [50]	1.76 [49]
CO* + H* → COH* + * → C* + OH*	1.63 [19]	3.26 [19]	
CO* + H* → HCO* + * → CH* + O*	0.79 [19]	0.95 [19]; 1.00 [50]	1.17 [49]
HCO* + H* → HCOH* + * → CH* + OH*	0.65 [19]	1.10 [19]	
HCO* + H* → H$_2$CO* + * → CH$_2$* + O*	3.29 [19]	1.63 [19]	

[a] For reaction routes including several steps, the energy corresponds to the highest barrier along the route.

The decomposition of the formyl species (Equation (4b)) has to compete with further hydrogenation of the latter species to hydroxymethylene, HCOH, followed by its decomposition,

$$HCO* + H* \rightarrow HCOH* + * \tag{5a}$$

$$HCOH* + * \rightarrow CH* + OH* \tag{5b}$$

or leading to formaldehyde, H_2CO, followed by its decomposition,

$$HCO* + H* \rightarrow H_2CO* + * \qquad (6a)$$

$$H_2CO* + * \rightarrow CH_2* + O* \qquad (6b)$$

While the formation of the formaldehyde species appears to be possible [19], the cleavage of the C–O bond in the latter species seems to be prohibited, because reaction 6b has to surmount quite large energy barriers with values 3.29 eV on Fe(110) and 1.63 eV on Co(0001) [19]. Thus, the fate of formaldehyde will be one of the following: (i) decomposition into HCO and H surface species (reverse of reaction 6a); (ii) desorption from the surface; or (iii) further reaction to other oxygenated compounds, e.g., formate, HCOO, species. However, based on the DFT computed barrier for the reverse of reaction 6a on Co(0001), i.e., 0.22 eV (0.13 eV with zero-point energy corrections) [70], formaldehyde desorption or further reaction to other oxygenated compounds seem difficult.

DFT-derived barriers for the reaction of Fischer-Tropsch synthesis on Ru catalysts show that CO predominantly reacts at (111) terraces through H-assisted reactions via the formyl route [51]. The formation of formyl species was experimentally detected by Eckle et al. [71] in time resolved in situ diffuse reflectance infrared Fourier transform spectroscopy (DRIFTS) measurements on Ru/Al_2O_3 catalyst in idealized reformate ($CO/H_2/N_2$) conditions (band at 1760 cm^{-1}). As it happens for Co and Fe, barriers on Ru catalysts for the unassisted reaction are much higher than those calculated after hydrogenation; this is the case of CO dissociation on the planar (111) terraces and also on low-coordination atoms at step-edge sites. The latter conclusions contrast with results arising from DFT/PW91 studies by van Santen and co-workers, concerning the CO dissociation on stepped Ru(10$\bar{1}$5) [52], and on open Ru(11$\bar{2}$1) [53] or Ru(10$\bar{1}$0)B [54] surfaces, where calculated barriers for CO direct dissociation on low-coordinated sites are only 0.92 eV, 0.67 eV and 0.49 eV, respectively. The barriers on these three Ru surfaces are significantly smaller than that calculated for the reaction on the planar Ru(0001) surface, i.e., 2.35 eV [52]; such barriers are also smaller than those required to cleave the C–O bond via the hydroxymethylidyne or the formyl routes. The latter conclusions can be extended to Co based catalysts [54]. However, formation of formyl species is predicted on the Co(10$\bar{1}$0)B surface but its decomposition seems to be unaffordable [54]. Interestingly, in the DRIFTS experiments of Eckle et al. [71] but on a Ru/zeolite catalyst, it was not observed the band at 1760 cm^{-1} attributed to adsorbed HCO species. These results are strong evidences that quite small changes in the structure of a typical catalyst for the Fischer-Tropsch reaction can change dramatically its activity. Similar conclusions are attained from recent DFT/PBE calculations by Liu et al., which demonstrate that the Fischer-Tropsch reaction on Co catalysts is structure sensitive [55]. In the case of

331

the flat Co(0001) surface, the mechanism through the formyl intermediate is the most favorable while on Co surfaces possessing less coordinated atoms, the mechanism based on the direct CO dissociation on the surface is also feasible.

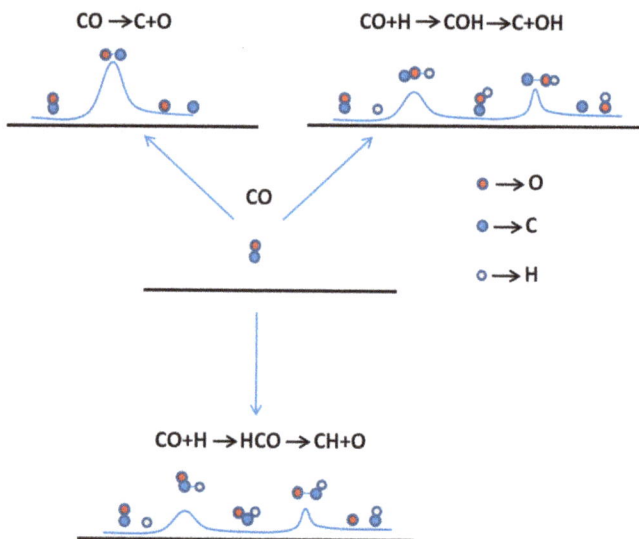

Figure 1. Representation of the carbide (**top left**), hydroxymethylidyne (**top right**) and formyl (**bottom**) routes for the CO dissociation in the Fischer-Tropsch reaction.

Recently, we considered the DFT/PW91 approach to study the mechanisms of methane and methanol formation on nickel based catalysts [56]. Nickel is active per se and has been found to promote the activity of iron and cobalt catalysts in the Fischer-Tropsch reaction [72] and references therein]. Therefore, it is not surprising to find in the literature several reports on the consideration of Ni-based catalysts for this reaction [73]. However, the number of studies with Ni-based catalysts is much smaller than with catalysts based on Co and Fe, maybe because the former catalysts are associated with formation of volatile carbonyls, which leads to deactivation and loss of the active phase [73]. As it happens in so many occasions, the catalyst formulation and the nature of the support seem to affect the conclusions about the influence of Ni species in the catalytic activity. In our computational study, our goal was to understand how the addition of a second metal (bimetallic catalyst) affect the activity of Ni in the Fischer-Tropsch synthesis reaction. With that aim we prepared flat (111) and crest (110) models of the Rh@Ni and Ru@Ni catalytic surfaces and compared the reaction mechanistic profiles on the latter with the profile calculated on a pure Ni surface [56]. We have found that surfaces possessing low coordinated atoms are more reactive than flat surfaces in the catalysis of dissociative reaction steps, which are in fact the rate determining steps of the process. Furthermore, it was found that the reactions toward methanol and methane formation preferentially evolve through the formyl intermediate on all the surfaces. Additionally, the reaction selectivity toward methane or methanol was affected by the surface doping and by the presence of low coordinated atoms on the surface. In fact, the routes leading toward methanol and methane on pure nickel surfaces and on flat surfaces doped with Rh or Ru atoms have similar activation energy barriers, while in the routes on crested doped surfaces the methane formation is more feasible than that of methanol [56].

DFT calculations were also used in studies of the mechanisms of hydrocarbon chain growth and chain termination, which are essential aspects in the synthesis of long chain hydrocarbons and oxygenated compounds through the Fischer-Tropsch process [57,58,74]. The reaction routes proposed are the carbide (also known as alkyl mechanism) and the CO insertion mechanisms, *cf.* Figure 2 [74]. In the carbide mechanism, the CO molecules dissociate to form CH_x intermediates. This process initiates the overall reaction, and the growth of the chains by reaction of these intermediates with other adsorbed C_nH_y intermediates eventually formed on the catalyst surface. In the CO insertion mechanism, the process is initiated by the formation of the CH_x species on the catalyst surface upon dissociation of CO, which is followed by (i) insertion of a CO molecule into the CH_x intermediate and by (ii) breakage of the CO bond in the resulting compound. This two-step mechanism will lead to the formation of long chain C_nH_y species. Thus, as it is well-illustrated in Figure 2, the chain growth in the carbide mechanism occurs through consecutive insertion of CH_x groups, while in the case of the CO insertion mechanism the chain growth occurs by sequential insertion of a CO molecule in C_nH_y species and cleavage of the inserted C–O bond. The reader must be aware of the mechanistic complexity of the Fischer-Tropsch process since, depending on the conditions, the reaction may go in the opposite direction, *i.e.*, to the steam reforming of alcohols or hydrocarbons [59,60,75], being Ni [76,77], Co [78–81], or Rh [78] the most common catalysts for the latter reaction. Although experimental knowledge acquired along the years allowed the proposal of different reaction routes, later analyses proved several of those to be less probable. An overview of such routes can be found in the work of Davis *et al.* [7].

Cheng *et al.* [57,58] investigated by DFT which are the CH_x intermediates that are preferentially coupled during the hydrocarbon chain growth on stepped Co, Rh, Ru, Fe and Re surfaces via the carbide mechanism. They found that despite the transition state structures are quite similar, the energy barriers for the different C–C coupling reactions differ considerably among the transition metals considered in their work, which leads to different preferential paths for the chain growth in each metal. In fact, C + CH and CH + CH paths are preferred on Ru and Rh, C + CH$_3$ and CH$_2$ + CH$_2$ on Co, C + CH$_3$ on Fe and C + CH on Re. Thus, with the exception of Ru and Rh surfaces, different coupling pathways are observed on different metal surfaces.

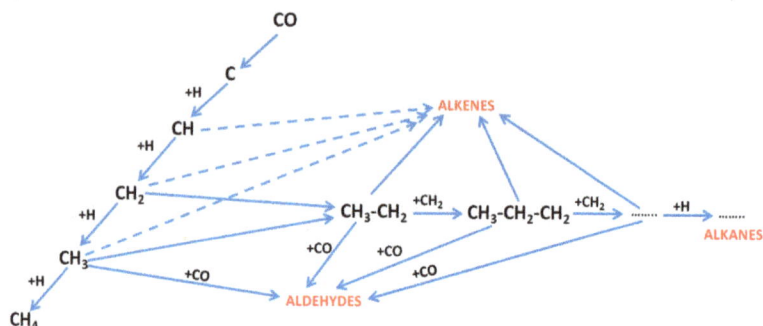

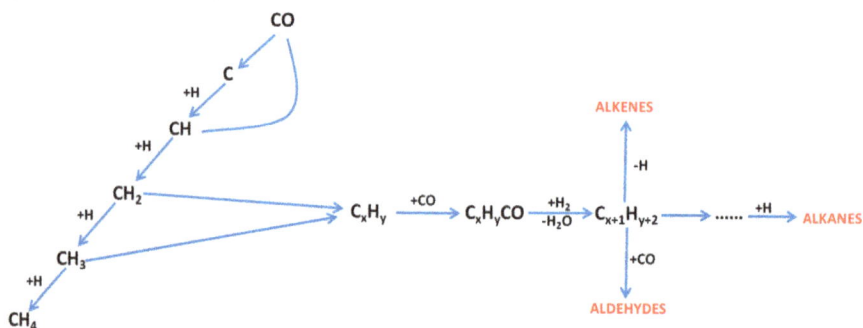

Figure 2. Representation of the Fischer-Tropsch (**a**) carbide and (**b**) CO insertion mechanisms steps (all the intermediates species represented are adsorbed on the surface).

DFT calculations have been also used with success in the explanation of the catalyst deactivation by carbon deposition or sulfur [61,62,82]. Within such works, Zhang et al. [61] studied the co-adsorption of CO and S species on the Rh(111) surface and found that the chemisorption of each species is not significantly influenced by the chemisorption of the other species. Thus, they concluded that the interaction between the CO and S species co-adsorbed on the metal surface is mainly short range type and, therefore, the suggestion for the responsibility of the long range electronic effect in the reduction of the CO methanation in presence of S is very likely incorrect. Czekaj et al. [82] studied also the poisoning of a Ni/γ-Al$_2$O$_3$ catalyst by sulfur with the RPBE approach and cluster models. They found several sulfur containing species to be stable on the catalyst, both on the metallic particles and on the oxide support, as for example, carbonyl sulfide, hydrogen sulfide and hydrogen thiocarbonates. The high stability of the sulfur containing adsorbates was suggested to deactivate the catalyst. Lee et al. [62] investigated by DFT the effect of adding Ni in the improvement of the sulfur tolerance of a Rh catalyst for the reaction of CO dissociation. In their calculations, they compared the adsorption and dissociation of CO on bimetallic RhNi(111) catalysts having compositions Rh_1Ni_2 and Rh_2Ni_1 with those on pure Ni(111) and Rh(111) surfaces. They demonstrate that the CO dissociation is less affected by sulfur poison on the bimetallic catalyst than on pure rhodium. This was attributed to the effect of Ni on the bimetallic catalysts which mitigates the repulsion of the adsorbed sulfur with the transition state structure during the CO dissociation. In fact, the repulsion between the adsorbed sulfur and the stretched CO molecule leads to higher reaction barriers than on the clean rhodium surfaces [62]. In the case of Fischer-Tropsch reaction poisoning, DFT calculations allowed to explain the effect of the Pt and Ru promoters in the prevention of the Co catalyst deactivation by carbon deposition at the catalyst surface in the reaction of CO hydrogenation. Balakrishnan *et al.* [32] considered pure Co and bimetallic models, where Co atoms were replaced by Pt or Ru, and found that the presence of the promoters in the catalyst decreased the carbon hydrogenation barrier and increased the carbon-carbon coupling barrier, but without changing the barriers for the diffusion of C atoms. These authors also found that the effect of Pt in the destabilization of adsorbed carbon structures was more significant than that of Ru. Very recently, it was unraveled how carbon deposition (together with CO adsorbates) can affect the surface roughness (surface reconstruction) of Co catalysts [33].

3. Conclusions

In this review it was analyzed how the inclusion of additional components in the traditional catalysts for the Fischer-Tropsch reaction catalysis changes the preference for a specific reaction mechanism and how they influence the catalyst activity and selectivity.

It was analyzed the energetics of the dissociation of the C–O bond in carbon monoxide. Generally speaking, the density functional calculations suggest that the direct dissociation of CO is less feasible than the dissociation of the bond in the products of its hydrogenation, *i.e.*, hydroxymethylidyne (COH) or formyl (HCO) species. In most cases, the formation of the latter species is much easier and the dissociation of its C–O bond is the less costly. Nevertheless, the breakage of the C–O bond in the formyl species has to compete with the formation of hydroxymethylene (HCOH) or formaldehyde (H_2CO) by further hydrogenation, and it was pointed out that the direct C–O bond rupture in CO competes with the dissociation of the bond in the formyl intermediate occurring on flat Co(0001) surfaces.

The chain growth was also explained with the aid of results from density functional theory calculations. Two alternative mechanisms, *i.e.*, the carbide and the CO insertion, arise as the most probable and are the most accepted nowadays. It is suggested that the occurrence of one or another is dramatically dependent on the nature of the catalyst and on the reaction conditions considered.

It was also reviewed the effects of poisoning and promoting species in the global catalytic activity by looking at the stability (energies of adsorption) of poisoning species such as sulfur containing molecules or carbon.

From what has been said above, the combination of density functional theory and realistic catalyst models provided very useful adsorption, diffusion, reaction and activation energies for the elementary steps in the Fischer-Tropsch process, which allowed the clarification of the preferential reaction mechanisms and also the understanding of atomic level structural details and their influence in the catalytic activity. Such information is crucial to design more efficient catalysts in the near future.

Acknowledgments: Thanks are due to Fundação para a Ciência e Tecnologia (FCT), Lisbon, Portugal, and to FEDER for financial support to REQUIMTE (projects Pest-C/EQB/LA0006/2013 and NORTE-07-0124-FEDER-000067-NANOCHEMISTRY) and to CICECO (projects Pest-C/CTM/LA0011/2013 and FCOMP-01-0124-FEDER-037271) and for Programa Investigador FCT. This work has been supported also by FCT through project PTDC/QUI-QUI/117439/2010 (FCOMP-01-0124-FEDER-020977) co-financed by *Programa COMPETE*. JLCF acknowledges FCT for the grant SFRH/BPD/64566/2009 co-financed by the *Programa Operacional Potencial Humano* (POPH)/*Fundo Social Europeu* (FSE); *Quadro de Referência Estratégico Nacional 2009–2013 do Governo da República Portuguesa*.

Author Contributions: José L. C. Fajín and José R. B. Gomes conceived the outline of the review and selected the materials. All authors contributed to the writing of the paper.

Conflicts of Interest: The authors declare no conflict of interest.

References

1. Fischer, F.; Tropsch, H. Synthesis of petroleum at atmospheric pressures from gasification products of coal. *Brennst. Chem.* **1926**, *7*, 97–104.
2. Anderson, R.B. *The Fischer Tropsh Synthesis*; Academic Press: New York, NY, USA, 1984.
3. Kusama, H.; Okabe, K.; Arakawa, H. Characterization of Rh-Co/SiO$_2$ catalysts for CO$_2$ hydrogenation with TEM, XPS and FT-IR. *Appl. Catal. A* **2001**, *207*, 85–94.
4. Schulz, H. Major and minor reactions in Fischer-Tropsch synthesis on cobalt catalysts. *Top. Catal.* **2003**, *26*, 73–85.
5. Lögdberg, S.; Lualdi, M.; Järås, S.; Walmsley, J.C.; Blekkan, E.A.; Rytter, E.; Holmen, A. On the selectivity of cobalt-based Fischer-Tropsch catalysts: Evidence for a common precursor for methane and long-chain hydrocarbons. *J. Catal.* **2010**, *274*, 84–98.
6. Guczi, L.; Stefler, G.; Koppány, Z.; Borkó, L. CO hydrogenation over Re-Co bimetallic catalyst supported over SiO$_2$, Al$_2$O$_3$ and NaY zeolite. *React. Kinet. Catal. Lett.* **2001**, *74*, 259–269.
7. Davis, B.H. Fischer-Tropsch synthesis: Current mechanism and futuristic needs. *Fuel Process. Technol.* **2001**, *71*, 157–166.
8. Tupabut, P.; Jongsomjit, B.; Praserthdam, P. Impact of boron modification on MCM-41-supported cobalt catalysts for hydrogenation of carbon monoxide. *Catal. Lett.* **2007**, *118*, 195–202.

9. Dorner, R.W.; Hardy, D.R.; Williams, F.W.; Davis, B.H.; Willauer, H.D. Influence of gas feed composition and pressure on the catalytic conversion of CO_2 to hydrocarbons using a traditional cobalt-based Fischer-Tropsch catalyst. *Energ. Fuel.* **2009**, *23*, 4190–4195.

10. Rønning, M.; Tsakoumis, N.E.; Voronov, A.; Johnsen, R.E.; Norby, P.; van Beek, W.; Borg, Ø.; Rytter, E.; Holmen, A. Combined XRD and XANES studies of a Re-promoted $Co/\gamma\text{-}Al_2O_3$ catalyst at Fischer-Tropsch synthesis conditions. *Catal. Today* **2010**, *155*, 289–295.

11. Bezemer, G.L.; Bitter, J.H.; Kuipers, H.P.C.E.; Oosterbeek, H.; Holewijn, J.E.; Xu, X.; Kapteijn, F.; van Dillen, A.J.; de Jong, K.P. Cobalt particle size effects in the Fischer-Tropsch reaction studied with carbon nanofiber supported catalysts. *J. Am. Chem. Soc.* **2006**, *128*, 3956–3964.

12. Wang, C.; Zhao, H.; Wang, H.; Liu, L.; Xiao, C.; Ma, D. The effects of ionic additives on the aqueous-phase Fischer-Tropsch synthesis with a ruthenium nanoparticle catalyst. *Catal. Today* **2012**, *183*, 143–153.

13. Inderwildi, O.R.; Jenkins, S.J. *In-silico* investigations in heterogeneous catalysis—Combustion and synthesis of small alkanes. *Chem. Soc. Rev.* **2008**, *37*, 2274–2309.

14. Stranges, A.N. A History of the Fischer-Tropsch Synthesis in Germany 1926–45. In *Fischer-Tropsch Synthesis, Catalysts, and Catalysis*, 1st ed.; Davis, B.H., Occelli, M.L., Eds.; Elsevier B.V.: Amsterdam, The Netherlands, 2007; pp. 1–28.

15. Davis, B.H.; Occelli, M.L. *Advances in Fischer-Tropsch Synthesis, Catalysts, and Catalysis*; CRC Press, Taylor & Francis Group: Boca Raton, FL, USA, 2010.

16. Jacquemin, M.; Beuls, A.; Ruiz, P. Catalytic production of methane from CO_2 and H_2 at low temperature: Insight on the reaction mechanism. *Catal. Today* **2010**, *157*, 462–466.

17. Park, S.-J.; Kim, S.-M.; Woo, M.H.; Bae, J.W.; Jun, K.-W.; Ha, K.-S. Effects of titanium impurity on alumina surface for the activity of $Co/Ti\text{-}Al_2O_3$ Fischer-Tropsch catalyst. *Appl. Catal. A* **2012**, *419–420*, 148–155.

18. Nawdali, M.; Bianchi, D. The impact of the Ru precursor on the adsorption of CO on Ru/Al_2O_3: Amount and reactivity of the adsorbed species. *Appl. Catal. A* **2002**, *231*, 45–54.

19. Ojeda, M.; Nabar, R.; Nilekar, A.U.; Ishikawa, A.; Mavrikakis, M.; Iglesia, E. CO activation pathways and the mechanism of Fischer-Tropsch synthesis. *J. Catal.* **2010**, *272*, 287–297.

20. Gual, A.; Godard, C.; Castillon, S.; Curulla-Ferré, D.; Claver, C. Colloidal Ru, Co and Fe-nanoparticles. Synthesis and application as nanocatalysts in the Fischer-Tropsch process. *Catal. Today* **2012**, *183*, 154–171.

21. Senanayake, S.D.; Evans, J.; Agnoli, S.; Barrio, L.; Chen, T.-L.; Hrbek, J.; Rodriguez, J.A. Water-gas shift and CO methanation reactions over $Ni\text{-}CeO_2(111)$ catalysts. *Top. Catal.* **2011**, *54*, 34–41.

22. Bundhoo, A.; Schweicher, J.; Frennet, A.; Kruse, N. Chemical transient kinetics applied to CO hydrogenation over a pure nickel catalyst. *J. Phys. Chem. C* **2009**, *113*, 10731–10739.

23. Vendelbo, S.B.; Johansson, M.; Mowbray, D.J.; Andersson, M.P.; Abild-Pedersen, F.; Nielsen, J.H.; Nørskov, J.K.; Chorkendorff, I. Self blocking of CO dissociation on a stepped ruthenium surface. *Top. Catal.* **2010**, *53*, 357–364.

24. Williams, C.T.; Black, C.A.; Weaver, M.J.; Takoudis, C.G. Adsorption and hydrogenation of carbon monoxide on polycrystalline rhodium at high gas pressures. *J. Phys. Chem. B* **1997**, *101*, 2874–2883.

25. Jenewein, B.; Fuchs, M.; Hayek, K. The CO methanation on Rh/CeO_2 and CeO_2/Rh model catalysts: A comparative study. *Surf. Sci.* 2003; 532–535, 364–369.

26. Bulushev, D.A.; Froment, G.F. A DRIFTS study of the stability and reactivity of adsorbed CO species on a $Rh/\gamma-Al_2O_3$ catalyst with a very low metal content. *J. Mol. Catal. A* **1999**, *139*, 63–72.

27. Panagiotopoulou, P.; Kondarides, D.I.; Verykios, X.E. Mechanistic aspects of the selective methanation of CO over Ru/TiO_2 catalyst. *Catal. Today* **2012**, *181*, 138–147.

28. Karelovic, A.; Ruiz, P. Mechanistic study of low temperature CO_2 methanation over Rh/TiO_2 catalysts. *J. Catal.* **2013**, *301*, 141–153.

29. Izquierdo, U.; Barrio, V.L.; Bizkarra, K.; Gutierrez, A.M.; Arraibi, J.R.; Gartzia, L.; Bañuelos, J.; Lopez-Arbeloa, I.; Cambra, J.F. Ni and Rh-Ni catalysts supported on zeolites L for hydrogen and syngas production by biogas reforming processes. *Chem. Eng. J.* **2014**, *238*, 178–188.

30. Pirola, C.; Scavini, M.; Galli, F.; Vitali, S.; Comazzi, A.; Manenti, F.; Ghigna, P. Fischer-Tropsch synthesis: EXAFS study of Ru and Pt bimetallic Co based catalysts. *Fuel* **2014**, *132*, 62–70.

31. Christensen, J.M.; Medford, A.J.; Studt, F.; Jensen, A.D. High pressure CO hydrogenation over bimetallic Pt-Co catalysts. *Catal. Lett.* **2014**, *144*, 777–782.

32. Balakrishnan, N.; Joseph, B.; Bhethanabotla, V.R. Effect of Pt and Ru promoters on deactivation of Co catalysts by C deposition during Fischer-Tropsch synthesis: A DFT study. *Appl. Catal.* **2013**, *462–463*, 107–115.

33. Weststrate, C.J.; Ciobîcă, I.M.; Saib, A.M.; Moodley, D.J.; Niemantsverdriet, J.W. Fundamental issues on practical Fischer-Tropsch catalysts: How surface science can help. *Catal. Today* **2014**, *228*, 106–112.

34. Bambal, A.S.; Guggilla, V.S.; Kugler, E.L.; Gardner, T.H.; Dadyburjor, D.B. Poisoning of a silica-supported cobalt catalyst due to presence of sulfur impurities in syngas during Fischer-Tropsch synthesis: Effects of chelating agent. *Ind. Eng. Chem. Res.* **2014**, *53*, 5846–5857.

35. Tian, D.; Liu, Z.; Li, D.; Shi, H.; Pan, W.; Cheng, Y. Bimetallic Ni-Fe total-methanation catalyst for the production of substitute natural gas under high pressure. *Fuel* **2013**, *104*, 224–229.

36. Tada, S.; Kikuchi, R.; Takagaki, A.; Sugawara, T.; Oyama, S.T.; Satokawa, S. Effect of metal addition to Ru/TiO_2 catalyst on selective CO methanation. *Catal. Today* **2014**, *232*, 16–21.

37. Wang, G.; Zhang, K.; Liu, P.; Hui, H.; Tan, Y. Synthesis of light olefins from syngas over Fe-Mn-V-K catalysts in the slurry phase. *J. Ind. Eng. Chem.* **2013**, *19*, 961–965.

38. Xu, J.-D.; Zhu, K.-T.; Weng, X.-F.; Weng, W.-Z.; Huang, C.-J.; Wan, H.-L. Carbon nanotube-supported Fe-Mn nanoparticles: A model catalyst for direct conversion of syngas to lower olefins. *Catal. Today* **2013**, *215*, 86–94.

39. Shi, B.; Wu, L.; Liao, Y.; Jin, C.; Montavon, A. Explanations of the formation of branched hydrocarbons during Fischer-Tropsch synthesis by alkylidene mechanism. *Top. Catal.* **2014**, *57*, 451–459.

40. Pendyala, V.R.R.; Shafer, W.D.; Jacobs, G.; Davis, B.H. Fischer-Tropsch synthesis: Effect of reaction temperature for aqueous-phase synthesis over a platinum promoted Co/alumina catalyst. *Catal. Lett.* **2014**, *144*, 1088–1095.

41. Jermwongratanachai, T.; Jacobs, G.; Shafer, W.D.; Pendyala, V.R.R.; Ma, W.; Gnanamani, M.K.; Hopps, S.; Thomas, G.A.; Kitiyanan, B.; Khalid, S.; *et al.* Fischer-Tropsch synthesis: TPR and XANES analysis of the impact of simulated regeneration cycles on the reducibility of Co/aluminacatalysts with different promoters (Pt, Ru, Re, Ag, Au, Rh, Ir). *Catal. Today* **2014**, *228*, 15–21.

42. Ning, W.; Yang, S.; Chen, H.; Yamada, M. Influences of K and Cu on coprecipitated FeZn catalysts for Fischer-Tropsch reaction. *Catal. Commun.* **2013**, *39*, 74–77.

43. Shimura, K.; Miyazawa, T.; Hanaoka, T.; Hirata, S. Factors influencing the activity of Co/Ca/TiO$_2$ catalyst for Fischer-Tropsch synthesis. *Catal. Today* **2014**, *232*, 2–10.

44. Azzam, K.; Jacobs, G.; Ma, W.; Davis, B.H. Effect of cobalt particle size on the catalyst intrinsic activity for Fischer-Tropsch synthesis. *Catal. Lett.* **2014**, *144*, 389–394.

45. Pendyala, V.R.R.; Jacobs, G.; Ma, W.; Klettlinger, J.L.S.; Yen, C.H.; Davis, B.H. Fischer-Tropsch synthesis: Effect of catalyst particle (sieve) size range on activity, selectivity, and aging of a Pt promoted Co/Al$_2$O$_3$ catalyst. *Chem. Eng. J.* **2014**, *249*, 279–284.

46. Ding, M.; Qiu, M.; Liu, J.; Li, Y.; Wang, T.; Ma, L.; Wu, C. Influence of manganese promoter on co-precipitated Fe-Cu based catalysts for higher alcohols synthesis. *Fuel* **2013**, *109*, 21–27.

47. Bakar, W.A.W.A.; Ali, R.; Toemen, S. Catalytic methanation reaction over supported nickel-rhodium oxide for purification of simulated natural gas. *J. Nat. Gas Chem.* **2011**, *20*, 585–594.

48. Carenco, S.; Tuxen, A.; Chintapalli, M.; Pach, E.; Escudero, C.; Ewers, T.D.; Jiang, P.; Borondics, F.; Thornton, G.; Alivisatos, A.P.; *et al.* Dealloying of cobalt from CuCo nanoparticles under Syngas exposure. *J. Phys. Chem. C* **2013**, *117*, 6259–6266.

49. Inderwildi, O.R.; Jenkins, S.J.; King, D.A. Mechanistic studies of hydrocarbon combustion and synthesis on noble metals. *Angew. Chem. Int. Ed.* **2008**, *47*, 5253–5255.

50. Inderwildi, O.R.; Jenkins, S.J.; King, D.A. Fischer-Tropsch mechanism revisited: Alternative pathways for the production of higher hydrocarbons from synthesis gas. *J. Phys. Chem. C* **2008**, *112*, 1305–1307.

51. Loveless, B.T.; Buda, C.; Neurock, M.; Iglesia, E. CO chemisorption and dissociation at high coverages during CO hydrogenation on Ru catalysts. *J. Am. Chem. Soc.* **2013**, *135*, 6107–6121.

52. Ciobica, I. M.; van Santen, R.A. Carbon monoxide dissociation on planar and stepped Ru(0001) surfaces. *J. Phys. Chem. B* **2003**, *107*, 3808–3812.

53. Shetty, S.; Jansen, A.P.J.; van Santen, R.A. Direct versus hydrogen-assisted CO dissociation. *J. Am. Chem. Soc.* **2009**, *131*, 12874–12875.

54. Shetty, S.; Jansen, A.P.J.; van Santen, R.A. Hydrogen induced CO activation on open Ru and Co surfaces. *Phys. Chem. Chem. Phys.* **2010**, *12*, 6330–6332.

55. Liu, J.-X.; Su, H.-Y.; Li, W.-X. Structure sensitivity of CO methanation on Co (0001), (10$\bar{1}$2) and (11$\bar{2}$0) surfaces: Density functional theory calculations. *Catal. Today* **2013**, *215*, 36–42.

56. Fajín, J.L.C.; Cordeiro, M.N.D.S.; Gomes, J.R.B. Methanation of CO on pure and Rh or Ru doped nickel surfaces. *J. Phys. Chem.* **2014**. submitted.

57. Cheng, J.; Hu, P.; Ellis, P.; French, S.; Kelly, G.; Lok, C.M. A DFT study of the chain growth probability in Fischer-Tropsch synthesis. *J. Catal.* **2008**, *257*, 221–228.

58. Cheng, J.; Hu, P.; Ellis, P.; French, S.; Kelly, G.; Lok, C.M. Chain growth mechanism in Fischer-Tropsch synthesis: A DFT study of C-C coupling over Ru, Fe, Rh, and Re surfaces. *J. Phys. Chem. C* **2008**, *112*, 6082–6086.

59. Van Grootel, P.W.; Hensen, E.J.M.; van Santen, R.A. The CO formation reaction pathway in steam methane reforming by rhodium. *Langmuir* **2010**, *26*, 16339–16348.

60. Mueller, J.E.; van Duin, A.C.T.; Goddard III, W.A. Structures, energetics, and reaction barriers for CH_x bound to the nickel (111) surface. *J. Phys. Chem. C* **2009**, *113*, 20290–20306.

61. Zhang, C.J.; Hu, P.; Lee, M.-H. A density functional theory study on the interaction between chemisorbed CO and S on Rh(111). *Surf. Sci.* **1999**, *432*, 305–315.

62. Lee, K.; Song, C.; Janik, M.J. Density functional theory study of sulfur tolerance of CO adsorption and dissociation on Rh-Ni binary metals. *Appl. Catal. A* **2010**, *389*, 122–130.

63. Perdew, J.P.; Chevary, J.A.; Vosko, S.H.; Jackson, K.A.; Pederson, M.R.; Singh, D.J.; Fiolhais, C. Atoms, molecules, solids, and surfaces: Applications of the generalized gradient approximation for exchange and correlation. *Phys. Rev. B* **1992**, *46*, 6671–6687.

64. Hammer, B.; Hansen, L.B.; Nørskov, J.K. Improved adsorption energetics within density-functional theory using revised Perdew-Burke-Ernzerhof functionals. *Phys. Rev. B* **1999**, *59*, 7413–7421.

65. Perdew, J.P.; Burke, K.; Ernzerhof, M. Generalized gradient approximation made simple. *Phys. Rev. Lett.* **1996**, *67*, 3865–3868.

66. Blöchl, P.E. Projector augmented-wave method. *Phys. Rev. B* **1994**, *50*, 17953–17979.

67. Kresse, G.; Joubert, D. From ultrasoft pseudopotentials to the projector augmented-wave method. *Phys. Rev. B* **1999**, *59*, 1758–1775.

68. Shustorovich, E.; Sellers, H. The UBI-QEP method: A practical theoretical approach to understanding chemistry on transition metal surfaces. *Surf. Sci. Rep.* **1998**, *31*, 1–119.

69. Storsæter, S.; Chen, D.; Holmen, A. Microkinetic modelling of the formation of C1 and C2 products in the Fischer-Tropsch synthesis over cobalt catalysts. *Surf. Sci.* **2006**, *600*, 2051–2063.

70. Luo, W.; Asthagiri, A. Density functional theory study of methanol steam reforming on Co(0001) and Co(111) surfaces. *J. Phys. Chem. C* **2014**, *118*, 15274–15285.

71. Eckle, S.; Anfang, H.-G.; Behm, R.J. Reaction intermediates and side products in the methanation of CO and CO_2 over supported Ru catalysts in H_2-rich reformate gases. *J. Phys. Chem. C* **2011**, *115*, 1361–1367.

72. Li, T.; Wang, H.; Yang, Y.; Xiang, H.; Li, Y. Study on an iron-nickel bimetallic Fischer-Tropsch synthesis catalyst. *Fuel Proc. Technol.* **2014**, *118*, 117–124.

73. Enger, B.C.; Holmen, A. Nickel and Fischer-Tropsch synthesis. *Catal. Rev.: Sci. Eng.* **2012**, *54*, 437–488.

74. Van Santen, R.A.; Markvoort, A.J.; Filot, I.A.W.; Ghouri, M.M.; Hensen, E.J.M. Mechanism and microkinetics of the Fischer-Tropsch reaction. *Phys. Chem. Chem. Phys.* **2013**, *15*, 17038–17063.

75. Birot, A.; Epron, F.; Descorme, C.; Duprez, D. Ethanol steam reforming over $Rh/Ce_xZr_{1-x}O_2$ catalysts: Impact of the CO–CO_2–CH_4 interconversion reactions on the H_2 production. *Appl. Catal. B* **2008**, *79*, 17–25.

76. Lv, X.; Chen, J.-F.; Tan, Y.; Zhang, Y. A highly dispersed nickel supported catalyst for dry reforming of methane. *Catal. Commun.* **2012**, *20*, 6–11.

77. Freni, S.; Cavallaro, S.; Mondello, N.; Spadaro, L.; Frusteri, F. Production of hydrogen for MC fuel cell by steam reforming of ethanol over MgO supported Ni and Co catalysts. *Catal. Commun.* **2003**, *4*, 259–268.

78. Karim, A.M.; Su, Y.; Sun, J.; Yang, C.; Strohm, J.J.; King, D.L.; Wang, Y. A comparative study between Co and Rh for steam reforming of ethanol. *Appl. Catal. B* **2010**, *96*, 441–448.

79. Lin, S.S.-Y.; Kim, D.H.; Ha, S.Y. Hydrogen production from ethanol steam reforming over supported cobalt catalysts. *Catal. Lett.* **2008**, *122*, 295–301.

80. Batista, M.S.; Santos, R.K.S.; Assaf, E.M.; Assaf, J.M.; Ticianelli, E.A. Characterization of the activity and stability of supported cobalt catalysts for the steam reforming of ethanol. *J. Power Sources* **2003**, *124*, 99–103.

81. Batista, M.S.; Santos, R.K.S.; Assaf, E.M.; Assaf, J.M.; Ticianelli, E.A. High efficiency steam reforming of ethanol by cobalt-based catalysts. *J. Power Sources* **2004**, *134*, 27–32.

82. Czekaj, I.; Struis, R.; Wambach, J.; Biollaz, S. Sulphur poisoning of Ni catalysts used in the SNG production from biomass: Computational studies. *Catal. Today* **2011**, *176*, 429–432.

Catalytic Glycerol Hydrodeoxygenation under Inert Atmosphere: Ethanol as a Hydrogen Donor

Efterpi S. Vasiliadou and Angeliki A. Lemonidou

Abstract: Glycerol hydrodeoxygenation to 1,2-propanediol (1,2-PDO) is a reaction of high interest. However, the need for hydrogen supply is a main drawback of the process. According to the concept investigated here, 1,2-propanediol is efficiently formed using bio-glycerol feedstock with H_2 formed *in situ* via ethanol aqueous phase reforming. Ethanol is thought to be a promising H_2 source, as it is alcohol that can be used instead of methanol for transesterification of oils and fats. The H_2 generated is consumed in the tandem reaction of glycerol hydrodeoxygenation. The reaction cycle proceeds in liquid phase at 220–250 °C and 1.5–3.5 MPa initial N_2 pressure for a 2 and 4-h reaction time. Pt-, Ni- and Cu-based catalysts have been synthesized, characterized and evaluated in the reaction. Among the materials tested, Pt/Fe_2O_3-Al_2O_3 exhibited the most promising performance in terms of 1,2-propanediol productivity, while reusability tests showed a stable behavior. Structural integrity and no formation of carbonaceous deposits were verified via Temperature Programmed Desorption of hydrogen (TPD-H_2) and thermogravimetric analysis of the fresh and used Pt/FeAl catalyst. A study on the effect of various operating conditions (reaction time, temperature and pressure) indicated that in order to maximize 1,2-propanediol productivity and yield, milder reaction conditions should be applied. The highest 1,2-propanediol yield, 53% (1.1 $g_{1,2-PDO}\ g_{cat}^{-1}\cdot h^{-1}$), was achieved at a lower reaction temperature of 220 °C.

Reprinted from *Catalysts*. Cite as: Vasiliadou, E.S.; Lemonidou, A.A. Catalytic Glycerol Hydrodeoxygenation under Inert Atmosphere: Ethanol as a Hydrogen Donor. *Catalysts* **2014**, *4*, 397–413.

1. Introduction

The catalytic conversion of renewable sources to fuels and chemical products has become particularly important, because of the global energy requirements and environmental problems caused by petroleum and its derivatives [1]. It has been reported that as biomass is the only renewable source of carbon, its upgrading to a value-added chemical product instead of energy, heat and fuels will offer the greatest advantages [2].

Among the renewable oxygenates formed from biomass processing, glycerol is identified as a building block intermediate that can be converted to target products.

Glycerol is produced as a by-product of various industrial processes, such as transesterification of oils and fats, soap manufacture, fatty acid production, *etc.* [3]. Glycerol is now largely available in the market mainly due to the rapid increase in bio-diesel production. For this reason, a great effort has been directed towards the conversion of glycerol to valuable products, such as 1,2- and 1,3-propanediols [4–10], acrolein [11,12], hydrogen [13] and fuel additives [14].

Among the various processes studied, the production of 1,2-propanediol (propylene glycol) is one of the most attractive. 1,2-Propanediol, conventionally produced via propylene oxide hydration, is an important chemical that finds applications in antifreeze fluids and in the polymer industry. This process has been already industrialized by companies, such as Archer Daniels Midland (ADM), which has already started the production of 1,2-propanediol from glycerol and announced a 61% reduction in greenhouse gas emissions compared to the conventional method [15].

The catalytic glycerol hydrodeoxygenation (HDO, also called hydrogenolysis) to 1,2-propanediol is feasible over metal catalysts and hydrogen pressures up to 8.0 MPa [16]. The majority of research studies and patents available so far have mainly focused on the development of suitable catalytic systems and on engineering issues [17,18]. Although this process has been effectively developed, there are still drawbacks that need to be solved. The requirement for an external H_2 supply is the main disadvantage of the method, as the cost of hydrogen production, distribution and storage is high and negatively impacts the process economics. In addition, hydrogen is industrially produced using fossil feedstocks, thus rendering the overall reaction petroleum dependent.

Recently, the idea of *in situ* hydrogen production and consecutive consumption for hydrodeoxygenation has been explored, aiming at overcoming the above-mentioned problems [19]. Within this concept, two different approaches have been investigated: the first one involves the use of aqueous solutions of glycerol, where hydrogen is formed via aqueous phase reforming (APR) of a part of glycerol [20–23]; and the second, the addition of a hydrogen donor molecule (alcohols, formic acid), through hydrogen transfer reaction. The idea of catalytic transfer hydrogenation (CTH) was firstly realized by Musolino *et al.* [24] for selective transfer hydrogenolysis of glycerol in the presence of palladium catalysts (Pd/Fe_2O_3) using ethanol and 2-propanol as H_2 donor molecules. The pathways proposed include alcohol dehydrogenation, glycerol dehydration to hydroxyacetone and subsequent hydrogenation of the latter, consuming the H_2 generated from the alcohols. The reaction sequence was performed in ethanol and 2-propanol under 0.5 MPa inert atmosphere at 150–200 °C. There was no remarkable difference between the different donor alcohols, and the best performance was observed at 180 °C after 24 h, where glycerol was fully converted to a mixture of propylene glycol and ethylene glycol (94 and 6% selectivity, respectively). In a more recent study by Xia et al. [25], ethanol was recognized as the most efficient H_2 source among various alcohols (methanol, ethanol, 1-propanol, butanol and iso-propanol) tested for their hydrogen donating ability over Cu:Mg:Al catalysts. The latter was attributed to the high dehydrogenation activity shown by copper in ethanol reactions. At 210 °C, 3.0 MPa N_2 and 10 h, glycerol was converted by 95% with 92% selectivity to 1,2-propanediol. Gandarias et al. [26–28] have significantly contributed to the field of CTH hydrogenation reactions, focusing mainly on 2-propanol and formic acid as hydrogen donor molecules. Formic acid was proven to be an effective donor over Ni-Cu/Al_2O_3 catalyst. Importantly, the authors proposed a direct mechanism for glycerol conversion to 1,2-propanediol. This mechanism was suggested to occur when hydrogen is generated in the proximity of the active sites and includes an intermediate alkoxide formation.

In our previous studies [29,30], a different approach of in situ H_2 formation has been explored. Specifically, the required hydrogen is formed in situ by the reaction of methanol and water (aqueous phase reforming-APR), which are already components of the crude glycerol stream after transesterification and consumed from glycerol to form 1,2-propanediol (glycerol hydrodeoxygenation). The tandem reaction cycle presents advantages, such as: it is performed in the liquid phase, in the presence of the same catalytic material and reactor set-up, thus allowing the production of 1,2-propanediol under inert atmosphere through a one-step process. A Cu:Zn:Al catalyst prepared by the oxalate gel technique was found to exhibit the best performance with a 45% yield to the target product at 220 °C 3.5 MPa N_2 and 4 h. These results correspond to productivity values ($g_{1,2-PDO}·g_{cat}^{-1}·h^{-1}$) up to five-times higher compared with previous reports [24,27]. Experiments with labeled $^{13}CH_3OH$ over this catalyst allowed us to quantify the H_2 formation origin (methanol and/or glycerol APR) and showed that ~70% of the total H_2 is indeed produced from the reformation of methanol.

The objective of the present study is to further develop the innovative process described above using alternative H_2 sources, such as bio-ethanol. The idea is based on the fact that bio-ethanol can be used instead of methanol for transesterification during biodiesel production. In addition, in contrast to methanol, bio-ethanol can be derived from renewable resources, such as starch crops and lignocellulosic biomass (see Scheme 1). Within this context, the synthesis, characterization and evaluation of Pt, Ni and Cu catalysts in the ethanol APR-glycerol hydrodeoxygenation reaction cycle under inert atmosphere are investigated. The catalyst formulations evaluated in this study were selected based on their performance in ethanol reforming [31] and glycerol hydrodeoxygenation reactions [16]. Furthermore, the stability of the best performing catalyst was examined for two consecutive reaction tests. In addition, in the presence of the most effective and stable catalyst, the impact of reaction time, reaction temperature and nitrogen initial pressure was examined.

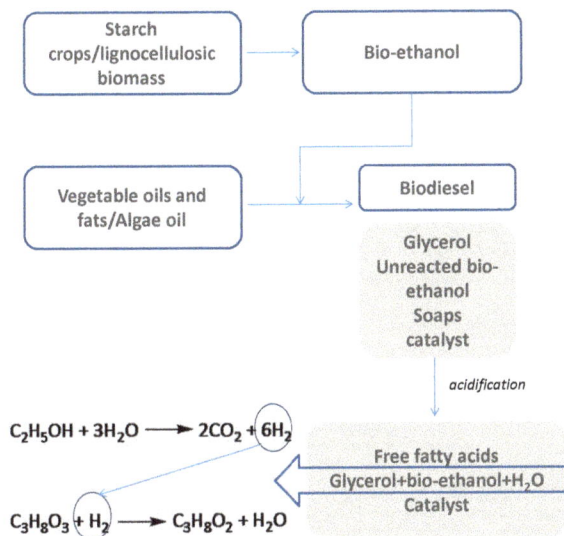

$$C_2H_5OH + 3H_2O \longrightarrow 2CO_2 + 6H_2$$

$$C_3H_8O_3 + H_2 \longrightarrow C_3H_8O_2 + H_2O$$

Scheme 1. Integrated process scheme for 1,2-propanediol production under inert conditions using bio-ethanol as the H_2 source.

2. Results and Discussion

2.1. Catalysts Characterization

The composition and the physicochemical properties of the supports and catalysts are presented in Table 1. Pt results in no significant changes when impregnated on Fe_2O_3-Al_2O_3 support. The impregnation of Ni on CeZrLa leads to a moderate decrease in the BET surface area. For the bulk Cu:Zn:Al catalyst, a comparison with literature data [30,32] for the corresponding materials prepared by the conventional carbonate co-precipitation method reveals the structural superiority of the oxalate gel prepared sample possessing a high BET surface area.

Table 1. Catalyst composition and porous characteristics.

Support	Composition (wt%)	BET surface area ($m^2 \cdot g^{-1}$)
FeAl	61 (Fe_2O_3), 39 (Al_2O_3)	80.0
CeZrLa	78 (ZrO_2), 17 (CeO_2), 5 (La_2O_3)	54.9
Catalyst		
Pt/FeAl	5.0 (Pt)	73.6
Ni/CeZrLa	10.0 (Ni)	37.2
Cu:Zn:Al	49.0 (Cu), 26 (Zn), 3.5 (Al)	71.5

The crystalline phases identified over all of the synthesized catalysts were investigated by X-ray diffraction and are presented in Figure 1. The XRD pattern of the Pt/FeAl catalyst is shown in Figure 1a. The dominant crystalline phase present is Fe_2O_3. The Al_2O_3 oxide was not detected, probably because of its amorphous nature. Platinum is highly dispersed on the surface, as no characteristic peaks were identified. The diffractogram of the Ni catalyst supported on CeZrLa is illustrated in Figure 1b. The characteristic peaks of the $Zr_{0.84}Ce_{0.16}O_2$ phase were identified, while no peaks corresponding to La_2O_3 were observed, most probably due to fine dispersion. The NiO phase was also present on the Ni/CeZrLa catalyst. Finally, Figure 1c shows the X-ray diffractogram of the Cu:Zn:Al catalyst. The catalyst exhibits broad CuO and ZnO peaks, which shows highly dispersed Cu and Zn phases, respectively. As previously for the Pt sample, Al_2O_3 was absent from the diffraction pattern.

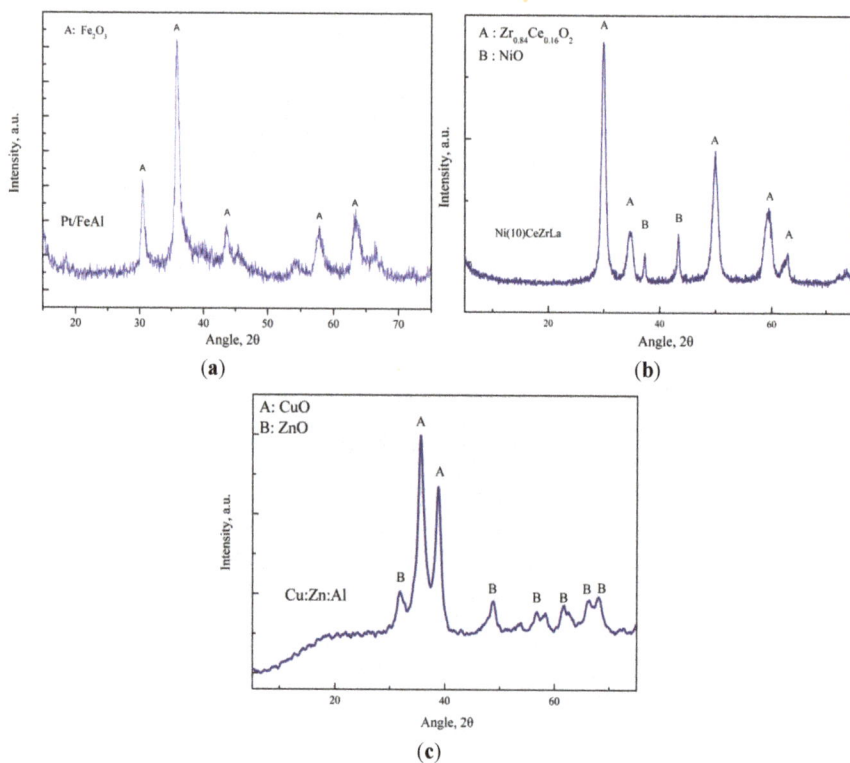

Figure 1. XRD patterns of (a) Pt/FeAl, (b) Ni/CeZrLa and (c) Cu:Zn:Al catalysts.

The reduction characteristics of the catalysts were studied by temperature-programmed reduction (TPR). The TPR profiles of all catalysts are illustrated in Figure 2. The TPR profiles of the Pt catalyst and the support are depicted in Figure 2a. The Fe_2O_3-Al_2O_3 support shows a H_2 consumption peak at

500 °C. This peak corresponds to Fe_2O_3 reduction to Fe_3O_4. The low temperature peak ($T < 100$ °C) that appears in the TPR profile of Pt/FeAl is attributed to Pt oxide reduction to metallic Pt^o. The impregnation of platinum leads to a shift at lower temperature (380 °C) of the peak observed for Fe_2O_3-Al_2O_3, as Pt facilitates the diffusion of hydrogen [33]. As seen from the H_2 consumption profile of the La_2O_3-doped CeO_2-ZrO_2 support (Figure 2b), a peak attributed to the partial reduction of CeO_2 appears in the temperature range between 300–400 °C [34]. In the TPR profile of the corresponding catalyst (Figure 2b), a main reduction peak is observed with maximum H_2 consumption at 445 °C, which corresponds to the reduction of NiO to metallic Ni^0. Moreover, a shoulder is also obvious in the temperature range of 300–400 °C, probably due to the reduction of the support. Figure 2c shows the reduction curves of unsupported CuO and the Cu:Zn:Al catalyst. The TPR profile of the catalyst exhibits a broad reduction peak at a temperature range of 200–300 °C, which suggests that the reduction of CuO to metallic Cu^0 proceeds through consecutive steps [35]. Compared with the unsupported CuO, it seems that the Cu:Zn:Al catalyst is more easily reduced.

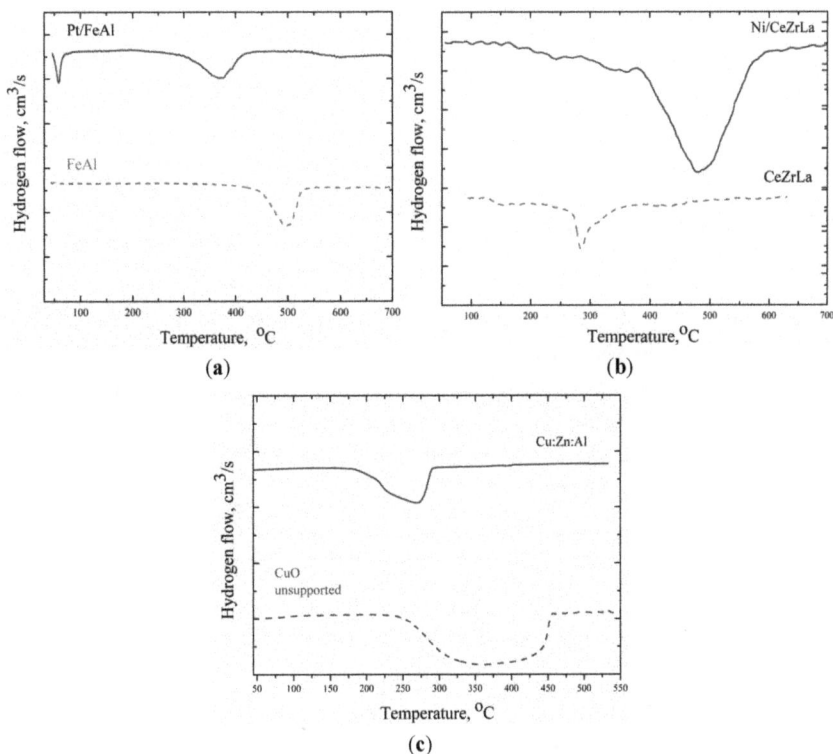

Figure 2. Reduction profiles of: (a) FeAl support, Pt/FeAl; (b) CeZrLa support, Ni/CeZrLa; and (c) CuO unsupported and Cu:Zn:Al.

347

2.2. Catalytic Performance in the Ethanol APR-Glycerol Hydrodeoxygenation Reaction Cycle

In order for the catalytic results to be clearly discussed, the possible reactions for individual compounds (glycerol and ethanol) are given below. The overall stoichiometric reaction leading to hydrogen formation via ethanol reforming is as follows:

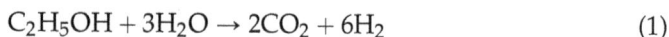

$$C_2H_5OH + 3H_2O \rightarrow 2CO_2 + 6H_2 \tag{1}$$

Apart from the reforming route, which requires a sufficient H_2O to ethanol ratio, ethanol dehydrogenation to acetaldehyde, dehydration to ethylene (the proposed precursor of coke formation), decomposition and subsequent reactions of decomposition products, as well as the water gas shift reaction may also take place [31]. Glycerol can undergo hydrodeoxygenation (Reaction 2), leading to 1,2-propanediol formation, while glycerol reforming (Reaction 3) is also an option under the present conditions.

$$C_3H_8O_3 + H_2 \rightarrow C_3H_8O_2 + H_2O \tag{2}$$

$$C_3H_8O_3 + H_2O \rightarrow 3CO_2 + 7H_2 \tag{3}$$

The three series of catalytic materials synthesized were tested under the standard reaction conditions. The catalytic results obtained over the different catalysts tested are summarized in Table 2. The experimental data show that 1,2-propanediol formation is possible under inert conditions with H_2 formed *in situ* from ethanol APR. Among the catalysts tested, the Pt-based sample showed the most promising performance in the tandem reaction cycle of ethanol APR and glycerol hydrodeoxygenation. Glycerol was almost fully converted in the presence of this catalyst, while for Ni and Cu, the conversion was somewhat lower. Of interest is the superior mass-specific rate exhibited by the Pt catalyst, which is over four-times higher compared with the Ni and Cu samples. Moreover, hydrogen conversion (defined as H_2 consumed for glycerol conversion to products that consume hydrogen/total H_2 formed) ranges between 75% and 85%, proving the potentiality of this cascade system. A comparison with our previous study [30], where H_2 is generated via methanol APR in the presence of the same Cu:Zn:Al catalyst, shows that methanol is a more efficient H_2 donor (see Table 2). The same performance was observed for the Pt/FeAl (Table 2), suggesting that methanol is much more reactive than ethanol under these conditions. It should be highlighted here that in this study, the ethanol/glycerol molar ratio corresponds to values lower than that of transesterification conditions (1.25 and 9.0, respectively [36]), which means that a part of the unreacted ethanol can be still recycled to the biodiesel reactor.

This is very important, as in previous studies, ethanol/glycerol molar ratios up to 46.0 have been used [24].

Table 2. Catalyst screening results under inert atmosphere, standard conditions: 3.5 MPa· N_2 initial pressure, 4 h, 250 °C, 7.1 wt% EtOH, 11.3 wt% glycerol (ethanol/glycerol molar ratio = 1.25) and water, catalyst/glycerol + ethanol weight ratio = 0.06 (for Pt) and 0.25 (for non-noble metal catalysts, *i.e.*, Ni and Cu).

Catalyst	Conversion (%)		Mass-specific rate (Integral) (MSR) (mmoles $g_{cat}^{-1}\cdot h^{-1}$)		Selectivity (%)				1,2-PDO productivity ($g_{1,2\text{-PDO}}\, g_{cat}^{-1}\cdot h^{-1}$)
	Glycerol	Ethanol/Methanol	Glycerol	Ethanol/Methanol	1,2-PDO	EG	AC	1-PrOH	
Pt/FeAl	97.4	9.5	26.4	3.6	32.2	2.1	4.5	7.3	0.65
Pt/FeAl [a]	96.4	14.4	26.2	12.2	41.7	2.6	5.3	5.4	0.83
Ni/CeZrLa	82.3	41.0	5.4	3.2	21.6	3.9	8.4	8.8	0.09
Cu:Zn:Al	87.7	3.6	5.8	0.3	32.9	1.9	7.5	5.6	0.15
Cu:Zn:Al [b]	88.8	14.1	6.7	1.9	39.2	1.9	5.4	6.1	0.20

[a] With methanol as the H_2 source; [b] with methanol as the H_2 source, from [30].

Under these conditions, glycerol was also subjected to reforming to some extent, generating additional hydrogen. Nevertheless, glycerol APR was not significant, especially over the Pt and Cu catalysts, while in the presence of Ni, glycerol reforming became more pronounced. The latter was expected, as Ni catalysts are among the most efficient materials for glycerol APR, aiming at renewable hydrogen production [37].

The target product, *i.e.*, 1,2-propanediol, was the main product formed in all cases, though in relatively low overall selectivity values of 21%–33%. It should be underlined that selectivity was based on the total C moles of glycerol reacted (see Section 3.3). Pt and Cu catalysts proved to be equally selective, while Ni, known for its C–C scission ability, also resulted in higher amounts of ethylene glycol and methanol (methanol selectivity Pt, 3.2%; Ni, 6.3%; and Cu, 2.1%) in the liquid phase. The undesirable sequential hydrodeoxygenation of 1,2-propanediol to propanols seems not to be promoted over the catalysts evaluated, as the selectivity to these mono-alcohols ranges between 5.5% and 9.0%. It should be underlined that the intermediate glycerol dehydration product, hydroxyacetone (acetol), was always detected in the product mixture at selectivity values of 4.5%–8.5%, supporting the proposed two-step dehydration-hydrogenation mechanism for the glycerol hydrodeoxygenation reaction [21,22].

For the present tandem process, ethanol APR is used as the hydrogen donor reaction. The activity order with respect to ethanol conversion was as follows: Ni > Pt > Cu (see Table 2). However, in terms of the mass-specific rate, the Pt-based catalyst shows better performance. Moreover, although Cu was the less active catalyst for ethanol APR, its high hydrogenation activity towards glycerol hydrodeoxygenation to 1,2-propanediol resulted in 1,2-PDO productivity higher than the Ni catalyst. The gas phase analysis showed CO_2 as the main product, while in the presence of Ni catalyst, CH_4 was also produced in considerable amounts (~48% selectivity based on gas phase products). Ethane originating from ethylene (an ethanol dehydration product) hydrogenation was additionally detected, though at very low concentrations (0.6%–2.9% selectivity in the gas phase). Due to the low reaction temperature, which favors the water gas shift reaction, CO formation is limited to selectivity values up to 0.8%–1.4%.

Comparing the performance of the catalysts on the tandem cycle of ethanol APR-glycerol HDO in terms of integral mass-specific rates, interesting observations can be obtained. The ratio between the two rates can be used as an indication for assessing the cascade nature of the system. For the noble metal catalyst, Pt/FeAl, this ratio equals 7.3, which means that the rate of glycerol conversion exceeds that of ethanol APR; however, the values are of the same order of magnitude. In the case of Cu catalyst, the mass-specific rates (MSRs) of glycerol conversion prevail over the ethanol rates (glycerol to ethanol MSR ratio = 19.3), as copper shows superior performance for glycerol hydrodeoxygenation and not for ethanol APR. The performance of the Ni/CeZeLa is clearly different, as glycerol and ethanol conversion proceeded in parallel (glycerol to ethanol MSR ratio = 1.7). However, Ni favors undesirable pathways, which include C–C bond scission reactions, leading to degradation product formation, like ethylene glycol, methanol and methane. Based on the above, it can be deduced that Pt-based catalysts are potential candidates for the ethanol APR-glycerol hydrodeoxygenation cycle.

2.3. Catalyst Stability

As deactivation phenomena are common in liquid phase reactions, the best performing Pt/FeAl catalyst was subjected to four reaction cycles without any pre-treatment between the tests. After every reaction cycle, the catalyst was collected by filtration and dried overnight. The performance upon reuse is illustrated in Figure 3. Our stability tests showed that there was no deactivation of the catalyst, as glycerol conversion is practically the same over the four catalytic tests. Moreover, hydrogen conversion was also the same, i.e., 76%. Furthermore, 1,2-propanediol selectivity and yield are somewhat improved compared with the first reaction cycle and stabilized at the initial performance during the third and fourth reaction cycles.

Based on previous reported studies on Pt/FeAl catalysts, we propose that the presence of iron is crucial for the catalyst's stable performance [38]. For this type of catalyst, Pt/FeAl, Pt sintering is the usual parameter causing deactivation. Iron addition improves the structural integrity, forming Pt-Fe alloy particles on Al_2O_3 under a reductive atmosphere. These particles have been found to segregate into Pt and Fe_2O_3 and form a Fe_2O_3 layer on the Pt particles, thus preventing them from sintering.

Measurement of Pt dispersion and particle size using the H_2-TPD technique (Table 3) showed a moderate dispersion decrease and particle size increment from 1.4 to 1.6 nm. However, this result seems not to significantly affect the performance upon reuse, as Pt still remains highly dispersed on the surface. Moreover, thermogravimetric analysis (Figure 4) of the used catalyst (after the fourth time) presents weight losses at low temperatures <480 °C, attributed to the loss of water and strongly adsorbed reactant or products onto the catalytic surface. From the thermogravimetric profile at higher temperatures, >480 °C, it is obvious that no carbonaceous products were formed, as no weight loss was observed.

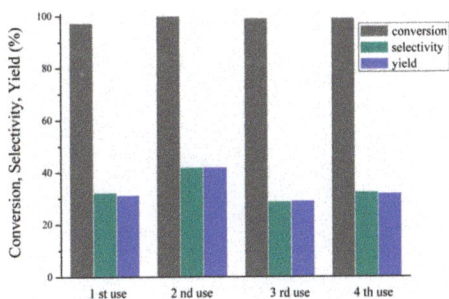

Figure 3. Reusability tests of the Pt/Fe-Al catalysts: 250 °C, 4 h, 7.1 wt% EtOH, 11.3 wt% glycerol (ethanol/glycerol molar ratio = 1.25) and water, catalyst/glycerol + ethanol weight ratio = 0.06.

Table 3. Dispersion and particle size of Pt/FeAl fresh and used samples.

Catalysts	Dispersion, %	Pt particle size, nm
Pt/FeAl-fresh	81	1.4
Pt/FeAl-used after 4th cycle	72	1.6

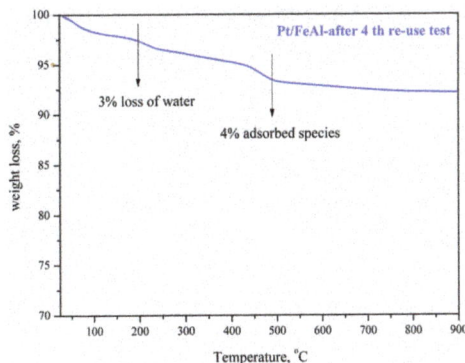

Figure 4. Thermogravimetric analysis in oxidative atmosphere of the fourth use of Pt/FeAl catalyst.

351

2.4. Effect of Various Operating Conditions

For the best performing catalyst, Pt/FeAl, the effects of reaction time (2 and 4 h), reaction temperature (220 and 250 °C) and system pressure (1.5 and 3.5 MPa N_2) were studied. Figure 5 illustrates the results obtained with reaction time variation. The increase of reaction time had no substantial effect on glycerol conversion, while the selectivity to 1,2-PDO was somewhat lower, due to sequential hydrodeoxygenation to 1-propanol. The above results demonstrate that the glycerol hydrodeoxygenation reaction is already completed at very short reaction times. This result enables efficient operation, as in such liquid phase reaction systems, prolonged reaction times are usually applied (>10 h) [19].

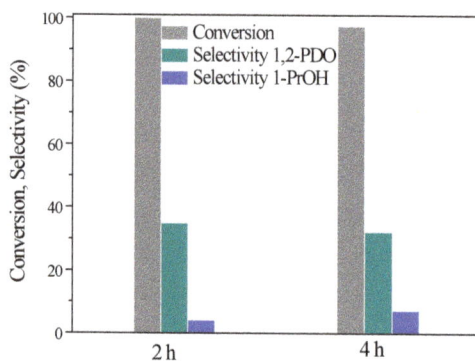

Figure 5. Effect of reaction time on glycerol conversion, 1,2-PDO selectivity and yield: 3.5 MPa N_2 initial pressure, 250 °C, 7.1 wt% EtOH, 11.3 wt% glycerol (ethanol/glycerol molar ratio = 1.25) and water, catalyst/glycerol + ethanol weight ratio = 0.06.

The influence of the reaction temperature on glycerol conversion, 1,2-PDO selectivity and yield is presented in Figure 6. The decrease of reaction temperature from 250 to 220 °C results in a decrease of glycerol conversion from 97.4% to 80.6%, as expected. The opposite trend is, however, observed for 1,2-PDO selectivity (from 32.2 at 250 °C to 66.0% at 220 °C), as it seems to be favored at the low temperature. The increased 1,2-PDO selectivity at 220 °C is mainly associated with the limitation of unidentified side product formation and secondarily with the decreased extent of its sequential hydrodeoxygenation to propanols (3% at 220 °C and 7.3% at 250 °C). The latter has been previously reported and is proposed to be favored under excess hydrogen and higher temperatures [30]. The operation at lower temperatures leads to a significant increase of the target product yield, from 31.3% at 250 °C to 53.3% at 220 °C, the highest yield obtained in this study. It is worth noticing here that at 220 °C, 1,2-PDO productivity equals 1.1 $g_{1,2\text{-PDO}}$ $g_{cat}^{-1} \cdot h^{-1}$, a value significantly higher compared with previous works using ethanol as the hydrogen donor ($0.13\ g_{1,2\text{-PDO}}\ g_{cat}^{-1} \cdot h^{-1}$ at $180\ ^{\circ}C$ and $0.62\ g_{1,2\text{-PDO}}\ g_{cat}^{-1} \cdot h^{-1}$ at $210\ ^{\circ}C$, both at almost complete glycerol conversion levels [24,25]).

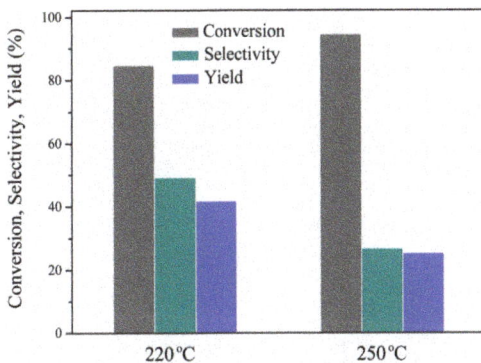

Figure 6. Effect of reaction temperature on glycerol conversion, 1,2-PDO selectivity and yield: 3.5 MPa N_2 initial pressure, 4 h, 7.1 wt% EtOH, 11.3 wt% glycerol (ethanol/glycerol molar ratio = 1.25) and water, catalyst/glycerol + ethanol weight ratio = 0.06.

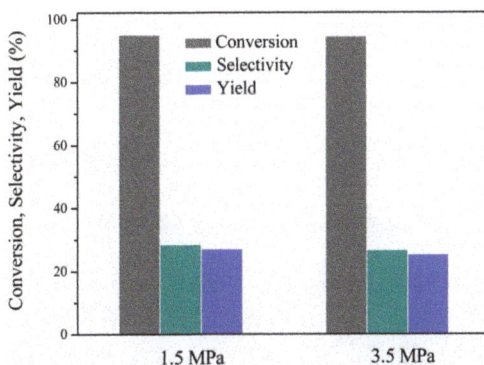

Figure 7. Effect of system pressure on glycerol conversion, 1,2-PDO selectivity and yield: 250 $^{\circ}C$, 4 h, 7.1 wt% EtOH, 11.3 wt% glycerol (ethanol/glycerol molar ratio = 1.25) and water, catalyst/glycerol + ethanol weight ratio = 0.06.

The effect of the initial nitrogen pressure on the conversion, 1,2-propanediol selectivity and yield is shown in Figure 7. It is quite clear that the variation of the N_2 system pressure between the range tested had no significant influence on the performance. This indicates that active hydrogen formed nearby the catalytic active sites readily reacts with glycerol, thus avoiding its escape to the gas phase. Moreover, it is also evident that hydrogen generation from ethanol APR and its consumption from glycerol proceed at comparable rates, in accordance with previous reported studies [28].

Taking into account the catalytic results discussed above, it is clear that the application of milder conditions is more effective. Short reaction times <2 h, low temperature (220 °C) and nitrogen pressure (1.5 MPa) positively affect the performance in terms of 1,2-propanediol selectivity and yield, exhibiting a maximum value of 53%. Therefore, operation at these conditions is beneficial as a means of reducing the energy requirements of the system.

3. Experimental Section

3.1. Catalyst Synthesis

3.1.1. Platinum Catalyst

The sol-gel technique was used to firstly prepare Fe_2O_3-Al_2O_3. The initial step of the sol-gel method was the preparation of a colloidal suspension (sol) of boehmite (γ-AlOOH), which was obtained by the hydrolysis of aluminum alkoxide (aluminum tri-sec-butylate, $C_{12}H_{27}AlO_3$) and stabilized in an acidic environment. The procedure was as follows:

(1) $C_{12}H_{27}AlO_3$ was dropwise and under vigorous stirring added into deionized water while heating at 80 °C.
(2)   After the addition of the alkoxide, the solution was maintained under stirring and heating for 30 min, so as to evaporate butanol (C_4H_9OH); then, $Fe(NO_3)_3 \cdot 9H_2O$ was added, and the solution was subjected to reflux conditions for 17 h at 80 °C.
(3)   The final sample was obtained after calcination at 600 °C for 3 h.

The supported platinum catalyst with 5 wt% Pt was prepared using the wet impregnation method. Platinum was deposited on Fe_2O_3-Al_2O_3 using an aqueous solution of $H_2PtCl_6 \cdot 6H_2O$. The impregnation was performed in a rotary evaporator at 70 °C for 1 h, followed by solvent removal drying at 120 °C for 17 h. The catalyst was then calcined under synthetic air at 450 °C for 3 h. Before the catalytic tests, the catalyst pre-reduced under continuous flow 10 v/v% H_2/N_2 for 2 h, at a temperature 90 °C.

3.1.2. Nickel Catalyst

The supported 10 wt% Ni catalyst was also prepared by the wet impregnation method. The support used was commercially available (CeZrLa-Mel Chemicals), and its composition is shown in Table 1. Ni was deposited on the carrier using a metal precursor compound, $Ni(NO_3)_2 \cdot 6H_2O$, in aqueous solution form. The impregnation took place in a rotary evaporator under stirring at 70 °C for 1 h, followed by removal of the solvent at 80 °C under vacuum and then drying at 120 °C for 17 h. The catalyst was then calcined in the presence of synthetic air at 450 °C for 3 h. The last step is the reduction in a specially-designed streaming unit. The reduction of the catalyst samples was carried out under continuous flow of 10 $v/v\%$ H_2/N_2 at a temperature of 500 °C for a period of 2 h.

3.1.3. Copper Catalyst

The synthesis method used was the oxalate gel co-precipitation. Details about the preparation and activation of this catalyst can be found in our previous publication [30].

3.2. Catalyst Characterization

Surface areas of the samples were determined by N_2 adsorption at −196 °C, using the multipoint BET analysis method, with an Autosorb-1 Quantachrome flow apparatus (Quantachrome Instruments, Boynton Beach, FL, USA). Prior to the measurements, the samples were dehydrated in a vacuum at 250 °C overnight.

X-ray diffraction (XRD) patterns were obtained using a Siemens (Munich, Germany) D500 diffractometer, with Cu-Kα radiation.

The reduction characteristics of the catalysts were studied by temperature-programmed reduction (TPR). These experiments were performed in a gas flow system equipped with a quadrupole mass analyzer (OMNIStar™, PFEIFFER, Asslar, Germany). Typically, the catalyst sample (100 mg) was placed in a U-shaped quartz reactor and pretreated in flowing He (30 cm^3/min) for 0.5 h at 250 °C, followed by cooling at room temperature. After pretreatment, the temperature was raised from room temperature up to 900 °C at a rate of 10 °C/min in a 10% H_2/He flow (30 cm^3/min).

Structural characterization and dispersion measurements of the Pt/FeAl catalyst were performed with hydrogen chemisorption and subsequent temperature-programmed desorption of the chemisorbed hydrogen. These experiments were performed in a gas flow system equipped with a quadrupole mass analyzer (OMNIStar™, PFEIFFER, Asslar, Germany). Typically, the fresh catalyst sample (100 mg) was placed in a U-shaped quartz reactor and pre-treated in flowing He (30 cm^3/min) for 0.5 h at 250 °C, followed by cooling at 90 °C (Pt/FeAl catalyst reduction temperature; see Figure 2a). The fresh sample was isothermally reduced at 90 °C with a mixture of 10% v/v H_2/He flow (30 cm^3/min flow). After reduction, H_2 was flushed with a 60 cm^3/min He flow, and then hydrogen chemisorption was performed at 90 °C with the same gas mixture as used for the reduction for 1 h. The sample was then cooled to room temperature under He flow and was kept at that temperature for 2 h in order to remove physisorbed H_2. The TPD experiment was performed by heating the sample at a rate of 10 °C/min from room temperature to 800 °C, under a He flow of 30 cm^3/min. The same procedure was followed for the used catalyst (after the 4th reuse test), but without the reduction step at 90 °C as described above for the fresh catalyst. Dispersion values were calculated from the quantification of the desorbed hydrogen and the assumption that the stoichiometry factor between chemisorbed hydrogen and surface Pt equals H:Pt = 1:1.

3.3. Catalyst Evaluation

The reaction tests were carried out in a 100-mL monel batch reactor (Parr Instrument Company, Moline, IL, USA) equipped with an electronic temperature controller and a mechanical stirrer (Scheme 2).

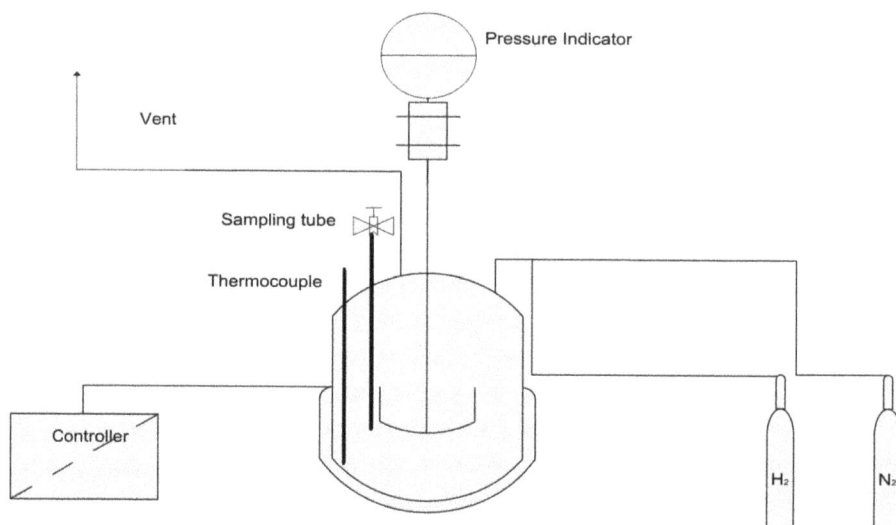

Scheme 2. Schematic representation of the experimental unit.

The reaction was typically conducted under the following standard conditions: 250 °C, 3.5 MPa initial N_2 pressure, 0.06–0.25 catalyst/glycerol + ethanol weight ratio, a feedstock mixture of: 7.1 wt% EtOH, 11.3 wt% glycerol (ethanol/glycerol molar ratio = 1.25) and water, 4-h reaction time and a 1000 rpm stirring rate. The effects of reaction time, temperature and nitrogen pressure were studied by varying the parameters, such as: 2 and 4 h, 220 and 250 °C and 1.5 and 3.5 MPa, respectively. The reactant conversion was calculated as follows:

$$\text{Conversion, \%} = \frac{\text{moles}_{in} - \text{moles}_{out}}{\text{moles}_{in}} \tag{4}$$

The product selectivity and yield were calculated using the equations:

$$\text{Selectivity of product } i, \% = \frac{\text{C moles product } i}{\text{C moles of glycerol reacted}} \times 100 \tag{5}$$

$$\text{Yield of product } i, \% = \frac{\text{moles of product } i}{\text{moles glycerol}_{in}} \times 100 \qquad (6)$$

Liquid samples were analyzed by GC (Agilent 7890A, Santa Clara, CA, USA, FID, DB-Wax 30 m × 0.53 mm × 1.0 μm). Acetonitrile was used as a solvent for the GC analysis. The multiple point internal standard method was used for the quantification of the results. The liquid compounds detected were: 1,2-propanediol-1,2-PDO, ethylene glycol-EG, hydroxyacetone-AC (acetol), 1-propanol-1-PrOH, 2-propanol-2-PrOH and ethanol-EtOH. Unidentified side liquid products were also detected by GC analysis. Gas analysis was performed in an Agilent GC (Santa Clara, CA, USA) 7890A, Molecular Sieve and Poraplot) equipped with a thermal conductivity detector. The gaseous products were: CO_2, H_2, CO, CH_4 and C_2H_6.

4. Conclusions

In this study, the formation of 1,2-propanediol using a glycerol by-product as a raw material without the need of external H_2 addition was realized. The required hydrogen is formed via the ethanol aqueous phase reforming over Pt, Ni and Cu catalysts. Ethanol was selected as a hydrogen source, because this alcohol is used instead of methanol for biodiesel production. All of the catalysts tested are able to catalyze the tandem reaction sequence (aqueous phase reforming and hydrodeoxygenation), producing 1,2-propanediol under initial N_2 pressure. Among them, the Pt-based catalyst (Pt/Fe_2O_3-Al_2O_3) exhibited the best results concerning 1,2-propanediol productivity and stability under reaction conditions. Moreover, it was found that milder operating conditions (reaction times <2 h, temperature 220 °C and nitrogen pressure 1.5 MPa) are essential for maximizing 1,2-propanediol productivity and yield.

Acknowledgments: This work was funded by the Aristotle University of Thessaloniki Research Committee in the frame of the Scholarships for Post-Doctoral Researchers 2013. Stylliani Sklari is acknowledged for the synthesis of the Fe_2O_3/Al_2O_3 support. Sofia Angeli and Vasileia-Loukia Yfanti are acknowledged for TPR experiments of the Ni and Pt catalysts, respectively.

Author Contributions: Efterpi S. Vasiliadou did the preparational work and wrote the paper; Angeliki A. Lemonidou supervised all the study.

Conflicts of Interest: The authors declare no conflict of interest.

References

1. Ruppert, A.M.; Weinberg, K.; Palkovits, R. Hydrogenolysis goes bio: From carbohydrates and sugar alcohols to platform chemicals. *Angew. Chem. Int. Ed. Engl.* **2012**, *51*, 2564–2601.

2. Vennestrøm, P.N.R.; Osmundsen, C.M.; Christensen, C.H.; Taarning, E. Beyond petrochemicals: the renewable chemicals industry. *Angew. Chem. Int. Ed. Engl.* **2011**, *50*, 10502–10509.

3. Bauer, F.; Hulteberg, C. Is there a future in glycerol as a feedstock in the production of biofuels and biochemicals? *Biofuels Bioprod. Biorefin.* **2013**, *7*, 43–51.

4. Vasiliadou, E.S.; Heracleous, E.; Vasalos, I.A.; Lemonidou, A.A. Ru-based catalysts for glycerol hydrogenolysis—Effect of support and metal precursor. *Appl. Catal. B* **2009**, *92*, 90–99.

5. Vasiliadou, E.S.; Lemonidou, A.A. Parameters Affecting the Formation of 1,2-Propanediol from Glycerol over Ru/SiO$_2$ Catalyst. *Org. Process. Res. Dev.* **2011**, *15*, 925–931.

6. Vasiliadou, E.S.; Lemonidou, A.A. Investigating the performance and deactivation behaviour of silica-supported copper catalysts in glycerol hydrogenolysis. *Appl. Catal. A* **2011**, *396*, 177–185.

7. Vasiliadou, E.S.; Eggenhuisen, T.M.; Munnik, P.; de Jongh, P.E.; de Jong, K.P.; Lemonidou, A.A. Synthesis and performance of highly dispersed Cu/SiO$_2$ catalysts for the hydrogenolysis of glycerol. *Appl. Catal. B* **2014**, *145*, 108–119.

8. Vasiliadou, E.S.; Lemonidou, A.A. Kinetic study of liquid-phase glycerol hydrogenolysis over Cu/SiO$_2$ catalyst. *Chem. Eng. J.* **2013**, *231*, 103–112.

9. Nakagawa, Y.; Shinmi, Y.; Koso, S.; Tomishige, K. Direct hydrogenolysis of glycerol into 1,3-propanediol over rhenium-modified iridium catalyst. *J. Catal.* **2010**, *272*, 191–194.

10. Huang, L.; Zhu, Y.; Zheng, H.; Ding, G.; Li, Y. Direct Conversion of Glycerol into 1,3-Propanediol over Cu-H$_4$SiW$_{12}$O$_{40}$/SiO$_2$ in Vapor Phase. *Catal. Lett.* **2009**, *131*, 312–320.

11. Martin, A.; Armbruster, U.; Atia, H. Recent developments in dehydration of glycerol toward acrolein over heteropolyacids. *Eur. J. Lipid Sci. Technol.* **2012**, *114*, 10–23.

12. Liu, L.; Ye, X.P.; Bozell, J.J. A comparative review of petroleum-based and bio-based acrolein production. *ChemSusChem* **2012**, *5*, 1162–1180.

13. Dou, B.; Song, Y.; Wang, C.; Chen, H.; Xu, Y. Hydrogen production from catalytic steam reforming of biodiesel byproduct glycerol: Issues and challenges. *Renew. Sustain. Energy Rev.* **2014**, *30*, 950–960.

14. Serafim, H.; Fonseca, I.M.; Ramos, A.M.; Vital, J.; Castanheiro, J.E. Valorization of glycerol into fuel additives over zeolites as catalysts. *Chem. Eng. J.* **2011**, *178*, 291–296.

15. ADM Propylene Glycol—Life Cycle Analysis. Available online: http://www.adm.com (accessed on 30 September 2014).

16. Nakagawa, Y.; Tomishige, K. Heterogeneous catalysis of the glycerol hydrogenolysis. *Catal. Sci. Technol.* **2011**, *1*, 179–190.

17. Mizugaki, T.; Arundhathi, R.; Mitsudome, T.; Jitsukawa, K.; Kaneda, K. Selective Hydrogenolysis of Glycerol to 1,2-Propanediol Using Heterogeneous Copper Nanoparticle Catalyst Derived from Cu–Al Hydrotalcite. *Chem. Lett.* **2013**, *42*, 729–731.

18. Akiyama, M.; Sato, S.; Takahashi, R.; Inui, K.; Yokota, M. Dehydration–hydrogenation of glycerol into 1,2-propanediol at ambient hydrogen pressure. *Appl. Catal. A* **2009**, *371*, 60–66.

358

19. Martin, A.; Armbruster, U.; Gandarias, I.; Arias, P.L. Glycerol hydrogenolysis into propanediols using *in situ* generated hydrogen—A critical review. *Eur. J. Lipid Sci. Technol.* **2013**, *115*, 9–27.

20. D'Hondt, E.; van de Vyver, S.; Sels, B.F.; Jacobs, P. Catalytic glycerol conversion into 1,2-propanediol in absence of added hydrogen. *Chem. Commun.* **2008**, *45*, 6011–6012.

21. Dasari, M.A.; Kiatsimkul, P.P.; Sutterlin, W.R.; Suppes, G.J. Low-pressure hydrogenolysis of glycerol to propylene glycol. *Appl. Catal. A* **2005**, *281*, 225–231.

22. Miyazawa, T.; Kusunoki, Y.; Kunimori, K.; Tomishige, K. Glycerol conversion in the aqueous solution under hydrogen over Ru/C + an ion-exchange resin and its reaction mechanism. *J. Catal.* **2006**, *240*, 213–221.

23. Mane, R.B.; Rode, C.V. Simultaneous glycerol dehydration and *in situ* hydrogenolysis over Cu–Al oxide under an inert atmosphere. *Green Chem.* **2012**, *14*, 2780–2789.

24. Musolino, M.G.; Scarpino, L.A.; Mauriello, F.; Pietropaolo, R. Selective transfer hydrogenolysis of glycerol promoted by palladium catalysts in absence of hydrogen. *Green Chem.* **2009**, *11*, 1511–1513.

25. Xia, S.; Zheng, L.; Wang, L.; Chen, P.; Hou, Z. Hydrogen-free synthesis of 1,2-propanediol from glycerol over Cu-Mg-Al catalysts. *RSC Adv.* **2013**, *3*, 16569–16576.

26. Gandarias, I.; Arias, P.L.; Requies, J.; El Doukkali, M.; Güemez, M.B. Liquid-phase glycerol hydrogenolysis to 1,2-propanediol under nitrogen pressure using 2-propanol as hydrogen source. *J. Catal.* **2011**, *282*, 237–247.

27. Gandarias, I.; Requies, J.; Arias, P.L.; Armbruster, U.; Martin, A. Liquid-phase glycerol hydrogenolysis by formic acid over Ni-Cu/Al$_2$O$_3$ catalysts. *J. Catal.* **2012**, *290*, 79–89.

28. Gandarias, I.; Fernández, S.G.; El Doukkali, M.; Requies, J.; Arias, P.L. Physicochemical Study of Glycerol Hydrogenolysis Over a Ni–Cu/Al$_2$O$_3$ Catalyst Using Formic Acid as the Hydrogen Source. *Top. Catal.* **2013**, *56*, 995–1007.

29. Vasileiadou, E.S.; Lemonidou, A.A. Catalytic process for the production of 1,2-propanediol from crude glycerol stream. Patent EP 2565175 A1, 31 August 2011.

30. Vasiliadou, E.S.; Yfanti, V.-L.; Lemonidou, A.A. One-pot tandem processing of glycerol stream to 1,2-propanediol with methanol reforming as hydrogen donor reaction. *Appl. Catal. B* **2014**, *163*, 258–266.

31. Ni, M.; Leung, D.Y.C.; Leung, M.K.H. A review on reforming bio-ethanol for hydrogen production. *Int. J. Hydrogen Energy* **2007**, *32*, 3238–3247.

32. Zhang, X.-R.; Wang, L.-C.; Yao, C.-Z.; Cao, Y.; Dai, W.-L.; He, H.-Y.; Fan, K.-N. A highly efficient Cu/ZnO/Al$_2$O$_3$ catalyst via gel-coprecipitation of oxalate precursors for low-temperature steam reforming of methanol. *Catal. Lett.* **2005**, *102*, 183–190.

33. Pendem, C.; Gupta, P.; Chaudhary, N.; Singh, S.; Kumar, J.; Sasaki, T.; Datta, A.; Bal, R. Aqueous phase reforming of glycerol to 1,2-propanediol over Pt-nanoparticles supported on hydrotalcite in the absence of hydrogen. *Green Chem.* **2012**, *14*, 3107–3113.

34. Angeli, S.D.; Monteleone, G.; Giaconia, A.; Angeliki, A. Low Temperature Methane Steam Reforming: Catalytic Activity and Coke Deposition Study. *Chem. Eng. Trans.* **2013**, *35*, 1201–1206.

35. Guerreiro, E.D.; Gorriz, O.F.; Rivarola, J.B.; Arrfia, L.A. Characterization of Cu/SiO_2 catalysts prepared by ion exchange for methanol dehydrogenation. *Appl. Catal. A* **1997**, *165*, 259–271.

36. Sales, A. Production of Biodiesel from Sunflower Oil and Ethanol by Base Catalyzed Transesterification. Master Thesis, Department of Chemical Engineering, Royal Institute of Technology (KTH), Stockholm, Sweden, June 2011.

37. Vaidya, P.D.; Rodrigues, A.E. Glycerol Reforming for Hydrogen Production: A Review. *Chem. Eng. Technol.* **2009**, *32*, 1463–1469.

38. Sakamoto, Y.; Higuchi, K.; Takahashi, N.; Yokota, K.; Doi, H.; Sugiura, M. Effect of the addition of Fe on catalytic activities of $Pt/Fe/\gamma-Al_2O_3$ catalyst. *Appl. Catal. B* **1999**, *23*, 159–167.

Valorisation of Vietnamese Rice Straw Waste: Catalytic Aqueous Phase Reforming of Hydrolysate from Steam Explosion to Platform Chemicals

Cao Huong Giang, Amin Osatiashtiani, Vannia Cristina dos Santos, Adam F. Lee, David R. Wilson, Keith W. Waldron and Karen Wilson

Abstract: A family of tungstated zirconia solid acid catalysts were synthesised via wet impregnation and subsequent thermochemical processing for the transformation of glucose to 5-hydroxymethylfurfural (HMF). Acid strength increased with tungsten loading and calcination temperature, associated with stabilisation of tetragonal zirconia. High tungsten dispersions of between 2 and 7 W atoms·nm^{-2} were obtained in all cases, equating to sub-monolayer coverages. Glucose isomerisation and subsequent dehydration via fructose to HMF increased with W loading and calcination temperature up to 600 °C, indicating that glucose conversion to fructose was favoured over weak Lewis acid and/or base sites associated with the zirconia support, while fructose dehydration and HMF formation was favoured over Brönsted acidic WO_x clusters. Aqueous phase reforming of steam exploded rice straw hydrolysate and condensate was explored heterogeneously for the first time over a 10 wt% WZ catalyst, resulting in excellent HMF yields as high as 15% under mild reaction conditions.

Reprinted from *Catalysts*. Cite as: Giang, C.H.; Osatiashtiani, A.; dos Santos, V.C.; Lee, A.F.; Wilson, D.R.; Waldron, K.W.; Wilson, K. Valorisation of Vietnamese Rice Straw Waste: Catalytic Aqueous Phase Reforming of Hydrolysate from Steam Explosion to Platform Chemicals. *Catalysts* **2014**, *4*, 414–426.

1. Introduction

The energy and atom efficient transformation of biomass waste feedstocks such as rice, corn and coconut husks, rice and wheat straw, corn cobs and palm kernels into sustainable gasoline and/or diesel drop-in biofuels, offers new routes to environmentally-benign renewable energy resources [1,2]. Furthermore, the attendant production of highly functionalised molecular intermediates affords synthetic pathways to diverse platform bio-derived chemicals which underpin the formulation of plastics and polymers with novel properties [3].

Lignocellulosic biomass comprises cellulose and hemicellulose carbohydrate polymers and lignin poly phenolics, and is susceptible to thermochemical decomposition into their constituent monomers. The cellulosic components are most readily activated and depolymerised into C_5 and C_6 monosaccharides, principally xylose and glucose, respectively, important building blocks for deoxygenation to chemical intermediates such as 5-hydroxymethylfurfural (HMF), levulinic acid and γ-valerolactone , and whose further deoxygenation yields fungible alkane, alkene and aromatic liquid fuels. The primary barrier to HMF utilisation in high-volume chemical and fuel applications is its high cost and correspondingly limited availability. To be commercially viable, large scale HMF production must be achievable at a comparable cost to that of petroleum-derived feedstocks such as para-xylene and terephthalic acid. Glucose or fructose derived from the cellulose component of biomass can be converted into HMF through its stepwise dehydration under mild aqueous conditions. A number of economically viable routes exist to these sugars [4–9], hence HMF production from such bio-derived feedstocks appears an attractive process in terms of the raw material supply and green credentials. Glucose conversion to HMF requires catalysts able to affect its isomerisation to fructose [10], and subsequent dehydration of this fructose intermediate, and poor HMF yields are consequently commonly reported in the literature.

Homogeneous mineral acid catalysed routes to HMF are well-known, however, heterogeneous (solid) variants offer numerous process advantages in terms of product separation, catalyst recycling, acid storage and handling, and opportunities for continuous flow operation. A number of solid acid catalysts have shown potential for this transformation [11–13], with a recent systematic investigation of sulfated zirconia highlighting a bifunctional pathway involving Lewis acid catalysed glucose isomerisation to fructose over the parent zirconia support, and subsequent Brönsted acid catalysed dehydration to HMF, conferring a yield of 2%–3% at 100 °C [14]. Other metal oxides, notably tungstates also exhibit solid acidity, particularly in conjunction with zirconia [11], but have never been applied to aqueous phase HMF production from rice straw.

Herein, we demonstrate the utility of tungstated zirconia for the aqueous phase transformation of hydrolysate sugar (obtained from rice straw lignocellulose hydrolysis) to HMF as a function of tungsten oxide loading and calcination temperature, achieving HMF yields as high as 10%–15% from steam exploded hydrosylate and liquid condensate.

2. Results and Discussion

2.1. Catalyst Characterization

The physicochemical properties of zirconia and WZs are summarised in Table 1 as a function of their calcination temperature and nominal W loading. There is good agreement between the nominal and surface W loadings observed by XPS, with the generally higher observed surface loadings consistent with the wet impregnation protocol employed and consequent concentration of tungsten at the catalyst surface relative to bulk. In almost all cases, tungsten addition increased the surface area relative to the corresponding parent ZrO_2, this effect being most significant at the lower calcination temperature of 500 °C, while the surface areas of all WZ materials decreased with increasing calcination temperature. There was no significant variation in either total pore volume or mean pore diameter, reflecting the non-porous nature of both parent zirconia and WZ, with measured porosity associated with interparticle voids and hence, was relatively insensitive to changes in particle morphology or crystallinity. Acid strength, as measured by the pH of an aqueous suspension of WZ, increased with W loading and calcination temperature (parent ZrO_2), reaching 3.31 for the 10 wt% WZ calcined at 500 °C. It is important to note that such moderate acidity is known to effect glucose dehydration/fructose isomerisation, but should be insufficient to drive further reaction of any HMF product to levulinic and formic acids observed when employing stronger acids [15,16].

Table 1. Physicochemical properties of ZrO_2 and WZ materials as a function of calcination temperature.

Catalyst	Calcination Temperature/°C	pH [a]	Surface W Loading [b]/wt%	Surface Area [c]/m².g⁻¹	Mesopore Diameter [d]/Å	Total Pore Volume [d]/cm³.g⁻¹	Surface W Density [e]/atoms.nm⁻²
ZrO_2		5.0	-	78	25	0.19	-
5 wt% WZ	500	5.6	7.7	122	17	0.18	2.1
10 wt% WZ		5.3	12.6	184	17	0.12	2.7
ZrO_2		-	-	44	50	0.15	-
5 wt% WZ	600	5.7	9.3	81	22	0.15	3.7
10 wt% WZ		4.6	8.4	116	17	0.16	3.0
ZrO_2		-	-	86	17	0.19	-
5 wt% WZ	700	5.5	11.7	64	25	0.14	6.0
10 wt% WZ		3.3	17.7	98	20	0.18	7.1

[a] Solution pH; [b] determined by XPS; [c] BET value from porosimetry; [d] Mesopore diameter and total pore volume from porosimetry; [e] Surface W density = W surface loading/[(100 × RMM_W × 6.023 × 10^{23}) × (BET surface area × 10^{18})], RMM_W is the Relative Molecular Weight of tungsten.

The nature of crystalline phases was also investigated via powder X-ray diffraction as a function of W loading and calcination temperature (Figure 1). Tetragonal zirconia (t-ZrO_2) is known to be thermodynamically unstable with respect to monoclinic zirconia (m-ZrO_2), and indeed the latter phase dominates the diffractograms of the parent support. The intensity of m-ZrO_2 reflections also increases with calcination temperature, accompanied by peak sharpening, indicative of an increase in crystallite size. Peak width analysis employing the Scherrer equation confirms the mean crystallite size rises from 10.2 nm after 500 °C calcination to 22 nm following 700 °C treatment, consistent with the decrease in surface area. In contrast, tungstated zirconias exhibited significant contributions from the t-ZrO_2 phase, with the proportion t-ZrO_2:m-ZrO_2 ratio proportional to the tungsten loading. For the

10 wt% WZ sample, only t-ZrO_2 was detectable even after calcination at 700 °C. Tungsten thus serves to stabilise the tetragonal phase with respect to the monoclinic, in accordance with previous reports for sulfated and tungstated zirconia [15,17,18]. Crystallite size for the t-ZrO_2 phase in the WZ materials were also far smaller than those of the parent m-ZrO_2, varying from 6.6 nm (500 °C) to 9.6 nm (700 °C). No reflections attributable to any tungsten oxide phase were observed for any WZ samples, indicating that tungsten must be highly dispersed, either as sub-2 nm particles or as a monolayer coating over the underlying t-ZrO_2 support. The latter hypothesis is in line with the surface W density, which increases with tungsten loading from 2.1 to 7.1 W atoms.nm^{-2}, but in all cases remains below that required to saturate a WO_3 monolayer (8.9 W atoms.nm^{-2}) [19].

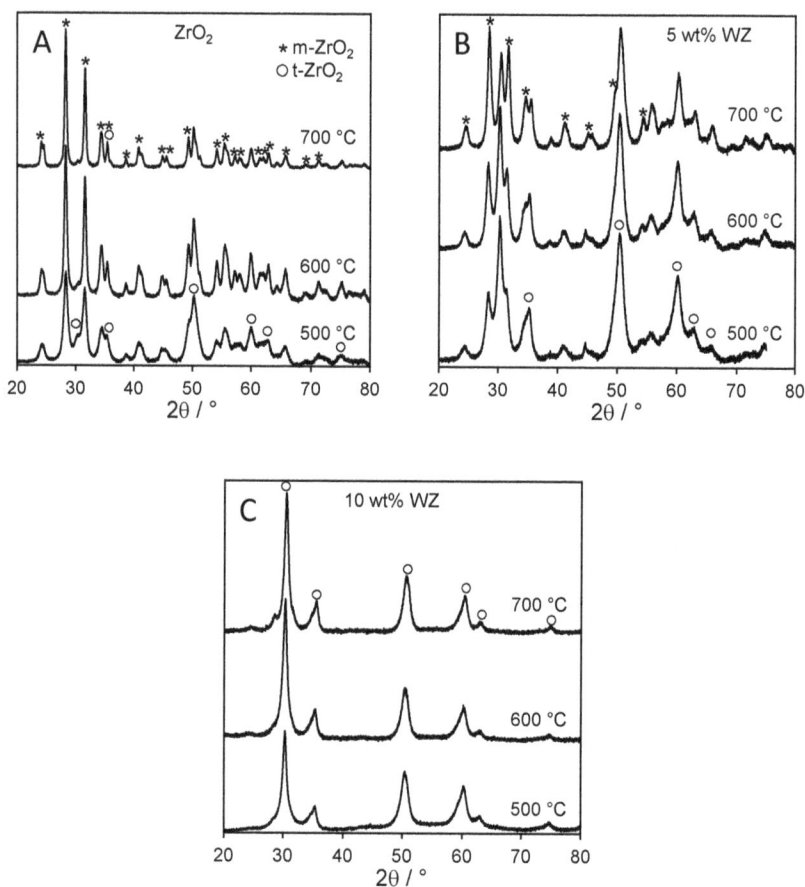

Figure 1. Powder XRD patterns for (**A**) parent zirconia, (**B**) 5 wt% WZ, and (**C**) 10 wt% WZ as a function of W loading and calcination temperature showing the thermal stabilisation of t-ZrO_2 by surface tungstate.

The nature of the surface acid sites was probed through pyridine chemisorption and subsequent DRIFTS analysis to identify the presence of Brönsted and Lewis acid sites and the relative Brönsted:Lewis character. Figure 2 shows the resulting DRIFT vibrational spectra for the parent and WO_x/ZrO_2 solids calcined at 500, 600 and 700 °C, revealing bands at 1540, 1490 and 1440 cm^{-1} typical of chemisorbed pyridine [19,20]. The band at 1540 cm^{-1} is attributed to a pyridinium ion bound to Brönsted acid sites, while those at 1580 cm^{-1} and 1438 cm^{-1} are attributed to molecular pyridine coordinated to Lewis acid sites. Since the 1488 cm^{-1} band is observed from pyridine adsorbed at both Brönsted and Lewis acid sites, the relative Brönsted:Lewis acidity can be quantified from the ratio of the 1540 cm^{-1} and 1438 cm^{-1} bands, as shown in Figure 2 as a function of calcination temperature and W loading. The Brönsted:Lewis ratio increased significantly with tungsten content, as anticipated upon impregnation of the parent zirconia support, which possesses oxygen deficient Lewis acid centres, with tungstate clusters. Calcination also increased the Brönsted character of all materials, possibly reflecting the formation of two-dimensional polytungstate clusters and attendant charge neutralisation by adsorbed protons [21].

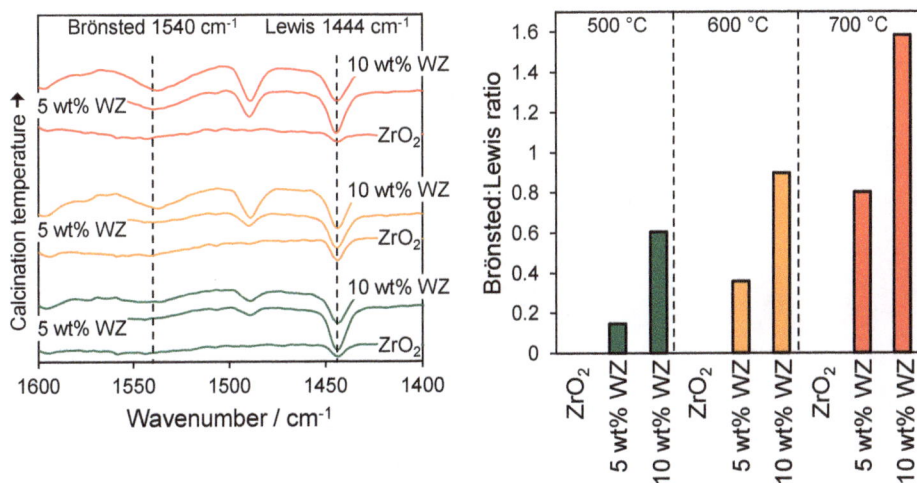

Figure 2. (left) DRIFT spectra of chemisorbed pyridine over zirconia and tungstated zirconias; and (right) ratio of Brönsted:Lewis adsorption bands as a function of W loading and calcination temperature.

2.2. Catalytic Conversion of Glucose

The impact of tungsten upon glucose isomerisation to fructose and subsequent dehydration to HMF was explored as a function of calcination temperature. In all cases, glucose conversion ranged between 5% and 20% under the mild conditions employed in this work, with Figure 3 revealing a decrease with increasing W loading and calcination temperature. These observations are consistent with another report on WZs, wherein glucose conversion is favoured over weak Lewis acid base sites associated with the bare zirconia support [14], while higher W loadings and calcination temperatures favour Brönsted acid sites which drive fatty acid esterification [19]. We can therefore infer a switchover in Lewis to Brönsted acid character resulting from the coalescence of isolated WO_x clusters as polytungstates and the resultant genesis of Brönsted acidity for the 5 and 10 wt% WZ samples. This observation is in excellent agreement with the pyridine titrations shown in Figure 2. The fall in conversion with temperature also likely reflects the corresponding decreases in surface area due to crystallite sintering. Normalising the final glucose conversion to the acid site density derived from the integrated chemisorbed pyridine DRIFTS intensity, yielded Turnover Frequencies of 4.7, 0.9 and 0.8 h^{-1} for the 700 °C calcined ZrO_2, 5 wt% and 10 wt% WZ materials, respectively. The superior activity of the parent zirconia is in accordance with its pure Lewis acid character.

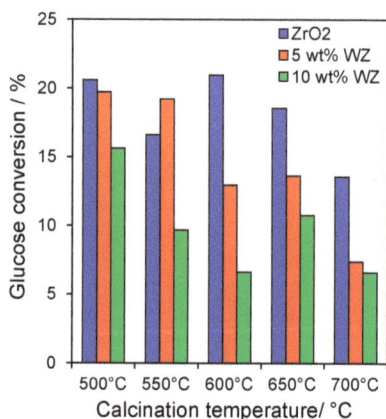

Figure 3. Glucose conversion after 6 h reaction at 100 °C over zirconia and tungstated zirconias.

Only two major reaction products were observed from glucose conversion over all WZ materials, fructose via glucose isomerisation, and HMF via dehydration of the reactively-formed fructose. This reflects the low reaction temperatures employed in this study relative to the literature [12] which disfavour HMF polymerisation and humin formation observed in other reports [22]. In all cases, fructose was overwhelmingly the dominant product. However, tungsten incorporation significantly enhanced HMF production, with the 10 wt% WZ sample twice as selective towards the furan as the parent zirconia support (Figure 4). High temperature calcination of WZ suppressed HMF formation at the expense of fructose, clearly indicating two competing reaction pathways, each requiring different active sites. Fructose appears favoured by Lewis acid sites present over the parent zirconia, whereas HMF is favoured by moderate surface tungstate densities of 2–3 atoms.nm^{-2}, significantly less than required to form a full monolayer (which impart Brönsted acidity), and calcination around 600 °C wherein surface areas remained around 100 m^2 g^{-1}. Fructose dehydration to HMF hence appears catalysed by isolated or dimeric WO$_x$ clusters.

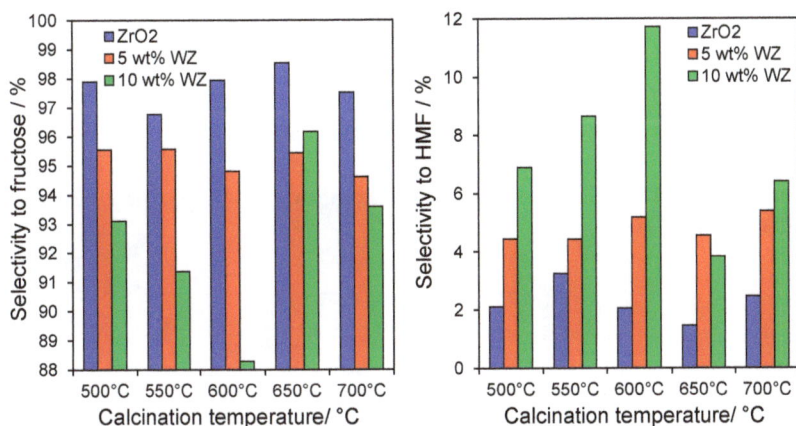

Figure 4. (**left**) Selectivity to fructose; and (**right**) HMF production from glucose after 6 h reaction at 100 °C over zirconia and tungstated zirconias.

2.3. Catalytic Conversion of Fructose

In order to confirm the preceding hypothesis, namely that glucose isomerisation occurs over Lewis acid sites whereas fructose dehydration is favoured by Brönsted acid sites associated with small tungstate clusters, the behaviour of our WZ series was also assayed for the direct dehydration of fructose (Figure 5). Activities were slightly higher than those observed for glucose, with fructose conversions spanning approximately 10%–30%. However, in contrast, regarding their reactivity to glucose, the WZ catalysts always outperformed the parent zirconia, consistent with their predominant Brönsted acidity observed by pyridine titration, and conferred significant conversion even at high calcination temperatures, despite the large decrease in BET surface areas; in contrast, calcination deactivated the pure zirconia. These observations are fully consistent with the notion that fructose dehydration occurs faster over WO$_x$ Brönsted acid sites than Lewis acid sites on the parent ZrO$_2$ support.

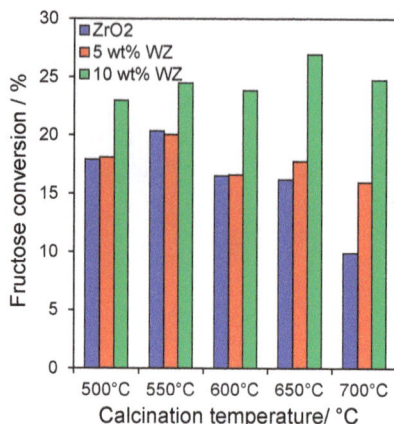

Figure 5. Fructose conversion after 6 h reaction at 100 °C over zirconia and tungstated zirconias.

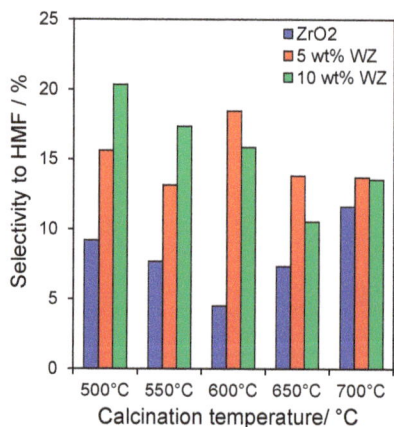

Figure 6. Selectivity to HMF production from fructose after 6 h reaction at 100 °C over zirconia and tungstated zirconias.

HMF (and trace oligomers of its condensation products), alongside glucose formed via the reversible isomerisation reaction, were the major products of fructose dehydration, with the WZ samples significantly more selective to HMF than ZrO_2 (Figure 6), with the highest selectivities of 15%–20% observed after moderate thermal processing of the 5 wt% and 10 wt% WZ catalysts, equating to a three-fold higher HMF yield, a consequence of the superior Brönsted acid character of the tugstated materials which favour rapid fructose dehydration over isomerisation back to glucose.

2.4. Catalytic Dehydration of Hydrolysate Sugar and Condensate

Having demonstrated the efficacy of WZ towards conversion of glucose and fructose model reactants, the behaviour of the 500 °C calcined materials was explored towards the transformation of hydrolysate sugar and condensate derived from Vietnamese rice straw. These catalysts were selected since they exhibited the optimum balance of glucose/fructose conversion and selectivity to HMF. The results in Figure 7 show similar poor levels of glucose conversion within both the hydrolysate and condensate at 100 °C for all three catalysts, with the parent zirconia the most active in accordance with Figure 3 and it superior Lewis acidity (Figure 2) which is required to drive initial glucose isomerisation to fructose. However, even a small increase in reaction temperature to 130 °C dramatically enhanced catalyst activity, with conversion climbing to around 30%–40% for the hydrosylate and 50%–60% for the condensate. At these higher reaction temperatures, the 10 wt% WZ also now outperformed zirconia towards both feedstocks.

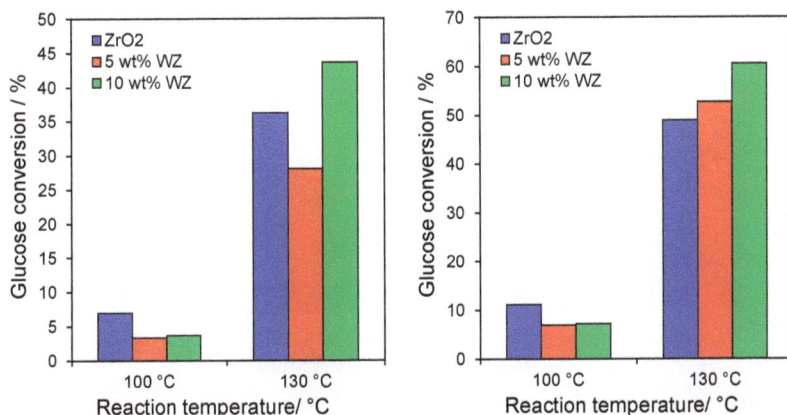

Figure 7. Glucose conversion after 6 h reaction at 100 °C from (**left**) hydrolysate; and (**right**) condensate over zirconia and tungstated zirconias.

The slight rise in reaction temperature enhanced overall product yields from ~10% to between 25%–35% from both hydrolysate and condensate. Fructose was the primary product in almost all circumstances, with the observed acetic acid reflecting that present in the starting feedstocks and remaining essentially unchanged during reaction (equating to ~3.5 g·L^{-1} in the hydrolysate and 2.5 g·L^{-1} in the condensate). HMF and furfural production (the latter presumably formed from trace hemicellulose via dehydration of xylose) increased dramatically at 130 °C, seen in Figure 8, particularly in the condensate wherein HMF actually becomes the principal product over the 10 wt% WZ catalyst with a yield of ~15%. This is a noteworthy achievement since it is difficult to obtain high HMF yields under aqueous reaction

369

conditions due to competing HMF hydrolysis and resultant levulinic and formic acid formation, coupled with further degradation and polymerisation reactions to humins.

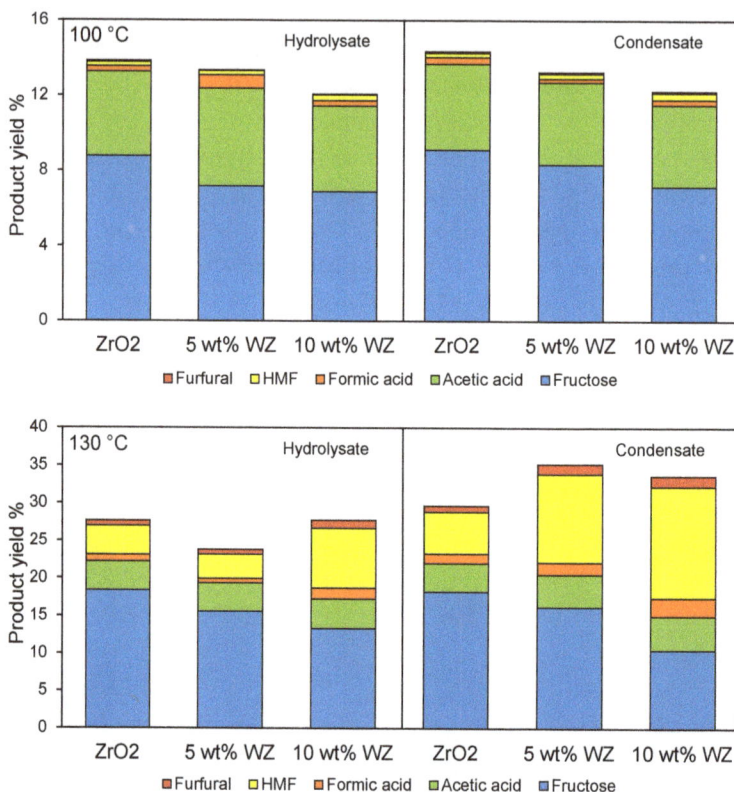

Figure 8. Yield of major products after 6 h reaction of processed rice straw at 100 °C and 130 °C.

3. Experimental Section

3.1. Catalyst Preparation

WO$_3$/ZrO$_2$ (WZ) catalysts were prepared with 5 and 10 wt% W loadings via wet impregnation of zirconium hydroxide (MelChem XZO 880/01, MEL Chemicals Company, Manchester, UK) with 40 cm^3 aqueous ammonium (para)tungstate hydrate solution (Sigma-Aldrich, Manchester, UK). The resulting slurry was stirred for 15 h at room temperature, and subsequently dried to evaporation overnight at 80 °C prior to calcination under flowing oxygen (10 cm$^3 \cdot$ min^{-1}) between 500 and 700 °C for 4 h. Pure zirconia controls were prepared by direct calcination of the parent zirconium hydroxide at different temperatures.

3.2. Catalyst Characterization

Textural properties of the WZ materials were characterised by nitrogen porosimetry using a Quantachrome Nova 1200 (Quantachrome, Oxford, UK) porosimeter and analysed with NovaWin software version 11. (Quantachrome, Oxford, UK) Samples were degassed at 120 °C for 2 h prior to analysis of adsorption/desorption isotherms by nitrogen physisorption at −196 °C. BET surface areas were calculated over the relative pressure range 0.01–0.2.

XPS was performed on a Kratos Axis HSi X-ray photoelectron spectrometer fitted (Kratos Analytical, Manchester, UK) with a charge neutraliser and magnetic focusing lens, employing Al K_α monochromated radiation (1486.7 eV). Spectral fitting was performed using CasaXPS version 2.3.14. (Casa Software Ltd., Teignmouth, UK) High-resolution O 1s, S 2p, Zr 3d and W 4f spectral binding energies were corrected to the C 1s peak at 284.6 eV and surface atomic compositions calculated via correction for the appropriate instrument response factors. pH measurements on aqueous catalyst suspensions were performed by adding 0.1 g of each catalyst to 20 mL of deionised water and stirring at room temperature for 30 min. Solution pH was subsequently measured using a Jenway 3305 pH meter (Jenway, Dunmow, UK). Powder X-ray diffraction patterns were recorded on a (PANalytical, Almelo, Holland) X'pert-Pro diffractometer (PANalytical, Almelo, Holland) fitted with an X'celerator detector (PANalytical, Almelo, Holland) with a Cu K_α (1.54 Å) source and nickel filter, and calibrated against a quartz standard. Brönsted/Lewis acid character was determined via Diffuse Reflectance Infrared Fourier Transform Spectroscopy (DRIFTS) on a Nicolet Nexus 479 FTIR spectrometer (Nicolet, MN, USA) via pyridine adsorption; approximately 1 mL of pyridine (99.8%, Sigma-Aldrich, UK) was dropped onto 200 mg of sample and dried by vacuum oven to remove physisorbed base.

3.3. Catalytic Reaction

Control experiments employing glucose and fructose as substrates were conducted employing 0.55 mmol dissolved in the 20 mL deionized water and 100 mg catalyst. Reactions were undertaken using a Radleys Starfish system within glass round-bottomed flasks at 100 °C for 6 h under stirring at 300 rpm. The reaction mixture was periodically sampled, with aliquots analysed with an Agilent 1200 series HPLC (Agilent, Oxford, UK) equipped with refractive index and UV diode array detectors and a Hi-Plex H column (Agilent, Oxford, UK) maintained at 65 °C under a 5 mM H_2SO_4 mobile phase flowed at 0.6 cm^3 min^{-1}.

Field grown rice (*Oryza sativa*, *cv.* KhangDan18) straw and husk was harvested at maturity in spring 2012 at the Ba Vi National Park, Hanoi, Vietnam. Both rice straw and husk were steam exploded into hot water using a bespoke CAMBI™ steam explosion facility with a 30 L reactor (Cambi AS, Asker, Norway). Before steam explosion, the rice straw was cut into 2–3 cm pieces for reactor loading. The

371

reactor was charged with 500 g feedstock, sealed and heated (husk: 220 °C; straw: 210 °C) with steam for 10 min. After this time, the contents were released (steam explosion), and deposited into 3.5 L hot water via a collection cyclone. The reactor was then pressurised again to 2–3 bar and released to dislodge any residue. The resulting pretreated slurry was collected and fractionated into solid and liquid phases by centrifuging through a 100 μm nylon mesh. The insoluble residue was rinsed extensively to ensure that any water-soluble material was removed. The residue was quantified and stored at −40 °C prior to further use or analysis.

Pilot-scale hydrolyses (5 L) were conducted in a bespoke high-torque bioreactor [23]. Digests were conducted at 20 $w/v\%$ substrate concentration using 1 Kg dry weight equivalent of steam exploded rice straw/husk suspended in sodium acetate/acetic acid buffer (4.1 g/L, pH 5.0). The buffer and substrate were initially heated to >85 °C to minimise the possibility of microbial contamination. The mixture was then cooled to 50 °C and an appropriate quantity of Cellic®CTec2 (Novozymes, Bagsværd, Denmark) was added (6.49 or 10 filter paper unit/g dry matter for straw and husk, respectively). The hydrolysate was agitated at 39 rpm for 4 days and the resulting aqueous solution contained 8.7 wt% glucose, while the steam exploded liquid condensate fraction contained 2.7 wt% glucose. Reactions were carried out as described above at 100 °C and 130 °C. Glucose and fructose conversion, product selectivity and yields are quoted ±2%, with mass balances >95%.

4. Conclusions

Here, we report the first application of a solid acid catalyst for the production of HMF from hydrolysate and condensate streams produced via the steam explosion of rice straw. Tungstated zirconia catalysts having 5 and 10 wt% W were prepared via wet impregnation and calcined at 500–700 °C to investigate the optimum WO_x surface density and Lewis:Brönsted ratio for the direct conversion of glucose to HMF under the mild conditions. In all cases, glucose conversion over WZ catalysts ranged between 5% and 20%, with HMF selectivity increasing with W loading and calcination temperature up to 600 °C. In comparison, reaction of pure fructose resulted in higher conversions of 10%–30%, for WZ, which always outperformed the parent zirconia, while being relatively insensitive to high calcination temperatures. These combined observations are consistent with the notion that glucose conversion to fructose is favoured over weak Lewis acid and/or base sites observed for the bare zirconia support, while fructose dehydration and HMF formation occurs preferentially over WO_x Brönsted acid sites, the latter being favoured for high W loadings and calcination temperatures. Application of the optimum 500 °C calcined 10 wt% WZ catalyst in the aqueous phase conversion of rice straw hydrolysate and condensate at 130 °C revealed fructose, HMF and furfural were the main products formed (the latter presumably formed from trace hemicellulose via dehydration of xylose). When using condensate from steam exploded rice straw, HMF actually becomes the principal product with a yield of ~15%, which is a noteworthy achievement under aqueous reaction since this is obtained on a crude feedstock containing impurities such as acetic acid, xylose and formic acid.

Acknowledgments: We thank the EPSRC for funding under (EP/K000616/1 and EP/G007594/4) and MEL Chemicals for the supply of Zr(OH)4, and BBSRC for funding (BB/J013838/1 and Institute Strategic Programme "Food and Health" BB/J004545/1). KW acknowledges The Royal Society for the award of an Industry Fellowship, and AFL thanks the EPSRC for the award of a Leadership Fellowship. The authors acknowledge financial support from CAPES (Coordenação de Aperfeiçoamento de Pessoal de Nível Superior), CNPq (Conselho Nacional de Desenvolvimento Cientifico e Tecnológico) and UFPR (Universidade Federal do Paraná). We also acknowledge Peter Ryden for assistance with rice straw/husk digestion.

Author Contributions: K.W. and A.F.L. conceived the research programme and drafted the manuscript; G.C., A.M. and V.C.S. synthesised, characterised and tested the catalysts; D.R.W. and K.W.W. supervised the steam explosion and digestion experiments.

Conflicts of Interest: The authors declare no conflict of interest.

References

1. Lee, A.F.; Bennett, J.A.; Manayil, J.C.; Wilson, K. Heterogeneous catalysis for sustainable biodiesel production via esterification and transesterification. *Chem. Soc. Rev.* **2014**, *43*, 7887–7916.
2. Wilson, K.; Lee, A.F. Rational design of heterogeneous catalysts for biodiesel synthesis. *Catal. Tech. Sci.* **2012**, *2*, 884–897.
3. Lee, A. Catalysing sustainable fuel and chemical synthesis. *Appl. Petrochem. Res.* **2014**, *4*, 11–31.
4. Pilipski, M. Saccharification of cellulose. U.S. Patent 4,235,968, 25 November 1980.
5. Onda, A.; Ochi, T.; Yanagisawa, K. Selective hydrolysis of cellulose into glucose over solid acid catalysts. *Green Chem.* **2008**, *10*, 1033–1037.
6. Pang, J.; Wang, A.; Zheng, M.; Zhang, T. Hydrolysis of cellulose into glucose over carbons sulfonated at elevated temperatures. *Chem. Commun.* **2010**, *46*, 6935–6937.
7. Zhou, C.H.; Xia, X.; Lin, C.X.; Tong, D.S.; Beltramini, J. Catalytic conversion of lignocellulosic biomass to fine chemicals and fuels. *Chem. Soc. Rev.* **2011**, *40*, 5588–5617.
8. Climent, M.J.; Corma, A.; Iborra, S. Converting carbohydrates to bulk chemicals and fine chemicals over heterogeneous catalysts. *Green Chem.* **2011**, *13*, 520–540.
9. Kobayashi, H.; Fukuoka, A. Synthesis and utilisation of sugar compounds derived from lignocellulosic biomass. *Green Chem.* **2013**, *15*, 1740–1763.
10. Moliner, M.; Román-Leshkov, Y.; Davis, M.E. Tin-containing zeolites are highly active catalysts for the isomerization of glucose in water. *Proc. Natl. Acad. Sci. USA* **2010**, *107*, 6164–6168.

11. Chareonlimkun, A.; Champreda, V.; Shotipruk, A.; Laosiripojana, N. Reactions of C_5 and C_6-sugars, cellulose, and lignocellulose under hot compressed water (HCW) in the presence of heterogeneous acid catalysts. *Fuel* **2010**, *89*, 2873–2880.

12. Nakajima, K.; Baba, Y.; Noma, R.; Kitano, M.; N. Kondo, J.; Hayashi, S.; Hara, M. $Nb_2O_5 \cdot nH_2O$ as a heterogeneous catalyst with water-tolerant lewis acid sites. *J. Am. Chem. Soc.* **2011**, *133*, 4224–4227.

13. Zeng, W.; Cheng, D.G.; Chen, F.; Zhan, X. Catalytic conversion of glucose on Al–Zr mixed oxides in hot compressed water. *Catal. Lett.* **2009**, *133*, 221–226.

14. Osatiashtiani, A.; Lee, A.F.; Brown, D.R.; Melero, J.A.; Morales, G.; Wilson, K. Bifunctional SO_4/ZrO_2 catalysts for 5-hydroxymethylfufural (5-HMF) production from glucose. *Catal. Tech. Sci.* **2014**, *4*, 333–342.

15. Qi, X.; Watanabe, M.; Aida, T.M.; Smith, R.L., Jr. Sulfated zirconia as a solid acid catalyst for the dehydration of fructose to 5-hydroxymethylfurfural. *Catal. Comm.* **2009**, *10*, 1771–1775.

16. Climent, M.J.; Corma, A.; Iborra, S. Conversion of biomass platform molecules into fuel additives and liquid hydrocarbon fuels. *Green Chem.* **2014**, *16*, 516–547.

17. Chen, F.R.; Coudurier, G.; Joly, J.F.; Vedrine, J.C. Superacid and catalytic properties of sulfated zirconia. *J. Catal.* **1993**, *143*, 616–626.

18. López, D.E.; Suwannakarn, K.; Bruce, D.A.; Goodwin, J.G., Jr. Esterification and transesterification on tungstated zirconia: Effect of calcination temperature. *J. Catal.* **2007**, *247*, 43–50.

19. Dos Santos, V.C.; Wilson, K.; Lee, A.F.; Nakagaki, S. Physicochemical properties of WO_x/ZrO_2 catalysts for palmitic acid esterification. *Appl. Catal. B* **2015**, *162*, 75–84.

20. Karim, A.H.; Triwahyono, S.; Jalil, A.A.; Hattori, H. WO_3 monolayer loaded on ZrO_2: Property-activity relationship in *n*-butane isomerization evidenced by hydrogen adsorption and IR studies. *Appl. Catal. A* **2012**, *433–434*, 49–57.

21. Barton, D.G.; Shtein, M.; Wilson, R.D.; Soled, S.L.; Iglesia, E. Structure and electronic properties of solid acids based on tungsten oxide nanostructures. *J. Phys. Chem. B* **1999**, *103*, 630–640.

22. Hu, X.; Lievens, C.; Larcher, A.; Li, C.Z. Reaction pathways of glucose during esterification: Effects of reaction parameters on the formation of humin type polymers. *Bioresour. Technol.* **2011**, *102*, 10104–10113.

23. Elliston, A.; Collins, S.R.A.; Wilson, D.R.; Roberts, I.N.; Waldron, K.W. High concentrations of cellulosic ethanol achieved by fed batch semi simultaneous saccharification and fermentation of waste-paper. *Bioresour. Technol.* **2013**, *134*, 117–126.

MDPI AG

Klybeckstrasse 64

4057 Basel, Switzerland

Tel. +41 61 683 77 34

Fax +41 61 302 89 18

http://www.mdpi.com/

Catalysts Editorial Office

E-mail: catalysts@mdpi.com

http://www.mdpi.com/journal/catalysts

www.ingramcontent.com/pod-product-compliance
Lightning Source LLC
Chambersburg PA
CBHW050345230326
41458CB00102B/6408